ISA HANDBOOK OF CONTROL VALVES

A comprehensive reference book containing application and design information written and compiled by control valve experts under the direction of the Final Control Elements Committee of ISA's Process Measurement and Control Division.

Edited by

J. W. Hutchison

Honeywell, Inc.

INSTRUMENT SOCIETY OF AMERICA
Pittsburgh, Pennsylvania

Library of Congress Catalog Card Number 73-163-121

SBN87664-151-6

PREFACE

The Instrument Society of America has prepared this handbook as part of an overall program to produce *in-depth* control valve technical material. The first phase was the production of a 16mm movie which gives a quick but broad look at control valve technology. Film strips which treat specific control valve subjects in detail are being planned. This handbook complements the films and is intended to permit both technicians and engineers to advance their control valve knowledge. The text emphasizes the whys of the variety of valve types, detail design features, manufacturing features, material selection, installation considerations and many other facets of control valve technology presented in an integrated text. The material ranges over a spectrum from fundamentals to advanced technical information.

This publication was prepared under the direction of the Final Control Elements Committee of ISA's Process Measurement and Control Division.

Participating authors were selected on the basis of their eminence in the particular topics which they covered. All are recognized authorities in the field of control valve technology and represent manufacturers as well as users. Their expertise plus many months of writing effort has resulted in a creditable compilation of man's present knowledge of control valves. To the individual authors and their companies we express our gratitude for the time, effort, and interest they have shown in the preparation of this book.

Hundreds of pages of manuscript contributed by nineteen individuals, all written and presented in different characteristic styles, presented a real challenge in having the book appear uniform from chapter to chapter, and yet retain continuity within each chapter. It is felt that this objective has been largely accomplished.

A cursory review of the technical information presented herein suggests that present knowledge of control valves has been developed to an advanced degree. Much that is now known has been developed in the relatively recent past. Both manufacturers and users have contributed to this store of information by working independently and by working together. This book, the first to gather so much of this data, presents some material which has heretofore been reported in the literature, but which still seems to be neither well known nor well understood by many people.

What may not be quite so evident from a cursory review is that there is a very extensive amount of investigative work still going on with respect to many different aspects of valve technology. This is notably true in the areas of sizing, noise, flashing, and cavitation. Considerable activity along these lines is presently under way within the ISA Standards and Practices Department Committee on Valve Sizing (SP39 Committee).

Over a period of many years, several specific systems of terminology and nomenclature have evolved with respect to control valves. This is true with valve components and parts and with the terms and symbols used in control valve sizing. This is not uncommon in the earlier stages of any technology, although it may not be a desirable situation. The advantages of standardization are both many and great, not the least of which is that everyone understands exactly what is meant by everyone else. However, getting people to agree to a new universal language is not easy, nor can it be accomplished quickly. This is quite understandable since everyone is reluctant to change from the system he knows best and is most familiar with, or in some cases in which he may even have a proprietary interest. It is, therefore, gratifying to be aware that standardization is at present actively being pursued with respect to control valves. The ISA Control Valve Sizing Committee (SP39) has worked long on the project and is still engaged in a great deal of activity in this direction. They have, for example, already come forth with the draft of a set of terms and symbols which they propose being used as standards for valve sizing terms.

With respect to "sizing" control valves, several methods of calculation have evolved, some of which have become well established. In many cases, those who work with control valves have become very familiar with, and very much attached to, one particular system and its accompanying set of symbols. Authors of this book, for example, have used (and in fact have insisted on using) symbols from one or another of these well known systems. Since each one defines within his chapter what the symbols mean, this should cause no problem in understanding the text.

During the planning stage it was found that many individuals expressed a desire to have an Appendix which would incorporate adequate reference material frequently used in control valve technology. The object, of course, was to eliminate the necessity of having to turn to several different books to find often used data.

The Appendix, therefore, has been expanded to be as complete as possible, with the intent of making the book more useful. The Appendix, however, is not the sole location of tables and other useful reference data. Much of this appears elsewhere within the framework of the associated chapters. This is particularly true with respect to Chapter 9, *Materials.*

More than two dozen individuals actively participated in the review of the material as it was prepared. Their comments, suggestions, and opinions were of special value in enabling the authors to revise their drafts for corrections and clarity. Reviewers' comments were of a particularly constructive nature and were well received by the individual authors. Practically every author participated in the review of chapters other than his own. In addition, there were others who were very active in the reviewing process. The names of all reviewers appear in a separate listing. To all who assisted in the review, we express appreciation for the quality and value of their comments and for the interest and cooperation they have shown all during the preparation of this book.

It may be noted that some specific material and information appears in more than one location in the book. It is expected that some readers may look for certain information within one chapter while others will expect to find it in another chapter; therefore, to facilitate useage of the book, and to eliminate the need of continual reference to several different parts of the book, some duplication has been allowed to remain.

Relatively minor differences of opinion among the authors may be either observed or construed in certain areas of the book. This is neither a totally unexpected situation to encounter in the still-developing technology of valves, nor is it a cause for undue concern. It merely points up the fact that further investigation and development should be, and in all probability, will be done within these areas. In all cases, the authors have experimental or engineering data to substantiate their statements.

Diagrammatic arrangements of control valve piping and manifolding are presented in Chapter 11 on "Installation." These cover a great number and a great variety of situations with corresponding satisfactory solutions. It is felt that many hours of drafting time may be saved in the preparation of engineering documents by simply referring to a page and figure number within this book.

The control valve is often the most costly element in a control loop. Also, since it is in the process stream, it is the element most likely to cause process downtime in the event of malfunction. For these reasons, how it is installed and how it is maintained are of vital concern to everyone who is in any way involved with control valves. Therefore, the information presented in Chapter 10, "Maintenance" and Chapter 11 "Installation" will be found to be of particular interest to all who use this book.

Throughout the book, authors have used illustrations and technical data from their own particular companies because of the availability of this material to them and because of their greater knowledge of it. In many cases, the data presented was developed and established by the author's company through specific research or experience. Therefore, to give the reader verification and credence for the statements he makes, the author refers specifically to his own company's findings. There is no intent on their part, or ours, that this be construed as advertising or sales promotional in nature.

James W. Hutchison
Editor in Chief
Honeywell, Inc.

REVIEWERS

E. E. Allen	Fisher Controls Company
E. A. Bake	Rockwell Manufacturing Co.
H. D. Baumann	Masoneilan International, Inc.
S. L. Davis	E. I. DuPont de Nemours & Co.
L. R. Driskell	Blaw-Knox Chemical Plants, Inc.
E. L. Dixson	Fisher Controls Company
M. L. Freeman	Masoneilan International, Inc.
D. J. Gaffney	Fisher Controls Company
A. J. Hanssen	Conoflow Corporation
M. W. Hott, Jr.	Jordan Valve Division, Richards Industries, Inc.
C. F. King	Fisher Controls Company
E. F. Kremer	E. I. DuPont de Nemours & Co.
C. A. Loflin	E. I. DuPont de Nemours & Co.
O. P. Lovett, Jr.	E. I. DuPont de Nemours & Co.
S. F. Luna	Westinghouse Electric Corporation
R. L. Moore	E. I. DuPont de Nemours & Co.
J. T. Muller	Leslie Company
P. L. Ochs	Atlas Valve Company
J. P. O'Connor, Jr.	Kieley & Mueller, Inc.
J. A. Parker	E. I. DuPont de Nemours & Co.
M. B. Rothstein	E. I. DuPont de Nemours & Co.
C. B. Schuder	Fisher Controls Company
A. T. Sherman	E. I. DuPont de Nemours & Co.
G. F. Stiles	Fisher Controls Company
C. R. Thornton	E. I. DuPont de Nemours & Co.
E. L. Vollmer	Honeywell, Inc.
D. L. Wassum	E. I. DuPont de Nemours & Co.
R. B. Weil	Jamesbury Corporation
P. Wing	Masoneilan International, Inc.
B. C. Zannini	ITT Hammel-Dahl, Inc.

CONTENTS

CHAPTER 1

INTRODUCTION

Orval P. Lovett, Jr.

E. I. DuPont de Nemours & Co.
Wilmington, Delaware

The *control valve* plays a very important part in the automatic control of modern plants, which depend on the correct distribution and control of flowing liquids or gases. Such control, whether for the exchange of energy, reduction of pressure, or simply to fill a tank, depends on some form of *final control element* to do the job. Final control elements may be considered the "muscle" of automatic control. They furnish the necessary power amplification between the low energy levels in controllers and the higher energy levels needed to perform their function in controlling flowing fluids.

The control valve is the most widely used type of final control element. Other final control elements include metering pumps, dampers and louvers (a variation of a butterfly valve), variable pitch fan blades, electric current control devices, and electric motors for positioning devices other than valves.

Despite the control valves wide use, there is probably no other element in the control system which receives less attention. In many systems, a control valve is subjected to more severe conditions of temperature, pressure, corrosion and contamination than other components, yet it must perform satisfactorily, with a minimum amount of attention, as it manipulates the flow of process fluid.

A control valve functions as a variable resistance in a pipeline. It provides a pressure drop by changing the turbulence in the process fluid or, in the case of laminar flow, the pressure drop is caused by the changed valve resistance or "drag". This pressure-drop process is often called *throttling*. For gases it approaches adiabatic, isothermal conditions. Deviations depend upon the degree of non-ideality of the gas (Joule-Thompson effect). In the case of liquids the pressure is dissipated by either turbulence or viscous drag. In either case, it converts the pressure energy to heat, resulting in a very small rise in temperature.

The usual control loop consists of three major parts. The first part is a *sensing element*, which usually is a transmitter. This is a device which is able to measure the process variable which is to be controlled, such as pressure, level, or temperature. The transmitter output is sent to a controlling instrument—*the controller*, which defines and measures the error between the set point or desired value, and the actual value of the process variable, and in turn sends a corrective signal to a final control element—the *control valve*. The valve varies the fluid flow to change the process variable to the desired value.

In pneumatic control systems, the pneumatic output signal from the controller may directly actuate the valve by either a spring-and-diaphragm actuator or a piston actuator. The positioning energy in this case is provided by compressed air which should be dry in outdoor installations to prevent freeze-up, and should be clean and filtered.

When a pneumatic valve is used with an electronic controller, either an electronic-to-pneumatic valve positioner or an electronic-to-pneumatic transducer is used. The same considerations for compressed air apply as in all-pneumatic control systems.

In control theory terminology, a control valve has both static and dynamic characteristics contributing to the overall success or failure of the loops controllability. The static or gain term, which is the flow characteristic of the valve, depends upon valve size, configuration of the plug and seat combination, type of actuator, valve positioner, upstream and downstream pressures, and to some extent fluid properties. Chapter V covers this subject in detail.

The dynamic characteristic is determined by the actuator or positioner-actuator combination. For slower processes, such as the control of temperature or liquid level, the valve dynamics are not usually a limiting factor in controllability. On faster systems, such as liquid flow control, the control valve may be the dominant lag and must be included in loop controllability considerations. Generally, it is only the control systems specialist who needs to be concerned with the valve's dynamics. Normal considerations for positioner applications, as discussed in Chapter VIII, will suffice in most control valve installations.

The history of the automatic control valve dates back to the self-actuated regulator which consisted of a globe-type valve with weight-loaded stem. The weights were used to balance plug forces and thus achieve some measure of control. Another early form of self-actuated regulator was the pressure balanced regulator. A process pressure was piped to the diaphragm chamber of a spring and diaphragm regulator. Either a pressure reducing valve, a back pressure regulator, or a differential pressure regulator could be built from variations of this basic valve.

The advent of pneumatic transmitters and controllers naturally led to pneumatic control valves. They were basically pressure reducing valves or back-pressure regulating valves, modified to utilize instrument air instead of process fluids. Many of the present day control valve manufacturers grew from companies which had been in the regulator business. Control valve applications continued to grow in number and complexity and many valve body variations as well as accessories became available to solve a variety of problems.

This handbook is intended to acquaint engineers with the factors of control valve design and application and to assist instrument engineers in the selection of the best valve body, actuator, and accessories for his applications.

CHAPTER 2

CONTROL VALVE COMPONENTS, DEFINITIONS, TERMINOLOGY AND SPECIFICATIONS

M. W. Hott, Jr.

Jordan Valve Division
Richards Industries, Inc.
Cincinnati, Ohio

PREFACE

This chapter presents the fundamentals that are basic to an understanding of the detailed discussions contained in the following chapters. Part 1 contains labeled illustrations identifying the parts of the control valve components —valve types, cage trim, body assemblies, bonnets and actuators. Part 2 presents the commonly accepted definitions of the components illustrated in Part 1. Part 3 includes the commonly accepted terminology used in discussions concerning control valve applications. Control valve specifications are covered in Part 4.

CONTENTS

Part 1
IDENTIFICATION OF CONTROL VALVE COMPONENTS

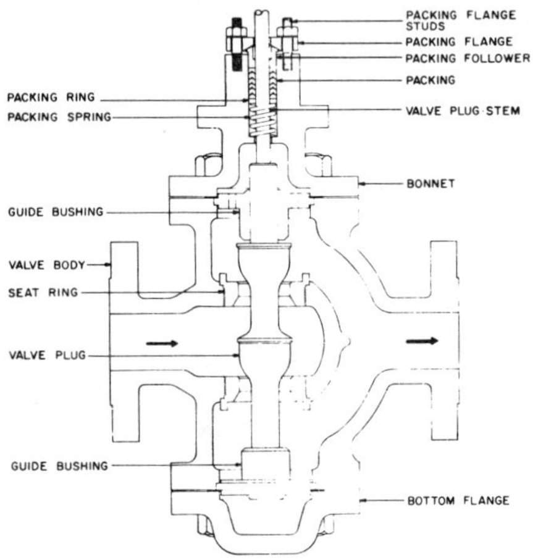

Figure 1. TYPICAL DIAPHRAGM CONTROL VALVE.

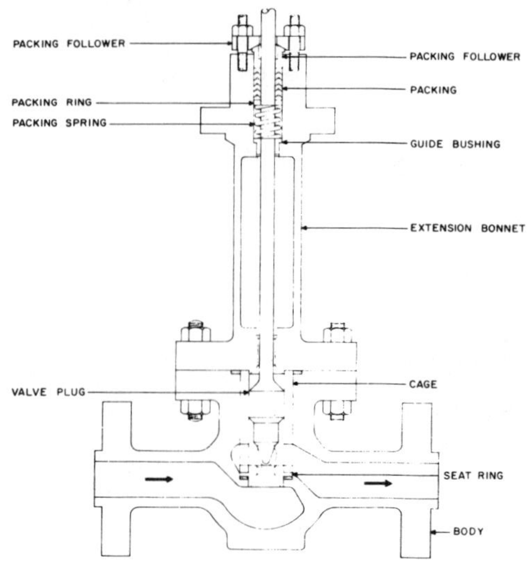

Figure 2. TYPICAL CAGE TRIM BODY.

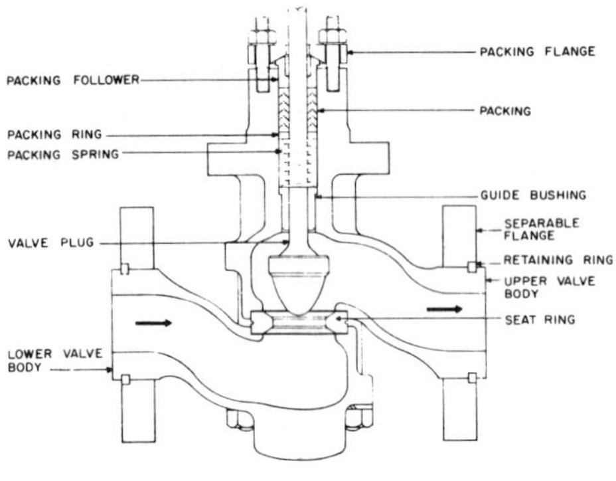

Figure 3. TYPICAL SPLIT VALVE BODY ASSEMBLY.

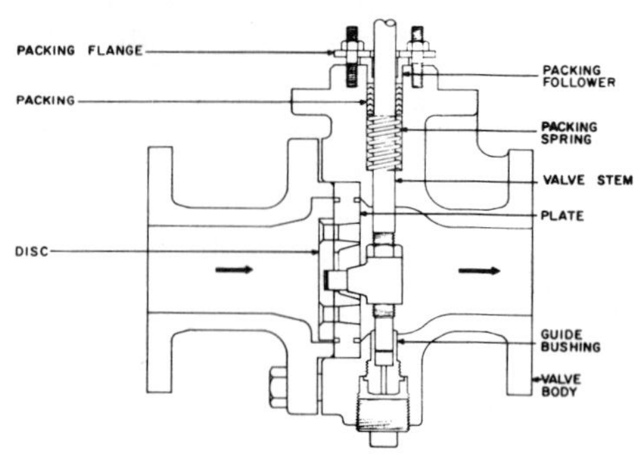

Figure 4. SLIDING GATE VALVE BODY.

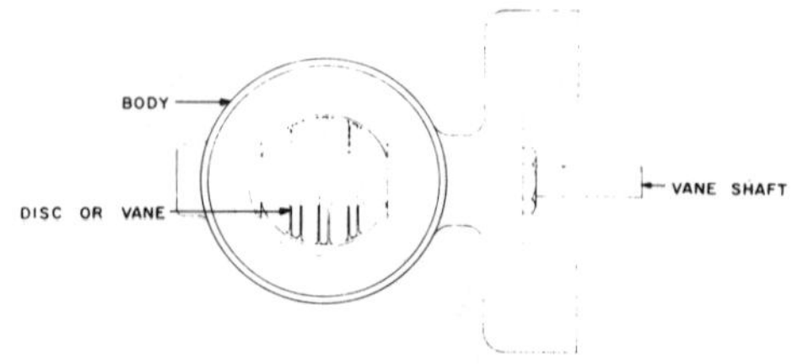

Figure 5. TYPICAL BUTTERFLY VALVE.

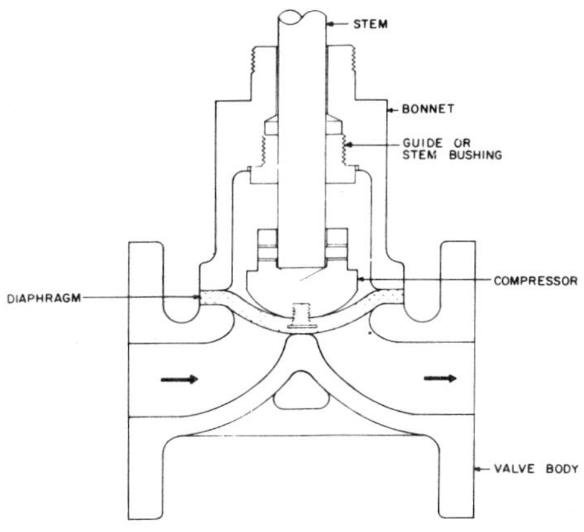

Figure 6. TYPICAL DIAPHRAGM VALVE.

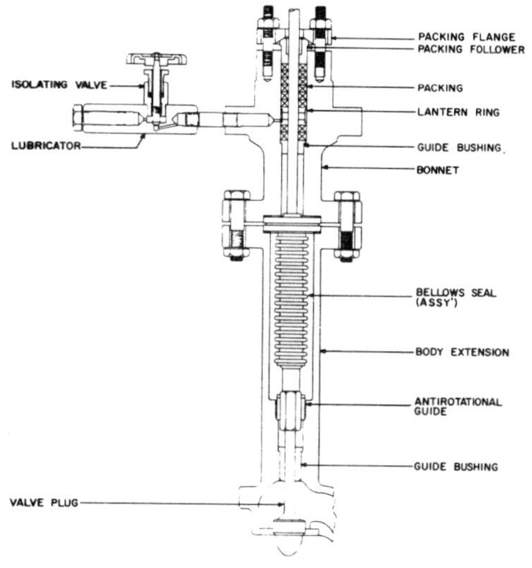

Figure 7. BELLOWS SEAL BONNET.

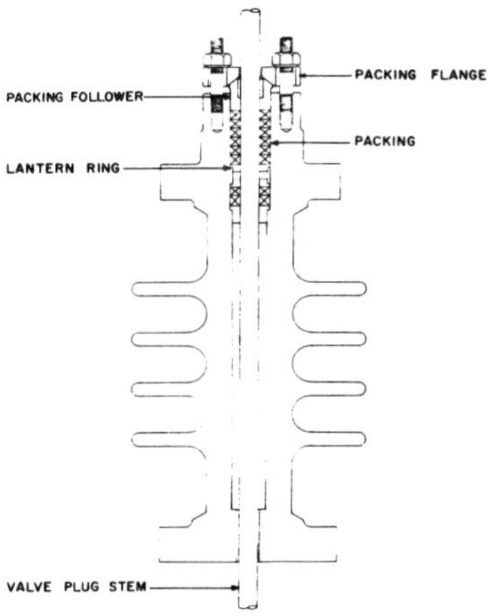

Figure 8. RADIATION FIN BONNET.

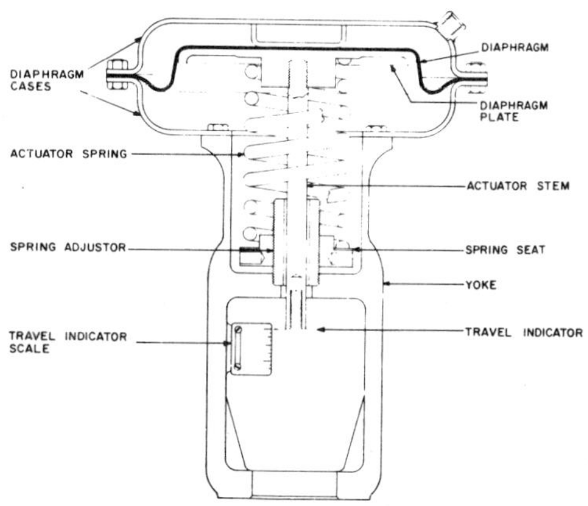

Figure TYPICAL DIRECT ACTING DIAPHRAGM ACTUATOR.

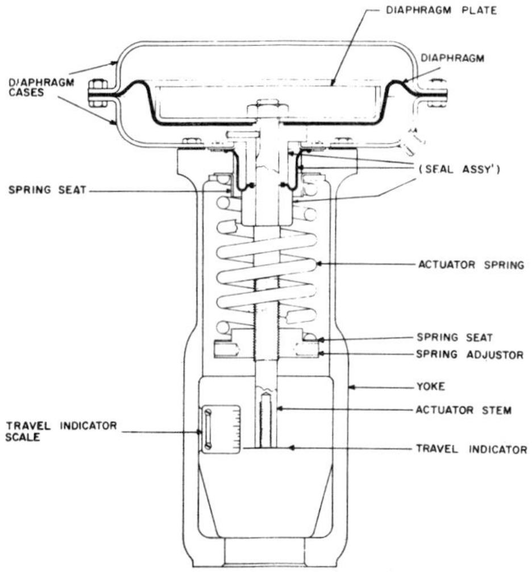

Figure 10. TYPICAL REVERSE ACTING DIAPHRAGM ACTUATOR.

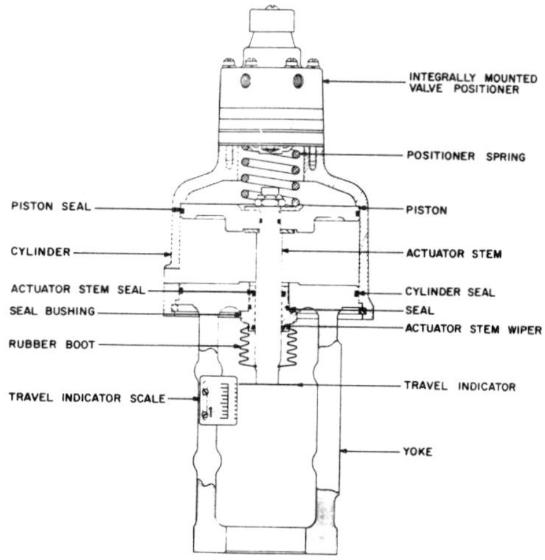

Figure TYPICAL PISTON ACTUATOR.

Part 2
DEFINITION OF CONTROL VALVE COMPONENTS

Actuator Spring: A spring to move the actuator stem in a direction opposite to that created by diaphragm pressure.

Actuator Stem: A rod-like extension of the diaphragm plate to permit convenient external connection.

Ball: The valve closure member in a ball valve.

Ball Valve: A control valve utilizing a rotatable ball shaped valve closure member that is bored, notched, or contoured to obtain various flow characteristics. This member is positioned to close, open, or to control the amount of flow by rotation.

Bellows Seal Bonnet: A bonnet which uses a bellows for sealing against leakage of controlled fluid around the valve plug stem.

Bonnet: The major part of the bonnet assembly, excluding the sealing means.

Bonnet Assembly: An assembly including the part through which a valve plug stem moves and a means for sealing against leakage along the stem. It usually provides a means for mounting the actuator. Sealing against leakage may be accomplished by packing or a bellows. A bonnet assembly may include a packing lubricator assembly with or without isolating valve. Radiation fins or an extension bonnet may be used to maintain a temperature differential between the valve body and sealing means.

Bottom Flange: A part which closes a valve body opening opposite the bonnet assembly. In a three-way valve it may provide an additional flow connection. It may include a guide bushing and, in a three-way valve, may also include a seat.

Butterfly Valve: A control valve that utilizes a rotatable disc or vane as a valve closure member.

Cage: A hollow cylindrical trim element that is a guide to align the movement of a valve plug with a seat ring and also retains the seat ring in the valve body. Often the walls of the cage contains openings which determines the flow characteristics of the control valve.

Center Valve Body: A housing placed between the upper and lower valve bodies to provide another flow connection as in a three-way valve.

Compressor: The part of the stem assembly in a diaphragm valve such as the Saunders patent diaphragm valve which effects flow restriction by forcing the diaphragm toward an internal web.

Control Valve: A valve with a power positioning actuator for moving the valve closure member to any position relative to valve port or ports in response and in proportion to an external signal. The energy for a control valve actuator is provided from an independent source.

Cylinder: The chamber in which the piston moves.

Diaphragm: A flexible pressure responsive element which transmits force to the diaphragm plate and actuator stem.

Diaphragm (As in Diaphragm Valve): The valve closure member in a diaphragm valve such as the Saunders patent diaphragm valve, which provides a variable restriction to flow.

Diaphragm Actuator: An assembly utilizing fluid pressure acting on a diaphragm to develop a force to move the actuator stem. It may, or may not, have a spring for positioning and return of the actuator stem.

Diaphragm Case: A housing, consisting of top and bottom sections, used for supporting a diaphragm and establishing one or two pressure chambers.

Diaphragm Control Valve: A control valve that is actuated by a diaphragm or one having a spring-diaphragm actuator.

Diaphragm Plate: A plate concentric with the diaphragm for transmitting force to the actuator stem.

Diaphragm Valve: A control valve that utilizes a flexible diaphragm as a valve closure member. The diaphragm is forced down toward an internal dividing web to restrict the flow.

Direct Actuator: A diaphragm actuator in which the actuator stem extends with increasing diaphragm pressure.

Disc: The valve closure member in a

slide valve which provides a variable restriction in a port.

Disc: The valve closure member in a butterfly valve which provides a variable restriction in a port. Sometimes called vane.

Extension Bonnet: A bonnet with an extension between the packing box assembly and bonnet flange for hot or cold service.

Globe Valve: A basic control valve type which gets its name from the globular shape of its body. It normally uses the basic valve plug as its valve closure member.

Guide Bushing: A bushing in a bonnet, bottom flange, or body to align the movement of a valve plug with a seat ring. Guiding of a valve plug may be accomplished by an internal part of a bonnet or bottom flange or by a seat ring or seat ring extension or by a cage.

Inlet Port: The port which is connected directly to the upstream side of a fluid system.

Isolating Valve: A hand operated valve between the packing lubricator assembly and the packing box assembly to shut-off the fluid pressure from the lubricator assembly.

Lower Valve Body: A half housing for internal valve parts having one flow connection. For example, the half housing of a split body valve.

Lubricator: See *packing lubricator assembly.*

Outlet Port: The port which is connected directly to the downstream side of a fluid system.

Packing Box Assembly: The part of the bonnet assembly used to seal against leakage around the valve plug stem, including various combinations of all or part of the following:

a. Packing
b. Packing Nut
c. Packing follower
d. Lantern ring
e. Packing spring
f. Packing flange
g. Packing Flange Studs or Bolts
h. Packing Flange Nuts
i. Packing Ring
j. Packing Wiper Ring
k. Felt Wiper Ring

Packing Lubricator Assembly: An optional part of the bonnet assembly used to inject lubricant into the packing box.

Piston: A movable, pressure responsive element which transmits force to the actuator stem.

Piston Actuator: A fluid pressure operated piston and cylinder assembly for positioning the actuator stem in relation to the operating fluid pressure or pressures.

Plate: A separate piece inserted in a slide valve body to form a valve body port.

Port Guided: A design on which the valve plug is aligned by the body port or ports only.

Radiation Fin Bonnet: A bonnet with fins to reduce heat transfer between the valve body and packing box assembly.

Retaining Ring: A split ring that is used to retain a separable flange on the upper and lower valve bodies.

Reverse Actuator: A diaphragm actuator in which the actuator stem retracts with increasing diaphragm pressure.

Rubber Boot: A protective device similar to a flexible bellows, which is used to prevent build-up of abrasive deposits on the valve stem.

Seat: That portion of a seat ring or valve body which a valve plug contacts for closure.

Seat Ring: A separate piece inserted in a valve body to form a valve body port.

Separable Flange: A removable flange which fits over a valve body flow connection. It is generally held in place by means of a retaining ring.

Split Valve Body Assembly: An assembly of an upper valve body, lower valve body, and trim elements. The assembly can also include a center valve body to provide another flow connection as in a three-way valve.

Spring Adjustor: A fitting, usually threaded on the actuator stem or into the yoke, to adjust the spring compression.

Spring Seat: A plate to hold the spring in position and also to provide a flat surface for the spring adjustor to contact.

Stem Connector: A clamp, in two pieces, to connect the actuator stem to the valve stem.

Stem Guided: A special case of top guided in which the valve plug is aligned by a guide acting on the valve plug stem.

Top Guided: A design in which the valve plug is aligned by a single guide in the body adjacent to the bonnet or in the bonnet.

Top And Bottom Guided: A design in which the valve plug is aligned by guides in the body or in the bonnet and in the bottom flange.

Top And Port Guided: A design in which valve plug is aligned by a guide in the bonnet or body and the body port.

Travel Indicator: A pointer, attached near the stem connector, to indicate the position of the valve plug.

Travel Indicator Scale: A graduated scale attached to the yoke for indication of valve travel.

Trim: Those parts of a valve body assembly, excluding the valve body, bonnet assembly, and bottom flange, which are exposed and in contact with the line medium usually consisting of but not limited to the seat ring, valve plug stem, valve plug, valve plug guide, guide bushing and cage.

Upper Valve Body: A half housing for a split body type valve. It contains internal valve parts and has one flow connection. It usually includes a means for sealing against leakage along the stem and provides a means for mounting the actuator.

Valve Body: A housing for internal valve parts having inlet and outlet flow connections. Several common body arrangements are employed, as follows:

a. Single ported having one port and one valve plug.
b. Double ported having two ports and one valve plug.
c. Two-way having two flow connections - one inlet and one outlet.
d. Three-way having three flow connections, two of which may be inlets with one outlet (for conveying or mixing flows), or one inlet and two outlets (for diverging or diverting flows).

Valve Body Assembly: An assembly of a body, bonnet assembly, bottom flange and trim elements. The trim includes a valve plug which opens, shuts or partially obstructs one or more ports.

Valve Closure Member: That part of a valve which is positioned to close, open, or to control the amount of flow.

Valve Plug: A movable part which provides a variable restriction in a port.

Valve Plug Guide: That portion of a valve plug which aligns its movement in either a seat ring, bonnet, bottom flange or any two of these.

Valve Plug Stem: A rod extending through the bonnet assembly to permit positioning of the valve plug.

Vane: The valve closure member in a butterfly valve which provides a variable restriction in a port. Sometimes called disc.

Yoke: A structure by which the diaphragm case assembly is supported rigidly on the bonnet assembly.

Part 3
CONTROL VALVE TERMINOLOGY

Actuator Stem Force: The net force from an actuator that is available for actual positioning of the valve plug.

Actuator Travel Characteristic: Relation between per cent rated travel and diaphragm pressure. This may be stated as an inherent or installed characteristic.

Capacity: Rate of flow through a valve under stated conditions.

Clearance Flow: That flow below the minimum controllable flow with the valve plug not seated.

C_v: See *valve flow coefficient.*

Dead Band: The amount the diaphragm pressure can be varied without initiating valve plug motion.

Diaphragm Pressure Span: Difference between the high and low values of the diaphragm pressure range. This may be stated as an inherent or installed characteristic.

Dynamic Unbalance: The net force produced on the valve plug in any stated open position by the fluid forces acting upon it.

Effective Area: In a diaphragm actuator, the effective area is that part of the diaphragm area which is effective in producing stem force.

Equal Percentage Flow Characteristic: An inherent flow characteristic which, for equal increments of rated travel, will ideally give equal percentage changes of the existing flow.

Flow Characteristic: Relation between flow through the valve and per cent rated travel as the latter is varied from 0 to 100 per cent. This is a general term. It should always be designated as either inherent flow characteristic or installed flow characteristic.

Inherent Diaphragm Pressure Range: The high and low values of pressure applied to the diaphragm to produce rated plug travel with atmospheric pressure in the valve body.

Inherent Flow Characteristic: Flow characteristic when constant pressure drop is maintained across valve.

Inherent Rangeability: Ratio of maximum to minimum flow within which the deviation from the specified inherent flow characteristic does not exceed some stated limits.

Installed Diaphragm Pressure Range: The high and low values of pressure applied to the diaphragm to produce rated travel with stated conditions in the valve body.

Installed Flow Characteristic: Flow characteristic when pressure drop across valve varies as dictated by flow and related conditions in system in which valve is installed.

Leakage: Quantity of fluid passing through an assembled valve when valve is in fully closed position under stated closure forces, with pressure differential and temperature as specified. Leakage is commonly measured and may be expressed as a percentage of capacity at rated travel or as a cumulative quantity over a specified time.

Linear Flow Characteristic: An inherent flow characteristic which can be represented ideally by a straight line on a rectangular plot of flow versus per cent rated travel. Therefore, equal increments of travel yield equal increments of flow at a constant pressure drop.

Modified Parabolic Flow Characteristic: An inherent flow characteristic which provides fine throttling action at low valve plug travel and approximately a linear characteristic for upper portions of valve travel.

Normally Closed Control Valve: One which closes when the diaphragm pressure is reduced to atmospheric.

Normally Open Control Valve: One which opens when the diaphragm pressure is reduced to atmospheric.

Quick Opening Flow Characteristic: An inherent flow characteristic in which there is maximum flow with minimum travel.

Rated C_v: The Value of C_V at the rated full open position of the valve plug.

Rated Travel: Linear movement of the valve plug from the closed position to the rated full open position.

Seat Load: The contact force between the valve seat and valve plug.

Spring Rate: Force change per unit change in length.

Static Unbalance: The net force produced on the valve plug in its closed position by the fluid pressure acting upon it.

Stem Unbalance: The net force produced on the valve plug stem in any position by the fluid pressure acting upon it.

Unit Sensitivity: The per cent increase in existing flow for an increase in movement of 1 per cent of rated travel at a stated operating point.

Valve Flow Coefficient (C_v): Number of U.S. gallons per minute of 60° F water that will flow through a valve with one psi pressure drop under stated conditions. Stated conditions include pressure and per cent of rated travel.

Part 4
CONTROL VALVE SPECIFICATIONS

The writing of control valve specifications is an extremely important segment of the complete design, purchase and fabrication of any control system.

When the wrong control valve is specified, due to either incomplete or erroneous specifications, the replacement of that valve becomes costly to both the user and the control valve manufacturer and often results in undue project delay.

Many control valve users have developed standard control valve specification forms that must accompany all control system project requisitions and they are designed to eliminate as much misunderstanding as possible regarding the valve specifications. These standard control valve specification forms are often based on the standardized specification sheet for control valves that is contained in the Instrument Society of America Recommended Practice RP 20.1, "Specification Forms for Instruments, Gages, Thermocouples, Orifice Plates and Flanges, Control Valves, and Pressure Safety Valves". This specification sheet is presently available from the Instrument Society of America, and will soon be available in revised format.

The following paragraphs contain most of the items that should be considered in any control valve specification form. There are undoubtedly some items not covered in this text that are peculiar and necessary to individual situations.

Refer to Parts 1, 2, and 3 of this Chapter for appropriate definitions and terminology; in addition, refer to Fluid Controls Institute's FC 58-2, "Standard Measurement Procedure for Determining Control Valve Flow Capacity," and FCI 62-1, "Standard Formulas for Sizing Control Valves"; and the Instrument Society of America's Recommended Practice RP 20.1, "Specification Forms for Instruments, Gages, Thermocouples, Orifice Plates and Flanges, Control Valves, and Pressure Safety Valves".

Specifications

Service: On-off, Proportioning, etc.

BODY

Type of Body: Globe, angle, three-way mixing, three-way diverting, butterfly, Saunder's patented, ball plug cock, gate, etc.

Body Size: Specify nominal body size in inches.

Port Size: Specify port size in inches.

Guiding: Specify type of plug or inner valve guiding such as top and bottom, trunnion, top, port, cage, etc.

Number of Ports: 1 - Single Port (SP); 2 - Double port (DP); 3 - Three way; etc.

Plug Form: Percentage, linear, parabolic, quick-opening.

Packing Material: Teflon V-ring, Teflon - Asbestos, Durametallic, Buna N "O"-ring, etc.

Lubricator: Indicate only when required.

Bonnet type: Standard, plain, or finned for high temperature, plain extension for low temperature, bellows seal, diaphragm seal, etc.

End Connections: Specify flanged and describe if raised face (R.F.); ring joint (R.T.J.), large male (L.M.); etc.; screwed; socket welded, butt welded, with pipe schedule.

Material, Body & Bonnet: Specify steel, cast iron, 316 stainless steel, etc.

Material, Plug: Specify plug (inner valve, ball or disc) material, such as 316 stainless steel, 416 stainless steel, monel, alloy #6, etc.

Material, Seat: Specify seat ring(s) material.

Material, Posts: Specify post(s) material.

Material, Guides: Specify guide(s) (cage, bushings or bearings) material.

Material, Stem: Specify stem (shaft) material.

Composition Seating: Specify composition seating ("O" rings, composition disc, liners in butterfly valves, diaphragms in Saunders patented) by material such as Teflon, Kel-F, Buna-N, if desired.

ACTUATOR

Type of Actuator: Specify type of actuator such as diaphragm, piston or cylinder, electric-hydraulic, electric motor, solenoid, etc.

Close at and Open at: Specify signal span (psi, volts, amps, etc.) at which valve shall be closed and open. Note: This is signal to valve operator and is not necessarily controller output signal if a positioner and/or transducer is used.

Push down To: Specify if valve is to "Open" or "Close".

Flow Action To: Specify if flow action tends to "Open" or "Close" valve.

Failure Position: Specify position valve is to take on loss of actuating power.

Power Supply: Specify source of power supply such as 115 volt, 60 cycle, 80 psig air, etc. Note: Normal *minimum* air supply should be specified if lock-up and fail safe devices are required. This is especially important when using piston operators.

Size and Travel: Specify size or effective area and/or travel in inches or degrees rotation of actuator.

Handwheel and Location: Specify type of handwheel, if required, such as top mounted, or side mounted on operator, or side mounted on valve body such as butterfly or ball valves.

POSITIONER

Type: Side mounted, Top mounted, etc.

Filter Regulator: Indicate if required.
Gauges: Indicate if required.
By-pass: Indicate if required.

Input Signal: Indicate positioner input signal span from controller.

Output Signal: Indicate corresponding output signal span to actuator.

TRANSMITTER OR CONTROLLER

Input Signal: Indicate input signal span (if controller indicate measured pressure span and material if bourdon tube or bellows. If temperature controller, indicate measured temperature span and class of thermal system.)

Output Signal: Indicate corresponding output signal span.

SERVICE CONDITIONS

Fluid Units: Specify units of fluid flow, such as G.P.M., B/H, #/Hr, SCFH, etc.

Fluid: Name of fluid and state if vapor or liquid if it is not apparent.

Maximum Valve Capacity: Maximum valve capacity when valve is 100% open calculated with sizing pressure drop. Note: Normally this information is supplied by the valve manufacturer and is important for relief valve sizing and/or future capacity.

Valve C_v: Valve coefficient at 100% open - supplied by valve manufacturers.

Quantity Maximum: Maximum quantity of fluid valve is required to pass.

Flow C_v: Calculated flow coefficient at maximum flow using sizing pressure drop.

Quantity Normal: Normal Quantity of fluid valve is required to pass.

Flow C_v: Calculated flow coefficient at normal flow using sizing pressure drop.

Inlet Pressure: Normal inlet pressure with units (psia, psig, inches H_2O press, etc.) Note: Must use *normal* inlet pressure for gas or steam sizing.

Delta P Sizing: Pressure drop (inlet pressure minus outlet pressure with units; psi, inches W.C., etc.). For all flows for sizing of valve port(s).

Maximum shut-off Pressure: The maximum pressure drop in the shut off position to determine proper actuator size [inlet pressure minus outlet pressure with units, psi, inches WC (water column) etc.] when valve is closed.

Temperature, Maximum: Show maximum temperature with units (°F, °C, °K) to establish valve body rating and trim selection.

Temperature, Normal: Normal temperature is required for gas and steam sizing.

Specific Gravity: Specific gravity of flow at normal operating temperature.

Molecular Weight: Molecular weight of fluid.

Viscosity: Show viscosity at flowing temperature with units (c.p.; S.S.U., etc.).

Percent Flash: Percent of maximum flow that will flash to the vapor state due to pressure drop across the valve.

ACCESSORIES

Also include details on any accessories; such as, solenoid valve, limit switches, travel limit stops, air tanks, lock-up in last position relays, etc.

REFERENCES

1. ASME Standard No. 112, 1961, *Diaphragm Actuated Control Valve Terminology,* 14 pp, American Society of Mechanical Engineers, New York, New York.

2. FCI 58-2, 1963, *Standard Measurement Procedure for Determining Control Valve Flow Capacity,* 4 pp, Fluid Controls Institute, Pompano Beach, Florida.

3. FCI 62-1, 1962, *Standard Formulas for Sizing Control Valves,* 8 pp, Fluid Controls Institute, Pompano Beach, Florida.

4. ISA RP20.1, 1956, *Specification Forms for Instruments, Gages, Thermocouples, Orifice Plates and Flanges, Control Valves, and Pressure Safety Valves,* 48 pp, Instrument Society of America, Pittsburgh, Pennsylvania.

CHAPTER 3

VALVE BODIES

H. D. Baumann

Masoneilan International, Inc.
Norwood, Massachusetts

with portions contributed by

Paul Wing, Jr.
Masoneilan International, Inc.

and

Rolf Weil
Jamesbury Corporation
Worcester, Massachusetts

CONTENTS

INTRODUCTION

A control valve consists of two major sub-assemblies, a valve body sub-assembly and an actuator. The valve body sub-assembly is the portion that actually controls the passing fluid. It consists of a housing, internal trim, bonnet and sometimes a bottom flange. The valve body sub-assembly is a pressure carrying part (a pressure vessel) which must meet all of the applicable pressure, temperature and corrosion requirements in the same manner as a normal pressure vessel.

Body sub-assemblies occur in many shapes and working arrangements depending upon the individual service conditions and piping requirements. Each type has certain advantages and disadvantages for given service requirements and should, therefore, be selected with care.

TYPES OF VALVE BODIES
(Reciprocating Stem)

Globe Styles

The most common control valve body style is in the form of a globe, i.e., it is similar in appearance to a hand valve. Such a control valve body can be either single- or double-seated. *Single-seated* valves, as illustrated by Figure 1, are usually employed when tight shut-off is required, or in sizes of 1 inch and smaller where the partial cancellation of unbalanced forces acting on the valve stem is unimportant as a factor in actuator selection. Tight shut-off in this case usually means that the maximum expected leakage is less than 0.01% of the maximum valve C_V when subjected to an air test with 50 psi upstream and 0 psig downstream pressure.

Single-seated valves usually have a top guided construction, i.e., the valve plug is guided within the lower portion of the valve bonnet or top closure. Such top guiding minimizes the effect of the weight of the valve plug and tends to increase resistance to trim vibration. It also allows a somewhat higher flow capacity than top and bottom guided valves for a given orifice size.

A *double-seated valve,* as shown in Figure 2, is generally top and bottom guided. Here the practical leakage figure approaches 0.5% of the rated C_V, because of the fact that it is nearly impossible to close the two ports simultaneously, particularly when thermal expansion provides additional distortion after the valve is installed. However, the advantage of double-seated construction lies in the reduction of required actuator forces, because the hydrostatic effects of the fluid pressure acting on each of the two seats, tend to cancel each other out. Double-seated valves have upper and lower ports of different diameters (to allow withdrawal of the smaller plug through the larger port), which contributes to the unbalanced condition. Additionally, complete cancellation of these forces is not possible, because of the hydrodynamic effects of the fluid that passes the plug contour at high velocity. Fluid passing the lower seat (tending to close the plug) has a tendency to "suck" the plug into the seat, thereby creating a dynamic imbalance between this force and the differential pressure acting across the upper plug area. The dynamic imbalance can reach 40% of the equivalent single seat hydrostatic force. Such a

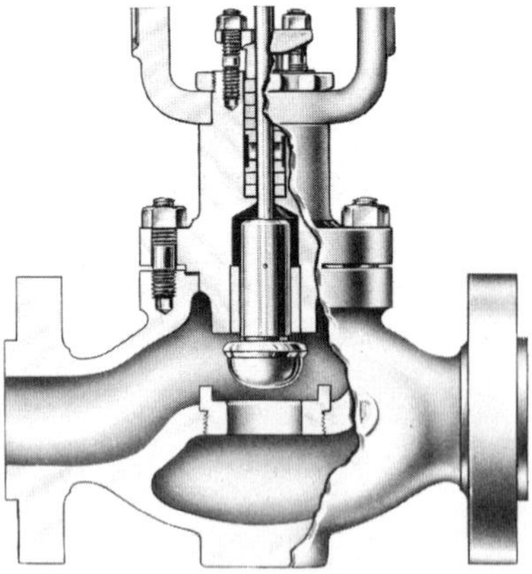

Figure 1. GLOBE TYPE SINGLE SEATED VALVE.

Courtesy Masoneilan International, Inc.

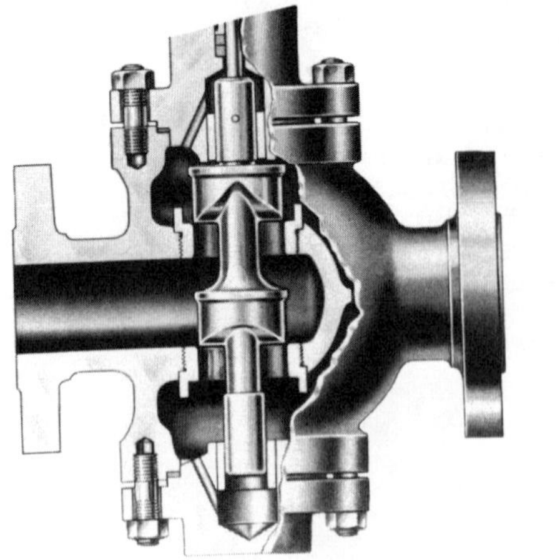

Figure 2. DOUBLE SEATED VALVE - AIR TO CLOSE.

Courtesy Masoneilan International, Inc.

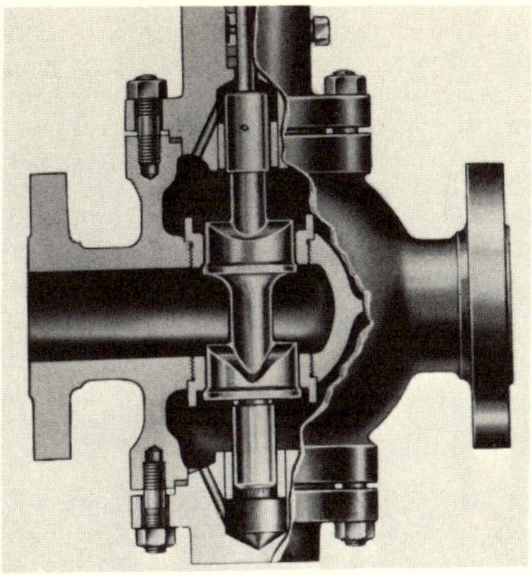

Figure 3. DOUBLE SEATED VALVE - AIR TO OPEN.
Courtesy Masoneilan International, Inc.

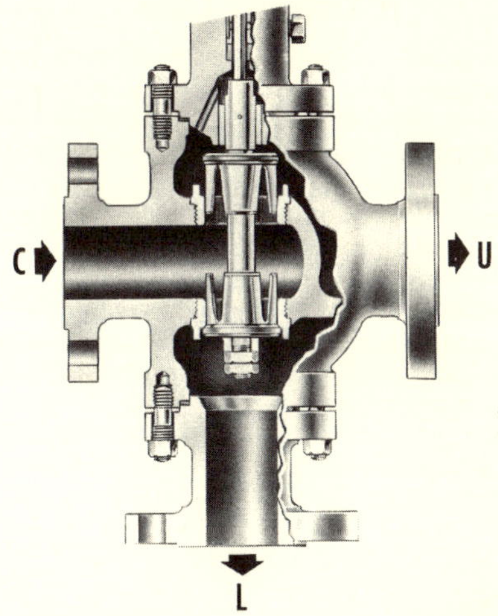

Figure 4. THREE WAY VALVE - DIVERTING SERVICE.
Courtesy Masoneilan International, Inc.

double-seated trim is therefore more justly called "semi-balanced".

With the top and bottom guided construction, the stroke direction can be reversed by simply inverting the valve body. This means that when a direct actuator pushes down, as a result of a signal increase, the valve will tend to close as shown in Figure 2. However, with the inverted body (body sub-assembly inverted 180 degrees), air pressure will tend to open the valve, as illustrated in Figure 3.

Three Way Valves

Three-way valves are a design extension of a typical double-ported globe valve. Here a distinction is made between a valve used for diverting service and one used for mixing or combining service. A typical three-way valve for diverting service as shown in Figure 4, has a modified double-ported body with the lower plug seating opposite from the normal shut-off position. Either a direct or a reverse actuator is used, again depending upon the desired fail-safe action. Note that the plug has additional rib guiding in the orifice to compensate for the omission of the lower guide post. The body has a internal bridge to separate the right hand and the lower outlet. A diverting valve might be used for a heat exchanger bypass where the heating medium enters port "C". Part of the fluid leaves port "U" to bypass the exchanger. The remaining portion of the fluid then goes to the heat exchanger, through port "L", to heat an independent process stream and, then rejoins the bypass fluid stream from port "U".

Figure 5 illustrates a three way valve used for combining service. Here, two separate fluids entering port "L" and port "U," respectively, are combined in a desired ratio and leave through the common port "C." The ratio between the amount of fluid coming through either port "U" or "L" is determined by the plug position. An upward movement of the plug decreases the flow passing port "U" and at the same time increases the flow area for port "L." Note again, that skirt guiding is employed to provide additional guiding in line with the upper guide post.

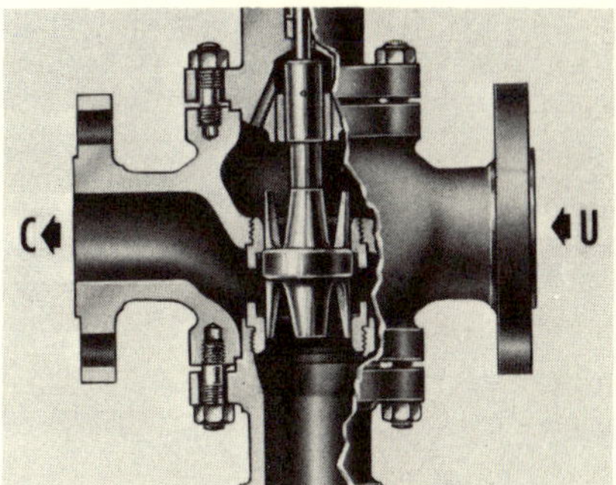

Figure 5. THREE WAY VALVE - COMBINING SERVICE.
Courtesy Masoneilan International, Inc.

Casual observation of Figures 4 and 5 might lead to the conclusion that hydrostatic forces acting on either plug are somewhat balanced as they are in valves having a double-seated construction. This conclusion is erroneous, however, since the pressure level in either port (Uand L) could vary considerably and, because the individual plug forces act in opposite directions. It is, therefore, prudent to treat a three-way valve in the same manner as two opposed, single-seated valves for purposes of actuator selection.

Angle Valves

Angle valves are single-seated valves with special body configurations to suit specific piping or flow requirements. A typical angle valve, with streamlined body interior, is shown in Figure 6. This valve was specifically designed for coking hydrocarbons. The streamlined interior passage tends to prevent an accumulation of solids on the body wall. This type of valve has been used, somewhat erroneously in the past, for high pressure-drop service. The underlying thought was to keep the turbulence, that is created by the throttling process, away from the valve's internal parts. However, in doing this, the energy conversion takes place in the downstream pipe, occasionally leading to severe pipe vibration and noise problems. Another disadvantage, in most applications, is the relatively high pressure recovery obtained with the streamlined flow pattern. High pressure recovery means a low cavitation index and severe cavitation under low or moderate pressure drops with liquid media (this is covered in Chapter 6).

In addition to the specific application listed above, angle valves may be used in cases where the piping layout does not allow the installation of a globe valve and may be used for handling certain erosive fluids, such as abrasive catalyst material. In the latter, a discharge into the outlet pipe (with flow direction tending to close the plug) prevents erosion of the inside of the valve housing. If cavitation or flashing is inevitable, an angle valve may be arranged to discharge directly into a vessel, or other enlarged fluid volume, thus avoiding damage.

An additional advantage of some angle valves is a self draining design which is of value when handling certain dangerous fluids such as radioactive liquids.

Y-Style Valves

Another valve which is self draining when installed at a certain angle is the Y valve. This design is commonly applied in molten metal installations and in cryogenic installations where the cool-down weight has to be kept to a minimum. Figure 7 illustrates the use of such a valve for liquid hydrogen. Here a special insulating jacket is placed around the valve body, to allow the maintenance of a vacuum around the valve housing, to prevent heat transfer. Note that bellows seals are employed to seal the end connection against similar vacuum jacketing on the adjoining pipe. The bellows compensates for thermal distortions and machining tolerances. Y valves have a relatively high flow capacity (C_V) and a low cavitation factor (F_L) as shown by Table 1.

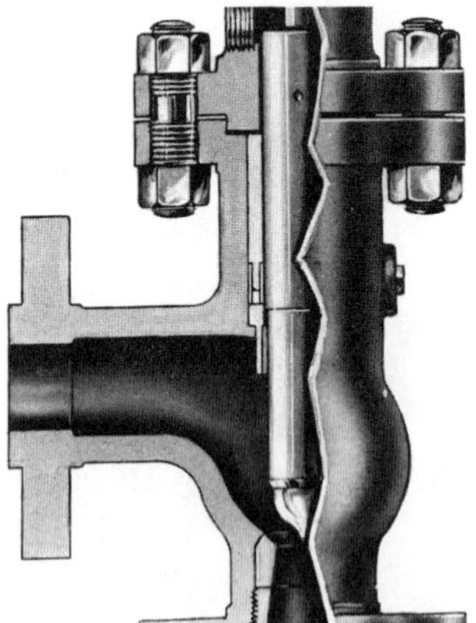

Figure 6. ANGLE VALVE - STREAMLINED BODY.

Courtesy Masoneilan International, Inc.

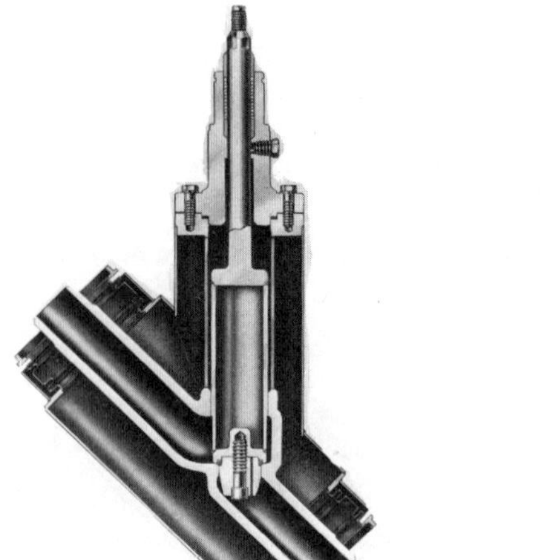

Figure 7. "Y" VALVE - VACUUM JACKETED FOR CRYOGENIC SERVICE.

Courtesy Masoneilan International, Inc.

Table 1. RELATIVE FLOW CAPACITIES OF CONTROL VALVES

Valve Type	C_d (1)	$C_d F_p$ (2)	$C_d F_L$ (3)
Double-seat globe	12	11	11
Single-seat top guided globe	11.5	10.8	10
Single-seat split body	12	11.3	10
Sliding gate	6-12	6-11	not available
Single-seat top entry cage	13.5	12.5	11.5
Eccentric rotating plug (Camflex)	14	13	12
60° open butterfly	18	15.5	12
Single-seat Y valve (300 & 600 lb.)	19	16.5	14
Saunders type (unlined)	20	17	not available
Saunders type (lined)	15	13.5	not available
Throttling (characterized) ball	25	20	15
Single-seat streamlined angle (flow-to-close)	26	20	13
90° open butterfly (average)	32	21.5	18

NOTE: This table may serve as a rough guide only since actual flow capacities differ between manufacturer's products and individual valve sizes.

(1) Valve flow coefficient $C_v = C_d \times d^2$ (d = valve dia. in inches)

(2) C_v/d^2 of valve when installed between pipe reducers (pipe dia. = 2 x valve dia.).

(3) C_v/d^2 of valve when undergoing critical (choked) flow conditions.

Split-Body Valves

Certain corrosive services in the chemical industry demand either frequent inspection or replacement of the valve trim. Such frequent maintenance service is difficult to accomplish with standard globe-type valves, having screwed-in seat rings, for example. Split-body valves were designed to provide easy maintenance. Here the seat ring is clamped between two halves of a valve body, as shown in Figure 8 and can be removed with relative ease. The weight and, therefore, the cost of exotic valve materials is further reduced by the use of slip-on flanges. These flanges, meeting standard ANSI dimensional requirements, can be made of ordinary carbon steel. Each flange locks behind a split-ring and pulls the body end connections against a gasket and the mating line flange.

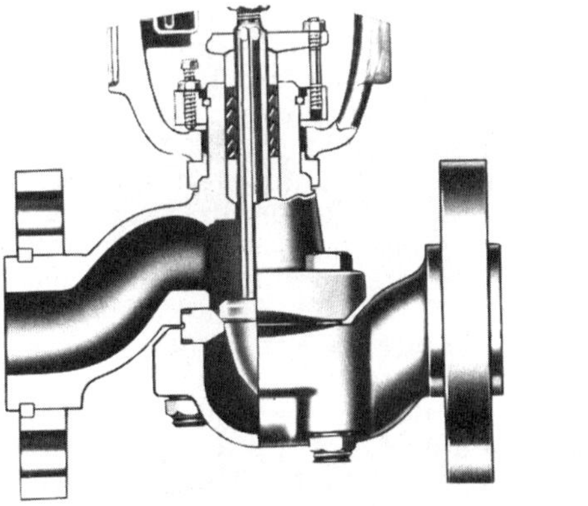

Figure SPLIT BODY VALVE WITH SLIP-ON FLANGES.

Courtesy Masoneilan International, Inc.

Note that the yoke connection near the stuffing box, has a similar split-ring connection, that is less affected by corrosion than the more universal threaded retainer, thereby facilitating maintenance.

Stem guiding of the plug, with valves handling slurries or crystalizing fluids, eliminates problems that would be caused by build-up on top of an ordinary plug guide.

One disadvantage of the split-body valves is that welded end connections, a common requirement in power plant service, are not practical because the body halves cannot be separated for removal of the seat ring.

Cage Valves

So-called "top entry" or *cage* valves have the advantage of trim removal. Figure 9 shows a typical top entry valve with unbalanced, single-seated trim. Valves of this type usually have streamlined body passages to permit increased flow capacity. The inner valve parts, often referred to as "quick change trim," can easily be removed after removing the bonnet, because of the absence of internal threads. Characteristic of all the various cage designs is the fact that a flexible metal gasket, with either asbestos or Teflon filler, is placed between the seat ring and the body bridge. This gasket in turn is compressed by the cage and serves as a spacer between the bonnet and the seat ring.

The cage and seat ring are sometimes designed as one piece. The valve plug then assumes the form of a flat disc, and the flow characteristic is determined by shaped windows, usually cast in the circumference of the cage.

The valve plug can also be made as a piston having a hollow internal passage, that permits the fluid pressure to communicate to both sides of the plug. A sliding seal, located in a groove near the top of the piston, seals the upper plug area against the outlet portion of the valve. This balanced design, as shown in Figure 10, tends to cancel out the hydrostatic forces acting on the plug and leads to a great reduction in the required actuator forces. However, as in double seated valves, a substantial increase in seat leakage must be tolerated.

As mentioned previously, the biggest advantage of the top entry valve is the ease of maintenance. The valve trim can be replaced, if necessary, without removing the valve body from the line. This is an important consideration in this age of trade unions, because maintenance men dealing with pipe flanges usually belong

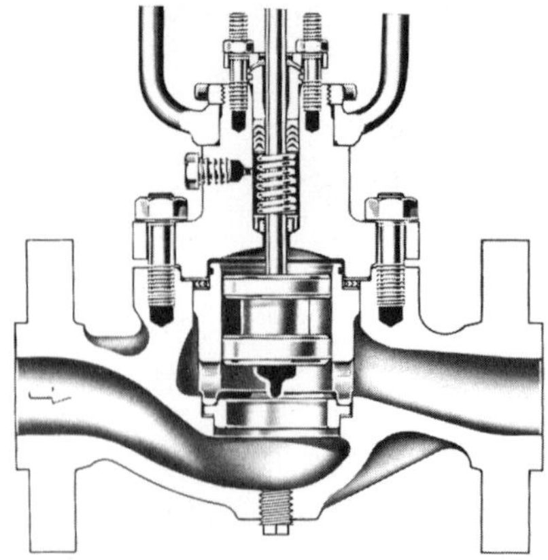

Figure 9. TOP ENTRY VALVE - UNBALANCED.
Courtesy Fisher Controls Company

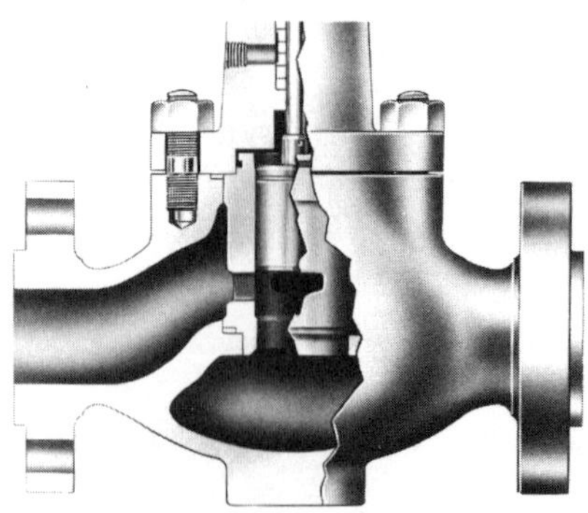

Figure 10. TOP ENTRY VALVE - BALANCED.
Courtesy Masoneilan International, Inc.

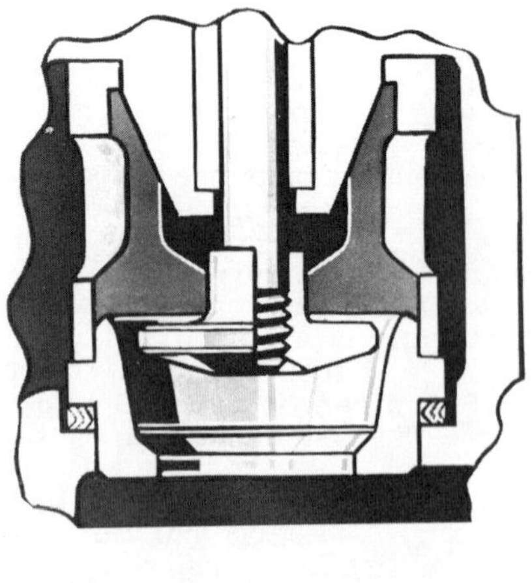

Figure 11. TOP ENTRY VALVE - CONTOURED REVERSIBLE SEAT RING.
Courtesy A. W. Cash Company

Figure 12. SLIDING GATE VALVE.
Courtesy Jordon Valve Division, Richards, Industries, Inc.

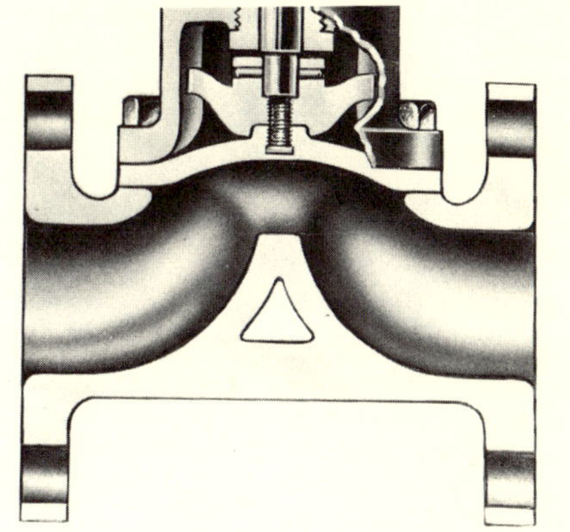

Figure 13. SAUNDERS VALVE.
Courtesy Masoneilan International, Inc.

to a different union than the mechanics that service the control valve. This is one of the reasons why this particular valve type has gained wide acceptance, since its quite-recent introduction.

A modified variation of an unbalanced top entry valve is illustrated in Figure 11. Here, the seat ring has an internal contour to provide for an equal percentage flow characteristic in conjunction with the disk type plug and can be inverted 180 degrees to convert from "fail-open" to "fail-close" stroke direction. This valve is stem guided only, which is an advantage for slurry applications.

Sliding Gate Valves

The slide valve approach is another way to reduce actuating forces in control valves. A typical design is shown in Figure 12, which consists of a two-piece housing, clamping a stationary gate that serves as a valve seat. Another "slideable" seal, pressed against the seat by the upstream fluid pressure, serves to adjust the flow rate by gradually uncovering the stationary slots in the seat ring gate. The valve is wide-open when the slots in the sliding member and the stationary gate are completely lined up. Both plates are lapped to light band flatness for tight shut-off. Standard material is chrome-plated, chrome-nickel type, stainless steel. This type of trim is relatively vibration proof, since the throttling element is pressed firmly against the seat at all lift positions.

Saunders Valves

The Saunders patent valves are used primarily for the control of corrosive fluids, heavy liquids, or liquids having solid particles in suspension. A typical valve, shown in Figure 13, consists of a housing, having a central raised weir and a moveable diaphragm, clamped between the body and the bonnet. This diaphragm is pressed against the weir or central body portion, to provide tight shut-off. The flow characteristic is determined by the distance between the weir and the diaphragm. Advantages are low cost, packless construction, and ease of maintenance. Disadvantages include a bad flow characteristic and high and non-uniform actuator force requirements. In addition, this valve is usually limited to fluid temperatures of below 180° F. If, the valve is to handle corrosive or dangerous fluids, it is prudent to specify a special bonnet with a packing box, to prevent leakage in the event of diaphragm failure.

The valve is available also with glass, plastic, lead or Teflon liners to provide extra corrosion resistance.

ROTATING TYPES OF CONTROL VALVES [1]

In recent years there has been a substantial increase in the use of valves broadly classified as "rotary." All types of rotary valves share certain basic advantages and disadvantages. Among the advantages are low weight, simplicity of design, high relative C_V, more reliable and friction-free packing, and generally low initial cost. They are generally not suitable in size, below 1 to 2 inches. Pressure-drop ratings are limited, particularly in the larger sizes, and the operating shaft must be designed to support a fairly heavy side-thrust that is not encountered in reciprocating stem valves. These points

[1] This section contributed by Paul Wing, Jr., Masoneilan International, Inc.

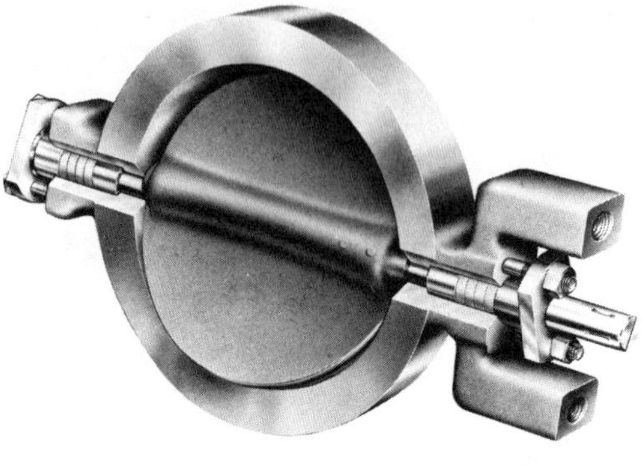

Figure 14. BUTTERFLY VALVE - UNLINED.
Courtesy Masoneilan International, Inc.

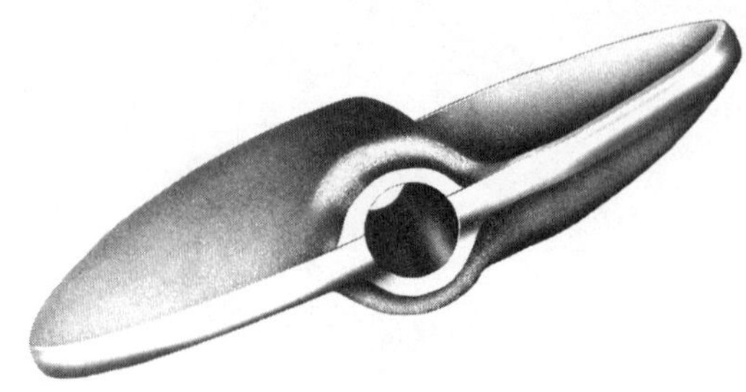

Figure 15. BUTTERFLY VALVE - LOW TORQUE VANE DESIGN.

Courtesy Masoneilan International, Inc.

Figure 16. BUTTERFLY VALVE - LOW TORQUE VANE DESIGN
Courtesy Fisher Controls Company

are brought out in more detail through the following discussion of some basic designs.

Butterfly Valves

The most common type of rotary valve used for control is the butterfly valve, as shown in Figure 14. The typical application range is in sizes from 2 inches through 36 inches or larger, for low or moderate pressures, or on unusual applications involving large flows at high static pressures, but with limited pressure-drop.

The typical valve, on throttling service, is limited to 60° opening resulting in a C_V rating of about 18 d^2. In all metal construction, the vane to body clearance in the closed position gives a leakage flow that is generally equivalent to the leakage of double-seated valves (0.5% of the rated C_V). Many design modifications are available which minimize leakage, including angle-seating, and the use of piston rings. Because of the simple body design, elastomer inserts can be easily adapted to give tight shut-off within the temperature limits of the insert material. Typical liner materials are neoprene, Buna-N®, or for high temperature and special chemical services, Viton A®. Actuator force requirements are dictated by a combination of two factors; the friction load imposed by side-loading on the shaft, due to the differential pressure; and the dynamic torque induced by the flow around the vane. The dynamic torque acts in the closing direction and, in the conventional design, reaches a maximum at about 70° open.

In designs recently introduced, special shaped vanes, shown in Figures 15 and 16, are used to reduce the dynamic torque and to permit 90° operation for increased capacity on certain installations. Another design modification involves an eccentrically-mounted vane to permit, for example, tight closure on cryogenic fluids using a seal of Teflon or Kel-F®, which may be mounted either within the body or on the circumference of the vane.

The most common body design is the flangeless "wafer" type for bolting between line flanges, as shown in Figure 14. ANSI body ratings are used, but the valves are also rated for maximum pressure-drop in the closed position and in the sixty degrees open position.

The rotary shaft and its supporting bearings are an important factor in assuring the success of butterfly valves in control service. Bearing materials range from reinforced Teflon for very low to moderate temperatures and pressures, up to combinations of the hard-facing alloys that permit use at temperatures well above 1000° F. For many years, outboard bearings were specified, almost exclusively. These were located outside the rotary packing box and were not in contact with the process fluid. This simplified the bearing design, but resulted in eccentric loading on the packing. The recent trend has been to inboard bearings, which are now practical, because of the availability of wear and corrosion-resistant materials.

Properly selected, the butterfly valve, generally in sizes 4 inch and above, offers the advantages of simplicity, low cost, light weight, and space saving, in combination with good flow-control characteristics. For moderate temperatures and pressures, the elastomer-lined valve includes the advantage of tight shut-off. In selecting the valve, consideration must be given to the possibility of cavitation on liquid flow, due to the low pressure recovery coefficient (F_L); the possibility of damage due to water hammer with fast closing on liquid service; and compensation for the effect of pipe reducers in computing the valve capacity, because of the high basic C_V rating.

Eccentric Cylindrical Plug Valves

The valve design shown in Figure 17, is a modification of the plug cock widely used for shut-off service. The eccentrically shaped cylindrical plug cooperates with an off-axis, seating-face within the body, permitting good shutoff, but easy operation, because there is no rubbing contact in the body. An elastomer facing is required for the shut-off feature.

The valve is available in sizes from 1/2 inch to 36 inches, and is available in a wide variety of materials. The temperature range is generally limited, and the pressure ratings are to 250 psi, decreasing as the valve size increases. Relative capacity is high and cost is low. It has been found useful on a wide variety of services, particularly for handling corrosive fluids, highly viscous liquids, or those containing solids in suspension, such as paper stock.

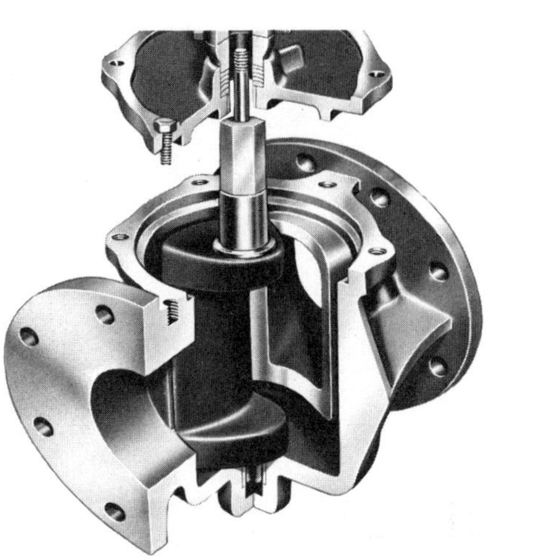

Figure 17. ECCENTRIC CYLINDRICAL PLUG VALVE.
Courtesy Dezurik Corporation

Eccentric-Rotating Plug Valve

A recently introduced valve, of interest, is the eccentrically-rotating plug valve shown in Figure 18. It is available in ratings through 600 lb. ANSI, in sizes from 1 inch to 12 inches. The normal rotary stroke is 50° and an essentially-linear, inherent flow characteristic is obtained with a C_V rating similar to that in the modern, high-capacity, top-entry, cage-guided globe valve. The eccentric motion of the spherical face of the plug reduces the operating torque requirements and permits stable operation when installed for either flow-to-open or flow-to-close. The seating action is positive and the sliding seal-friction associated with ball valves is eliminated. Tight shut-off is obtained with relatively low applied force, because minor misalignment is compensated by flexing of the plug support-arms.

Flangeless construction is used to reduce weight and to allow all ratings of up to 600 lb. ANSI, with a single body machining. A short, plain, bonnet extension, extends the standard temperature range from -320° F to 750° F.

In contrast to other body sub-assemblies, this valve requires a special long-stroke, spring-opposed, rolling-diaphragm actuator. The valve with no modifications, other than an optional stainless steel body, can cover a large percentage of all process control requirements.

BALL VALVES[1]

History

Since 1960, the acceptance of a ball valve as a modulating device has grown rapidly. Although the ball valve dates back many years in concept, it was not until the early fifties that an alternative to

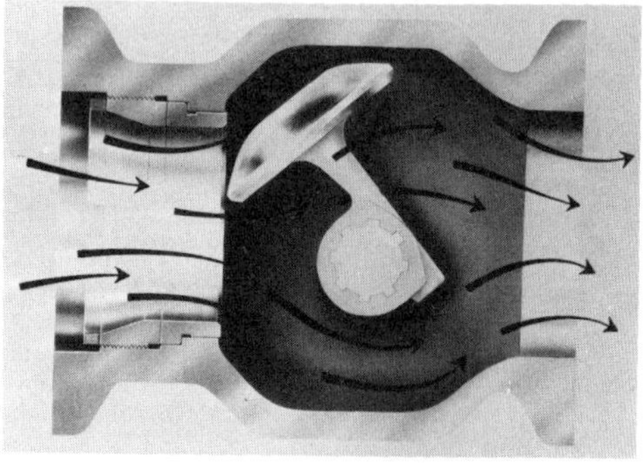

Figure 18. ECCENTRIC ROTATING PLUG VALVE.
Courtesy Masoneilan, Inc.

[1] This section contributed by Rolf B. Weil, Jamesbury Corporation.

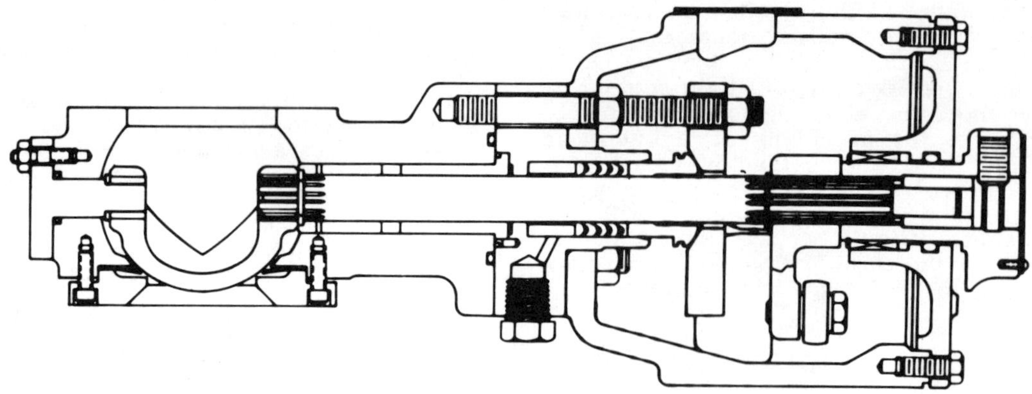

(a) V-shaped notch.

Courtesy Fisher Controls Company

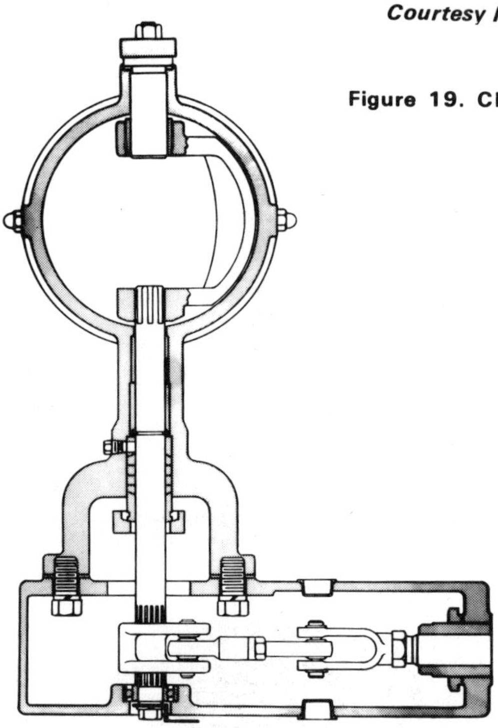

(b) Parabolic port.

Courtesy Masoneilan, Inc.

Figure 19. CHARACTERIZED BALL VALVES.

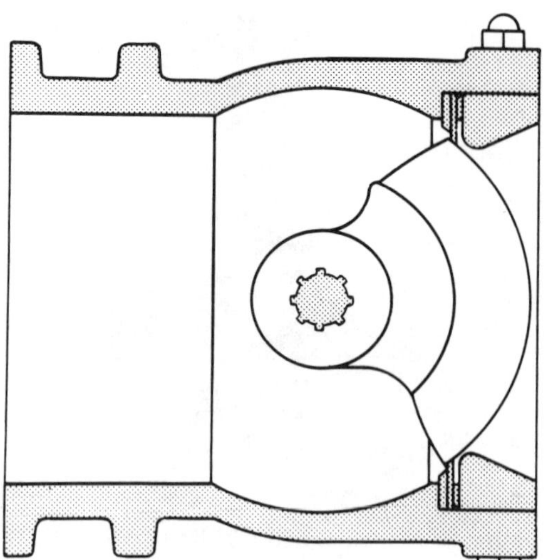

Figure 20. CROSS-SECTION OF A CHARACTERIZED BALL VALVE.

Courtesy Masoneilan, Inc.

either metal or rubber seats was generally available. The advent of elastomers and flurocarbons, in particular TFE, allowed the development of resilient seats which give today's ball valve its unique sealing properties. Starting with a single manufacturer in 1950, more than fifty companies now make ball valves, although most of these do not make control valves.

Initially, the ball valve was marketed as a tight shut-off hand valve, but manufacturers soon realized that a quarter turn device could readily be automated either electrically or pneumatically. Early attempts to adapt positioners to the ball valve were relatively crude, but the unique ability of the valve to handle slurries or fibrous materials, without straining or settling, stimulated interest in it. As a result, many improvements were made to the initial ball control valve and a number of modified designs evolved. As the realization came that the ball valve could capture a considerable portion of the control valve market, the major control valve manufacturers began to add ball valves to their line.

This section discusses the various design concepts employed in ball valves and outlines some of their advantages and limitations.

Construction

Two distinctly separate concepts developed in the evolution of the ball control valve. One type which evolved from the hand valve, utilizes a full ball, supported entirely by the seats. The second concept which is developed along the lines of the concentric plug valve, utilizes a hollowed-out, spherical segment, supported by two stub shafts running in bearings. For purposes of clarity, this latter design will be referred to as a characterized ball valve.

Both designs are quarter-turn, rotary valves, which eliminates alternate wetting and exposure to air of the valve stem. Consequently the valve stem, in the stuffing box area, does not become as severely corroded as it does in a lift type valve. This permits the use of stuffing boxes which are of relatively simple design. The valves are available with threaded, flanged, or flangeless bodies in a large variety of alloys, and may be equipped with either resilient or metal seats.

The characterized ball valve is so named, because a spherical segment with a naturally shaped convex leading edge, when used as a plug of the control valve, will produce an essentially linear characteristic. In order to obtain other characteristics, the leading edge of the segment must be contoured as in the V-notch and parabolic port designs, as illustrated in Figures 19a and 19b. The plug consists of a 120° spherical segment which is hollowed out and supported at either side by a stub shaft running in either bushings or bearings. Sealing is effected by a flexible metal or synthetic seal ring, which is pressed against the outside surface of the segment by a retainer ring. Shut-off capability is assisted by line pressure behind the seal ring, while the stub shafts hold the ball segment tightly in place. The interior of the valve is somewhat larger in diameter than the entrance port in order to allow free rotation of the plug. Because of this, flangeless characterized ball valves are restricted to port diameters no larger than 75% or 80% of nominal pipe size. A cross-section of a characterized ball valve is illustrated in Figure 20.

The full ball valve as shown in Figure 21 differs from the characterized ball valve in that its plug is made from a solid sphere with a waterway[1] drilled through

Figure 21. EXAMPLE OF A FULL BALL VALVE.
Courtesy Jamesbury Corporation

[1] Editors Note: The term 'waterway' is commonly used to describe the hole in the ball of full ball valves. This differentiates it from the term 'port' which is used for characterized ball valves and is the orifice formed between the ball and the seats.

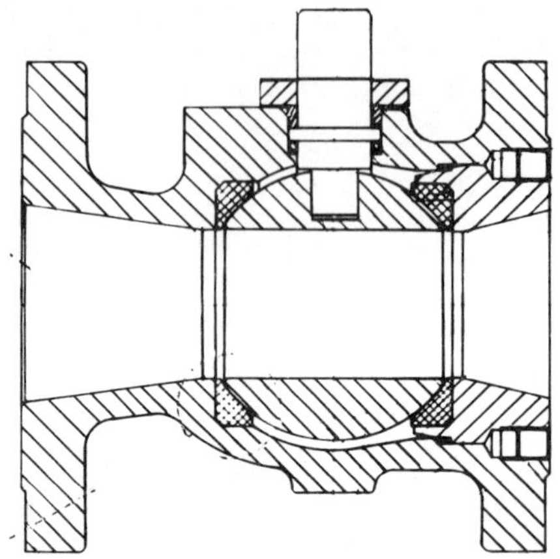

Figure 22. CROSS-SECTION OF A FULL BALL VALVE.
Courtesy Jamesbury Corporation

it. The ball is normally supported entirely by a pair of seat rings, and stem to ball design allows the ball to move between them when the valve is in the off position. This movement assists sealing as the ball is pressed into the downstream seat by line pressure. A cross-section of a full ball valve is shown in Figure 22. Accurate control of angular tolerances between ball and stem, which are a slip fit, must be maintained to prevent hysteresis in the valve.

A number of ball valve designs have evolved to handle higher pressure differentials. These utilize full balls but support them by means of trunnions or stub-shafts for additional strength (see Figure 23). "Full port" ball valves are available where the diameter of the ports is identical to the bore of the pipe size. This type is seldom used as a modulating device, except where lines must be "pigged" or rodded out.

Figure 23. EXAMPLE OF A TRUNNION-MOUNTED FULL BALL VALVE.
Courtesy Fisher Controls Company

Generally, the ball valve consists of a relatively small number of parts, but it must be removed from the line to replace soft or hard trim.

Actuator Requirements

The ball valve is a device in which the imbalance forces are mostly offset by the frictional forces between the ball and the seal ring. Because the frictional forces are relatively high, most ball valves are provided with piston type actuators to provide the necessary power. These may be either double-acting or spring-return types. The long stroke of a piston type actuator allows the use of a long-moment arm, thus providing efficient conversion of thrust to torque. Piston actuators also allow the use of higher operating air pressures, thus reducing the required diameter of the cylinder. For modulating service, a piston actuator requires a positioner.

The piston actuators that are used vary widely in design, from those which are basically linear and use external conversion to rotary motion, to those which are specifically designed for rotary service and utilize internal conversion of linear to rotary motion by means of a crank arm, rack and pinion, or scotch yoke mechanism. Rotary vane actuators have also been used for this service. The various designs include pistons which are rod or skirt guided, and piston seals which may be O-rings, U-cups, or rolling diaphragms.

Spring and diaphragm actuators, utilizing low pressure instrument air, have not been widely used, because their relatively short stroke allows only a small-moment arm with resulting inefficient conversion of thrust to torque. Their use has been limited to relatively small ball valves. As frictional forces utilize the greater part of the torque requirement of the ball valve, it is advisable to use a positioner with a spring and diaphragm to avoid excessive hysteresis.

Electric actuators have been used for two-position operation, almost from the inception of the ball valve. Because of the lack of imbalance forces, these electric actuators require only a relatively simple, spur gear train, making them inexpensive. Electronic positioners have been developed for use with these actuators to provide total electric modulating ball valves.

Characteristics Of Ball Valves

In discussing characteristics, it is necessary to treat the full ball valve and the characterized ball valve separately, because the characterized ball valve is a single orifice device, whereas the full ball valve has two orifices in series.

As the full ball valve is moved from open to closed position, the shape of each orifice goes from circular to eliptical. In the full-open position, each orifice is the same size as the waterway in the ball, making the effect similar to that of a single nozzle. As the valve is closed, the orifices become smaller in area than the waterway through the ball, and begin to have the effect of two independent orifices in series, with the waterway acting as a partial recovery area. The resultant inherent characteristic is essentially equal percentage. Changing the shape of the waterway in a full ball valve is not a practical method of changing the valve characteristic; therefore, where a different characteristic is desired, it is normal practice to alter positioner response by using a series of interchangeable cams in the feedback mechanism.

In the characterized ball valve, the leading edge of the ball segment is contoured to achieve the desired characteristic. A segment cut directly from a sphere, without any modification, would provide a convex leading edge to the plug, resulting in a port which is circular in the open position and is crescent shaped in the partially-closed position. The basic characteristic of this shape of plug is linear; however, as an equal percentage characteristic is usually more desirable, some form of characterization is required. The most common approach is to contour the leading edge of the plug, making it concave rather than convex. In Figures 19a and 19b, examples are illustrated with a V-shaped notch and a parabolic shaped contour in the leading edge of the plug.

Capacity Of Ball Valves

Ball valves have the highest flow capacities of any commonly used modulating valves. Their flow coefficient is in the order of two to three times that of an equivalent size globe body. It is common to use a globe valve that is one pipe size smaller than the line in which it is installed; however, a ball valve for the same service could be one-half of the line size. While this offers some obvious economic advantages, somewhat greater care must be given to system design when applying ball valves.

The ball valve will begin to show signs of cavitation and will reach choked flow at considerably lower differentials, relative to the inlet pressure, than will a globe type body. Its high flow capacity and the well defined vena contracta produced by a straight-through valve design are contributing factors. Typically, a wide-open ball valve, in service where vapor pressure is insignificant, may cavitate at a differential roughly equal to three-tenths of the inlet pressure, expressed in psia. As the valve is closed, the critical cavitation index increases, allowing the use of differentials as high as eight or nine-tenths of the inlet pressure. As a consequence, in some systems it may be necessary to install a ball valve of somewhat larger size than that indicated by calculations made for noncavitating conditions. This allows the use of the higher differentials, because the larger valve would be operating closer to the closed position. Obviously, in this type of system, some of the high capacity advantage is lost. Whereas reduced trim is normally available in globe pattern valves, it is not normal practice to provide this capability in ball valves.

Applications And Limitations Of Ball Valves

The ball valve found its initial application primarily in the pulp & paper industry where its straight-through flow path, tight-shut-off capability and ease of operation were distinct advantages in the handling of fibrous material such as pulp stock. It has remained the most suitable type of valve for handling this and many other types of slurries, because it mimimizes the tendency of these materials to either settle or strain. It has been accepted wherever tight shut-off is a necessity or where the economics of its high flow capacity justify its use. It is available in a wide variety of materials, with either resilient or metal seats.

The limitations of the valve depend largely on the type of seating material used. Resilient seats have essentially no leakage, but limit the temperature capability to that at which the resilient material begins to soften under load. Metal seats increase the temperature capability considerably, but have a higher leakage rate. Current designs are available with allowable differentials in modulating service of up to 1,000 psi.

SPECIAL CONTROL VALVES[1]

In a basic control valve handbook, it is impractical to do more than suggest the enormous variety of body and actuator configurations used on special applica-

[1] This section contributed by Paul Wing, Jr., Masoneilan International, Inc.

tions. Two classifications which are of general interest, are discussed in this section; namely, low-flow and low-temperature control valves. A brief treatise on some other valve designs concludes this section.

Low Flow Control Valves

While there is not a sharp line of a demarcation, the term *low-flow* will be used in this discussion for C_V ratings of 0.05 or less. There are many applications in industrial systems calling for either small, accurately controlled, quantities of material at ordinary pressure levels or somewhat larger quantities having very high pressure-drops, where the calculated C_V is very small. These applications either may be in pilot plants where a process is being conducted in miniature, or they may be in a commercial plant where, for example, a small amount of some reagent is being added to a large amount of process fluid. The theoretical C_V in the literature ranges upward from 0.00001 and as the accompanying Figure 24 shows, covers a wider range than the equivalent for valve sizes 1 inch through 16 inches.

In considering these low flow applications, the selection criteria must be quite different from those used for conventional control valves. For example, there is seldom any friction loss in series with the control valve nor is there an unfavorable pump-head characteristic. There is no real requirement for an equal percentage characteristic. The principal design problem is one of getting down into the desired flow range in a practical manner. C_V ratings become more and more theoretical as they become smaller. They are useful as a capacity index, but are of doubtful accuracy or repeatability, because in many cases, the flow through the principal restriction may be either laminar or a combination of laminar and turbulent and this flow mixture may change radically with valve stroke or with flowing conditions.

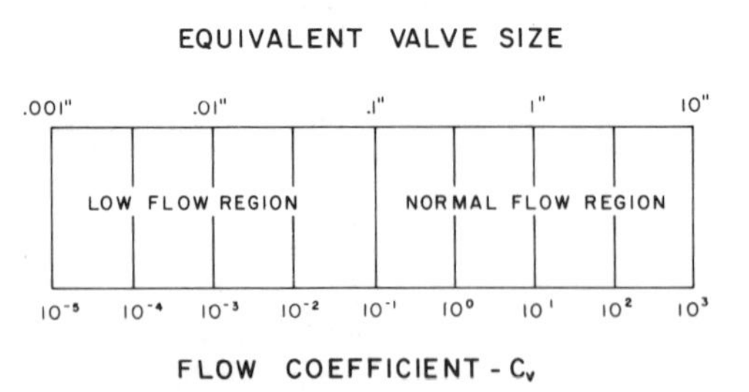

Figure 24. LOGARITHMIC PLOT OF CONTROL VALVE C_V RANGE SHOWS WIDE COVERAGE REQUIRED ON LOW FLOW APPLICATIONS.

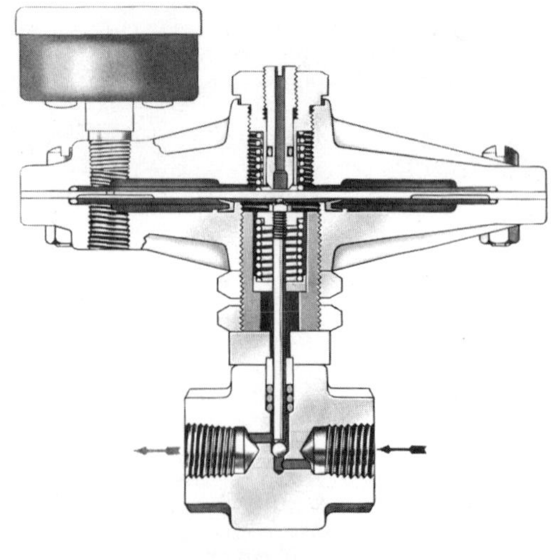

Figure 25. LOW FLOW VALVE WITH BALL PLUG AND ACTUATOR WITH VARIABLE STROKE FOR WIDE C_V RANGE.

Courtesy A. W. Cash Company

From the design standpoint, the physical dimensions of the required flow restriction is very small indeed. An additional problem is imposed by the fact that a great many low-flow applications involve high pressure-drop with the accompanying erosion problems.

In standard valve design, capacity is largely determined by the diameter of the orifice. As the capacity is reduced, a practical minimum diameter of about 1/16 in. is reached. Such an orifice would have a theoretical rated C_V of approximately 0.06 or near the upper limit of the low flow range. All further reductions must be obtained either by radically reducing the plug stroke or by restricting the port opening.

An interesting example of the short stroke approach is shown in Figure 25. In this design, a synthetic sapphire ball modulates the flow through a small hardened metal orifice. A specially designed actuator, with variable pneumatic feedback, permits the stroke to be adjusted so that a 3-15 psi signal will cover a very wide C_V range.

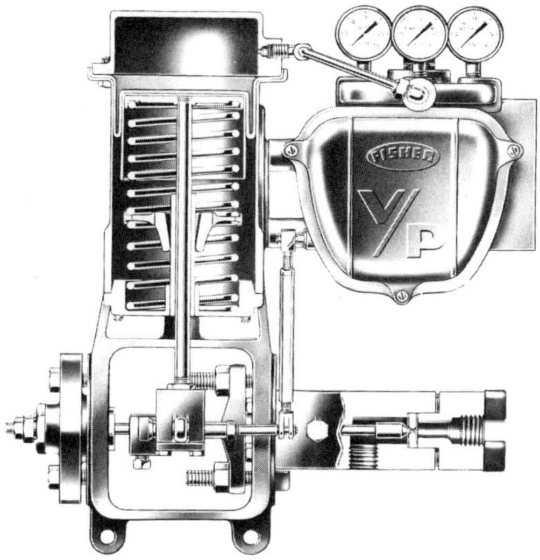

Figure 26. SPECIAL ROTARY ACTUATOR GIVES LOW STROKE BY USING DIFFERENTIAL THREAD DESIGN.

Courtesy Fisher Controls Company

For a short-stroke actuator, which is suitable for smooth throttling action on very high pressure service, another approach has been to rotate a threaded plug stem through part of a revolution, by means of an arm driven by a pneumatic actuator. Two threads of different pitch may be used in such a system to further limit the stroke, which may be reduced in this manner to as little as 0.005 in. An example of this design is shown in Figure 26. Another version using this basic approach, but with a mechanical variable ratio linkage, is shown in Figure 27. In this unit, the maximum stroke of the valve plug is 0.188 in., a 4 to 1 reduction over the 3/4 in-rated actuator stroke. This may be reduced to a minimum of 0.010 in., by adjustment of the linkage. This allows field adjustment for a wide range of capacity, as well as providing a high operating force for high pressure-drop operation. Both of these specialized valves may use a simple cone-type plug. For very low flows, a precision-made, tapered pin is used, operating in a tapered hole.

For less exacting services, tapered needles or pistons with milled slots or notches are generally used in orifices ranging from 1/8 to 1/4 in. in diameter. Valve stroke

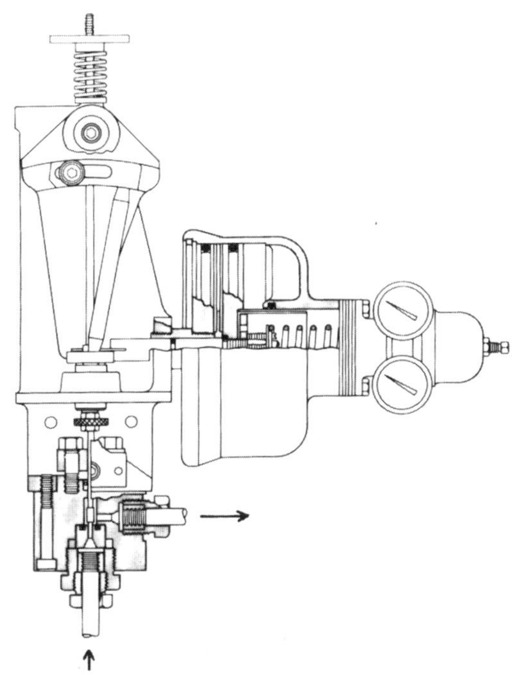

Figure 27. VARIABLE MECHANICAL LINKAGE GIVES MORE THAN 10 TO 1 STROKE RANGE FOR APPLICATION FLEXIBILITY.

Courtesy Masoneilan International, Inc.

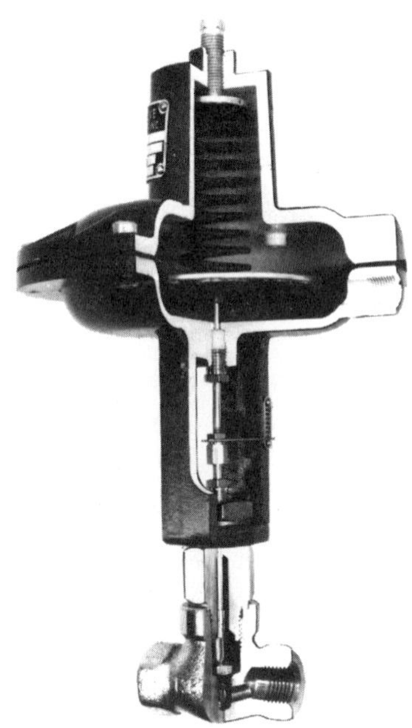

Figure 28. COMPACT VALVE AND ACTUATOR WITH PISTON OR NEEDLE TRIM IS IDEAL FOR PILOT PLANT WORK.

Courtesy Precision Products and Controls, Inc.

for this type is in the order of 1/2 in. Such low flow trim may be installed in conventional globe valve bodies for industrial applications, but it is possible to reduce both the body and actuator size substantially as shown in **Figure 28.** These valves are light in weight, compact, and ideally suited for pilot plant work.

As already mentioned, the C_V rating on low flow trim has less and less meaning as the size decreases. Fortunately, the useable flow range obtained in the valve orifice is very large. In other than adjustable stroke valves, valve trim can be changed easily and, if a number of similar valves are in service, a few sets of trim can be maintained without excessive cost since all are interchangeable in a given body.

The required C_V is calculated by the conventional formulas and then the next largest-size trim, available in the particular type of valve, is chosen. If the trim is too small in actual operation, a capacity increase may be possible either by simply increasing the valve stroke or by slightly modifying the plug.

Cryogenic and Cold Service Control Valves

When control valves are applied at operating temperatures that are substantially below the freezing point, special precautions must be taken and, in certain cases, special designs are required for satisfactory operation. The principal problem is the proper selection of materials of construction, particularly on moderately cold service (to -150° F). Cryogenics is the term applied to processes operating at temperatures below -150° F, where a number of additional factors must be considered.

In the temperature range from -20° F to -150° F, special impact-resistant, carbon steels are used for the pressure containing parts. The commonly used materials are carbon steel (grade LCB) from -20 to -50° F and 3.5% nickel steel (Grade LC3), to -150° F. These valves are generally equipped with a plain extenstion bonnet. In this temperature range, the primary objective is to reduce the influx of heat to the system to a reasonably low value, and to prevent the packing box area from getting so cold that frost and ice form on the upper part of the bonnet and on the valve stem above the packing box. The simple extension bonnet is generally installed in the upright position to minimize heat transfer in the operating fluid itself.

At cryogenic temperatures, materials of construction now exclude the carbon steel group and include austenitic stainless steels, bronze, and monel. Valves operating at this temperature range are generally handling extremely cold gases or liquefied gases. Among those commonly encountered are air, nitrogen, oxygen, hydrogen, and helium. Properties of some of these fluids are tabulated in **Table 2.** For these services, a number of factors must be considered in addition to the material of construction of the valve's pressure containing parts.

Table 2. PROPERTIES OF CRYOGENIC FLUIDS

Gas	N_2	O_2	CH_4	H_2	He	F_2
Density, NTP* lb./cu. ft.	0.0724	0.0827	0.0415	0.0052	0.0103	0.0982
Boiling Point 1 atm. ° F.	-320.4	-297.4	-258.7	-423.2	-452.1	-306.6
Vapor Density at B.P. lb./cu.ft.	0.288	0.296	0.111	0.084	1.06	—
Liquid Density at B.P. lb./cu.ft.	50.41	71.27	26.46	4.37	7.62	94.2
Heat of Vaporization at B.P. BTU/lb.	85.2	91.7	219.2	193	8.8	74.1
Critical Temperature ° F	-232.8	-181.1	-116.5	-400.3	-450.3	-200.2
Critical Pressure, psia	492.3	736.9	673.1	187.7	33.2	808.3

*NTP - Normal Temperature and Pressure, 70° F and 14.696 psia.

Very special attention must now be paid to the design of the extended bonnet. Even though the body material may be bronze, the extension would generally be made of austenitic stainless steel to minimize heat transfer. Care must be taken that liquefied gases cannot become trapped in the bonnet space, where subsequent vaporization on shut-down could generate dangerously high pressures.

In all cases, the valve must be heavily insulated. The process equipment piping and valves in the cold section of the plant are often installed in a "cold box." The valve in this case may have an exceptionally-long, plain, extension bonnet which may be installed in the horizontal position. A special valve design, such as that shown in Figure 29, is often used in this case. A large-diameter bonnet section is fabricated from stainless steel and brought out through the wall of the cold box. This permits removal of the valve trim, including the plug and seat-ring, without disturbing the valve body, which may be welded into the system to minimize the possibility of leakage within the cold box itself. Within the extension at the lower end is a Teflon, chevron seal which acts as a check valve, minimizing leakage of the cryogen into the bonnet section, yet permitting the release of any gas that is generated. In this way, the heat flow into the system is minimized.

A mounting flange is usually provided at the outer end of the extension bonnet, which cooperates through a flexible diaphragm with a larger flanged opening in the cold box itself.

Figure 30 illustrates a very special cryogenic valve that has a vacuum insulating jacket enclosing the entire body and body extension. Stainless steel expansion bellows are provided on the jacket to eliminate the piping strain that is caused by differential expansion. Note that a thin-walled cylinder is used instead of the usual stem extension to provide a long conductive path and to minimize the vapor space between the stem and the body cylinder. Radiation losses become an important factor and are minimized by using an elec-

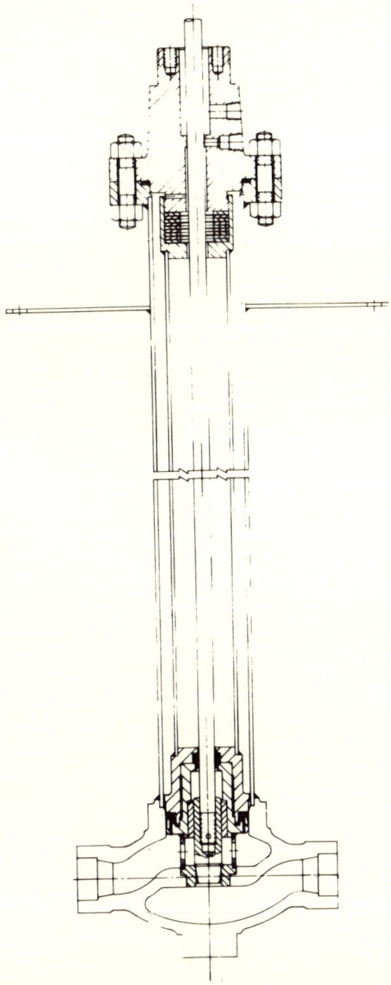

Figure 29. "COLD BOX" VALVE WITH PLUG AND SEAT RING REMOVABLE THROUGH THE BONNET.

Courtesy Masoneilan International, Inc.

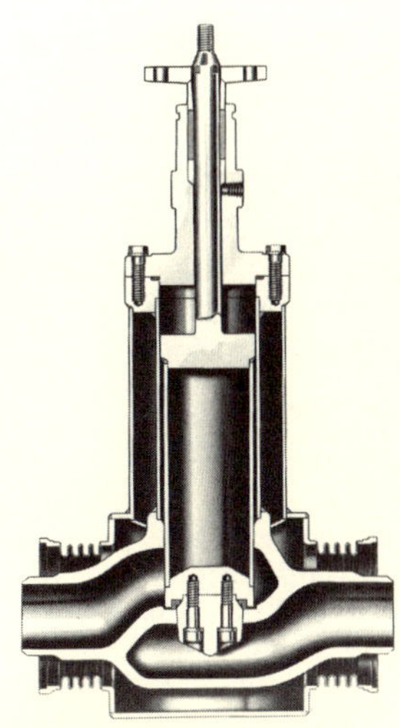

Figure VACUUM JACKETED CRYOGENIC VALVE FOR SERVICE ON LIQUID H2, He OR OTHER VERY COLD FLUIDS.

tro-polished valve exterior and an aluminum foil barrier. In this kind of valve, it is important to hold the weight to a minimum, but at the same time, to provide a high relative capacity, because low mass minimizes the boil-off on cooldown. Similar designs are available in a "Y-pattern" configuration to further reduce the head loss, particularly for on-off service.

Rotary valves, also equipped with an extension bonnet, are often specified for cryogenic service where tight shut-off is required. Figure 31 shows a special design, with an offset vane and a spherical seating surface, which gives good shut-off using a special Kel-F® lip-seal insert and yet it minimizes the torque required for breakout. In the larger sizes, the low mass of the valve is an important factor.

Figure 31. OFFSET VANE CRYOGENIC BUTTERFLY VALVE, WITH ELASTOMER LIP SEAL FOR TIGHT SHUT OFF.

Courtesy Royal Industries

Generally, requirements for cleanliness must be more rigid for cryogenic valves, during manufacture and, particularly, during assembly.

Special attention is required in the selection of lubricants, which may be used when assembling parts, to prevent galling. Many lubricants either solidify or become brittle and many may also be completely incompatible with the flowing fluid. A number of compounds have been developed for this service and are readily available, commercially. Very special attention is required, for example, when handling liquid fluorine, which is self igniting on contact with any hydrocarbon. A more common problem exists in the handling of oxygen, either at ambient temperature, or at cryogenic temperatures. In addition to the possibility of self-ignition in contact with certain contaminents, the oxygen supports combustion and a fire, if started, will consume the entire piping system. Most companies, that use cryogenic liquids, have detailed specifications on cleaning and degreasing procedures.

Low Noise Control Valves

Control valves designed for quiet operation are of special and rapidly growing interest. The Walsh-Healey Act, in the United States, defines the permissible noise level for continuous human exposure at 90 dB. Most control valve noise is generated on compressible fluid service as a result of the conversion, into sound, of a portion of the energy dissipated in the throttling process. A general measure for estimating purposes is that, if the product of the C_V and the inlet pressure exceeds 1000, the valve will be excessively noisy under critical pressure drop.

The two principle factors in designing for low noise, are fluid velocity and noise frequency. Noise energy level increases at up to the 8th power of the velocity below sonic, indicating the great importance of this factor. An increase in average frequency at the same noise energy level can also be beneficial, because of greater atteuation of the higher frequencies through the valve body and the downstream pipe wall.

A low noise design is shown in Figure 32, using a series of steps which force the flowing fluid to change direction continously causing high turbulence and friction energy loss. Compensation for change of fluid density with pressure is provided by increasing the plug and seat diameter toward the downstream section. The resultant cone shaped assembly produces a nearly-constant fluid velocity through the entire throttling process. The maximum velocity attained by this assembly is seldom higher than 300 ft/s, giving a substantial reduction in noise level and improved service life.

To give fail-safe action and, to reduce actuator thrust requirements, partial internal balancing can be provided. This, in combination with short valve stroke and a conventional spring diaphragm actuator, allows application at pressure drops to 6000 psi. An interesting version of this

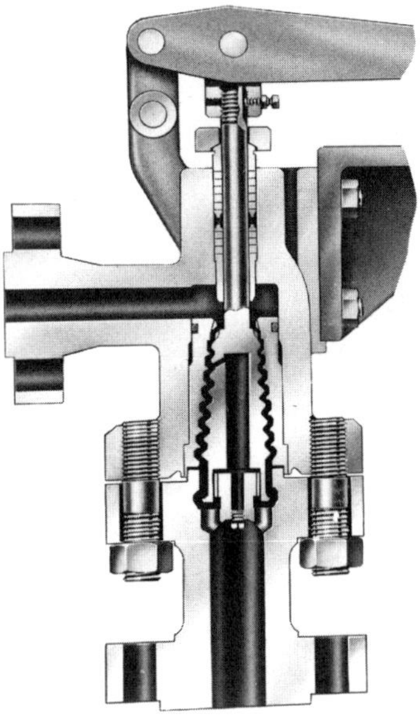

Figure 32. HIGH PRESSURE LOW NOISE, ANGLE VALVE WITH LABYRINTH TYPE SEMI-BALANCED CONE PLUG.

Courtesy Masoneilan, Inc.

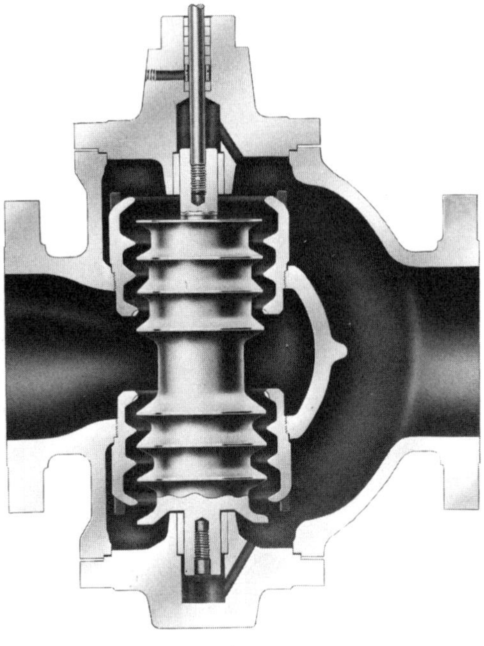

Figure STANDARD DOUBLE-SEAT VALVE WITH MODIFIED PLUG AND SEAT FOR LOW NOISE SERVICE.

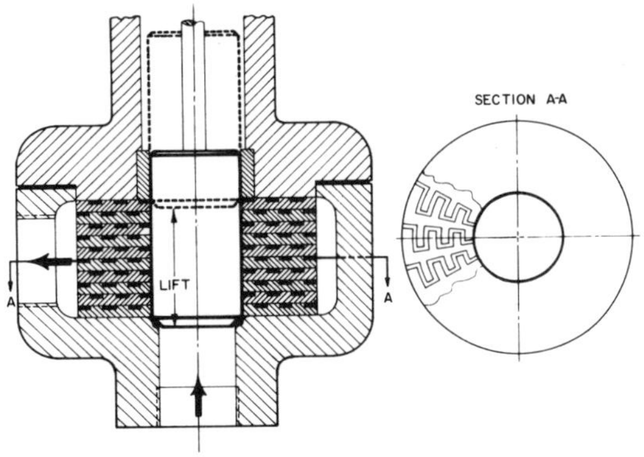

Figure 34. STACKED DISC, LONG-STROKE VALVE DESIGN REDUCES VELOCITY AND NOISE.

Courtesy Control Components, Inc.

design using a conventional double-seat, globe body is shown in Figure 33. Valve capacity of approximately 40 percent of the normal rated maximum is attained.

In yet another design, shown in Figure 34, the throttling passages through the valve are produced by stacking a number of annular discs, which are either etched or punched to produce flow passages. The fluid path is tortuous, similar to that in Figure 32, and compensation for expansion on compressible fluids is handled by an increase in the number of flow channels. The valve plug is of simple cylindrical shape and a long valve stroke is required.

Conventional single-seat globe valves, particularly those of ported cage designs, can be modified to increase peak noise frequencies resulting in less audible noise. In all designs, if the valve outlet velocity either reaches or exceeds the speed of sound, objectionable noise may be generated in the down-stream piping. On high-drop applications, therefore, special down-stream treatment is required. This may take the form of one or more special throttling plates, designed to both reduce the velocity in successive stages and to increase frequency. Various commercial silencers may also be used. Acoustical insulation is beneficial to limiting noise in a specific area; however, objectionable noise may still be produced beyond the insulated zone.

Other Valve Designs

Special valve designs evolve either through the ingenuity of mechanical designers or in a trial and error process to meet a particular field requirement.

Many, such as the pneumohydraulic,

axial-flow valve shown in **Figure 35** are quite unconventional in the approach to the problem. Others, such as the butterfly valve shown in **Figure 36**, with a two-piece body and an offset vane shaft to permit the use of an inflatable seal in the body itself for critically tight shutoff, are specialized adaptations of the basic types already shown.

One classification generally labeled *special* includes the valves designed for high-static and high pressure-drop service. These may be divided into two major classes:

1. Valves for the process industries in service such as ammonia letdown, polyethylene letdown, and urea letdown.
2. High pressure-drop, steam and water valves on large power boilers.

For the process industries, the most common valve configuration is shown in Figure 37. For most services such as ammonia letdown and urea letdown, temperatures are moderate and pressures are in the range of 3000 to 10,000 psi. Valve body-size seldom exceeds 3 inches and the orifice diameter 1 to 1.5 in. The design problem is one of pressure integrity and material compatability with the process fluid.

The angle-style body is used either flow-to-open or flow-to-close, depending on the process requirements. In the manufacture of polyethylene, for example, a control

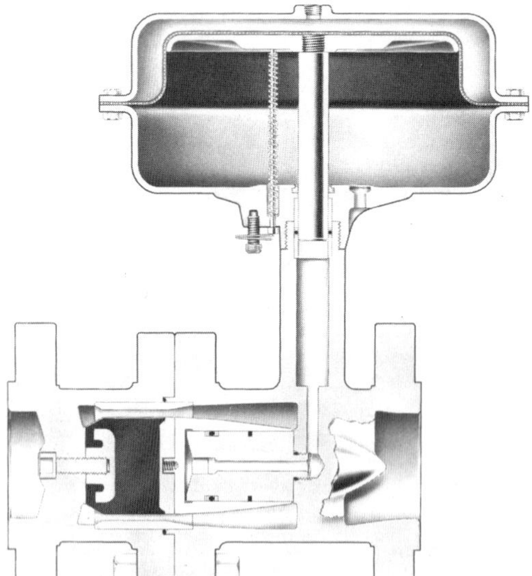

Figure 35. PNEUMOHYDRAULIC VALVE WITH AXIALLY MOUNTED PLUG.

Courtesy American Meter Company

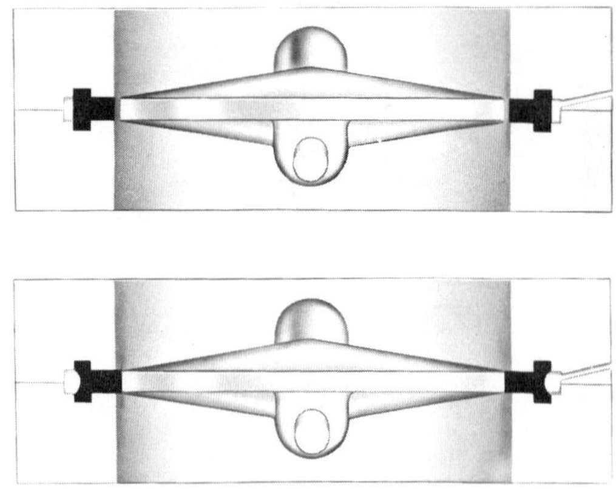

Figure 36. DIAGONALLY SPLIT BUTTERFLY BODY PERMITS USE OF INFLATABLE SEAL FOR CRITICALLY TIGHT SHUT OFF.

Courtesy Fisher Controls Company

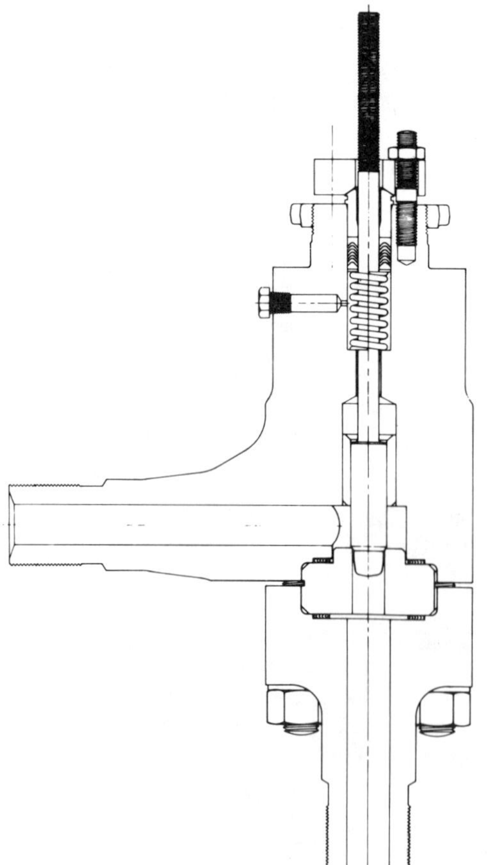

Figure 37. FORGED ANGLE VALVE FOR MODERATELY HIGH PRESSURE PETROCHEMICAL SERVICE.

Courtesy Fisher Controls Company

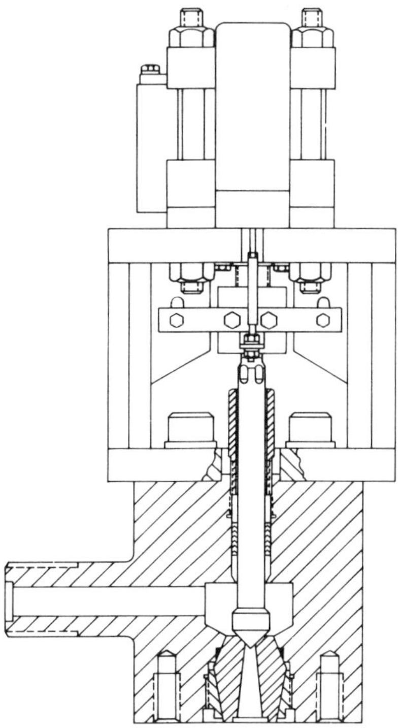

Figure 38. HIGH PRESSURE FORGED ANGLE VALVE WITH PRESSURES UP TO 50,000 PSI.

Courtesy Masoneilan International, Inc.

Figure 39. LARGE HIGH PRESSURE SINGLE SEAT VALVE ESPECIALLY DESIGNED FOR POWER PLANT SERVICE.

Courtesy Bailey Meter Company

valve is required, operating at an inlet pressure of 30 to 50,000 psi, and extremely high pressure-drop. A typical design is shown in **Figure 38**, again using the angle body which permits the use of hammer forgings or barstock. Such valves often incorporate special safety features, such as a back seat in the bonnet section, to minimize the hazard of a packing failure. Body-joint, end-connections are generally established on a custom basis. This is particularly true in operating pressure ranges above 10,000 psi.

On high-pressure power-plant applications, the valves are much larger and the temperatures are higher (up to 1100° F).

On many applications, the valve may be used only during start-up or emergency; therefore, the prevention of seat leakage is an important consideration. The boiler feed-pump recirculation valve, used to maintain a minimum flow through the pump, is generally required to remain shut over long periods of time with differential pressures across the seat of up to 6000 psi. Temperatures are moderate, but the prevention of seat leakage, which rapidly increases due to erosion of even the best hard trim alloys, presents a real design challenge. Such valves are generally high-pressure versions of the conventional cage-trim design shown earlier in this chapter.

As a last example, Figure 39 shows a large, high-pressure single-seat valve designed especially for once-through boiler start-up service. The design exceeds 2500 lb. ANSI rating and has a special breech-block type, body closure that uses internal pressure to aid in the sealing. Special attention is paid to body contouring, to minimize distortion due to thermal cycling. Exceptionally rugged construction is required throughout, because the valve normally sits in the closed position and may be required to fully stroke in as little as 2 seconds, in the event of a serious plant upset.

VALVE BODY END CONNECTIONS

Most flanged control valve bodies conform to the standardized face-to-face dimensions listed in ISA Recommended Practice, RP 4.1 or Fluid Controls Institute Standard, FCI 65-2, with the exception

Table 3. STANDARDIZED FACE TO FACE DIMENSIONS FOR FLANGED CONTROL VALVE BODIES

	ANSI RATINGS (LB.)		
Pipe Size (Inches)	125 Iron 150 Steel (Inches)	250 Iron 300 Steel (Inches)	600 Steel (Inches)
1/2	-	7-1/2	8
3/4	-	7-5/8	8-1/8
1	7-1/4	7-3/4	8-1/4
1-1/2	8-3/4	9-1/4	9-7/8
2	10	10-1/2	11-1/4
2-1/2	10-7/8	11-1/2	12-1/4
3	11-3/4	12-1/2	13-1/4
4	13-7/8	14-1/2	15-1/2
6	17-3/4	18-5/8	20
8	21-3/8	22-3/8	24
10	26-1/2	27-7/8	29-5/8
12	29	30-1/2	32-1/4
14	35	36-1/2	38-1/4
16	40	41-5/8	43-5/8

of some of the newer flangeless body styles, Saunders types, angle valves, and rotary valves (see Table 3). End connections can be classified as follows: threaded; welded, i.e., either butt-weld or socket weld; and flanged; or flangeless.

Threaded connections (female NPT thread) are standard in valve sizes smaller than 1 inch and are frequently used for control valves up to 2-inches. Steel control valve bodies having threaded ends are generally rated at, and limited to, 600 lb. ANSI. Their use in corrosive service is somewhat limited.

Welded ends are quite common where high pressure, high temperature fluids are encountered, particularly in steam and water service in power plants. Care should be taken to see that the valve body material specified is compatible with the adjoining pipe material. High chrome or carbon alloys should be avoided, because they have a tendency to air-harden and, therefore, require special heat treatment. Butt weld-end connections as illustrated by Figure 40, are standardized in ANSI B16.5. However, socket-welding ends, per

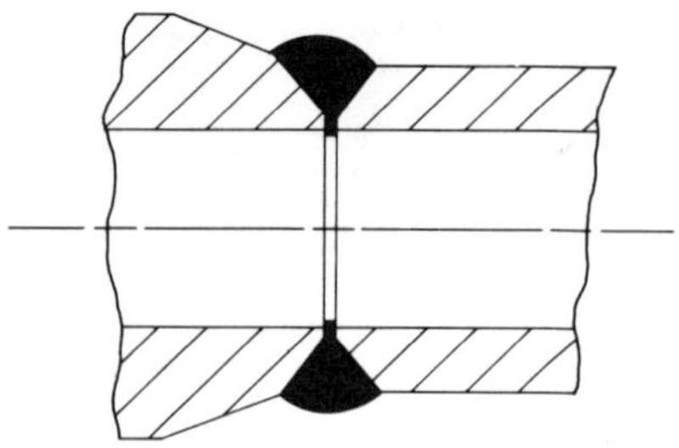

Figure 40. BUTT WELD END.

ANSI B16.11, are usually specified in valve sizes of up to 2 inches.

Steel or steel alloy flange dimensions conform to ANSI B16.5. Similar standards govern flanges for cast iron, bronze or other materials.

Cast iron flanges have no raised face to prevent bending moment around the gasket. Steel flanges below 600 lb. ANSI rating have a 1/16-inch high raised face, and above 600 lb., they have an ANSI 1/4 inch raised faced dimension. This simplifies identification.

The flange rating is determined by the type of the service, required material, presure, and the maximum fluid temperature. In contrast to certain European standards, an ANSI flange rating defines the maximum working pressure at an elevated temperature, i.e., 300 psi at a temperature of 850°F for carbon steel, 300 lb. ANSI rating. Such a flange is rated at 720 psi at room temperature and is tested at 1100 psi hydrostatic pressure.

Flanged valve bodies can be removed quite easily from the pipe; however, they are subject to either thermal distortion or thermal shock. Welded construction should be considered if wide temperature fluctuations are to be encountered.

The flange facing is determined by the type of joint desired. A raised face requiring a flat gasket is the most popular joint for service of up to 600 lb. ANSI. Ring-joint facings are more popular above these ratings. For very high pressure, (10,000 to 50,000 psi) a lens gasket or other special design will be necessary. In a typical design, as shown by Figure 41, extra heavy flanges are threaded onto the adjoining pipe ends to clamp a solid-metal, lens-ring gasket.

Another type of valve body has no flange connection at all and is simply clamped between the adjoining line flanges. This is a method originally used with butterfly valves and other rotary-type valves and is now enjoying increased popularity, because of the decrease in weight and the reduced space requirements.

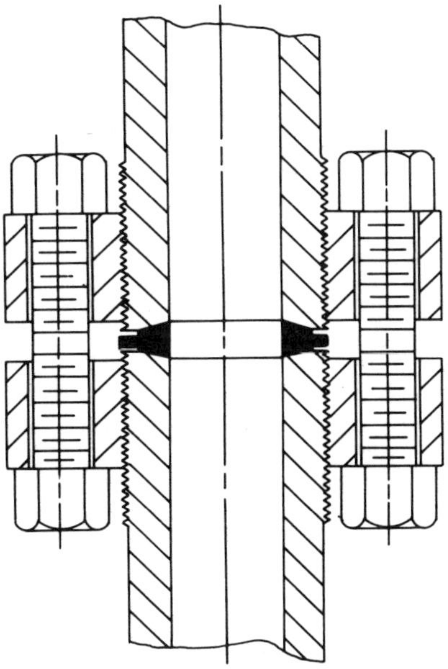

Figure 41. HIGH PRESSURE CONNECTION - LENS RING GASKET.

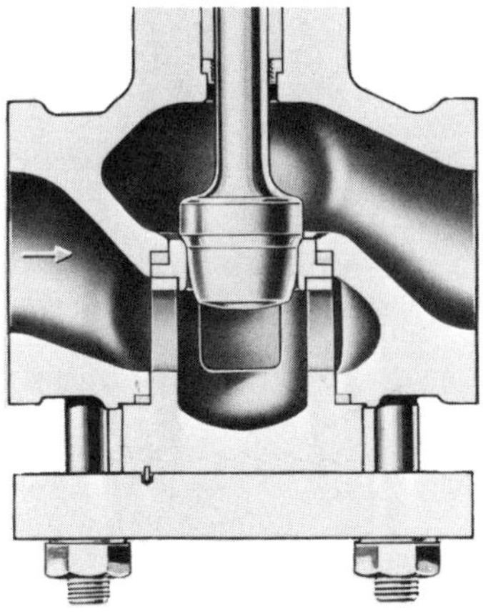

Figure 42. FLANGELESS VALVE - BOTTOM ENTRY.

Such a flangeless valve, as shown in Figure 42, is also somewhat simpler to install, because alignment between the line flange and the body flange is less critical. Long tie rods must be made from a high-strength, bolting material. The imposed strain on these tie rods, when properly installed, is sufficient to compensate for any thermal expansion of the valve body. Good design practice is to specify a flexible gasket with a controlled-compression, metal shoulder, between the line flange and the body end-connection to prevent crushing of the gasket during thermal cycling. Flangeless valve bodies are usually available for flat gaskets only; the body end connection having the same diameter as the adjoining flange, raised-face surface.

VALVE BONNETS

The valve bonnet or top closure is the removable upper portion of the valve body sub-assembly and is normally connected to the body by high strength bolting. It is a pressure-carrying part and is, therefore, subject to the same design requirements as the valve housing. The bonnet normally contains suitable mounting means for the actuator; for the packing, or other similar stem sealing means; and for an upper stem guide. Removal of the valve bonnet generally provides access to the valve trim; however, there are certain valve designs, such as eccentrically-rotating plug valves and sliding-gate valves, where the bonnet is a cast-on part of the valve body. In this case, trim removal is possible, either through the inlet or outlet flange connection or through another separable body connection.

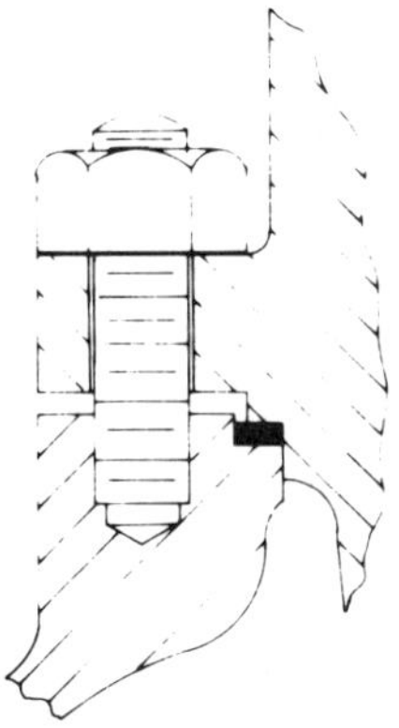

Figure 43. BOLTED BONNET JOINT.

The conventional bonnet joint contains a machined recess to provide guiding and containment of a flat gasket, as illustrated by Figure 43. The joint is usually stud bolted; however, through bolting is common on high pressure valves. Both the bolt size and the flange thickness should be calculated in accordance with applicable ASME Code requirements, such as ASME Pressure Vessel Code, Section 8, paragraph UG-34.

Some low-pressure valves, particularly in sizes below 2 inches, have a threaded bonnet connection which is more economical to produce than a flanged joint; however, it may be difficult to remove the bonnet after extended service, particularly when corrosive fluids are involved. The alignment between the stem, plug, and seat ring is less accurate.

Finally, there is the pressure-seal type bonnet connection, which employs a pressure activated metal seal ring and a bridgelock type of thread engagement, as illustrated in Figure 44. The main advantage of this type of seal, usually employed in high pressure and high temperature service valves, is the absence of a large bonnet flange which can be subjected to high internal stresses with frequent thermal shocks or thermal cycling.

The plug guide bushing, when employed, is usually made of a hard material such as heat treated ANSI type 440-C stainless steel or 17-4PH stainless; however, special guide materials, such as glass filled Teflon are also available for corrosive or cryogenic applications. These guide bushings are usually either pressed-in or permanently held by a suitable retainer.

The upper portion of the bonnet contains either the valve packing or the stem seal. This packing area is relatively near the bonnet flange connection for fluid temperatures of up to 450° F. A tubular extension separates the packing area from the flange in case cryogenic temperatures, to 450° F, are encountered. This extension bonnet places the packing far enough away from the cold area of the valve to prevent freeze-up of the packing, as illustrated by Figure 45.

For high fluid temperatures (above 450° F) it has been customary to cast either horizontal or vertical cooling fins onto the bonnet extension to provide sufficient radiating heat loss. The representative design shown in **Figure 46** is called a "Radiation Fin Bonnet."

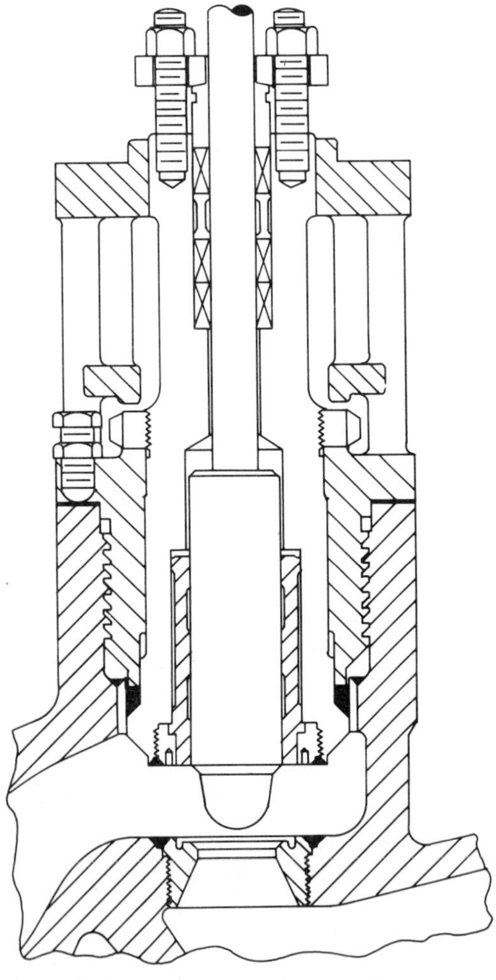

Figure 44. PRESSURE SEALED BONNET.

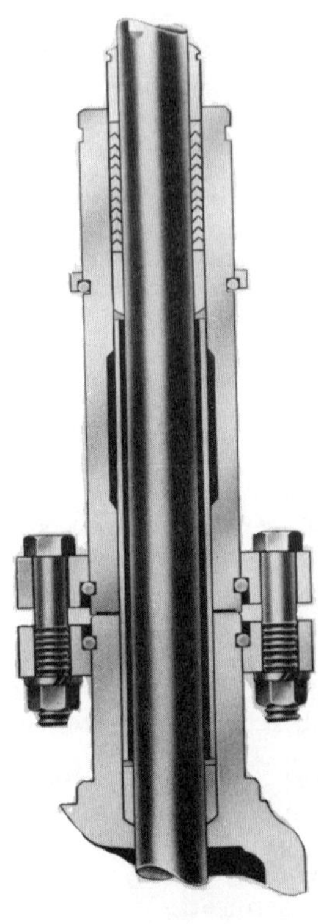

Figure 45. EXTENSION BONNET.

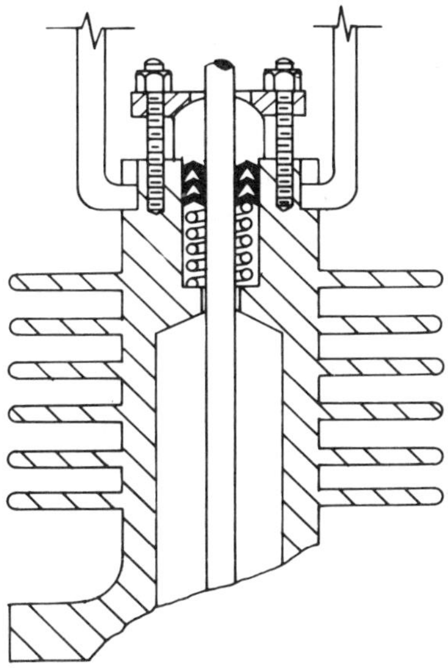

Figure 46. RADIATING FIN BONNET.

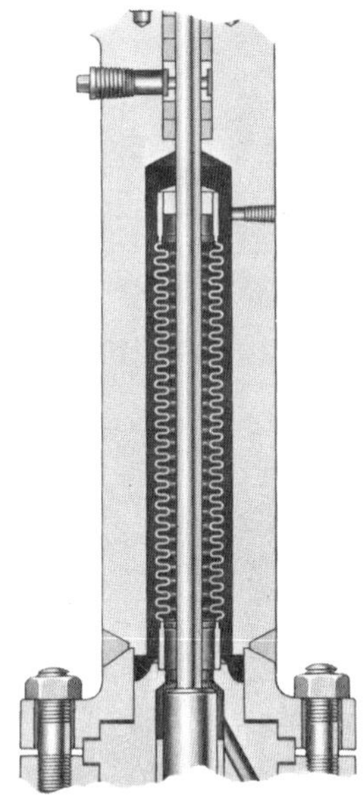

Figure 47. INTERNALLY PRESSURIZED BELLOWS SEAL.

Courtesy Fisher Controls Company

Recent designs omit these radiation fins, because they have been proven unnecessary. In other words, a well designed, plain extension bonnet will provide sufficient radiating surface, in combination with a limited cross-section, for heat conduction to keep the packing within acceptable temperature limits. In no case should such a bonnet be covered with thermal insulating material.

For service, in which radioactive fluids, highly-toxic fluids, or very rare and costly media are encountered, it is customary to specify a *bellows seal.* A bellows seal usually consists of either a hydroformed metal bellows or a welded leaf-type bellows, surrounding the valve stem and forming a completely enclosed, metal seal for the moving valve stem. Bellows seals are of either internal pressurized or external pressurized design. In the internal pressurized version, as shown in **Figure 47**, only the inner surface of the bellows is subjected to the fluid pressure. This saves cost in the case of corrosive fluids, because the outer-extension bonnet can be made of carbon steel. On the other hand, the external pressurized bellows can withstand a somewhat higher pressure rating. A well-designed bellows seal has as anti-rotation device preventing bellows damage during assembly and disassembly.

One basic disadvantage of bellows seals is their relatively-low pressure rating. The pressure ratings decrease with an increase in valve size. Typical ratings are 150 psi at 600°F. Higher ratings to about 500 psi can be obtained by the use of special bellows, having increased wall thickness. Unfortunately, this is done at the expense of the number of obtainable life cycles. Another way of increasing the pressure rating is to load the exterior side of the bellows with a pressurized inert gas, but in most cases, this method is too costly and cumbersome to be practical.

The connection between the bellows and the stem or the lower flange is usually tested with a mass spectrometer that will measure leak rates of less than 1×10^{-6} cc/s of helium, from atmospheric pressure to high-vacuum.

A more economical substitute for the bellows seal is the double packing, described below.

PACKING BOX ASSEMBLY

The purpose of the packing box assembly is to contain an elastic means for preventing the leakage of a process fluid, past the surface of the reciprocating or rotating valve stem. Suitable adjustments should be provided for varying the compression of the packing material against the surface of the stem. The ideal packing material should be elastic and easily deformable, but yet, cause virtually no fric-

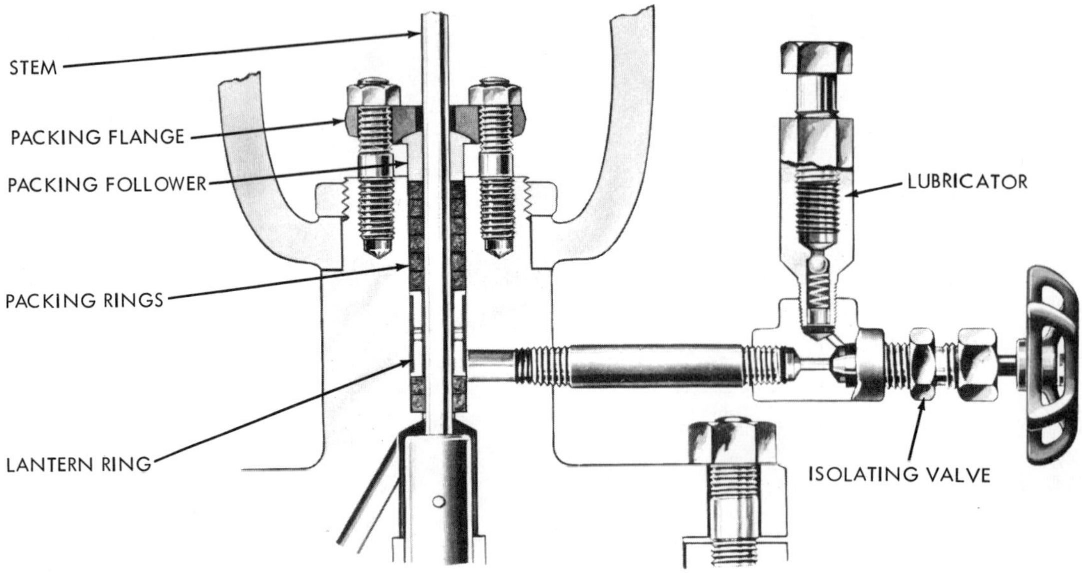

Figure 48. PACKING BOX ASSEMBLY (BOLTED) WITH LUBRICATOR AND ISOLATING VALVE.

tion. In addition, the packing should be as chemically inert as possible, and should be able to withstand high pressures and high temperatures. Existing packing materials will meet all of these requirements; therefore, selection must be made from a variety of materials to suit the specific service conditions.

Figure 48 is a cross-sectional view of the most-popular, bolted, packing box assembly. The assembly consists of a packing flange, packing follower, lantern ring, and a number of equally spaced packing rings. The packing flange contacts the follower, over a ball-shaped surface, to allow for unintentional misalignment of the packing flange, without side-loading on the stem. The lantern ring provides a space for the insertion of lubricating grease through a separate packing lubricator shown here in combination with an isolating valve. The isolating valve effectively seals the stuffing box from the lubricator to prevent the process pressure from blowing the grease out to atmosphere. This is a safety feature, particularly required on high pressure installations. Valve stem lubricants (lubricating greases) are available for different services and for temperatures of up to 550° F. The viscosity varies with different grades, to provide maximum lubricity at given service temperatures. Another variation in the composition of the grease is dictated by the corrosion requirements. The valve manufacturer should be consulted for his recommendation.

Certain low pressure valves for less critical service, such as water or steam, sometime employ screwed packing retainers such as that shown in Figure 49. In this design, a threaded cap compresses the packing rings inside the packing box. This type of construction is not permissible with highly corrosive fluids. Corrosion, either by atmosphere or by leaking process fluid, either could make the packing adjustment unserviceable after a certain time or could lead to the possibility of a packing blowout.

A well designed packing box permits repacking of at least one additional split packing-ring without disassembling the upper valve portion.

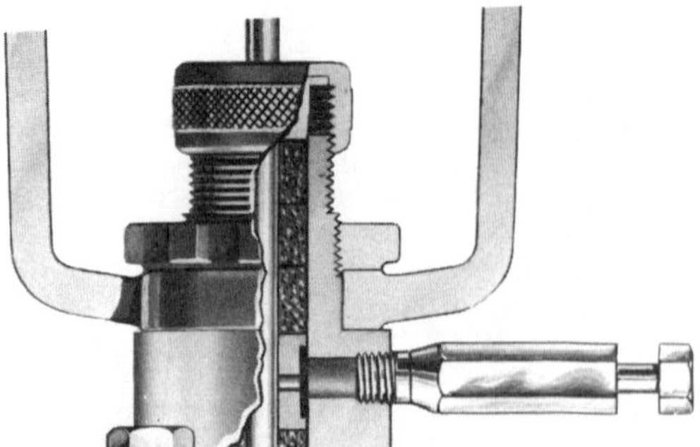

Figure 49. SCREWED STUFFING BOX.

A special variation in the packing box design is shown in Figure 50. This is a *double packing,* i.e., an assembly having two stem packings, in series. Such a packing arrangement is popular on atomic reactor valves that handle pressurized water. It can also be used on high vacuum service, but in this case the V rings have to be inverted. Any dangerous fluid, that leaks around the first set of packing rings, can be vented through the side connection to a special waste tank. Another approach is to pressurize the area between the two packings with inert gas at a pressure level, slightly higher than the fluid pressure inside the valve. In this case, the inert gas might leak into the valve, but no fluid is allowed to leak by the stem. This packing arrangement is more reliable and less expensive than a bellows seal.

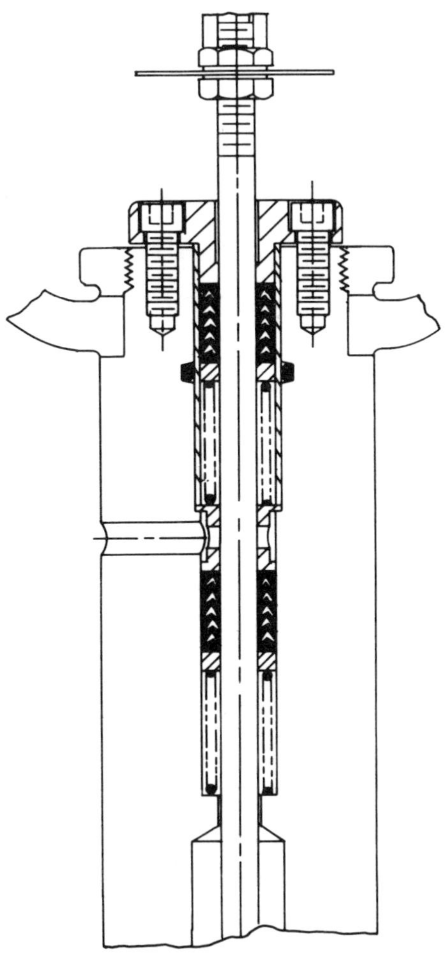

Figure 50. DOUBLE PACKING ARRANGEMENT.

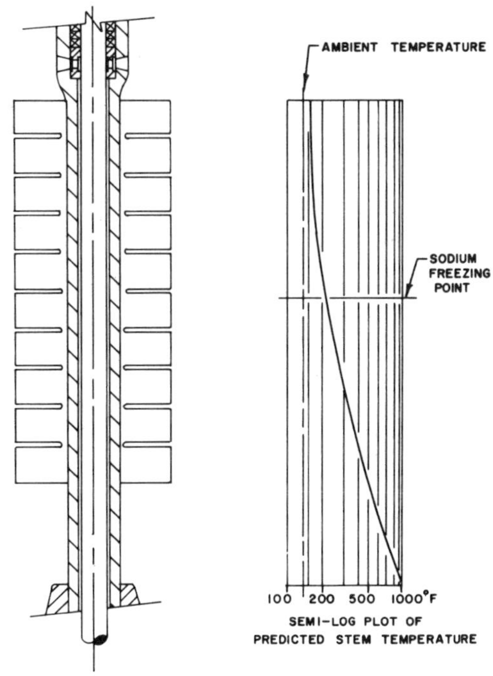

Figure 51. FREEZE SEAL PACKING - MOLTEN METAL SERVICE.

Finally, there are special stem seals such as the so-called "freeze-seal," employed with valves for molten metal (see Figure 51). In this design, a fine annular clearance passage is placed between the stem and the bonnet (inside diameter), ahead of the packing, and long enough to allow sufficient heat absorption to freeze liquid sodium. Solid sodium, of course, cannot leak through the packing box. The disadvantage of this type of seal is the high static friction, i.e., the actuator has to overcome the shear force of the frozen sodium everytime the valve moves.

A freeze-seal is more successful with rotary stem design, because it will reduce the amount of sodium oxide formation near the follow-up packing.

TYPES OF VALVE PACKINGS

The most popular valve packing material is Teflon, because of its excellent chemical inertness and its good lubricating properties. Teflon can be used in solid-molded or turned form (chevron rings), as shown in Figure 52; or it can be used as a lubricant for an asbestos packing. Solid chevron rings should be spring-loaded to provide a minimum initial pressure against the stem. Disadvantages of the solid Teflon packing are its high coefficient of thermal expansion, particularly near room temperature, and the require-

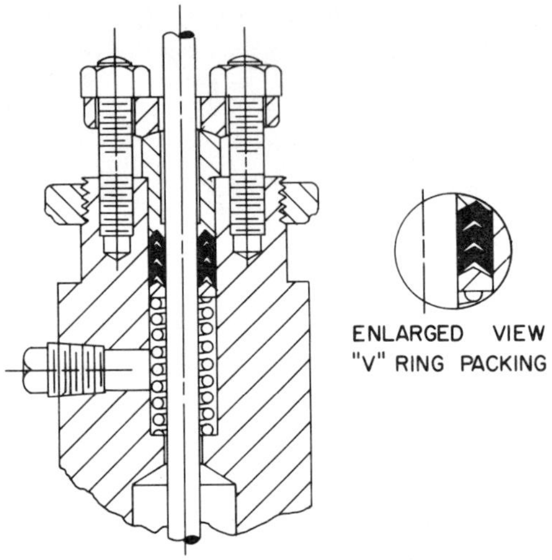

Figure 52. CHEVRON "V" RING PACKING.

ment for extra-fine surface finishes. Surface finishes of 8RMS (roughness in micro-inches) on the stem surface and 16RMS at the inside of the packing box are commonly specified to prevent undue friction and wear of the chevron rings. Solid Teflon rings are non-replaceable, once the actuator is mounted on the valve.

Braided asbestos is still a popular packing material, because it can be made as split rings, which can be wrapped around the valve stem; therefore, facilitating maintenance when the valve is installed. This type of packing usually employs additives, such as mica or graphite, for lubrication, particularly in high temperature service. The maximum temperature limit of asbestos is approximately 1000° F; however, the use of radiation bonnets lowers the local packing temperature sufficiently, to allow its use in valves handling fluid temperatures higher than 1000° F.

A more popular lubricant for asbestos packing, particularly for temperatures below 450° F is Teflon. Teflon is used either as a suspensoid or as a pure braided cover around an asbestos core. The latter is preferred, because it combines the elasticity and deformability of asbestos with the lubricity of Teflon, i.e., low friction coefficient of Teflon against the sliding stem surface.

A recent addition to the list of available packing materials is Grafoil®. Grafoil is an all-graphite product that is flexible and has direction-dependent properties, similar to pyrolytic graphite. It is essentially, chemically-inert, except when strong oxidizers are handled. The coefficient of friction is low and the packing can be used for quite-high temperature applications (sublimation point is 6,600° F). Care has to be taken in adjusting this packing, because it has a high density and over-tightening could lead to a lock-up of the valve shaft.

"O" rings, or chevron rings, made out of elastomers such as Neoprene® or Buna-N® can be used for certain low-pressure valves, handling mild fluids at below 180° F. This type of packing material is found on certain specialized valve applications, such as temperature control valves for air conditioning units.

Table 4 is a selection guide for various packing materials.

Table 4. SELECTION GUIDE FOR VALVE PACKING TEMPERATURE LIMIT

			TEMPERATURE LIMIT	
TYPE	*SERVICE*	*LUBRICATION RECOMMENDED*	*W/STANDARD BONNET*	*W/EXTENSION*
Neoprene or Buna-N O-rings or V-rings	Non-corrosive liquids	No	180° F	—
Braided asbestos with suspensoid Teflon	All except molten alkali and hot hydrofluoric acid	Yes	0-450° F	-450 to 800° F
Braided asbestos with solid Teflon cover	All except molten alkali and hot hydrofluoric acid	No	0-450° F	-450 to 800° F
Solid Teflon V-rings	Limited to fluids that will not attack the 316 SS packing spring	No	0-450° F	-450 to 800° F
Braided asbestos with inconel wire, graphite or mica lubricated	Steam and petroleum	Yes	0-450° F	0 to 1000° F
Grafoil	All except for strong oxidizers	No	0 to >2000° F	

BOTTOM FLANGE CONNECTIONS

As mentioned before, double-seated valves can be made reversible by inversion of the valve body. This necessitates two equally-dimensioned openings in the valve body; the first covered by the valve bonnet and the second covered by a bottom flange. The bottom flange contains the lower stem guide in the case of top- and bottom-guided valves, and sometimes contains a drain connection. Single-seated, top-entry valves do not need the bottom flange, because the lower section of the body is part of the cast housing. However, an interesting variation of the single-seated cage valve is the bottom entry valve in which the cage is contained against the valve bridge by the bottom flange and the bonnet becomes the cast-on portion of the valve housing. A typical design with flangeless configuration was shown in Figure 41. The main advantage of the bottom entry valve is the ability to remove the valve trim, without disconnecting the actuator from the bonnet; however, it may be necessary to provide additional space beneath the valve.

An added benefit of the bottom flange is the ability to remove sediment or other debris from the valve housing. A simple pipe-plug can be substituted for a complete blind flange at the bottom of the valve casting for this purpose, as illustrated by Figure 9.

Another purpose can be served by having the blind flange contain an adjustable-limit stop as illustrated by Figure 53. This limit stop consists of an adjustable screw, locked into postion by a locknut. It prevents the valve plug from completely closing or opening, depending upon on the fail-safe direction. A threaded cap, with a suitable gasket, seals the process fluid. This limit stop is available as an accessory with top- and bottom-guided globe valves. Other valve types have limit stops built into the actuating device.

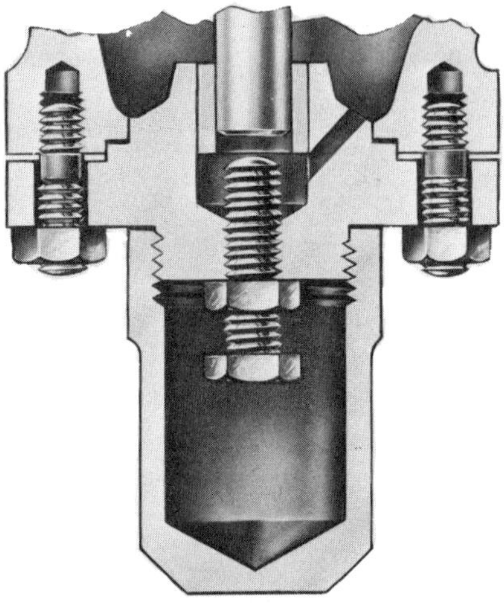

Figure 53. LIMIT STOP IN LOWER BOTTOM FLANGE.

BOTTOM FLANGE AND BONNET GASKETS

For ordinary valve styles, at ratings of 300 lb. ANSI and below, one can use rubber bonded asbestos gaskets, approximately 1/16 inch thick, as bonnet and blindhead gaskets. For higher service pressures and for temperatures above 250° F, metal-clad asbestos gaskets are used. A metal-clad asbestos gasket requires, of course, a higher bolt load; therefore, the bonnet bolting must be designed with this in mind.

Certain top-entry valves, such as the design shown in Figure 9, rely on elastic deformability to make up for the difference in thermal expansion between the cage material and the body. This is provided by spiral-wound stainless steel gaskets with asbestos or Teflon filler. These gaskets are known under brand names such as Spirotallic or Flexitallic. Spiral-type gaskets with Teflon filler are limited to about 450° F, while the ones with asbestos are useable for up to approximately 1000° F. Grafoil or solid metal gaskets can be used on applications exceeding 1000° F.

It is a good design practice to limit the amount of compression of the flexible bonnet and seat-ring gaskets, to a predetermined value, to prevent crushing the gasket. This can be done, for example, by having the bonnet contact the housing material after a certain amount of compression.

REFERENCES

1. ANSI B16.5, 1968, *Steel Pipe Flanges and Flanged Fittings,* American National Standards Institute, New York.
2. ANSI B16.11, 1966, *Forged Steel Fittings, Socket-Welding and Threaded,* American National Standards Institute, New York.
3. ASME *Pressure Vessel Code,* 1968, Section VIII, "Pressure Vessels," American Society of Mechanical Engineers, New York.
4. FC165-2, 1965, *Standards for Face-to-Face Dimensions of Control Valves,* Fluid Controls Institute, Pompano Beach, Florida.
5. ISA RP4.1, 1951, *Uniform Face-to-Face* Dimensions for Flanged Control Valve Bodies, Instrument Society of America, Pittsburgh.

CHAPTER 4

VALVE TRIM

Mathew L. Freeman*

Rockwell Manufacturing Company
Pittsburgh, Pennsylvania

CONTENTS

*The author was employed at Rockwell during the time he wrote this chapter, but he is now employed by Masoneilan International, Inc., Norwood, Massachusetts.

INTRODUCTION

Trim is the heart of the valve with a primary function to proportion the valve orifice area in such a manner that a prescribed relationship exists between flow capacity and valve plug lift. A secondary function may be to shut off tightly. These functions are accomplished in conjunction with body shape, actuator, stem packing, and line fluid forces acting upon the trim. The overall valve design and service application must properly relate all of these parameters to obtain good valve performance.

Valve trim consists of the removable internal parts of the valve contacting the flowing fluid; namely, plug, seat(s), stem plug guide(s), bushings, and cages. Stuffing box components considered as trim are the packing follower, spring, lantern ring, and packing retaining ring. Secondary trim parts are stem-to-plug attachments, seat retaining rings, seat to body seals, spacers and the like. Parts excluded are packing, bonnet, bottom flange and gaskets sealing bonnet and bottom flange to the valve body. (See Figure 1)

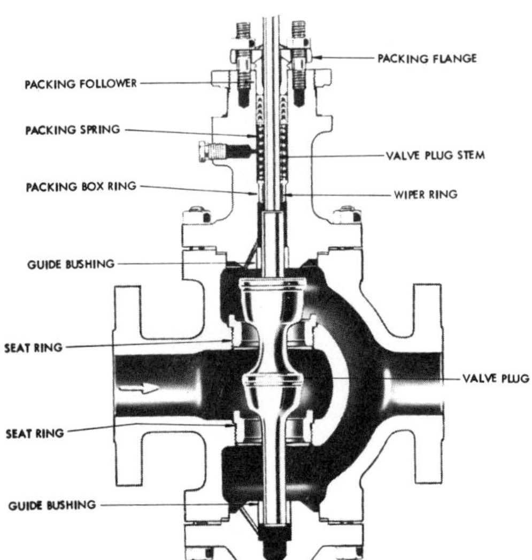

Figure 1. VALVE TRIM COMPONENTS
Courtesy Fisher Controls Company

The plug is the moving component of the valve which throttles flow by positioning itself within the seat orifice and shuts off flow by contacting the seat. The plug is moved against dynamic fluid flow forces by stem force transmitted from the actuator. Throttling may be done by a contoured or V-port plug or shaped opening in a close fitting cage guide.

The plugs (or cage slots) are shaped to throttle flow with a given characteristic such as quick opening, linear, parabolic or equal percentage. Graphs of these flow characteristics are shown in Chapter 5.

Line fluid and flow dynamics of the particular service application may subject trim to an erosive and/or corrosive environment plus static and dynamic forces acting along the plug axis, normal to the axis and rotational to the axis. Vibration is also a factor.

Temperature, pressure, flashing fluids, cavitation, fluid viscosity and solids content all affect trim design. Control of flow requires a pressure drop across the trim and the energy dissipated is absorbed by the fluid and in part by the valve trim and body.

Each valve has a given flow characteristic: i.e., for a given percentage of total lift of the plug off the seat, flow is a given percent of the full open flow. This is true only when the pressure drop across the valve remains constant as in the manufacturer's flow test; it is known as the "inherent flow characteristic."[1]

The installed[2] flow characteristic may differ considerably at much higher pressure drops. In many moderate to low pressure applications the differential pressure across the valve falls off as flow rate increases thus causing a difference between the inherent and installed flow characteristics. This frequently occurs with limited pump capacity or long upstream lines.

Selection of valve trim in regard to flow control is based upon: 1) knowledge of service pressure and flow conditions; 2) the manufacturer's inherent flow characteristic curves for the different trim shapes (these shapes are different profiles on: - plug noses, "V" slots in plug cylinders or cage) and 3) how the inherent characteristic is altered by varying flow conditions. This allows a prediction of the installed flow characteristic for each trim. From these, choose the trim which gives the closest required installed flow characteristic. Beyond this point most controllers will correct discrepancies.

Maximum flow capacity depends upon seat port sizing, body design, and plug

[1] Service applications such as dump valves removing liquid bottoms from vessels do have a fairly constant pressure drop over a wide range of flow. This also occurs in gas reducing service with a large capacity upstream.

[2] Flow characteristics of a specific valve installed in a specific service application.

lift and type, and whether or not the full open flow characteristic must exactly match a straight flow characteristic curve.

Shut off and opening capability are determined by the diameter of the mean seat-to-plug seal contact, pressure differential and available actuator force, provided the stem design is adequate and allowance is made for seating force and packing friction. Not all control valves are required to shut off "drop tight" and manufacturers often state the shut off leakage flow as less than 0.01 to 0.10% of the full, open flow capacity. Upstream block valves are used where necessary.

Double ported valves were introduced many years ago to gain increased flow capacity in a given body size with low lifts. These were necessary for air to close, spring opened diaphragm actuators and also to give a high degree of pressure balance in the closed position thus reducing actuator requirements.[1] Seating tightly on two ports simultaneously is difficult to achieve, except in heavy duty and expensive designs. The balanced cage design valve overcomes much of the double seating leakage, as will be described later.

"Drop tight" or "bubble tight" sealing may be obtained in single seat valves either by using narrow seat joints and high unit compressive seating stress, or by going to soft seats of elastomer materials. Modern actuators possess the power and sensitivity needed for tight shut off combined with good throttling while valve positioners, pressure boosters, and controllers greatly improve performance of simple diaphragm and cylinder actuators to meet tight shut off requirements.

Cage design, incorporating a balance of pressure forces across the closed plug, reduces the differential pressure load by 98% or more, thus allowing a much smaller actuator and greater seat joint compressive load for tight sealing.

Globe valve trim is discussed in detail in this chapter along with a brief description of trim used in non-globe type valves.

TRIM DESIGN FOR MAXIMUM FLOW

Maximum flow capacity in globe valve design is achieved by using the largest possible flow orifice combined with a quick opening plug form. The orifice should not be so large that valve body flow resistance becomes an appreciable part of the total valve pressure drop and changes the flow characteristic. Discharge flow area is equal to the orifice area at lifts of 1/4-seat diameter. (See Figure 2)

Turning fluid past the plug nose causes flow resistance and an additional 10 to 15% lift is required to obtain maximum flow. Thus total lift required is about 35% of the orifice diameter. Y-bodies require lifts of 100% of orifice diameter to clear

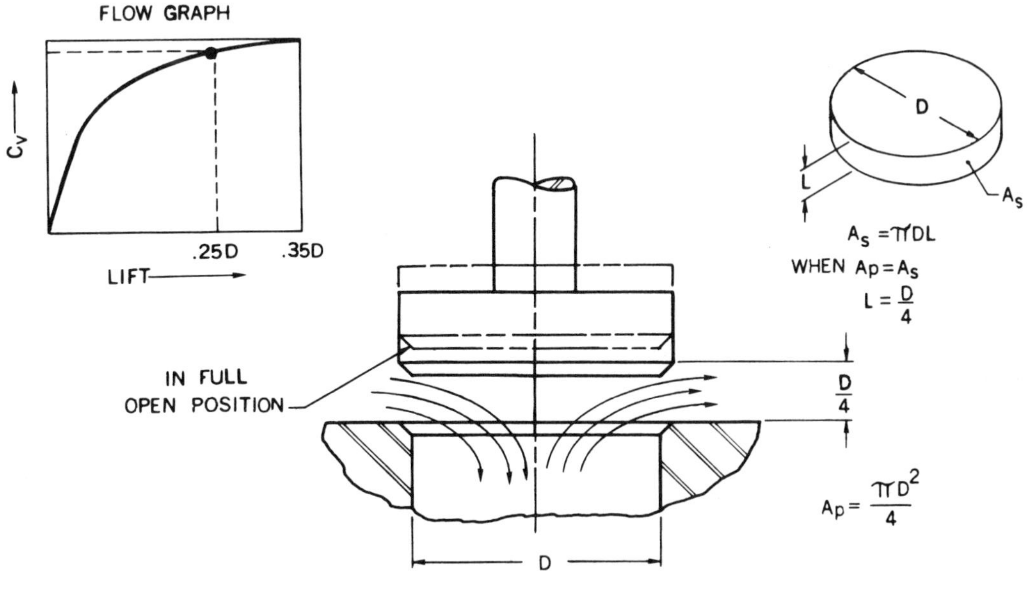

Figure 2. QUICK OPENING PLUG FLOW AREA.

[1] Double ported valves are far from being dynamically balanced in throttling position with flow at moderate pressure drop.

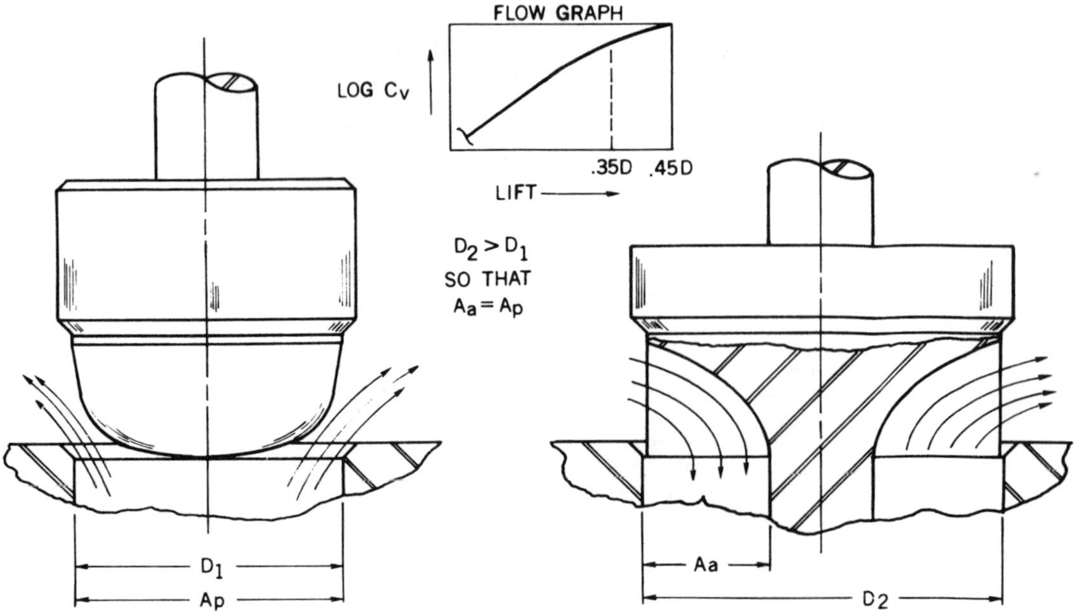

Figure 3. FLOW AREAS OF CONTOURED AND V-PORT PLUGS
Note: A larger port diameter is required for a plug with guide post to equal the port area of a top-guided plug.

the plug out of the flow stream, but have a much higher flow capacity than globe types with an equal seat orifice diameter.

Linear or equal percentage flow characteristic requires a plug nose extending into the seat orifice and this must also be withdrawn to get full flow. In this case the contoured or V-port shape aids in turning flow gradually and the lifts for maximum flow are similar or slightly greater than for quick opening plugs; usually, about 45% of orifice diameter for top guided plugs. (See Figure 3)

PLUG SHAPE EFFECT ON MAXIMUM FLOW CAPACITY

The choice of plug type effects maximum obtainable flow capacity as indicated below for two inch Single seated, full port valves with top and bottom guiding:

Plug Characteristic and Type	C_v *Maximum**	*%Flow*
Quick Opening	59	118%
Linear (contoured)	57	114%
Parabolic (contoured)	56	112%
Equal Percentage (contoured)	52	104%
Equal Percentage (V-Port)	50	100% (Base)

*Typical Values

MAXIMUM FLOW CAPACITY OF VALVES HAVING SINGLE AND DOUBLE SEATED PLUGS

Plug type has an important effect on maximum flow capacity in addition to body type. Note how the double seated flow advantage rapidly increases in the large sizes. The cage designs for double seated valves have oversize orifices with flow throttling within the cage. Typical values are given below:

SIZE (inches)	*DOUBLE SEATED* (Parabolic)	*SINGLE SEATED* Top and Bottom Guided	Cage Type	
		(Parabolic)	Equal Percentage)	(Linear)
1	20	17	18	21
2	65	55	65	75
3	170	120	130	150
4	220	165	200	235
6	500	375	410	465
8	860	590	650	735

Single seated valves without plug guide posts have the highest flow for a given body weight or flow passage area. The post adds turbulence and reduces the orifice flow area. Angle bodies turn fluid one half as much as globes and thus have higher flow capacities for equal orifice diameters. Body flow passages should be streamlined and have no sudden contractions or expansions; this applies only to maximum flow capacity designs used at moderate pressure drops.

The cage trim of single seated valves uses a larger seat orifice than double seated trim, since the principal restriction to flow is the cage flow ports. Some designs have a 2-5/16 in. seat orifice in a 2 in. valve to prevent orifice flow resistance from altering the flow characteristic governed by the contoured cage ports.

"Y" bodies obtain increased flow capacity by streamlining flow, but require much higher lift than globe or angle bodies. This either increases the diaphragm-spring actuator cost or calls for a springless piston type actuator and valve positioner. Stem, plug-guide and packing wear are increased slightly when compared to a vertical valve installation. "Y" bodies are primarily used for on-off high flow capacity service with intermittent throttling.

FACTORS LIMITING MAXIMUM FLOW

The break point in a flow curve, where increasing lift of the plug out of the seat orifice does not increase flow, may be due to either:

1) Plug is completely lifted out and the orifice is the primary flow restriction; or
2) Body flow resistance is becoming an appreciable part of the total valve resistance. (See Figure 4)

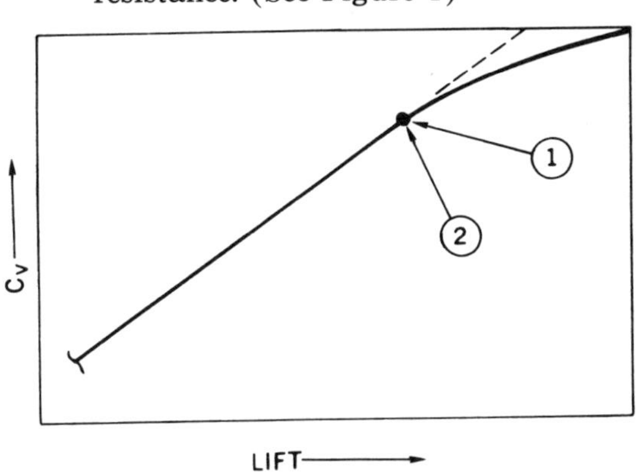

Figure 4. EFFECT OF ORIFICE AND BODY RESISTANCE ON FLOW CHARACTERISTIC.

A break point in the curve, where increasing pressure drop does not increase flow, may be due to either:

1) Critical flow at sonic velocity after pressure drop becomes one half of upstream pressure for gas service; or
2) Flashing flow chokes the body downstream of the seat joint with vapor bubbles in liquid service. In cavitating service no further reduction in pressure can be obtained at the vena contracta to increase flow, since the fluid is vaporizing. (See Figure 5)

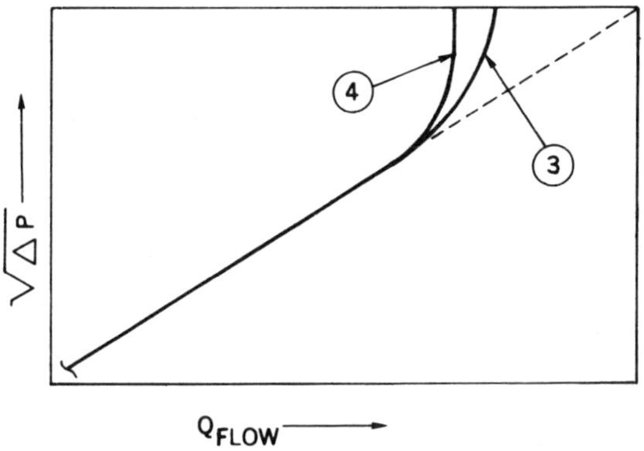

Figure 5. FLOW STAGNATION AND FLASHING EFFECT.

SHAPE OF FLOW CHARACTERIZED PLUGS AND CAGES (See Figure 6)

How closely the manufacturer can get the valve to accurately follow a design flow curve depends upon: 1) valve body type; 2) plug style; 3) available lift in relation to seat diameter; 4) ratio of maximum to minimum flow control as required by the process, called rangeability[1]; and 5) seat joint design. (Refer to Chapter 5)

Maximum flow capacity is usually higher with flow under the contoured plug. Flow-over may be only 80% of flow-under for full port valves and 96% or higher for 40% capacity trim. Flow-under more closely follows the slope or gain of the design flow characteristic resulting in improved control stability. Refer to Chapter 5 for a discussion on gain. V-port plugs show much less difference in flow capacity, over and under.

Positioner Characterizations may be used to attain a special characteristic not available in a standard plug. The positioner cam is shaped to modify the lift for a

[1] No standard of deviation from the specified characteristic has been established; therefore, this must be included in the definition of rangeability.

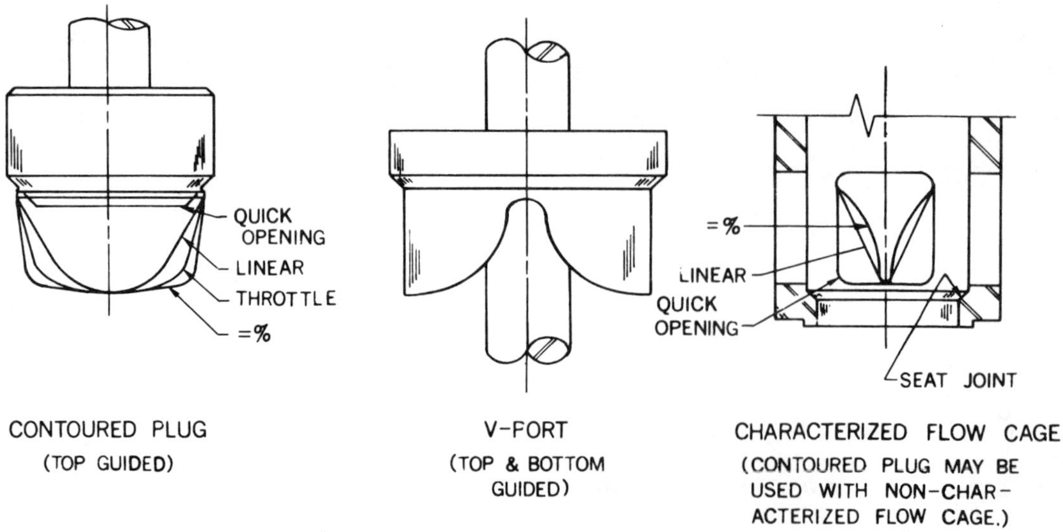

Figure 6. SHAPE OF FLOW CHARACTERIZED PLUGS AND CAGES.

given controller signal to accomplish the special characteristic. There is some limitation on cam shape and follower contact so it is best to start with the nearest plug characteristic, sometimes a parabolic. (See Figure 7)

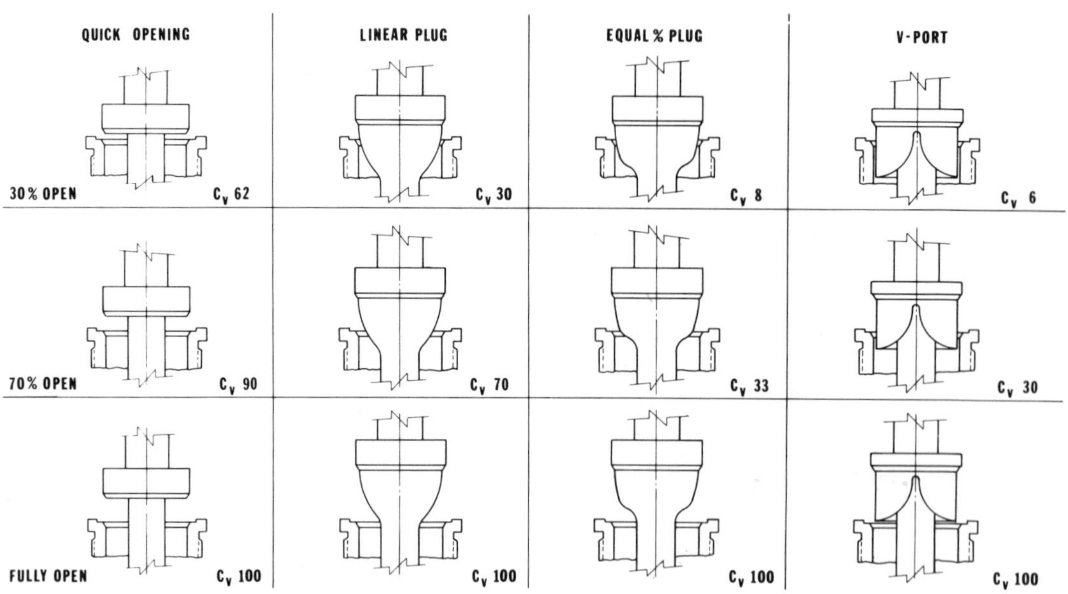

Figure 7. RELATION OF PLUG SHAPE, FLOW CHARACTERISTIC & FLOW CAPACITY AT EQUAL LIFT.

Shown here are the four basic inner valve contours, typified with scalular representation of equal lifts and given flow capacities for each lift. Courtesy WKM Valve Division, ACF Industries

Plug Types

Ultra Low Flow

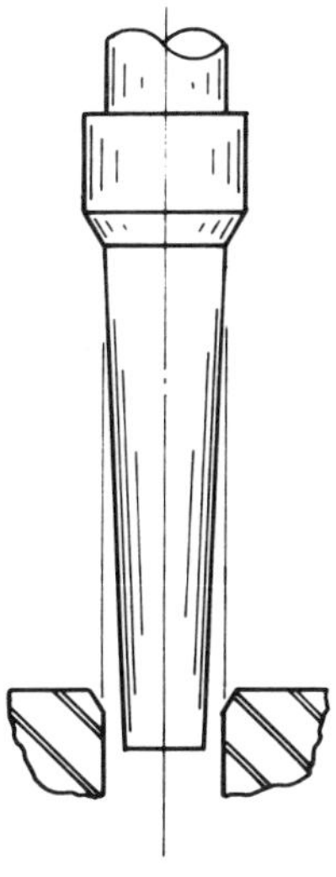

(a) Metering Pins

Pin sizes 1/16, 1/8 and 3/16 in. (Pin does not lift out of orifice). Long pins ground with a small taper angle. Used in precise flow control with constand upstream pressure. Linear characteristic. Used in valve body sizes 1/4 to 1 in.

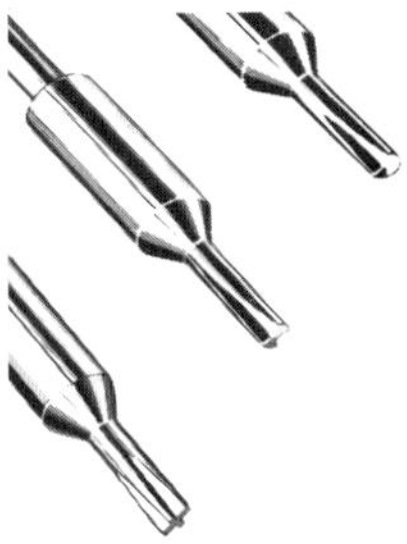

(b) Splines

Plug diameters 1/4 to 5/8 in. (spline does not lift out of orifice). Top and port guiding. Linear Characteristic. Cv range from 0.00001 to 0.6. Add splines up to four to increase flow capacity in a given size. Used in valve body sizes 1/4 to 1 in.

Courtesy of Kieley & Mueller, Inc.

Low Flow

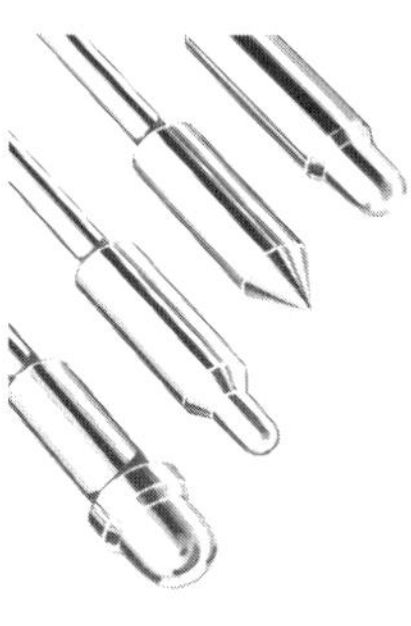

(c) Contoured

Size (in.)	Nominal C_V
3/16	1.0
1/4	1.6
3/8	2.5
1/2	4.0
3/4	6.3
1	12.0

Characteristic: equal percentage. Top guided. Used in 1/4 to 1 1/4 in. body sizes. Plug may lift clear of orifice for max. flow capacity and handling of slurries with less erosion.

Courtesy of Kieley & Mueller, Inc.

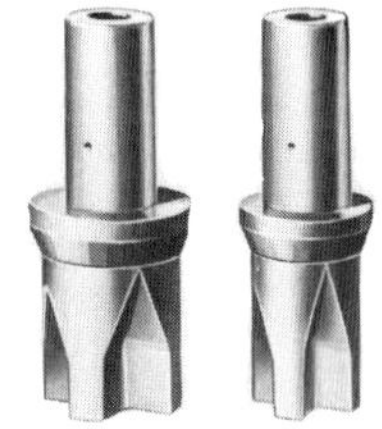

(d) Fluted

Sizes 1/4 to 1-1/4 in. Characteristic - normally equal percentage. Top and port guided. Rangeability, 40:1.

Courtesy of Honeywell, Industrial Division

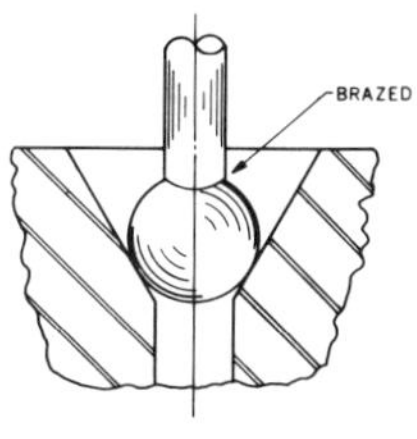

(e) Ball

Tungsten carbide ball brazed to stem, conical seat. For abrasive service. Sizes 1/8, 3/16, 1/4, 3/8 in.

Figure 8. ULTRA LOW FLOW AND LOW FLOW PLUG SHAPES

Regular Flow

Single Seated, 1 in. and larger. Double Seated, 1 in. and larger. Characteristics: linear, parabolic, equal percentage and throttle. Top guided, top and bottom guided, cage guided. Better than solid V-port in entrained particle service. For extreme load variations in high pressure service.

(a) Contoured,
l. Double Port,
r. Single Port

Courtesy Fisher Controls Company

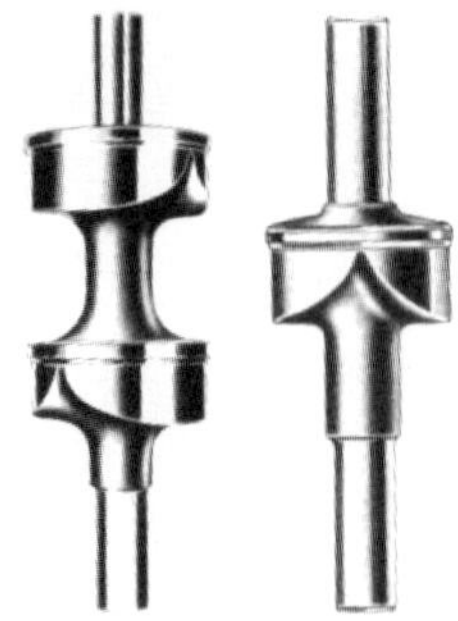

(b) Solid - V[1],
l. Double Port,
r. Single Port

Sizes 1 in. and larger. Characteristic: equal percentage. Skirts of plug are cast solid. Top and bottom guided. Rangeability, 50:1. Greater stability with flow either under or over and less difference between directional C_V than contoured plugs. Best *not* used in entrained particle service. For moderate load variations in high pressure service and extreme load variations to 100 psi ΔP.

Courtesy Fisher Controls Company

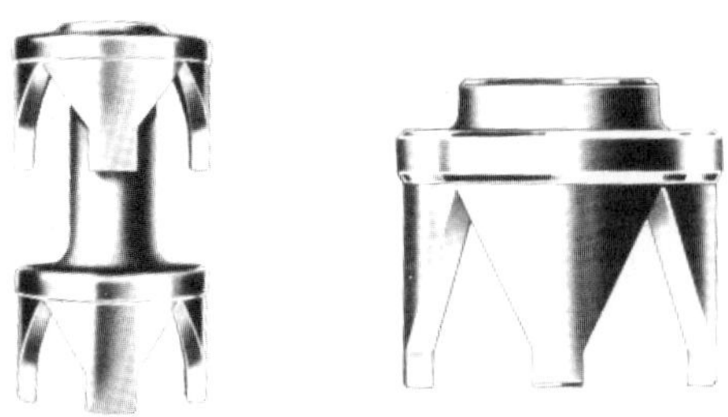

(c) V - Port, l. Double Port, r. Single Port
Courtesy Fisher Controls Company

Sizes 1 in. and larger. Characteristics: modified parabolic and equal percentage. Port guided. Skirt is hollow. Moderate pressure drops to 100 psi. Also used in three-way valves. May be obtained with O-Ring seal at plug seat joint. Best *not* used in entrained particle service. May cause high rotary stem torque with large flows.

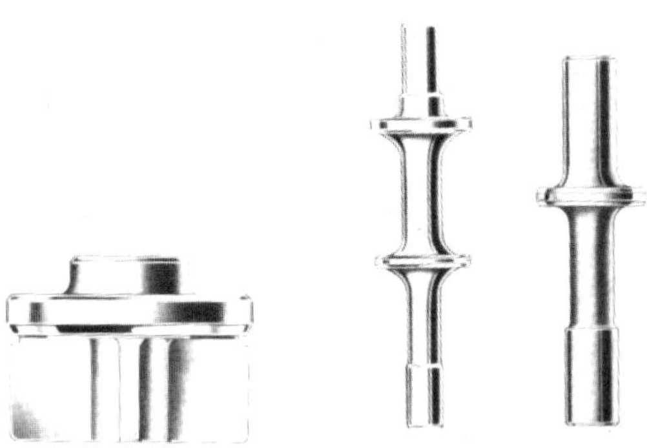

(d) Flat Disc - Quick Opening
l. Port Guided, c. Top and Bottom Guided Double Port, r. Top and Bottom Guided Single Port
Courtesy Fisher Controls Company

All sizes. Short lift. Characteristic: quick opening. Soft seal on plug available. May be used for moderate load changes with 100 psi ΔP. Add very short nose to prevent slamming on higher ΔP applications.

Figure 9. REGULAR FLOW PLUG SHAPES (continued next page).

[1]Note dissimilar top and bottom "V" shape to characterize flow (Also see Figure 27 in Hard Facing Section).

Dissimilar top and bottom shape to reduce dynamic unbalance fluid forces.

(e) Mixed Types (Double Ported)
Courtesy Fisher Controls Company

Sizes to 1.5 in. Plug acts as a sharp edge orifice in contoured seat ring which characterizes flow. Seat ring may be reversed to change direction of flow.

(f) Contoured Seat Ring and Sharp Edge Plug
Courtesy A. W. Cash Company

Sliding gate with multiple orifices. (Refer to section on non-globe valve trim.)

(g) Sliding Gate With Multiple Orifices
Courtesy Jordan Valve

Figure 9. REGULAR FLOW PLUG SHAPES.

MINIMUM CLEARANCE FLOW

Linear, parabolic, and equal percentage plugs are designed with a short cylindrical section just under the seat joint taper as a diameter for plug characterized contour run out and to establish a definite plug operating clearance in the seat orifice. V-port plugs and splined design must also have a clearance fit. (See Figure 10.)

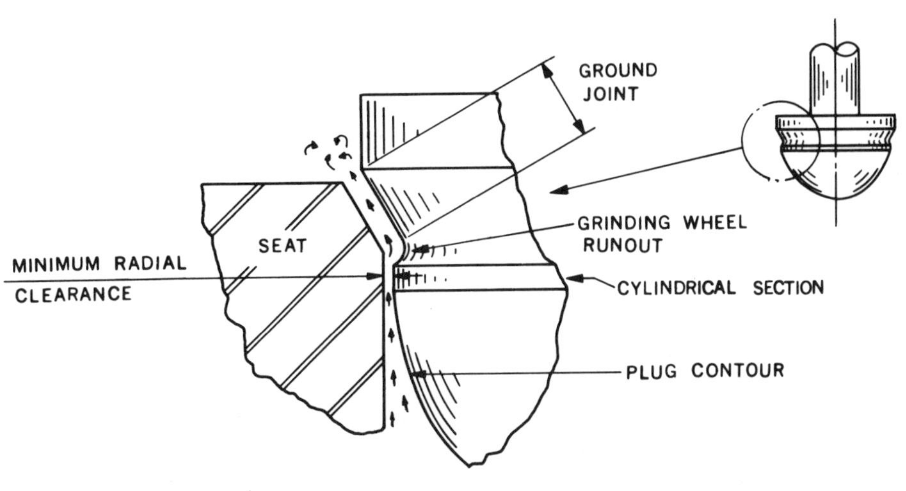

Figure 10. MINIMUM CLEARANCE FLOW OF CONTOURED PLUG.

When the plug is first cracked open, there is a linear characterized flow controlled by the distance across the opening of the joint. Soon the joint opening exceeds the annular flow area of the clearance between the cylindrical plug section and the seat orifice diameter, temporarily causing a choked flow which increases as the plug rise decreases the length of the close clearance flow path. True characterized flow should begin after the plug-contour run-out into the cylindrical section reaches the bottom of the seat-angle run-out in the orifice. Actually this does not occur until the plug makes a slight additional lift to eliminate the affect of the previous resistance.

Minimum clearance flow, as a percent of total flow, decreases as valve size increases, because it varies with the seat orifice diameter, whereas total flow varies with orifice area (diameter squared); therefore, the rangeability of plugs normally increases with size. Double-seated valves have larger clearance flows than single seated valves for the same flow capacity, because the length of the seat circumference of an area divided between two ports is longer, and clearance flow area is greater than for single ports. Guide posts also cause the seat diameter to be larger because they block part of the flow path. Rangeability of top and bottom guided valves, both single and double seated, is thus restricted below 3/4 in. size. Top guided valves should be used in sizes 3/4 in. and smaller.

Minimum flow in cage valves is controlled by the diametral clearance between the cage and plug cylinder, and by the length of the annular flow passage before it reaches the "V" at the bottom of the cage port. The plug-to-seat joint, flow area is similar to that shown in Figure 11.

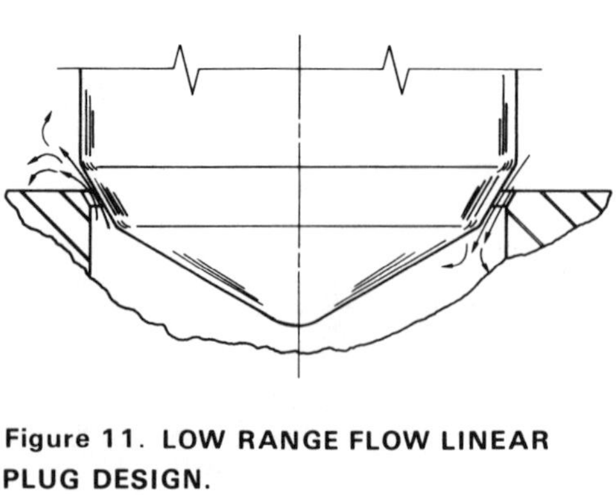

Figure 11. LOW RANGE FLOW LINEAR PLUG DESIGN.

RANGEABILITY

The parameters previously discussed, control the maximum low flow rangeability of linear, parabolic and equal percentage plugs. Quick opening plugs with flat joints have much less of a problem. Eliminating most of the minimum clearance flow is feasible in short-stroke, *linear* plugs, resulting in increased rangeability. The design, illustrated by Figure 11, functions without a cylindrical contour, run-out diameter. Rangeability of 800:1 is readily obtained by this design; however, rangeabilities of over several hundred are meaningless, because no process approaches this variation and, because plug positioning accuracy to obtain them is beyond the precision limits of most actuators.

Plug Trim vs Nominal Rangeability

Characteristic	*Plug Shape*	*Rangeability*
Quick Opening	Flat Disc	100 and up (Higher in larger sizes with lower seat angles)
Linear	Contoured	30:1 to several hundred
Parabolic	Contoured V-Port Skirt	
Equal Percentage	Contoured	30:1 to 60:1
	Solid V-Port	50:1

So much of the equal-percentage plug's lift is taken up by low flow, that in a contoured plug, the nose-to-control low flow often projects into the conical clearance needed at higher lifts for larger flows. As a result, the maximum flow of the equal percentage plug is less than a linear plug; therefore, with the same minimum clearance flow the equal percentage rangeability is lower.

DESIGNING A PLUG TO A SELECTED FLOW CHARACTERISTIC

The theoretical C_V of a perfect orifice having one square inch area is 38.1. This means that an orifice with a flow efficiency of 100% would pass 38.1 gpm (gallons per minute) with a pressure drop of 1 psi. No orifice has such an efficiency, of course, and the addition of a valve plug restriction further reduces the flow efficiency.

When designing a flow characteristic for a plug, the designer must attain a specific flow, corresponding to a specific lift. To do this, he needs to know the efficiency of the annular seat-to-plug orifice area (not the horizontal annular area, but the truncated conical area, normal to the plug contour at that lift). The efficiency varies with the flow rate and the length of the restricted flow path; the latter being predominant at low lifts. Also, the body-flow resistance is a factor which must be accounted for at high lifts and each type of body has a different affect on the resistance.

One method of plug design is to make a preliminary flow test using a rough-shaped plug with each new body design to determine the flow efficiency curve from C_V minimum to C_V maximum for that body. The plug is designed to a given flow characteristic using this data, then it is flow-tested and, if necessary, reshaped slightly to follow the design flow curve more closely. (See Figure 12.)

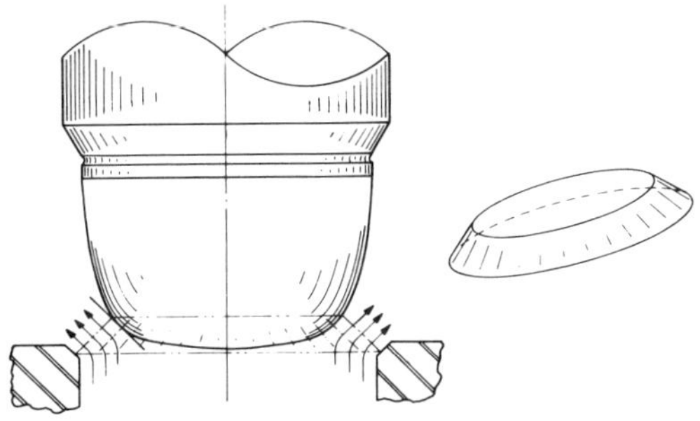

Figure 12. TRUNCATED CONICAL FLOW AREA.

REDUCED CAPACITY TRIM

Reduced capacity trim is used for the following reasons:

1) To achieve precise control at low flow in a process which may soon be expanded in capacity to a much higher flow rate. Valve body change-outs are eliminated. Much of the full capacity stroke length is retained in designing the reduced capacity plug, thus increasing the rate of change of lift to flow, therefore, increasing the precision of flow control.
2) To absorb vibrations and thermal energy in the relatively larger, heavier walled body, having a smaller flow port and much heavier guiding. This occurs in flashing liquid service, high velocity liquid flow, high pressure drop and/or cavitating service. The larger body downstream of the seat orifice in cavitating service, allows the seat wall to be well away from the body wall so that bubbles collapse more in mid stream than against the body wall. By keeping the same guide diameter, the plug is a much stiffer design than the same flow capacity in a smaller orifice valve. Thus, more metal surface is available to dissipate energy for high pressure drop service.
3) To provide ample feeding and gradual turning of fluid into the throttling zone resulting in reduced velocities in order to reduce abrasive erosion by slurry. The body size is relatively larger than the trim.
4) To reduce gas outlet velocity to a subsonic level in the valve body, downstream of the seat orifice.
5) To avoid using pipe line-reducers as a cost savings or to avoid using them where piping strength requires a full size body.
6) To correct initial oversizing errors or to handle flow when the process capacity is reduced.

Most manufacturers offer *standard reduced capacity trim* at 40% of full flow trim capacity for top and bottom guided valves; however, other sizes are available in top guided designs. For example, some 2 in. valves may be fitted with seven sets of trim having seat port sizes of 1/4, 3/8, 1/2, 3/4, 1, 1.5, and 2 in. Top and bottom guided valves and double-seated valves are restricted to fewer sizes by the bottom guide post.

The seven sets of trim are not truly *reduced capacity trim* for high energy dissipation, unless guide post diameters are equal to the full capacity trim posts. If this not the case, they are referred to as *restricted trim.* Reduced capacity trim is illustrated graphically in Figure 13.

In the following example, trim capacity is determined for a 2 in. valve.

Full capacity port of 2 in. diameter has a 3.14 in. flow area (with top guiding).

Reduced capacity port of 1-1/4 in. diameter has a 1.23 in. flow area (with top guiding); therefore,

$$\frac{1.23}{3.14} \times 100 = 39\% \text{ of area}$$

With a lower flow, valve body resistance is less; therefore, reduced trim flow capacity would be approximately 40%.

Plug lift for some manufacturers' 2 in. full port valves is 1-1/4 in. for linear or equal percentage plugs; therefore,

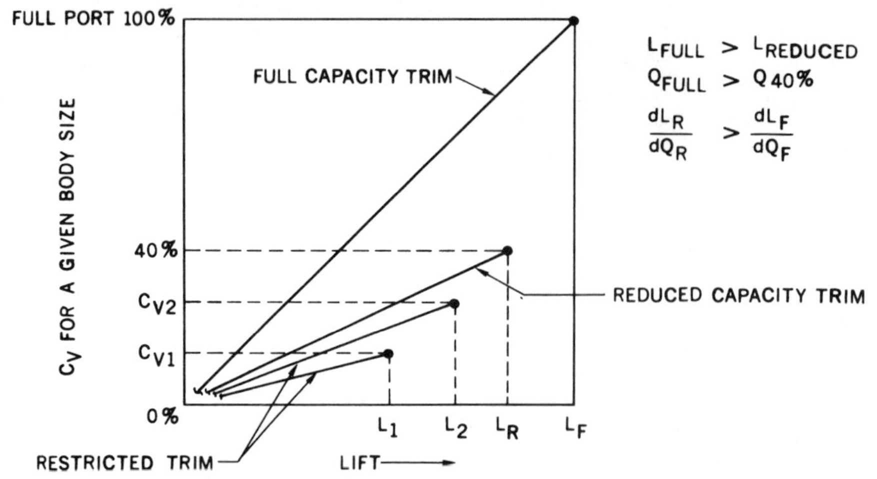

Figure 13. GRAPHICAL REPRESENTATION OF REDUCED CAPACITY TRIM.

the lift: diameter (L/D)ratio is:

$$L/D = \frac{1.25}{2.00} = 0.62$$

Plug lift for some manufacturers' 2 in. reduced port valves is 1 in. for linear or equal percentage plugs; therefore:

$$L/D = \frac{1.00}{1.25} = 0.80$$

As a result, the reduced port plug lift is:

$$100 \times \frac{0.80}{0.62}$$

or relatively 30% longer; therefore, dL/dQ (Q is flow rate) is greater and the flow control precision increases.

Changes Required By Installing Reduced Capacity Trim

In standard valves, plug and seat ring changes occur. The seat orifice is smaller and the plug has a smaller outside diameter, but the same guide post diameter. The plug lift-to-diameter ratio may be increased causing increased sensitivity by a more gradual profile. Body, guide post bearings and seat ring thread in the bridge wall remain the same.

In cage guided valves only the cage is changed as follows:

1) The new cage has smaller ports, yet seats with the old plug, or the cage may have fewer ports. The same size actuator is used since the seat port area does not change.
2) Non-characterized cage designs, which have contoured plugs, require the replacement of both plug and cage. A smaller actuator may be used with a smaller seat port area.

Some manufacturers use additional supporting parts for undersized trim so it will fit into a larger body. They are usually transfer pieces, fitting the standard, *restricted trim* seat to the larger seat ring thread in the body.

Top guided, single seated valves give a greater precision of low-flow control with reduced capacity trim (greater low flow rangeability) than do top and bottom guided, or double ported valves, because the clearance flow of the latter two is higher.

Reducing the number of cage ports in high pressure service may cause side loads and plug vibration; therefore, a reduction in port size is preferred. A long undercut on the plug just above the seat joint helps to eliminate 60-80% of the side thrust in a non-uniform, flow cage.

SELECTION OF TRIM MATERIALS

Over the years, valve trim has resolved into a limited number of materials which have the necessary balance of physical and chemical properties to give good performance in a wide range of applications. The material selection is coupled with a practical means of fabrication.

Property Considerations

Physical	*Thermal*	*Environment*	*Corrosion*	*Fabrication*
tensile	creep	non-galling couples	chemical resistance	castability
compressive	rupture	erosion resistance	electrolytic potential with body	machinability
torsional				
shear	scaling (oxidation)			finishing surfaces
yield	hot hardness			suitable for hard facing
ductility	cold impact			method of hard facing
Modulus of elasticity				
hardness				
density				

Relative Cost Of Trim

(Relative to a 2 in. single seated valve, including plug, stem and seat)

bronze	0.5X
Type 316 SS	1.0X (base)
Type 440-C SS, hardened	1.4X
Type 17-4 PH SS, hardened	1.4X
Stellite hard faced joint on both plug and seat (316 SS base material)	1.4X

Application Of Common Trim Materials

Application	*Materials*
mild service	bronze
general service	Type 316 SS
severe service	17-4 PH SS Type 410 SS Stellite Colmonoy
very erosive service	Type 440-C SS hardened tool steel tungsten carbide
very corrosive service	Nickel Monel® Inconel® Hastelloy-B® Hastelloy-C® Durimet® titanium

TRIM EROSION

Control valves are active components of flow control systems, constantly throttling and frequently shutting-off flow; therefore, in severe service, trim wear is to be expected. Because of the wear, ease of trim replacement is an important consideration in valve design.

Wearing away of plug and seat material causes leaky valves, alters flow characteristics and, in cases of severe erosion, may create a new flow path and may actually penetrate the valve body wall.

Particle impingement erosion is related to the kinetic energy of the abrasive particles rather than to particle momentum. High viscosity, carrying fluids reduce erosion by creating a "cushioning effect" between the particles and trim. Trim hardness is not the sole criterion in evaluating erosion resistance to particle impingement. Small abrasive particles cause less erosion than larger particles and *erosion damage increases roughly as the square of the flow velocity.*

There are four principal types of erosion: 1) abrasive particle; 2) cavitation; 3) erosive-corrosive action; and *4) high liquid velocity impingement.* Methods of elimination, reduction, or transferring of erosion to a non-critical area differ with the type of erosion.

Abrasive Particle Erosion

Billions of tiny particles, harder than the trim surface, are carried at high velocity in the fluid stream and impinge upon and scour away the trim metal. Slurries have such a high concentration of abrasive particles, that the plug and seat joint literally grind together on each closure. In the latter case the seat join is often propped open by crushed particles and wire drawing occurs.

Abrasive erosion may be reduced by:

1) Selecting trim material or trim with a facing which is harder than the abrasive particles; however, it must not be so hard as to be brittle to impact. This will extend the wear period and restrict abrasion to a few critical zones.
2) Streamlining flow to prevent direct particle impingement upon the trim. To do this, fluid flow should be parallel to the critical seat joint and plug contour surfaces. Flow must be turned gradually and flow attachment and detachment principles must be used to protect plug-to-seat joint areas. These factors are handled by valve designers, but the user and manufacturers' application engineers play important roles in the selection of trim types and body style to aid in achieving these principles.

Valve designers use several techniques to develop and improve flow patterns to reduce erosion. Half- and full-model flow studies using plastic see-through bodies are often used. Flow patterns are shown in air by smoke or attached filaments, while in water either an aluminum flock or a colored fluid injection is used. Trim and body shapes are altered to reduce direct impingement, to remove eddies, and to reduce pressure drops. Once this is done, a final erosion locating method may consist of circulating a high-sand-content fluid through a valve in which the trim has been coated with machinists' layout-bluing. The erosion patterns are visible in a few

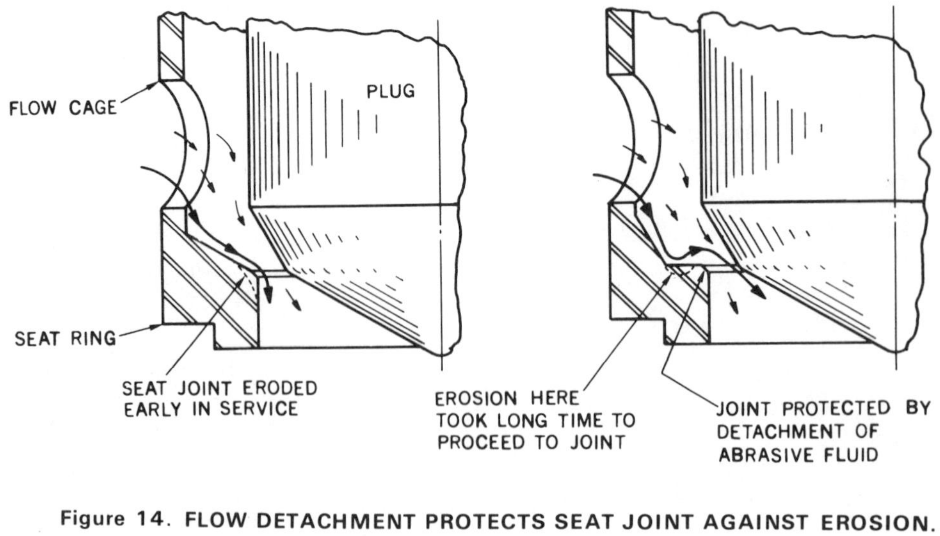

Figure 14. FLOW DETACHMENT PROTECTS SEAT JOINT AGAINST EROSION.

minutes and further changes may be needed to create a uniform wear rate and to reduce the wear at the seat joint. Once this is accomplished, several days of testing are required to erode the metal to prove the design changes. High pressure drop blow-down dynamic throttling tests may include a range of abrasive particle sizes including pea gravel. This really proves impact resistance.

1) In Figure 14, flow attachment and detachment are utilized to greatly reduce the rate of seat joint wear.
2) Figure 15 shows the effect of the removal of an eroding flow eddy in the discharge of a three-way valve. Eddy was necking stem in an area where it passed through the stem seal.

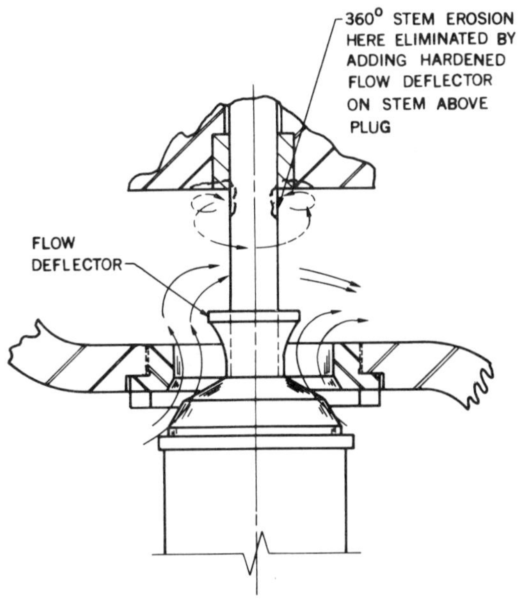

Figure 15. DEFLECTOR ON STEM OF 3-WAY VALVE STOPS EROSION.

3) In Figure 16, the flow cage more uniformly distributes flow around the seat to give longer life and more uniform wear.

Abrasive erosion may be further reduced by using narrow seat joints to limit the particle hang up on the valve closure and to allow sufficient seating force to crush large particles which do hang up. For example in a seat port diameter of 1/4 to 2 in., use a joint width of 0.015 to 0.030 in.

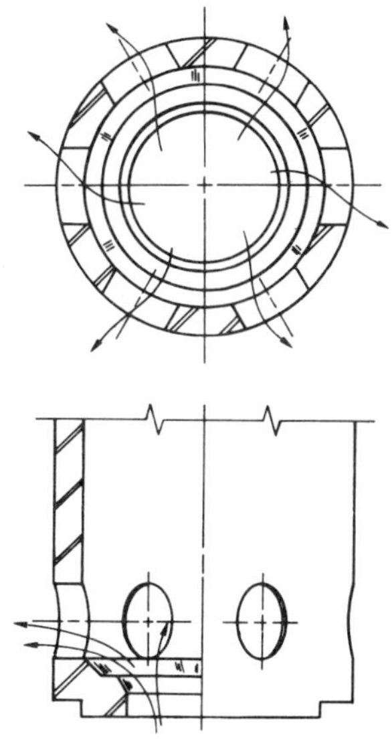

Figure 16. FLOW CAGE DISTRIBUTES EROSIVE FLUID AROUND SEAT JOINT INCREASING SEAT LIFE.

Recommended Pressure Drop Limits For Trim Handling Slurries (Figure 17.)

Material	*Throttling Pressure Drop Limitations*
17-4 PH	125 psi
No. 6 Stellite	125 psi
No. 6 Colmonoy	150 psi
Type 440-C SS	200 psi
tungsten carbide	Above 200 psi

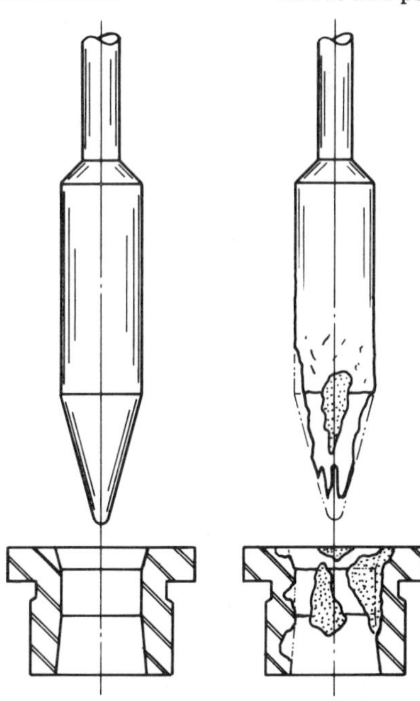

Figure 17. FLUID ABRASIVE & IMPINGEMENT EROSION.

Cavitation Erosion of Trim

Cavitation may occur any time the pressure of a high velocity liquid flowing in restricted passages of a valve becomes less than the vapor pressure of the liquid at that temperature.

When this happens, vapor bubbles form downstream of the restriction. The enlarging flow passage downstream of the restriction reduces flow velocity and the resultant pressure recovery collapses the bubbles. The impact of bubble implosions causes severe damage to trim. This is covered in greater detail in Chapter 6.

No known material can withstand continuous, severe cavitating conditions without failure, because the impact forces approach or exceed the yield strength of the hardened trim or the binding alloys used for extremely hard tungsten carbides. Ease of trim replacement is, therefore, essential in service under cavitating conditions. Even mild cavitation, if it is continuous, will eventually damage regular trim. Hardened trim is required to withstand a higher intensity of cavitation; whereas, severe cavitation will require a scheduled replacement of even the hardest trim.

The following trim materials are used for cavitating service, listed in the order of increasing resistance to cavitation effects, R_c (hardness).

Material	*Hardness at 70° F*
No. 6 Stellite	40 / 45 Rc
Type 440/C	50 / 60 Rc
No. 6 Colmonoy	56 / 61 Rc
hardened tool-steel	60 / 63 Rc
tungsten carbide (Sintered and held in nickel binder)	70 Rc[1]

Cavitation erosion of the selected trim may be reduced by:

1) Selecting the hardest trim feasible, without the chance of cracking from the impact of repeated valve closure and thermal shock.
2) Maintaining sufficient downstream pressure to exceed the fluid vapor pressure; therefore, preventing bubble formation in the high velocity low pressure regions. Often 50 to 100 psi downstream pressure, on pressure drops of 3000 psi for water at 120° F, will greatly reduce cavitation damage. This may be done by installing downstream of the valve, either a back pressure regulator or a very hard choke orifice that is smaller than the valve seat.
3) Using a sharp edge seat port orifice to keep the discharge away from the body wall in high pressure angle valves with flow over the disc. Collapse of bubbles within the fluid stream is less damaging than collapse against a rapidly diverging seat port or body wall. If this is not feasible, a gradually diverging seat orifice may be used.
4) Limiting the valve pressure drop by using two valves in series and distributing the drop between them, with a greater drop in the upstream valve.
5) Using special valves that incorporate a multiple series of restrictions and turns in the flow path, thus limiting the pressure drop across any one restriction.

Special trim designs are available for cavitating service. Some valve manufacturers resort to unique designs which greatly reduce or eliminate cavitation in very high-energy dissipation service. These designs cause the fluid to impinge upon itself, creating a high degree of turbulence within the plug-to-seat passage. The upstream potential energy, represented in any throttling valve by inlet pressure, is converted to heat energy in fluid friction (massive turbulence). Pressure reduction is accomplished with essentially no pressure recovery, thus reducing low-pressure, bubble-forming areas. Other designs have large plenums downstream of the seat orifice to allow bubble collapse in the fluid stream rather than against the body wall. Five designs are illustrated in Figures 18, 19, 20 and 21.

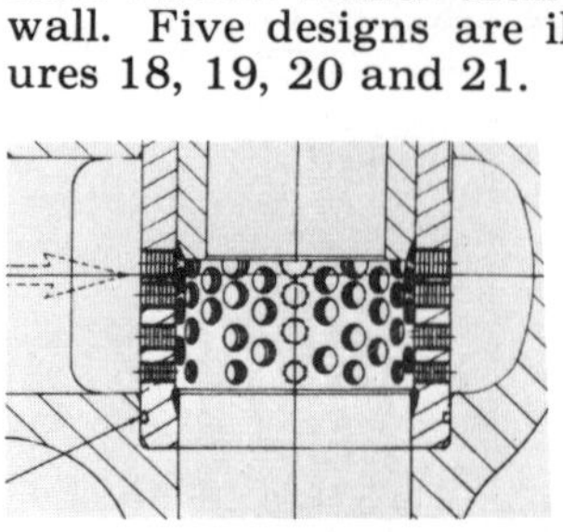

(a) Cage Design With Tapped Holes

Tapped holes cause fluid turbulence in hole orifices and also jet impingement from opposite holes. Seat ring and outlet piping has same internal diameter to prevent pressure recovery downstream of seat. Design successfully used in 1 to 18" sizes.

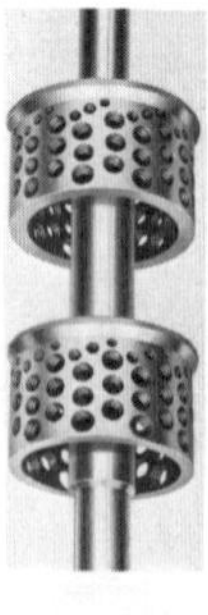

(b) Plug With Tapped Holes

Figure 18. TAPPED HOLE DESIGNS REDUCE CAVITATION.

Courtesy ITT Hammel - Dahl

[1]Some tungsten carbides are harder, but are brittle and do not have sufficient impact strength for repeated valve closure, particulary when cold. Tungsten carbides of 82 Rc may be used in throttle orifices, below seat joints, as shown in Figure 23. Boron carbide also falls in this category.

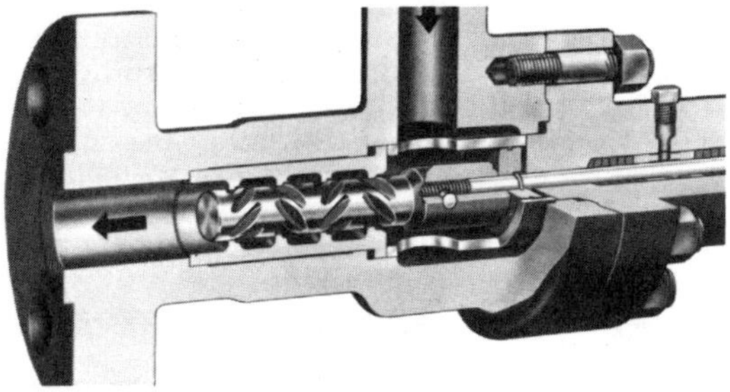

(a) Turbo-cascade® Control Valve

Fluid vortexes are produced within each stage which may take up to 1000 psi pressure drop. Fluid is turned 90 as it passes through the flutes from one stage to the next, and flutes spiraling around the plug force fluid against the containment wall by centrifugal force. Cavitation vapor bubbles tend to remain in the main stream away from the walls since they are lighter than the fluid. This design has been applied to regulating boiler feed water pump bypass.

Courtesy Yarway Corporation

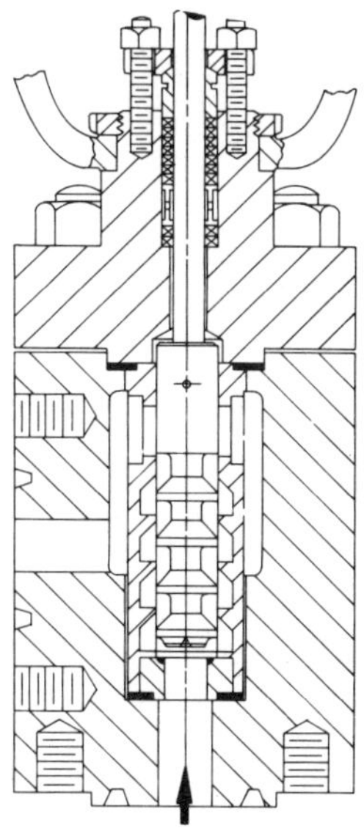

(b) Cascade Flow Restricting Orifices

Abrasive and cavitation erosion in high pressure drop liquid service are practically eliminated in this valve design which reduces pressure in multiple small steps thus maintaining much lower velocity than conventional valves. Conventional single plug orifice valves exhibit an average velocity head loss coefficient of 0.8 whereas this design has a K value of 20 due to the continuous turns the fluid is subject to when passing through. (Head loss $H = V^2/2g$ Each port of the plug sees only 5% of the pressure drop experienced in a conventional valve. The throttling process is nearly identical to a constant enthalpy flow with friction much like pressure drop in a long pipe line. Long guiding is restricted to 4 corners of the circumference which prevents jamming when flowing particle entrained liquids. Working pressures range to 6000 psi in sizes 1/2 to 2". Flow capacity is 1/16 to 1/4 that of a conventional low pressure drop valve.

Courtesy Masoneilan International, Inc.

Figure 19. **TWO DESIGNS FOR CONTROLLING FLOW BY CASCADING.**

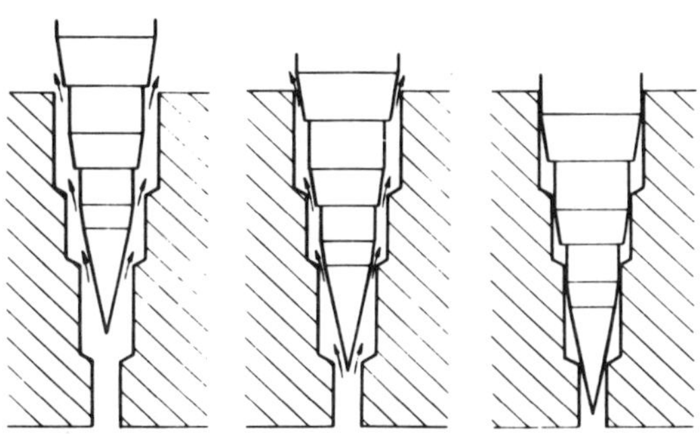

Figure IV - 20. MULTIPLE VELOCITY HEADLOSS TRIM.

Courtesy A. W. Cash Company

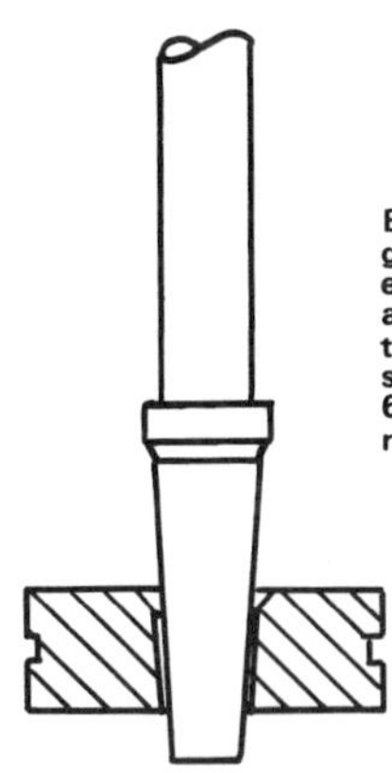

Both plug and seat orifice are gradually tapered. Friction generated in the long flow path absorbs energy, reducing cavitation erosion. The design is stable and has operated to 60,000 psi pressure drop. The rangeability is up to 250.

Figure IV - 21. REDUCED FLOW VELOCITY TRIM FOR HIGH PRESSURE DROP.

Courtesy Masoneilan International, Inc.

Erosion-Corrosion

Many metals do not corrode, because they form a protective surface film. In erosive valve service this film is rapidly worn away by abrasive particles, fluid impingement, or cavitation erosion, thereby exposing a fresh surface to the corrosive fluid. Under these conditions erosion-corrosion may proceed at a very rapid rate. Keep in mind that corrosion is a very complex phenomenon and simple guide rules are no substitute for actual valve usage history in the corrosive media.

Erosion-corrosion may be reduced by using more noble metals for trim such as: Monel®; nickel; titanium; Hastelloy®; or Durimet®. Refer to the trim material table given in the appendix for general application and for corrosion resistance to process fluids.

TFE (teflon) Trim in FEP Lined Bodies is the most completely corrosion resistant valve available in the 2 in., ANSI 150 lb. class for -20 to 300°F service (FEP is fluorinated ethylene propylene). The FEP lining is usually locked into grooves cut into a carbon steel body. The plug and seat are composed of TFE which may be glass-filled to increase stability. Alloy valve stems, compatible with the process fluid, are furnished by the manufacturer.

The following list is the relative galvanic corrosion resistance of trim materials, presented in the order of most corroded (anodic) to least corroded (cathodic). Copper and Platinum are shown for reference.

carbon steel	body material
cast iron	body material
Ni-Resist	body material
Type 440-C SS	trim
17 - 4 PH SS	trim
Type 316 SS	trim
Stellite and Colmonoy	trim
nickel	trim
Inconel	trim
copper	reference
bronze	trim
Monel	trim
platinum	reference

The electrolytic potential between trim and body material is higher in proportion to the greater their separation in the above table. The rate of corrosion is, therefore, higher from galvanic action. The larger area of the body, available to corrode, prevents this from being a serious problem because the body is the sacrificial component. Specification codes call for an allowance in body wall thickness for this corrosion.

High Velocity Fluid Impingement Erosion

Extremely high velocity fluid jets sometimes occur in the flow path, causing the fluid to turn abruptly, bouncing off one surface to impinge and erode the adjacent part. Impingement erosion may be a form of erosion-corrosion, whereby, the high velocity fluid jet, blasts away the protective surface coating as rapidly as it forms. Average fluid velocity through the seat-to-plug flow area must exceed several hundred feet per second for impingement erosion to occur.

Boiler feedwater pump bypass valves are a typical example of high velocity erosion. The high pressure drops, to 4600 psi, taken in this service, often cavitates the seat joint and fluid impingement may either wash away the contoured plug nose or cut peculiar flow patterns in its surface. Cage trim helps, but may not completely eliminate, impingement erosion.

Liquid droplets entrained in a vapor stream also cause impingement erosion; however, the distribution is spread over a much wider area. Dry gases and superheated steam cause little wear.

Impingement erosion as illustrated in Figure 22, may be reduced by: 1) streamlining flow paths; 2) using hard trim; and by 3) separating entrained droplets from the fluid stream before it enters the valve.

The following is the relative erosion resistance of various plug types and body combinations, in order from the "best" to the "poorest:"

1) *Single-seated, contoured plug, angle body, flow over plug.* Ample all around fluid feed to seat inlet and gradual turning of the fluid into the throttling zone reduces the inlet velocity and abrasive erosion. The discharge pattern results in the least cavitation damage. Long sweep angle body styles may be used to advantage.
2) *Single-seated, contoured plug, globe*

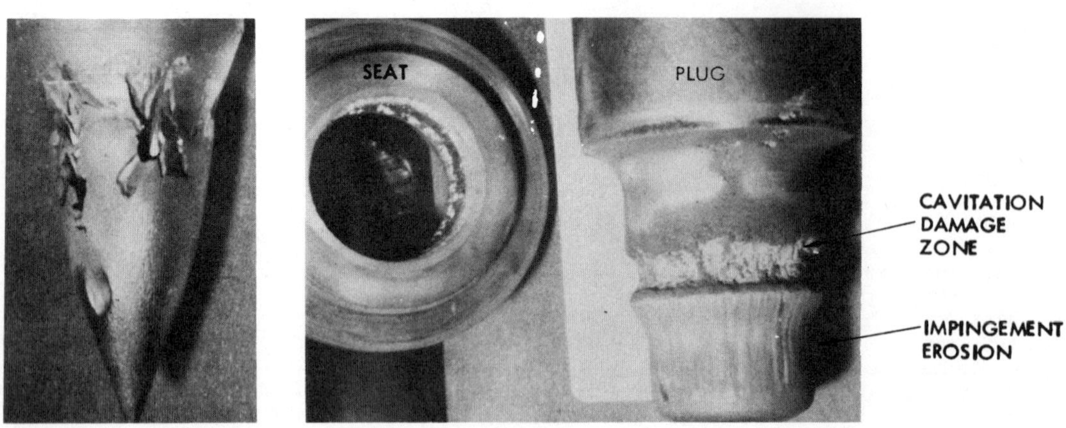

Figure 22. IMPINGEMENT EROSION DAMAGE.

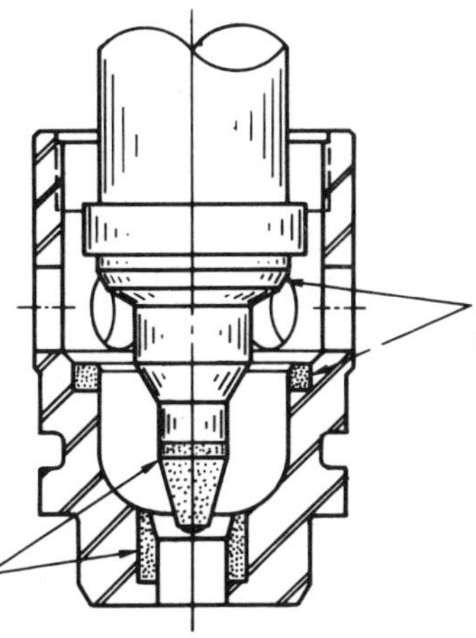

Figure 23. CAGE TRIM WITH SEPARATE THROTTLING AND SEATING ZONES.

body. Pressure recovery within the body, aggravates cavitation damage.

3) *Single-seated, V-port plug, globe body.* Jet action of the fluid at low lift may erode the body wall.
4) *Double-seated, contoured plug, globe body.* Plug guide post is subject to severe erosion.
5) *Double-seated, V-port plug, globe body.* Same results as items 3 and 4 above.
6) *Splined plugs* concentrate seat wear in a small segment and are unsatisfactory.

Elastomer lined bodies and plugs give excellent results at moderate pressure drops. Ceramic trim is also very resistant, and ceramic discs with orifice ports are used to advantage in sliding gate type control valves.

Separation of the throttling and seating areas leads to a multifold increase in trim life and is often used in severe services. By means of heavy guiding, the plug nose cylinder makes a very close fit in the seat orifice without contacting the orifice side wall. The seat-to-plug joint is located above the orifice and has a larger diameter. It is made of "tough" materials while the plug nose and orifice which do not impact on closure may be of much harder and more brittle materials. These parts take the throttling wear. The cylindrical section of the plug extends well into the orifice, past its entrance taper. On opening, the plug-to-seat joint is cracked and causes little flow restriction. In a 3/8 in. valve orifice, the seat joint offers little flow restriction after lifting 0.001 in. Flow at this point is through the annular clearance of the plug cylinder and orifice, and the valve does not begin to throttle above the minimum flow until the lift reaches 0.030 in. To limit seat abrasion and cavitation, the first portion of the lift is accomplished very rapidly, and then normal lift speed is resumed for throttling. On cracking the first 0.001 in., pressure drop is divided between the seat joint and the plug annular clearance in the throttle orifice, and the major portion of drop rapidly shifts to the orifice. In effect, two valves are momentarily in series, thus reducing cavitation as illustrated in Figure 23.

The cascade trim illustrated and explained in Figure 24, is another means of reducing cavitation.

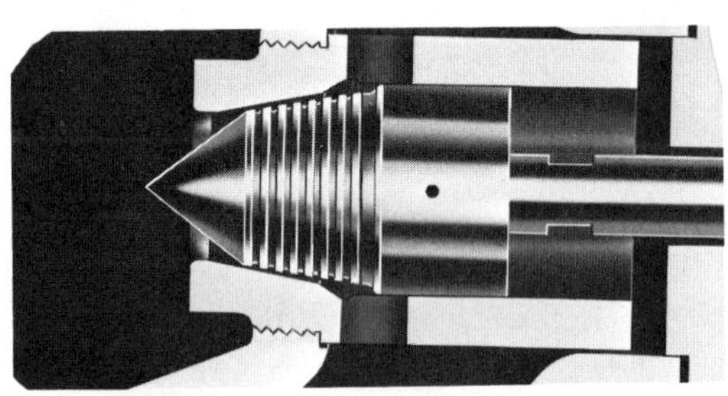

Figure 24. CASCADE TRIM.

Provides a series of pressure drops prior to seating surface, thus reducing the velocity across the seat joint. The gap between plunger and cage in the labrinth groove section is held to 0.002 in. for proper cascade effect. A differential seating angle of 20° plug mating a 22° seat, allows seating approaching 100,000 psi, causing a slight deformation of the 440-C SS trim hardened to 53-56 R_C. Used in liquid service above 1500 psi pressure drop and steam service above 2500 psi ΔP. One year of service, throttling 300°F water at 4000 psi pressure drops and 0.050″ lift did not damage the seating surfaces.

Courtesy Copes-Vulcan, Inc.

In a full port valve, the cylindrical section of the plug above the contour run-out, and below the seat joint, may be lengthened so that the seat is removed from the joint before large flow occurs.

Hard Facing

Trim is hard faced to increase erosion and/or corrosion resistance, and also to reduce galling at high temperatures. Stellite®, Colmonoy®, and tungsten carbide are the principal materials, although the latter has limited corrosion resistance and is subject to thermal shock fracture. The corrosion resistance considered here is a mild-pitting type, which might occur in the plug-to-seat joint contact areas. Hard facing is also used on the plug-to-seat joint to maintain a tight seal. The hard-faced areas are illustrated in Figure 25.

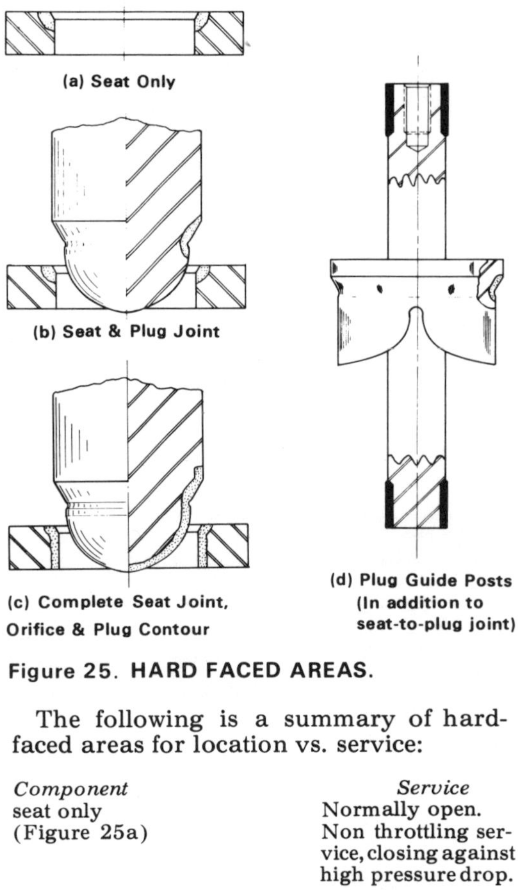

Figure 25. HARD FACED AREAS.

The following is a summary of hard-faced areas for location vs. service:

Component	*Service*
seat only (Figure 25a)	Normally open. Non throttling service, closing against high pressure drop.
seat and plug joint (Figure 25b)	Throttling service. Moderate pressure drop and abrasion. Normally Closed / opening to high pressure drop.
Complete Seat Joint, Orifice and Plug Contour (Figure 25c)	Tough throttling service. High pressure drop and severe erosion.
Plug Guide Posts, In addition to seat-to-plug joint (Figure 25d)	High temperature service.

Stellites® are patented alloys of hard tungsten and chromium particles in a softer cobalt matrix, which make up a group of materials for valvetrim. Number 6 Stellite is used on valve seats, while the slightly harder, but more brittle Number 12 may be used on plugs. The erosion resistance of Stellites is higher than indicated by their surface hardness, which is a measure of the matrix rather than micro particle hardness.

Stellite parts of small valves of 1/4 to 2 in., may be cast solid, cast as rings and projection welded, or applied as hard facing by welding to critical erosive areas of base trim material. Projection welding is a form of electric resistance welding where sharp ridges of the two parts are fused together. The degree of hard face coverage required, as well as economics, determine the method selected. Complete coverage of plugs above 4 in. size is not practical, due to cost and welding difficulties; however, there is no limit to hard facing seat joints of both plug and seat. Twenty four inch control valves with Stellite facings, have been made and used successfully. Figure 26 shows the automatic welding of a Stellite face to a large, quick-opening plug.

Figure 26. STELLITE - FACING A QUICK OPENING PLUG, USING AUTOMATIC EQUIPMENT.

Courtesy of Rockwell Manufacturing Company

Colmonoy® hard facing (for erosive service) is an alloy of very hard chromium boride crystals in a nickel base, which is also corrosion resistant (similar to inconel). It is applied to all trim shapes by the spray welding process and is then fused to give a non-porous surface. It has a high, hot-hardness and holds this hardness with thermal cycling. A plug with Colmonoy hard facing is shown in Figure 27.

Figure 27. COLMONOY® COATED AND FUSED V-PORT DOUBLE SEATED PLUG.

Ready for grinding seat joint, skirt, and guide post diameters.

Courtesy Wall Colmonoy Corporation.

Tungsten carbide trim is generally made by pressing a powder mixture, sintering, and grinding to size. Various alloy binders of cobalt, chromium and nickel are used to give the desired properties of impact resistance, compressive strength and ability to withstand some degree of thermal shock; however, these vary the apparent hardness.

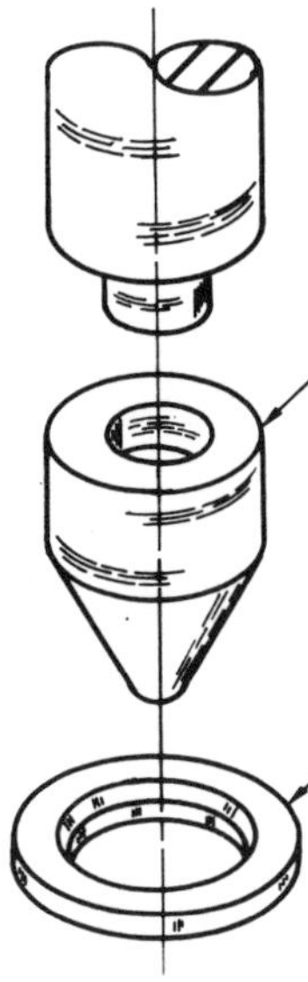

Figure 28. TUNGSTEN CARBIDE INSERTS FOR HARD FACING VALVE TRIM.

Finish ground after attachment by brazing or epoxy bonding of cylinderical fits.

Tungsten carbide is attached to valve parts by brazing, epoxy cement bonding and interference fits. Brazes have been developed for all temperatures, while the bonded trim thermal limit is about 200°F with a bond tensile strength of approximately 8000 psi as compared to 70,000 psi for brazing. Service temperature and thermal shock must be given careful consideration in its application. Valve seat loading must be even and impact of closure should be low, to prevent cracking this material. Cost is generally three to four times that of Type 316 stainless steel trim. (See Figure 28.)

HIGH TEMPERATURE TRIM MATERIAL AND DESIGN

(For Service Above 450°F)

The material properties, considered for establishing maximum operation and peak temperature limits are: tensile; yield; creep and rupture (above 800°F); hot hardness; impact strength; and aging. Other factors to consider are: scaling resistance; heat treating temperature; and galling. The design considerations include: clearances of moving parts and fitted parts (loosening) as related to the differential rate of thermal expansion of their respective materials; and galling of material combinations related to temperature, loading and degree of contact such as repeated impacts from closing or vibration. Also to be considered are the thermal cycling effects on: valve sealing; seat gasket sealing; and loosening of guide bushings.

Materials

Yield, tensile and compressive strengths are lowered by high temperatures, and above 800°F, creep and rupture become factors. In high temperature service, trim undergoes an initial elastic deformation and then continues to deform or "creep" with time, under load. Plastic deformation occurs at stresses lower than the yield strength for a given temperature. This applies to stem-to-plug attachment loading and the compressive seat, stem, and plug loads. Also, the compressive load, holding the cage trim in place, is reduced.

Parts, designed with such low stress levels that creep does not occur, are heavy and uneconomical; therefore, the designer, knowing the creep rate of a material, must select a stress, such that the total creep over the normal service life will not extend into the rupture region or allow deformation to cause interference of moving parts.

Hot hardness is necessary to prevent damaged seating surfaces and to prevent galling. Aging and the accompanying changes in physical properties such as toughness, grain size, etc., must be considered.

Scaling resistance is the ability of a material to withstand oxidation on thermal cycling without repeated scaling or flaking of the surface, exposing new metal.

Maximum temperature limits must be chosen, with a buffer zone to prevent a change of metal hardness, when service conditions approach or exceed the heat treating temperature which would anneal the plug, seat joint and erosion-resistant surfaces.

Differential Thermal Expansion of components is carefully considered in high temperature valve design. When hot fluid enters a cold valve, it surrounds the plug and rapidly brings it to line temperature with plug heat-loss limited by conduction through the, much smaller, cross-section stem and/or plug guide posts. The seat ring, due to its low mass, heats as rapidly as the plug but the heat is initially conducted through the body bridge wall. Since the linear coefficient of thermal expansion of the body is often less than the seat, it acts as a restraining member, limiting the seat expansion in a radial direction. Plug post and guide bushings are similarly affected. These conditions act to reduce the working clearances designed into the standard valve; therefore, the room temperature clearances of standard valves are increased when they are manufactured for hot service. Thus, at maximum working temperature, clearances are nearly normal. The plug post makes a tighter fit in the guide bushing than the plug cylinder-to-seat orifice clearance, in order to prevent galling and drag. As the guide clearances are increased, the plug seat orifice clearance must also be increased. For cold service, or flow testing, this has the effect of reducing the rangeability or control of low flows.

Double seated valves have problems, because the linear expansion of the plug between the two seat joints may exceed the body expansion which supports these seat rings. This problem is avoided in balanced cage valves, which use two separated metal seats, by making the plug and cage of the same material.

The following example is given for a 2 in., double seated valve:

Full port seats separated by a 3 in. elevation.

Plug material, 316SS, 10.3 x 10^{-6} in/in/° F coefficient of linear expansion.

Body material, Carbon Steel, 6.3x10^{-6} in/in/° F coefficient of linear expansion.

The difference is 4x10^{-6} in/in/° F coefficient linear expansion.

Plugs lapped to fit both seats simultaneously at 70° F.

At 270° F, ΔT =200° F

(200° F)x(4 x 10-6 in/in/° F)x(3 in.)
= Differential Expansion of 0.0025 in.

Thus with the lower plug seated at 270 F, the top plug would be 0.0025 in. off its seat, and a high leak rate would result.

To overcome the above problem, the plug of high quality valves is often made of the same alloy steel as the body, and the seat joint is hard-faced.

Thermal Cycling causes seat rings and guide bushings to loosen, so they must be either seal-welded or tack welded, respectively. For very high temperature, in large valves it is better to use a hard facing applied directly to the bridge-wall seat opening; this is known as an integral seat. Since seat gaskets must be loaded beyond yield to seal, severe thermal cycling causes them to creep and leak; therefore, they are best avoided for high pressure thermal cycling service.

Trim Design vs. Service Temperature

The following design changes must be considered for the corresponding temperature ranges.

Above 450° F	- The bonnet extension requires a longer stem to keep the packing cool.
Above 600° F	- Clearances must be increased. The plug and seat sealing areas must be hard-faced.
Above 750° F	- All threaded seat rings must be seal-welded to prevent loosening which will cause leakage and undercutting.
Above 900° F	- All guide bushings, plug guides and posts must be hard-faced. All guide bushings must be tack-welded.
Above 1050° F	- Hard-faced, integral seat joints must be used.

An additional consideration is that the fitted parts are subject to increased compressive stresses and creep due to differential thermal expansion, adding strain to the bonnet bolting. This involves the cage trim held by the body-bridge wall and the bonnet.

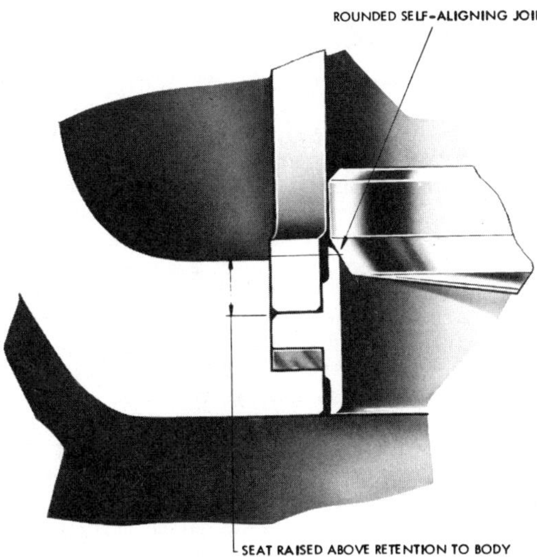

Figure 29. FLEXIBLE SEAT FOR THERMAL CYCLIC SERVICE.

Courtesy Fisher Controls Company

Seat Joint Designs for Thermal Cycling Service

A flexible seat-lip design, as shown in Figure 29, will allow for variations in the expansion rates of parts, for strains causing out-of-roundness and for seat wear caused by many closings. This design has been proven satisfactory in: gas dehydration at 60 to 600° F, 8 times/day for 40 psi pressure drops; and liquid sodium service at 300 to 1200° F, for 50 psi pressure drop.

Figure 30 is a comparison of the hot-hardness of the hard-facing materials, Stellite® and Colmony®. The 400 series, stainless steel-hardened trim materials are shown for reference. Note the rapid softening of the 400 series, above 700° F.

The following are the affects high temperature has on the physical properties of Type-316 stainless steel trim base material:

Physical Property	*70° F*	*1200° F*
ultimate tensile	85 ksi [1]	56 ksi
yield	38.5	20 ksi
Creep (1% in 100,000 hrs. or 11 yrs.)	(see [2])	6 ksi
Rupture (in 100,000 hrs.)	(see [2])	12 to 16 ksi

Galling

Galling is related to temperature, material pairs, surface finish, hardness, and loading; plus the line-fluid effect. One metal surface will score, weld-to, and/or ball-up material from another to: 1) jam the valve stroke; 2) ruin the seat joint; or 3) increase the operating force, causing jerky plug positioning. High temperature will either anneal or soften the metals, in-

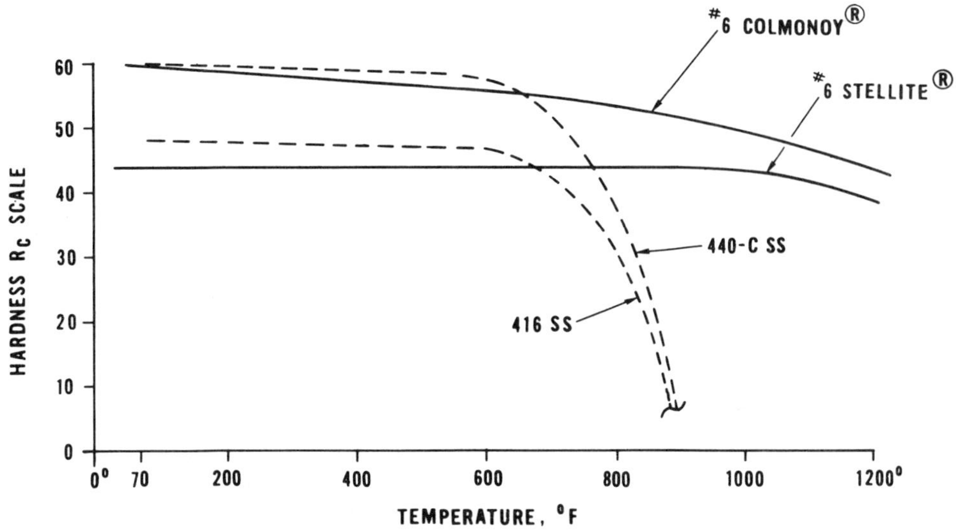

Figure 30. HOT-HARDNESS OF HARD FACING MATERIALS.

[1] ksi-units of 1000 psi

[2] Creep and rupture below 800° F are insignificant.

creasing their galling potential.

Large abrasive particles in the line fluid will sometimes rough-up the soft trim, initiating galling. Vibrational impact may create a "fretting" condition, leading to guide-bearing surface breakdown and eventual galling.

Liquid metals, such as sodium and potassium, clean the protective oxide films from metal surfaces so effectively, that galling is a serious problem at and above service temperatures of 400° F. Materials must be carefully selected and loading must be kept at low levels for this kind of service (for example, heat exchange in nuclear power applications).

Galling may be prevented by the following:

1) Using hard materials for one, or preferably both, parts;
2) Selecting pairs of material with low galling potential. (See the list given below).
3) Using different materials for both parts rather than the same material.
4) Assuring a 5- to 10-Rc difference in the hardness of the materials.
5) Specifying either a required surface finish and surface hardness or special coating.
6) Designing a reasonable loading.
7) Designing adequate operating clearances.
8) Breaking in the valve at low loads and with a lubricant.

The following metal pairs have low galling potentials:

Plug		*Guide Bushing*	*Rating*[1]
312 SS[2]	vs	17-4PH SS	Good
		400 - Series SS	Good
		Stellites	Good
		Colmonoys	Good
		Inconel	Good
304 - L SS	vs	Inconel	Good
440 - C SS	vs	440-C SS	Good
		17-4 PH SS	Good
Tool Steel	vs	17-4 PH SS	Good
Monel	vs	K-Monel	Good
K - Monel	vs	400 Series SS	Good
Stellites	vs	Stellites	Excellent
		Inconel	Excellent
Colmonoys	vs	Colomony	Excellent
Durimet 20	vs	Carpenter 20	Good
Hastelloy "B"®	vs	Hastelloy "B"®	Good

LOW TEMPERATURE TRIM MATERIALS AND DESIGN

The trim design for low temperature and cryogenic service is based on the following requirements: 1) bubble-tight sealing at low temperature; 2) minimum heat leak; 3) minimum cool-down mass; 4) quick change design; 5) simple design; 6) adequate impact strength of materials at low temperature; and 7) differential thermal contraction of materials. These factors have been resolved in valve design and material selection which have performed very satsifactorily.

Bubble-tight sealing, a must in cryogenic service, is obtained by using TFE or Kel-F on the plug seal. These materials become much harder at low temperature, but still retain enough yield to seal tightly. Kel-F has sealed 225 psi, bubble-tight, at 10° F above absolute zero. Both materials may be used for the guide bushings and the seat joint gaskets.

The stems must pass through extended bonnets to prevent the freezing of atmosphere moisture, resulting in binding of the stem at the packing exit, and to clear the thick insulation of cold boxes.

Hollow plugs are frequently used to prevent rapid heat transfer, yet allow a large-diameter plug for guiding and throttling stability. The void in the plug may be evacuated and/or filled with insulating material or reflectors to reduce radiant or convective heat transfer. Stainless steel is an excellent material due to its poor thermal conductivity; however, cross-sectional metal area is kept to a minimum. The clearance between plug extension and guide is close enough to prevent convection currents and it enlarges, above the guide, to a larger annular area to prevent pumping of liquid out of the body into this space.

Valves are normally installed either vertically, or up to 45° from vertical, to maintain a low-convective vapor-lock in the extended bonnet. The design shown in Figure 31 is applicable to valves which must be installed horizontally. The seal "seals" on downstroke and bypasses on upstroke.

The weight of the parts should be kept to a minimum, to reduce cool-down of the mass and boil-off loss of the liquified gases that are entering the piping system. To reduce weight, the large plugs are hollow and the seats may be integral with the body.

Quick-change trim, with removal through the bonnet, is a must, because

[1] When used within recommended temperature limit and proper hardness and loading.

[2] 300-Series stainless steel readily galls with itself when in sliding contact, but the plug-to-seat sealing does not usually involve this type of contact.

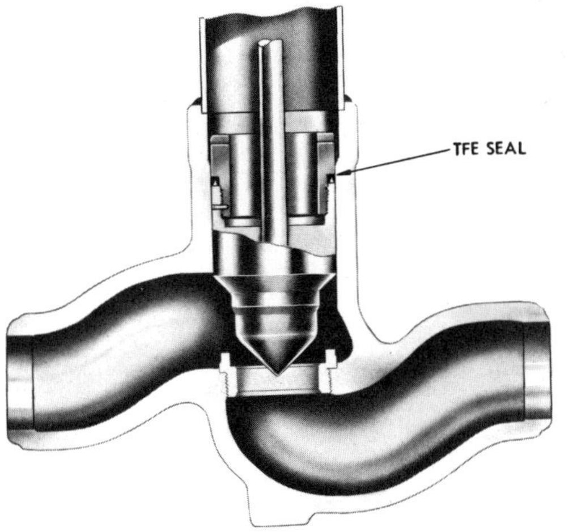

Figure 31. SEAL TO PREVENT PUMPING OF CRYOGENIC FLUID INTO VALVE BONNET WITH VALVE STROKING.

Courtesy ITT Hammel-Dahl

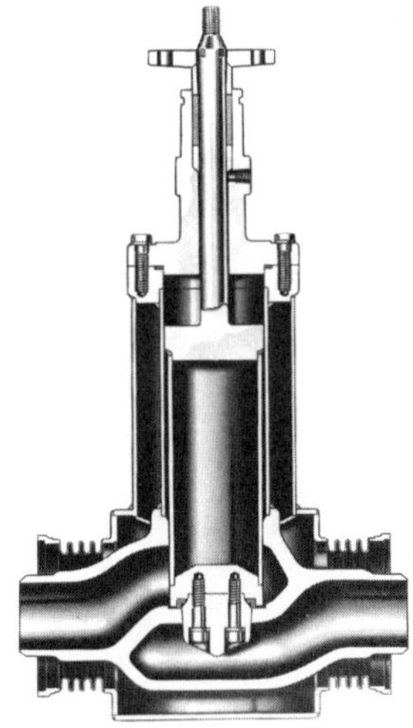

Figure 32. CRYOGENIC VALVE WITH VACUUM JACKET.

Courtesy Masoneilan International, Inc.

it is impractical to remove a vacuum-jacketed valve body from the line, or even to remove the insulation to get to the bonnet. Therefore, the bonnet is always located well above the normal position, requiring a longer plug in order to use a short, stiff stem having a small cross-sectional area necessary to limit heat leakage into the valve.

Stuffing boxes must be deep enough to accommodate a second or partial set of packing, installed back-to-back, to prevent leakage on cool-down if a vacuum should occur in the line.

One type of valve, designed for cryogenic application is shown in Figure 32. Note the vacuum jacket and the soft seat that is retained by the bolted nose plug.

Cold Impact Strength

Trim materials must possess adequate cold impact strength to prevent fracture and complete shattering. Carbon and chromium-alloy steels rapidly lose impact strength below 0° F; therefore, they are limited to -20° F and -50° F service, respectively. 3.5% nickel steel may be used to -150° F, and 9% nickel steel may be used to -320° F. Austenitic 300-Series stainless steels, nickel, Monel, Hastelloys, Durimet, titanium, aluminum alloys and bronze may be used to -460° F. These fall into the group of face-centered, cubic metals. Cooling increases their yield and tensile strength, while gradually reducing ductility. The impact strength of these materials may either remain the same or increase.

The ignition point of austenitic stainless steels in liquid oxygen, is so high that very few applications exist where they are not entirely satisfactory. If they are not satisfactory, Monel or bronze trim should be specified if very high-energy shock impacts are probable.

The following modifications are used for low temperature trim.

Low Temperature Ranges	*Trim Design Requirements*
Cool Valve Service	
0° to -50° F	The extension bonnet requires a longer stem. Sometimes the addition of ethylene glycol in the packing follower, annular clearance around the stem will prevent freezing and will permit operation of a standard length bonnet valve to about 0 F.
Cold Valve Service	
-50° to -150° F	The extension bonnet requires a longer valve stem to prevent stem freezing, from the condensation of atmospheric moisture at the exit of the stem from the packing.

Refrigeration Service (Typical operating temp.)	
Freon (-92° F) Ammonia (-84° F) Propane (-106° F)	The guide bushing may be tack-welded to prevent loosening from the differential thermal contraction of the bushing and bonnet materials. The seat rings, of screwed-design, are seal-welded to prevent loosening and leakage.
Cryogenic Service	
-150° F to -460° F	A double extension bonnet is used with a length of 12 inches for -320° F nitrogen service, and longer for -450° F He and H_2 Service.
Liquified Gases Service[1] (Typical operating temp.)	
Methane (-259° F) Oxygen (298° F) Nitrogen (-320° F) Hydrogen (-422° F) Helium (-452° F)	The guide bushings may be either welded or lock-rolled in place. The operating clearances for moving parts may be increased by 50%, to prevent binding from differential thermal contraction. A hollow plug extension is used - sealed void, with a low cross-section metal area. Extension in large valves is carried high enabling the use of a short and stiff stem due to the requirement of a long bonnet extension. The seat rings, of screwed-design are seal-welded to prevent loosening and leakage.

The following trim materials are used for plugs and seats, and bushings in cold service.

Plugs and Seats

Bronze
Type-316 SS
K-Monel
Stellite or Collmonoy hard facing on 316 SS
TFE or Kel-F is used as plug-seat joint material and seat gaskets. They are attached to the plug with strain-hardened 300 SS bolting or stem nuts.

Guide Bushings

Hard Aluminum Bronze
K-Monel
Hard Chrome Plated Materials
17-4 PH SS and Type 440-c SS are uses for guides only where brittleness is not a factor.

[1] Boiling points at 14.7 psia.
[2] Ksi = units of 1000 psi.

The affect of low temperature on the mechanical properties of type 316 stainless steel trim material is given below:

	-70° F	-423° F
yield strength (0.2% offset)	35 ksi[2]	84 ksi
ultimate tensile strength	84	210
elongation	70	55
impact strength (Izod, "V" notch test)	110	110 (at -320° F)

The following are the cold properties of the soft seat materials (TFE and Kel-F).

Material	*Tensile Strength (psi)*	*Thermal Contraction (from 70° F)*
TFE	1,500 (100° F), 22,000 psi (-460° F)	2.2%
Kel-F	4,000 (100° F), 16,300 psi (-300° F)	1.2%

TRIM EFFECTS ON NOISE

Trim contributes to valve noise by the following: 1) mechanical vibration; 2) natural frequency vibration; 3) throttling instability; 4) flow media - the cavitation of liquids or aerodynamic effects of gas flow; and 5) water hammer-shock on valve closure. These effects are discussed in detail, in Chapter 6.

Mechanical vibration may be reduced by: 1) holding close diametral clearances; 2) using heavy guiding to distribute the impact load and to dampen vibration; 3) selecting materials that will withstand heat and reduce wear, preventing the enlargement of clearances; and 4) using an elastomer dampening ring on the heavy plug guides of cage valves, which may also serve as the seal on pressure balanced cage designs;

Natural frequency vibration may be eliminated by: 1) breaking up symmetry by using solid-cast plugs and parts rather than using cylindrical skirts welded to posts made from bar stock; 2) changing from a skirted plug to a contoured plug, or vice versa; 3) reversing flow direction; 4) changing stem diameter; and 5) using single-seated valves with heavy plug guides (not guide posts), which are less susceptible to vibration because of greater plug stiffness.

Throttling Instability is the vertical oscillatory movement of the assembly, including plug, stem, and moving actuator components. Both single and double ported non-pressure-balanced valves are unstable when throttling to a high pressure drop, at low lifts, as explained in the "dynamic flow effects" section. Tremendous up-and-down forces, caused by flow impingement upon the plug, rapidly reverse their direction and/or magnitude, as the plug is moved by the actuator. This effect may be amplified by an actuator with positioner, whose combined frequency of response may fall out of phase with the required control action. This causes pressure fluctuations in the flowing media, generating a rumble with a frequency of around 30 Hz. Vibration depends upon the mass-spring rate relation-ship of the plug-stem-actuator moving parts mass. Seats, plugs and stems are damaged by the vibration and either leakage or broken stems result. In addition, the stem-packing wear rate is increased.

Throttling instability may be reduced by: 1) using a stiff actuator (high spring rate); 2) having a pulsation damper, also known as a "hydraulic snubber," installed on the actuator stem; 3) designing a pressure-balanced cage which reduces the magnitude of the unbalancing forces, therefore, improving stability; and 4) maintaining a fast rate of frequency response for the controller-positioner-actuator combination.

Flow media noise includes: 1) cavitation noise, caused by the impact of collapse of vapor bubbles formed in the passages of the seat-to-plug throttling annular clearance at high pressure drops; and 2) Aerodynamic noise, caused by high-pressure gas entering and exiting at the valve ports, is extremely annoying. Aerodynamic noise is also caused by pressure recovery, accompanied by a reduction in velocity, in the downstream passages, causing sonic shock waves.

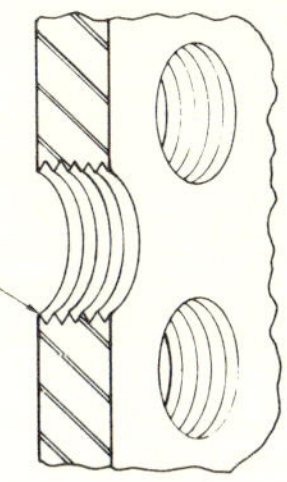

Figure 33. MULTIPLE, HIGH VELOCITY ORIFICES USED IN FLOW CAGES TO REDUCE NOISE

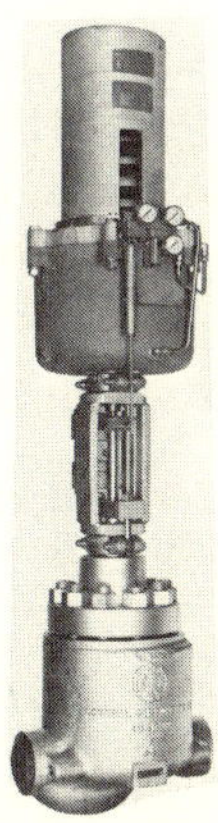

(a) 4 in., 2500 lb. Boiler Feedwater Recirculation Valve

(c) Flow Disc Stack for Gas Service

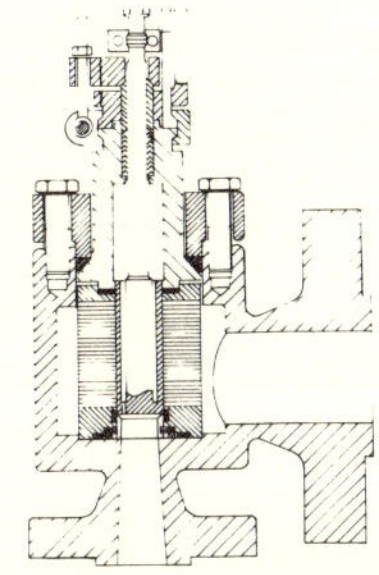

(b) Flow Disc Stack in Sectionalized Valve

In (c), the multiple, right angle turns in the staggered flow paths of each disc affect numerous velocity head losses. This alters the basic velocity equation from $V = \sqrt{2gh}$ to $V = \sqrt{2gh/N}$, where N represents the number of turns in series in each flow passage. A proven pre-determined number of discs with multiple flow passages etched into their faces is stacked around the plug. The number of fluid passages, open to flow, increases with valve plug lift. Noise and erosion are greatly reduced in high-pressure-drop service. Designs are also available for anti-cavitating liquid service.

Figure 34. PRESSURE REDUCTION AT CONSTANT VELOCITY REDUCES NOISE.

Courtesy Control Components, Inc., Patents Issued and Pending

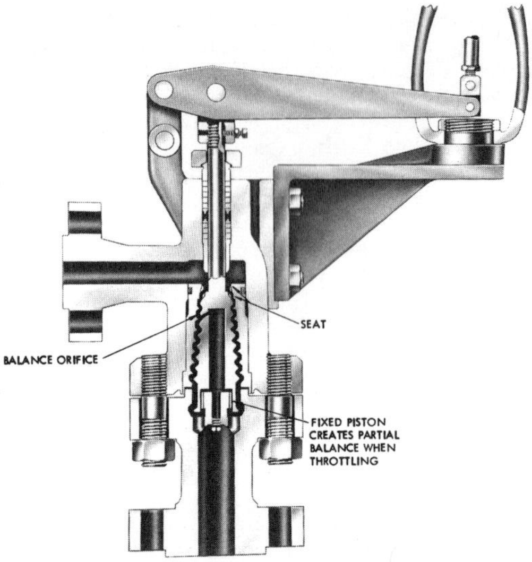

The tapered plug allows gas to expand along a tortuous path with little increase in velocity, since the area of flow-path is expanding; thus, greatly reducing noise. Adiobatic flow, with molecular friction, occurs at a nearly constant level of enthalpy, as pressure drop occurs in steps across a series of orifices. After opening, the plug is partially pressure balanced, but when closed the full pressure drop load is applied to the seat. Partial balance acts to close the valve for safety and for stable throttling. Maximum flow velocity is approximately one-fourth that of a conventional valve; therefore, trim erosion is not excessive, even though particles may impact the throttling surfaces.

Figure 35. SPECIAL LOW NOISE VALVE DESIGN.

Courtesy Masoneilan International, Inc.

Aerodynamic noise may be reduced as illustrated in Figures 33, 34 and 35 by:

1) Using trim that has many, small high-velocity ports which operate at a higher frequency, thus, increasing the attenuation through the body wall.
2) Using trim that has sharp edges and a distorted flow pattern which will reduce gas velocity and vortex size. Streamlining the flow has a negative effect, because it permits further isentropic expansion of the sonic or supersonic jet flow, causing greater shock waves upon pressure recovery.
3) Using in-line valves with expandable elastomer stem plugs rather than standard globe valves, because of the dampening effect of the plug sleeve.
4) Application of acoustical barrier treatment is nearly always necessary for standard trim.

Water hammer-noise and shock are caused by the sudden closure of valves that are stopping a moving column of liquid and by the reflection of shock waves within the liquid and piping system. Intensity varies directly with the density, velocity, and rate of deacceleration of the flowing liquid. In severe cases, flanges may be parted, causing temporary leaks or blown gaskets, line supports may be ripped loose, or piping ruptured in iron pipe systems. Linear and equal percentage plugs usually cause no problem with normal controller-closing speeds; however, with on-off snap acting controller settings, they may give some trouble. Quick opening plugs may require a throttling contour on their last 10 to 15% of travel before closure; this is advisable in the case of high pressure drops, having high flow rates. Hydraulic "snubbers" installed on the actuator stem, may be designed to slow valve closure just before shut off; beyond this, pulsation dampers should be installed in the piping system (these will require "tuning" of their gas charging pressure to match the system characteristics).

Two-phase (liquid and gas) "slug" flow causes some shock when passing through a control valve; however, the gas cushion in the flow line usually limits this to a minor shock.

TYPES OF PLUG GUIDING

Guiding is the means by which the plug is aligned with the seat before the contact for tight seating by the normal stem force. The guiding must resist all side thrust on the plug, from dynamic fluid flow conditions.

Guiding design is divided into the following four general types:

1) *Top Guiding*, illustrated by Figure 36(a), has a single, extra-long bushing which guides the plug from above. This is used with a contoured, single-seated plug. It is used for high pressure drop in sizes from 1/4 in. to 1 in. Valves above 1 in., use top and bottom guiding for globe bodies.
2) *Top and bottom guiding*, as shown in Figure 36(b) uses contoured and V-port plugs with post extensions that are guided above and below the single or double plug in bodies with bottom covers. This design allows plug reversal to change the closing action and the flow direction. It should be specified for pressure drops greater than 100 psi.
3) *Port guiding* is shown in Figure 36(c). In this design, the plug is guided in the seat ring by a skirt on the plug. It is used in bodies with V-port plugs and is used in some three-way valve designs. Port guiding is also called skirt guiding.

It is used for the moderate pressure drops under 100 psi.

4. *Cage guiding,* as illustrated by Figure 36(d), the cage guides the plug. This design is for moderate to very high pressure drop.

Other types include: (a) combined top and skirt guiding; (b) Stem guiding of large stems, as shown in Figure 36(e); and (c) Bottom guiding of the stem in extension bonnets, as shown in Figure 36(f).

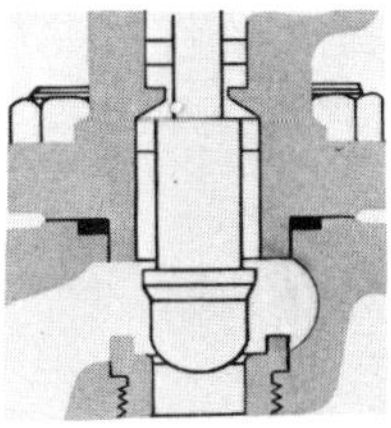

(a) Top Guiding

Courtesy Honeywell Industrial Division

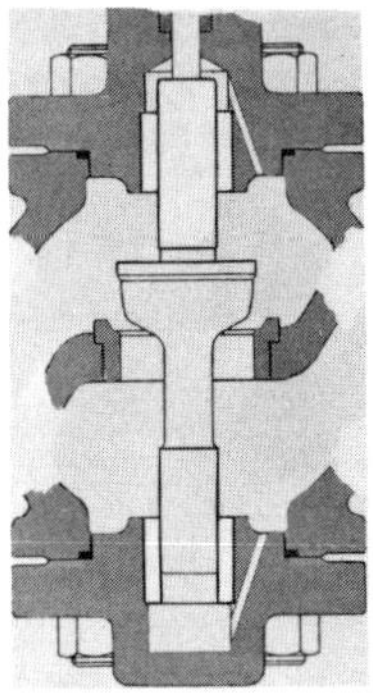

(b) Top & Bottom Guiding

Cortesy Honeywell Industrial Division

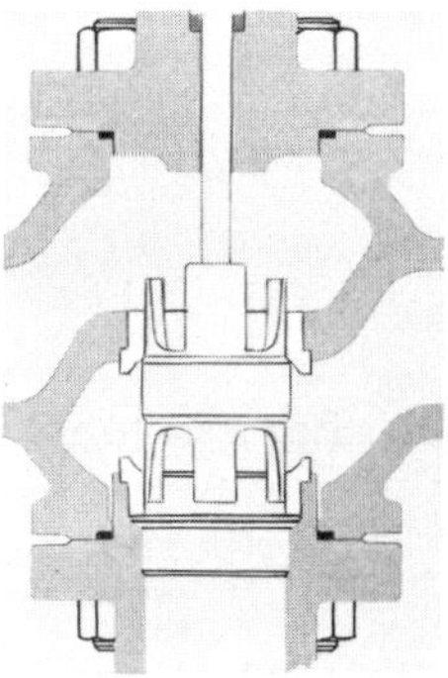

(c) Port Guiding

Courtesy Honeywell Industrial Division

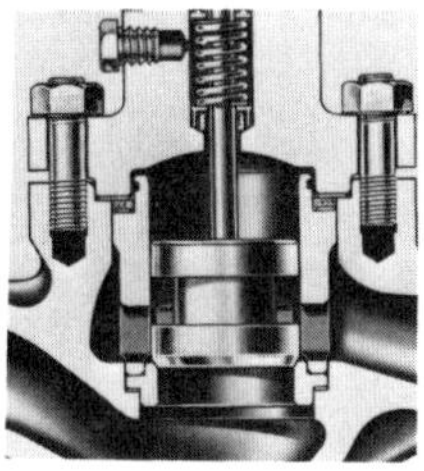

(d) Cage Guiding

Courtesy Fisher Controls Company

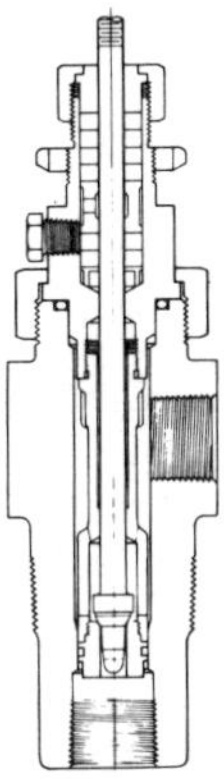

The body design illustrated is used to extend the seat joint into the vessel interior to prevent freezing on high pressure gas, pressure-reduction service.

(e) Stem Guiding

Courtesy Fisher Controls Company

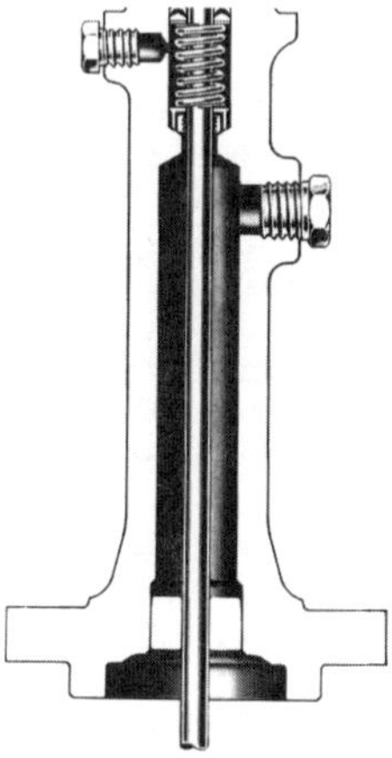

(f) Bottom Guiding of Stem in Extension Bonnet

Courtesy Fisher Controls Company

Figure 36. EXAMPLE OF GUIDING DESIGNS.

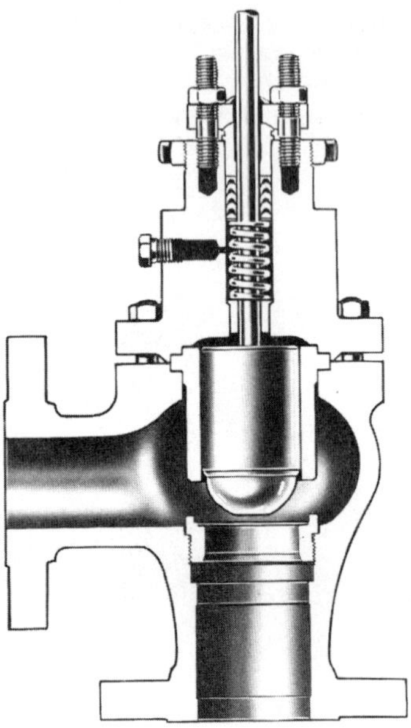

37. HEAVY DUTY CYLINDER GUIDE IN ANGLE BODY VALVE FOR HIGH PRESSURE DROP SERVICE.

Courtesy Fisher Controls Company

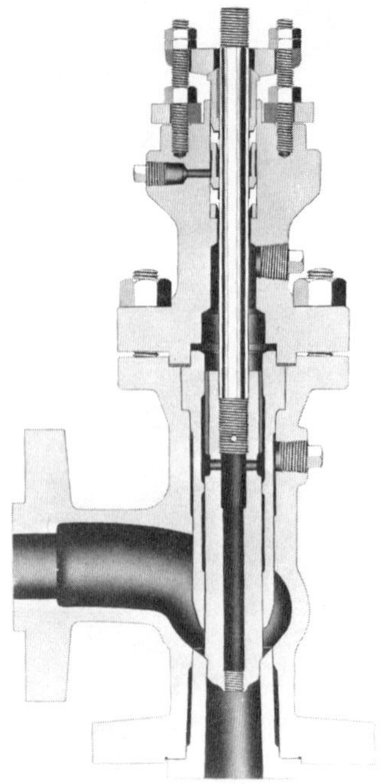

Figure 38. FLUSHING TYPE TOP EXTENSION GUIDING.

Courtesy ITT Hammel-Dahl

Extra heavy guide bushings, shouldering in the body bonnet neck, are used in high pressure drop angle valves of size larger than 1 in. as shown in Figure 37. This design is ideally suited for flow under, thus allowing larger flows with the smaller seat orifices and lower actuator forces that result from the smaller seat. The "stubby" plug deadens high frequency vibrations, giving quiet operation in steam service.

Figure 38 is a valve design that has flushing type, top guiding. This type is used in "dirty fluid" service. The guide may be continuously flushed by supplying a clean fluid into the plugged opening above the bonnet flange. Note the pressure balanced plug (the stem is not balanced).

Redundant guiding should be avoided for accurate plug-to-seat alignment. Cage guiding, followed by top guiding, best meets this requirement. Top and bottom guiding is required for plug side thrust stability in some designs; however, the two guides and one seat must be in perfect axial alignment after assembly. Skirt guiding maintains concentric alignment, but offers little restraint to cocking, because of the low ratio of skirt guide length to seat ring length at relatively large diameters. (The L/D ratio is too small.)

Guide fit clearances, materials, hardness, etc. are discussed throughout the chapter.

The characteristics and requirements of stem post guide bushings include:

1) They are press-fitted for normal service.
2) They are tack welded for thermal cycling at service below -100° F [1] and above 900° F. [2]
3) They are hard-faced for use at temperature above 900° F.
4) They may be screw-fitted with a pilot diameter providing axial alignment and a shoulder that will prevent cocking, thus offering fast maintenance advantage. This design, as illustrated by Figure 39, enables simple field replacement of the plug guide post or the stem guide bushings.

[1] The bushing materials will contract more than the bonnet materials on cool down; thus, the bushing might loosen and fall out if it is not tack-welded or lock-rolled in place.

[2] The bushing material expands more than the bonnet bore and could yield the bonnet, thus loosening the interference fit.

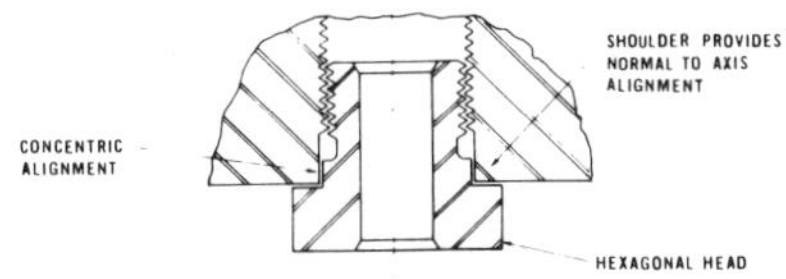

Figure 39. SCREWED GUIDE BUSHING.

The cage guiding technique requires a hardened plug, a Stellite® or Stellite®-faced plug, or a hard chrome-plated plug and guide cage.

DYNAMIC FORCES ACTING ON TRIM AND AFFECTING THROTTLING CONTROL ACTION

Valve plugs, when throttling flow, are subject to forces that are axial along the stem, at 90° to the stem axis and rotational to the plug and stem axis. The side forces cause bending moments and vibration.

Improved flow control can be obtained by valve trim design that will reduce the magnitude of the forces tending to unbalance the plug position and this will reduce the operator force requirements, making a more sensitive valve.

The following considerations will increase valve stability:

1) Select a sensing and controller system that has a high frequency response.
2) Use a relatively stiff actuator, one which has a high spring rate, so that plug force changes must be large to cause a substantial change in plug position.
3) Use flow under the plug when feasible, to eliminate the large "suck downward" force as the plug throttles near the seat. Fluid forces acting on the stem should always be compressive, forcing the plug out of its seat for best stability.
4) Use a small diameter stem to reduce the effects of line pressure change thrusting on the stem area, but keep the stem rigid enough to prevent bowing (packing wear and drag) and to reduce the natural frequency vibration.
5) Select proper trim designs for high energy dissipation to reduce plug drag.
6) Streamline the liquid flow to reduce the variation of flow patterns within the valve, which cause different pressure drops as the flow cycles from one to the other.
7) Use a hydraulic snubber on the actuator stem for severe pulsating service (such as a bypass on duplex or triplex, pump discharge pressure control).
8) The use of cage trim increases stability over regular trim by improving the the flow pattern and reducing side thrust friction and vibration.
9) Use balanced cage trim to counterbalance most of the forces across the plug so they are never transmitted to the actuator stem. Balanced cage trim is most stable of all.
10) Balance the valve stem thrust for extremely high line pressure (10,000 psi and up).
11) "Stubby" plug shapes with heavy guides may be used to help reduce vibration.

Stem Thrust

Stem Thrust varies with the body fluid pressure acting on the unsupported, cross-sectional area of the stem in the packing, thus tending to force the stem out of the bonnet. Figure 40 presents a comparison of stem thrust with various ratios of stem-to-seat diameter. It is usually a small part of the total plug load and may be discounted except for high pressure operation. Conversely, in very small orifice valves, the stem may be equal or larger in size than the port. Thrust also varies with the plug nose position in the orifice (for plugs without guide posts); here the vena contracta, at high flow rates, may cause a low pressure area with pressures below the downstream body pressure for flow over the plug. For flow under the plug, the same conditions apply, plus the velocity head impact of the fluid on the plug nose cross section.

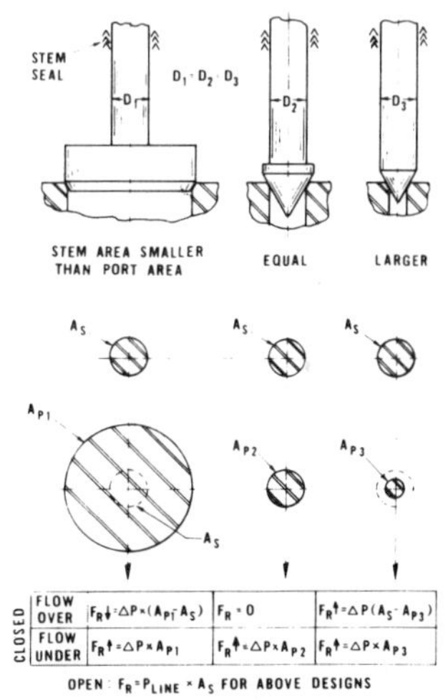

Figure 40. STEM THRUST.

Stem thrust may be minimized by:

1) Using the smallest diameter stems that are practical for strength requirements.
2) Removing stem thrust from the build-up of pressure, when the valve is closed, by flowing over the plug. This applies only where the stem diameter is equal to or less than the orifice diameter and is effectively used in angle valves. This keeps line pressure on the packing at all times.
3) Balancing the stem for use in extremely high line pressure, or by building-up of the shut-in pressure acting on the stem area. This can be accomplished by one of two means:
   a) By using a double extension stem, that goes through both the bonnet and the bottom flange with a second packing box, as shown in Figure 41. This has been very effective for full port valves where a large stem diameter is required to handle pressure drop and seating loads of 20,000 psi WP for 2 in. valves or 15,000 psi WP for 4 in. valves.
   b) By designing a balancing piston on the stem with an annular area equal to stem area, as shown in Figure 42. This has been effectively used on plug-balanced angle valves having small stems (for example, 1 in. orifice - 10,000 and 20,000 psi WP).

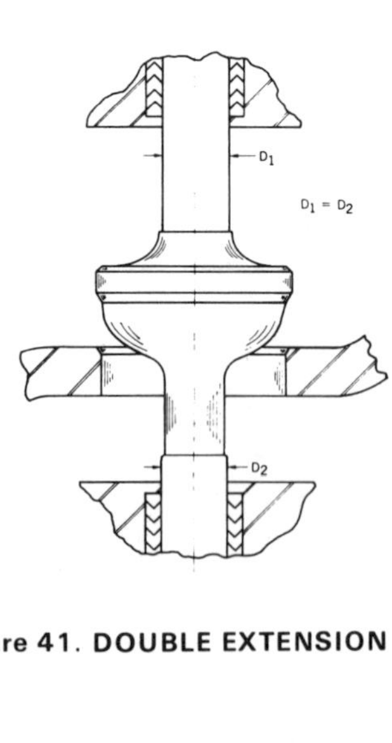

Figure 41. DOUBLE EXTENSION STEM.

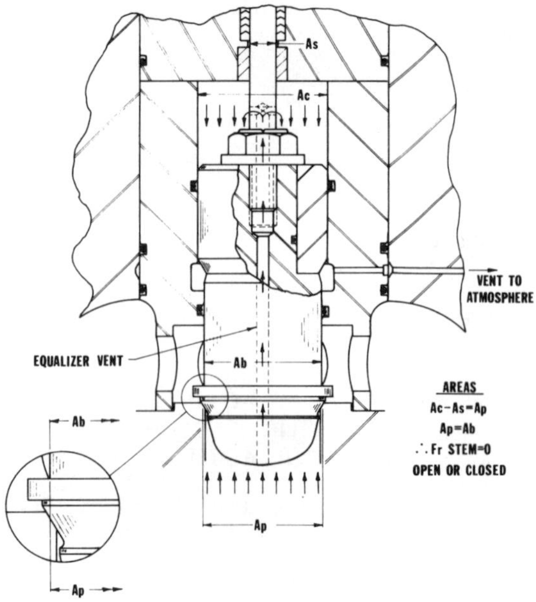

Figure 42. PISTON STEM BALANCE

Both plug and stem are balanced. A two piece plug is required for assembly. The design is limited to small valves because the plug and balance piston seals always sense the pressure drops between the line and atmospheric pressures. This design has been effectively used on small stem plug balanced angle valves, 1 in. orifice-10,000 and 20,000 psi WP.

Side Thrust

Side Thrust of fluid, forces the plug to one side of the guide bushing(s) increasing the operating friction and the "dead band" for the actuator. This may cause "jerky" plug positioning. Thrust effect depends upon trim type, flow velocity, fluid density, pressure drop and the flow pattern entering the valve body. Plug guide posts add to the side thrust by interrupting the natural flow patterns. Natural frequency vibration of the plug-stem assembly may occur as the plug is moved horizontally in the guide bushing clearances by fluid pulsations and by change of flow pattern. It becomes a problem only with high pressure drop services or with large valves.

The following considerations will reduce side thrust:

1) Select single ported valves with plug guides (not posts).
2) Streamline flow into the valve body and provide adequate inlet and exit passageways around the seat.
3) Use cage trim that has several orifices to uniformly distribute flow around the plug.
4) For double-seat valves, divide flow into the center cavity so that the impact will not be on the stem, midway

between the two plugs. Stem bending, binding and severe guide bushing wear has been greatly reduced in high pressure applications by flowing in this direction.

Y-Body valves inherently have higher side thrust components from fluid flow forces than globe or angle valves which have a larger flow passage around the plug. At low lifts and high pressure drop they impose a high drag on the plug and guide bearing surface, which often causes galling at high temperature. For example, throttling drops of 3500 psi at 1050° F.

This may be eliminated by preventing the fluid flow from getting behind the plug, thus giving a reduced side thrust as illustrated in the design of Figure 43. The plug is closely guided in the body bore by upper and lower, hard-faced guide rings, with an annular leak-off channel between the rings. For high temperature high pressure service, the bore is also hardfaced.

Fluid Swirl Rotational Forces

Fluid Swirl Rotational Forces in the orifice react strongly on large V-port plugs (4 in. and up) and can cause disengagement of the valve and actuator stem connection, unless this joint is secure. Diaphragm twist may also occur in springless actuators. In the later case it is advisable to attach an anti-rotation clamp and guide to the valve stem extension.

Rotational forces may be reduced by:

1) Streamling flow into the valve body.
2) Using contoured, rather than V-ported plugs.
3) Using cage designs that reduce this force by distributing flow, more uniformly around the plug.

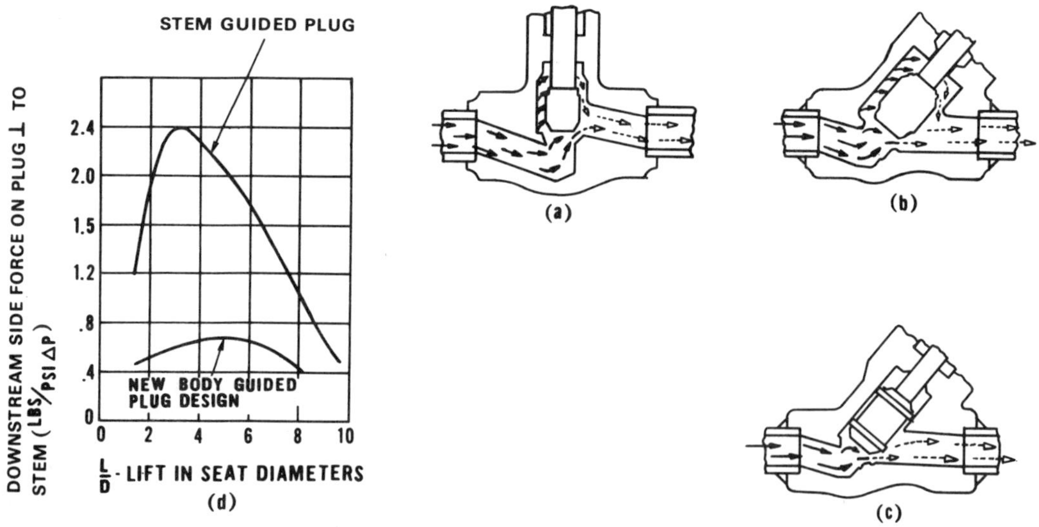

Figure 43. DESIGN FOR REDUCTION OF SIDE THRUST.

Courtesy Rockwell Manufacturing Company

(a) Vertical stem, globe valves are subject to side-thrust under high pressure drop conditions. This illustration shows how upstream pressure can slip past the stem-guided plug and impart a thrust toward the downstream side of the valve. Tests have proven that this thrust causes plug-seat misalignment plus galling and scoring.

(b) Inclined stem Globe Valves of the stem-guided type are also subject to side-thrust under the same conditions. This illustration shows the path of pressure through the valve.

(c) The new design, with Stellite® anti-thrust rings, has a lower guide which eliminates 90% of the flow upward and behind the plug. Both guide rings maintain perfect alignment. This effectively eliminates all side-thrust problems in both globe and Y-body valves.

(d) Illustrates the relationship of side-thrust in a conventional stem-guided globe valve and in a new design with body-guided plug.

Torsional Force Reactions

Torsional Force Reactions on a 4 in. V-Port Dougle Seated Valve Plug are covered in reference 36. Measured stem torque, varying from 8 to 20 ft.-lb., and oscillating at a frequency of about 30 Hertz, has been made for a 4 in. valve, flowing 220 GPM, with a pressure drop of 170 psi. Most of the torque results from an impulse reaction on the bottom skirt caused by the impingement of incoming fluid against the projected area of the edges of the "V" notches. Torque is proportional to the mass velocity of the fluid at the body inlet and impingement area presented by the lower skirt.

The torque may be reduced to ±4 ft.-lb. by changing from four to two "V's" in each skirt and by lifting the bottom skirt completely out of the orifice to achieve maximum flow. No sacrifice need be made in flow characteristic to accomplish this.

Orienting the "V's" in relation to the pipe axis may reduce torque to zero, but this not practical in economic valve design, because the angle of the "V"-to-pipe axis varies for different bodies.

Fluid Forces Acting on Plugs

When the valve is closed, differential pressure acting on the mean-seat joint area tends to open the valve for flow under or to close it for flow over. If the pressure drop is high, these forces build up quickly as the plug approaches its seat and may cause slamming and bounce-back for the flow-over direction. If throttling near the seat is continued, chattering results, which quickly damages the seat joint. This slamming action creates fluid pulsations which cause the plug-stem-actuator spring-mass to oscillate between 8 to 30 Hertz.

Flow capacity of globe bodies is generally 5 to 10 per cent higher with flow under and this goes hand-in-hand with improved throttling stability. Angle valves also have a higher flow capacity with flow under; however, they are usually installed with flow over for abrasion and cavitation reasons.

Flow under removes stem packing as a source of leakage from the build up of a high shut-in pressure.

The main advantage of flow over, is that the seating force rises with the increased pressure drop build up of the shut-in pressure. Sealing is tighter and safety shut-in is assured, even with low actuator thrust. To assure positive safety shut-in from the open position, the actuator must be spring loaded to close the valve against an opening stem thrust from line pressure. A second advantage is that stem area is subtracted from the mean seat joint area in calculating the actuator force required to open against the pressure drop. This applies when stem-to-seat area ratio is equal to, or less than 1.0. Larger stem ratios would create a self-opening action, as in the case of many small splined plugs.

The unity ratio creates a pressure balanced condition for flow over *only* and may be effectively used on small port, high pressure dump valves (3/8 in. port, 3000 psi pressure drop, and up). In this case the valve will always open at the same controller signal, regardless of change in pressure drop, since this change has no effect on actuator stem force in the closed position. This improves level control and *eliminates spring adjustments.* The need for a positioner is eliminated. Stem packing friction increases with pressure in this application.

Vena Contracta Forces

Flow through the contoured plug-to-seat annular area occurs at a much higher velocity, and lower pressure occurs in this vena contracta zone of the flow stream, than upstream or downstream of the orifice zone. This creates a force normal to the plug surface at the vena contracta which tends to draw the plug into the seat. The magnitude of the force acting on the plug, as illustrated in Figure 44, depends upon the following:

a) The downstream pressure recovery distance.
b) The width of the seat joint (length of low pressure zone) and the extent of the vena contracta along plug surface contour.
c) The plug surface angle (to horizontal at this point.
d) The pressure drop across the vena contracta zone.

The horizontal forces are balanced as shown by Figure 44. The magnitude of the vertical force, relative to other forces acting on the stem, is small for moderate size valves, except when used for high pressure drop service. The relative magnitude of the vertical force increases, however, for larger valves throttling low pressure.

Resultant Force

The resultant of the plug stem forces should always tend to open the valve, or if this is not feasible, the forces should never change direction to any great extent as the plug strokes. Otherwise, the rate of change of force with lift (dF/dL), is so rapid that the actuator-controller system will lose stable control and cycling will occur. This would result in chatter, excessive packing wear, noisy valves and damaged seat joints. The addition of a positioner may aggravate this situation.

$F_v = F_r \cos\theta$

$$F_r = \sum_{s_1}^{s_2} (P_d - P_{vc})\,A$$

A = Plug surface area sensing a difference in P_d and P_{vc}

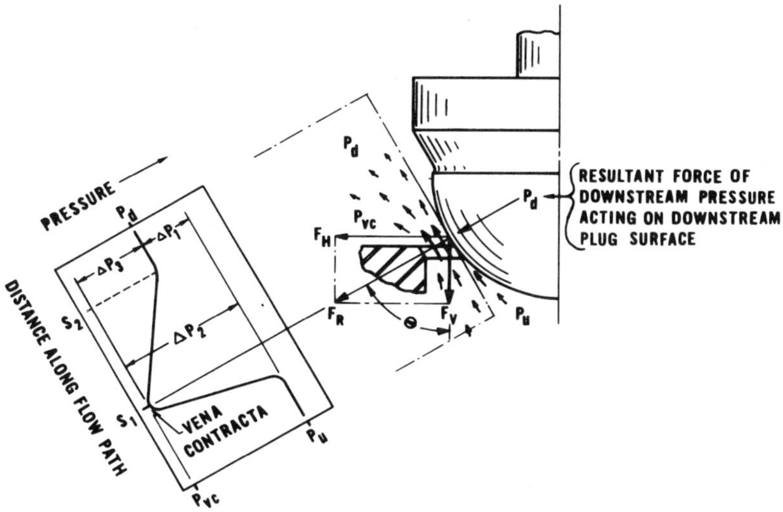

Figure 44. VENA CONTRACTA FORCES ACTING ON PLUG.

Dynamic Flow Forces

The dynamic fluid flow forces acting on several plug types are illustrated in Figures 45-50.[1] The curves given in the Figures, show forces which act axially on the plug when the pressure drop is held constant, as the valve opens. The magnitude of force change is less for V-Ports than for contoured plug shapes; and is less for double-seated than single-seated types. The reason the magnitude is less for V-ports than for contoured plugs is that the force difference between the downstream and vena contracta zone of the V-port acts on a smaller area of plug surface.

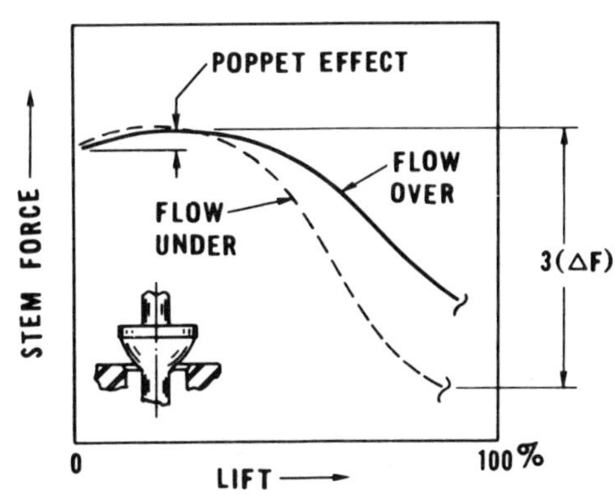

Forces act in same direction except for the poppet effect near the seat, which also occurs with flow-over due to the vena contracta effect moving downstream of the seat joint. Flow-over has the more gradual change (dF/dL); however, chatter will occur near the seat. The poppet effect of the valve on cracking open, exposes more area to the upstream pressure with flow-under, causing a force increase. Narrow seat joints and the least excessive plug diameters reduce this effect.

Figure 45. PARABOLIC.

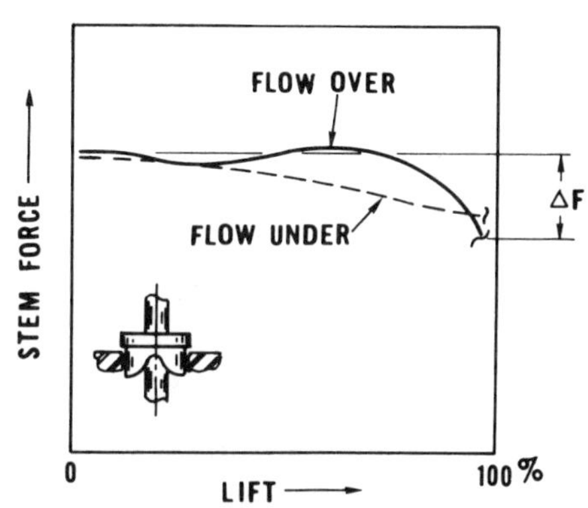

Flow-under forces are constant in direction and have a more gradual change (dF/dL), giving the plug greater stability. V-port plugs have approximately one-third the force change with lift, compared to parabolic plugs.

Figure 46. V-PORT.

[1](Figures 45, 46, 47-a, 48) This information was originally presented in a different form in a paper, "Unbalanced Force Reactions in Control Valve Plugs", R. A. Rockwell, *Instruments,* Dec. 1950, Reference No. 37. Flow tests were conducted at Masoneilan International, Inc.

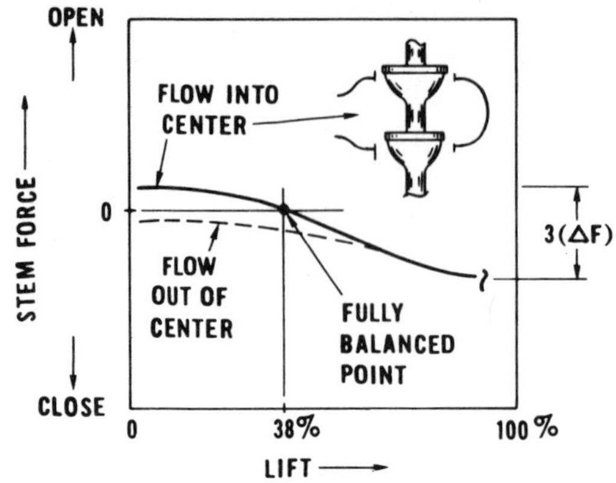

(a)

In (a) complete balance-closed is not obtained, since the upper seat orifice is larger than the lower, to pass the bottom plug. Note the much smaller change of force with lift compared with the single-seated plug of Figure 45. Flow into the center reduces slamming, but high flow rates may bow the plug, causing drag in the stem post guide bearings. Flow out of center, does not change the direction of the stem force with stroking and should give a stable plug.

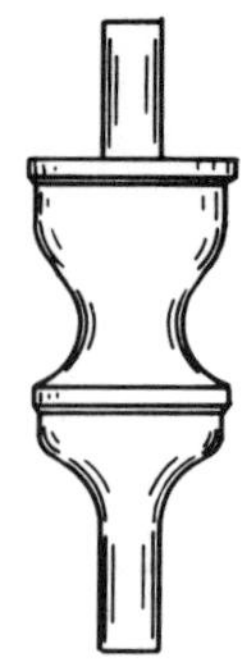

(b)

In (b) the degree of dynamic unbalance is reduced by using a different top and bottom contour form. Actual tests have shown a reduction of force to 90% by this method, applied to 4 in. valves.

Figure 47. DOUBLE SEATED PARABOLIC PLUGS.

Courtesy Fisher Controls Company

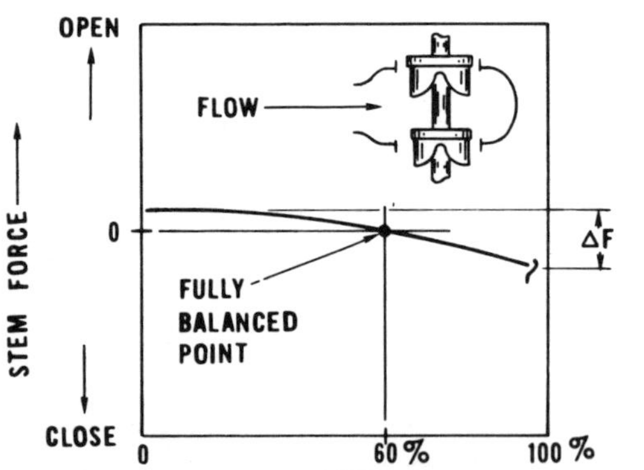

Force change is one-third that of a contoured, double-seated plug (Figure 47(a)), but does change direction; however, (dL/dF) is very gradual.

Figure 48. DOUBLE SEATED V-PORT.

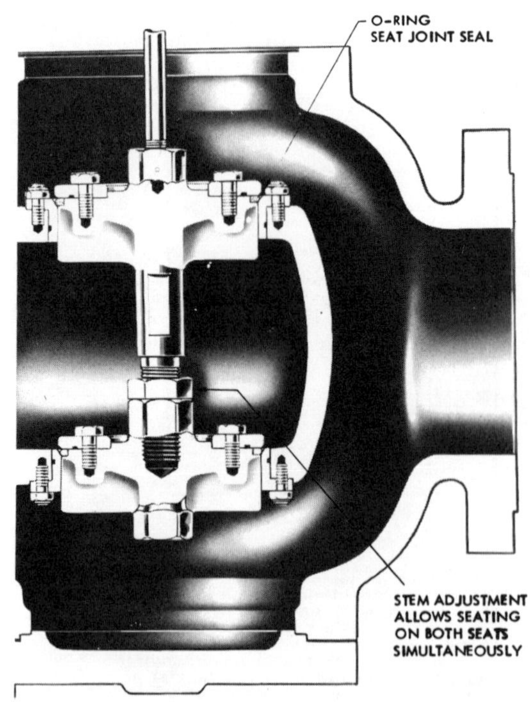

Some double ported gas regulators have a side entry through the body wall between the seats so that by using a two-piece plug, the ports' areas may be exactly balanced when closed.

Figure 49. STATICALLY BALANCED, DOUBLE SEATED VALVE.

Courtesy Fisher Controls Company

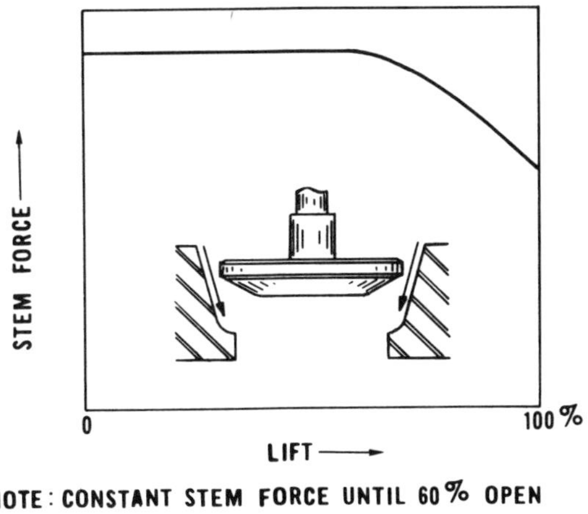

This design minimizes fluid separation along the throttling surfaces, which reduces pressure recovery in this region. Cavitation, if any, occurs downstream and valve throttling stability is improved.

Figure 50. SHARP EDGE PLUG IN CONTOURED SEAT RING
Courtesy W. A. Cash Company

CAGE GUIDED AND CAGE GUIDED, BALANCED TRIM

History

Flow cages have been used for many years as seat hold-down devices and as spacers in a stack of quick-change trim valve parts. Balanced designs appeared in early steam pressure regulators and have been effectively used in valves for high pressure oil and gas production, since the late 1940's. Characterized flow ports in cages have been common to some high-pressure, boiler feed-water, control valves for many years.

Cage trim, as we know it today, made a major market entry into the general control valve field in 1963. Now, nearly all manufacturers offer it, and in 1968 sales reached 40 to 60 percent of the intermediate and best quality control valve lines. This rapid acceptance indicates some definite advantages over regular trim, plus increased demands of process control.

The present massive cage guiding of the plug, combined with pressure balance across the plug and uniform distribution of flow around the plug by the cage ports, results in four principal advantages:

1) Stable throttling at any pressure drop.
2) Reduction in side load and friction.
3) Shut-off of large flows at high pressure by much smaller actuators.
4) Longer seat joint life.

Seat joint life is increased by eliminating chatter at low lifts and by a more uniform distribution of flow through the seat orifice, resulting in less channeling and erosive wear than experienced with standard trim. The characterized cage design reduces erosion by separating the throttling and seating areas, thus removing the seat from the high velocity flow zone.

Description and Functioning

The cage is a cylindrical spool piece surrounding the plug. It has a number of flow ports which uniformly distribute flow around the plug and, in most designs, serves as a massive plug guide. It also may carry the seat joint. Cage trim lends itself to a quick-change design. The two most common designs of cage trim are "cage guided" and "cage guided, balanced."

Cage Guided Trim

Cage Guided Trim (without balance), as shown in Figure 51, uses the cage as a massive plug guide, while excellent distribution of flow balances the horizontal side loads on the plug. Contoured cage openings, or a contoured plug establishes the flow characteristics. With a contoured plug the cage flow ports are simply openings which cause little flow restriction. The flow ports have a larger area than the seat orifice where throttling occurs. The

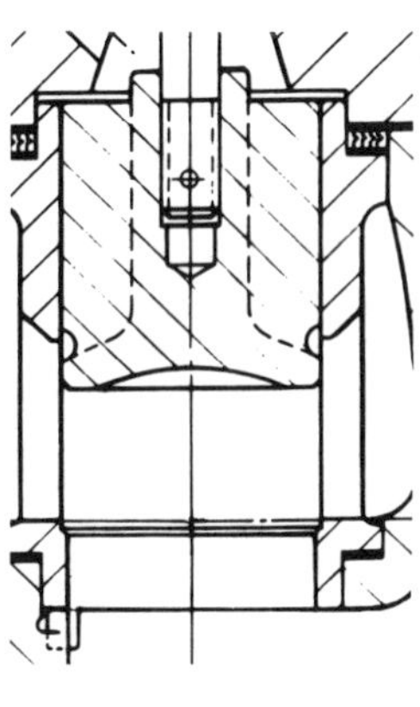

(a) Plug with Guide Ribs

Courtesy ITT Hammel-Dahl

Slots allow pressure above the seat to freely communicate with the enclosed cage cylinder space above the plug.

(b) Solid Plug with Pressure Vent Slot(s)

Courtesy Honeywell, Industrial Division

Figure 51. CAGE GUIDED TRIM.

plug is grooved along its sides to equalize pressure in the valve body, above the seat, with pressure above the plug, which fits snugly into the cage cylinder. One metal-to-metal seat joint is used for single port valves and this design shuts off drop-tight, if the actuator force is sufficient to overcome the differential pressure load acting on the mean seat joint area. Double-ported valves are also offered with two cage guides. Leakage is 0.1% C_V maximum for them.

The primary advantage of cage guided over regular trim is the improvement in flow stability. It may be used for high pressure drops when the actuator has sufficient power and is dampened by a hydraulic snubber.

Cage Guided Balanced Trim

Cage Guided, Balanced Trim, as shown in Figure 52, achieves pressure balance across the plug by vent holes passing through it. The plug, depending on trim design, makes either: a piston type seal; a partial sliding seal on the cage wall; or when seated, simultaneously makes a double seal on the regular orifice seat joint and on the upper cage seat joint (used for all-metal construction). In the latter, the plug makes a close sliding fit in the cage when throttling; therefore, the upward force caused by high differential pressures acting under the plug is practically eliminated. For flow over, this force cannot get behind the plug to slam it closed nor act to hold it closed.

The greatly improved throttling stability of cage guided, balanced trim, results in a much smaller actuator size than needed for either standard trim or cage guided trim. Tight shut-off is obtained by using elastomer seal rings on the plug piston; otherwise a slight leakage would occur.

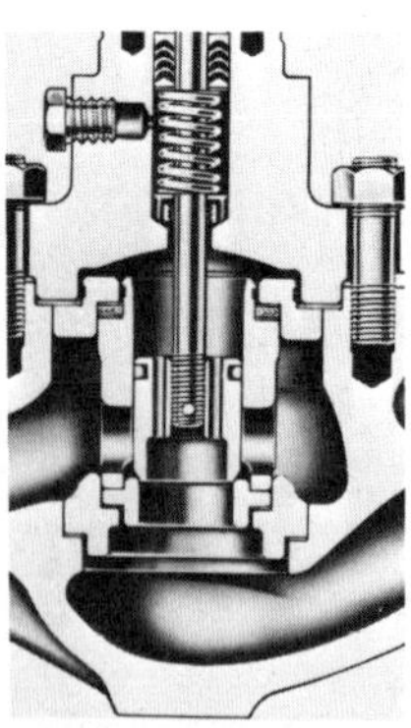

Note pressure equalizing holes in plug to create balance and the seal between plug and cage cylinder for tight shut-off. The reduced port design illustrated uses a smaller cage and plug with transfer pieces to fit in a full port body.

Figure 52. CAGE GUIDED BALANCED TRIM.

Courtesy Fisher Controls Company

Seating and/or sealing of cage guided, balanced trim occurs at two points:

1) *The seat joint between the plug and the seat.* Regular metal-to-metal joints are used and some manufacturers also offer a soft seat on the plug. Joint widths are made very narrow to effect a better pressure balance design between the orifice seat joint diameter and the cage seal or joint diameter.
2) *The seal between cage guide and plug, which may be of three designs:*
   a) Metal-to-metal seat joints, as shown in Figure 53, are used for temperature service above 450° F and may seal to a leakage of 0.1% C_V maximum (in effect, a double seated valve). The cage and the plug should be made of the same materials for uniform expansion and for retention of sealing at the operating temperature.
   b) Elastomer seal rings are used for bubble-tight shut-off at moderate temperatures, usually -20° F to 350° F.
   c) Loose fitting, piston ring type seals of glass or carbon filled TFE may be used to 450° F. Graphite rings are used above 450° F. All of these rings have leakage rates to 0.5% C_V maximum.

The primary advantages of cage-guided, balanced trim are greatly improved throttling stability at low lifts and highest pressure drops, tight shut-off (with elastomer seals), and smaller valve actuator requirements.

Summary Of Design Features

The following is a summary of the features of the three cage styles used (cage guided and cage guided, balanced as described above, plus erosive service cage trim):

1) *Cage Guided Trim* features both characterized and noncharacterized (non-restricting) flow ports; massive plug guiding over the entire stroke; and non-pressure balanced.
2) *Cage Guided Balanced Trim* features: characterized flow ports; massive plug guiding over the entire stroke; and pressure balanced across plug.
3) *Erosive Service Cage Trim* features: The cage does not guide plug but allows an annular flow area between the plug and the cage wall. The plug is characterized. The massive guide and plug cylinder sealing spool extends from the top of the cage to the bonnet. Erosive service trim is either non-balanced or balanced across the plug.

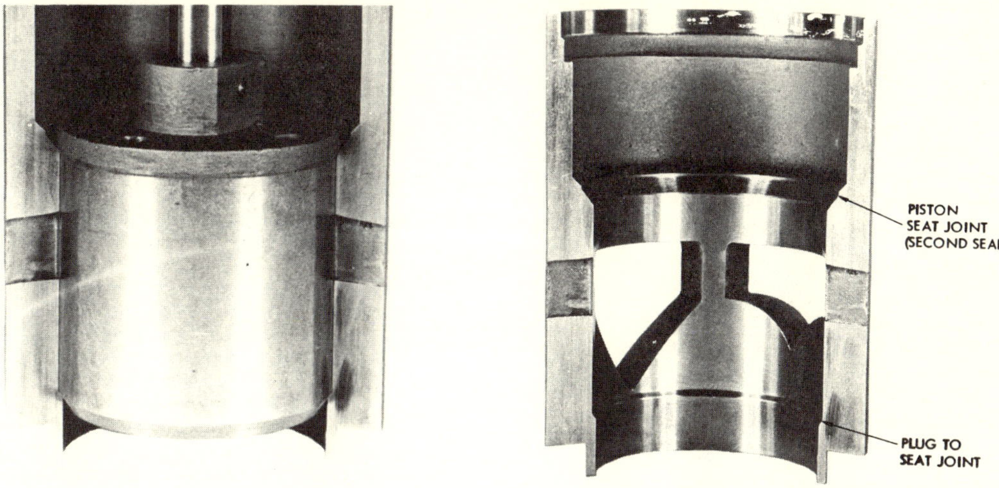

Figure 53. DOUBLE SEAT JOINT CAGE, Metal-to-Metal Sealing.

Courtesy Honeywell, Industrial Division

Note narrow seat joints, characterized cage ports, and degree of plug guiding. Pressure equalizing vent holes are visible in top of plug.

Note: Close fitting cages with flow-under, cause flow jets which will readily penetrate the body wall at low lift, in highly erosive service. On flow-over direction, the plug will be severely eroded in line with the "V" tips of the cage flow ports. At low lifts, flow is concentrated over small areas of the seat joint. Flow must be given room, and guided into turning, to eliminate direct impingement for either flow-over or flow under. This is accomplished in erosive service cage trim design by the annular flow area between the plug and cage and the plug contour, as shown in Figure 54.

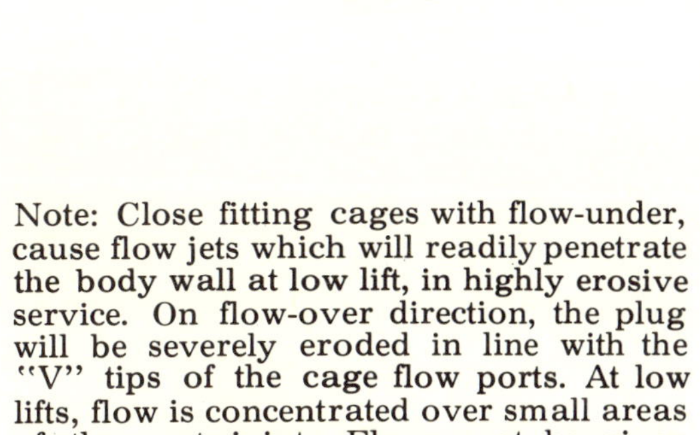

Figure 54. EROSIVE SERVICE CAGE TRIM (Characterized Plug Standard).

Cage Guided, Balanced Trim Simplifies Valve Application

For comparative purposes, a typical non-balanced valve application chart is given below for a 2 in., 3000 psi WOG valve (body rating) with a 1 in. seat port and a *medium* size diaphragm actuator.

Pressure Drop Opening and Closing Limits

Air To Close		Air To Open	
Flow Over	Flow Under	Flow Over	Flow Under
1400 psi	2200 psi	1100 psi	130 psi

The application problem would be apparent if one were required to: 1) enlarge this chart by all port sizes 1/4 in. to 2 in.; 2) make new charts for each actuator size (usually seven); and 3) add the various springs and settings to give all the variations, without exceeding valve stem, load limits.

In contrast, a cage balanced application chart for the same valve with a *small* actuator be as follows: The pressure drop limit equals 3000 psi body rating for all conditions: flow under, flow over, air to open, and air to close. One actuator size and spring covers the entire range. (See Figure 55.)

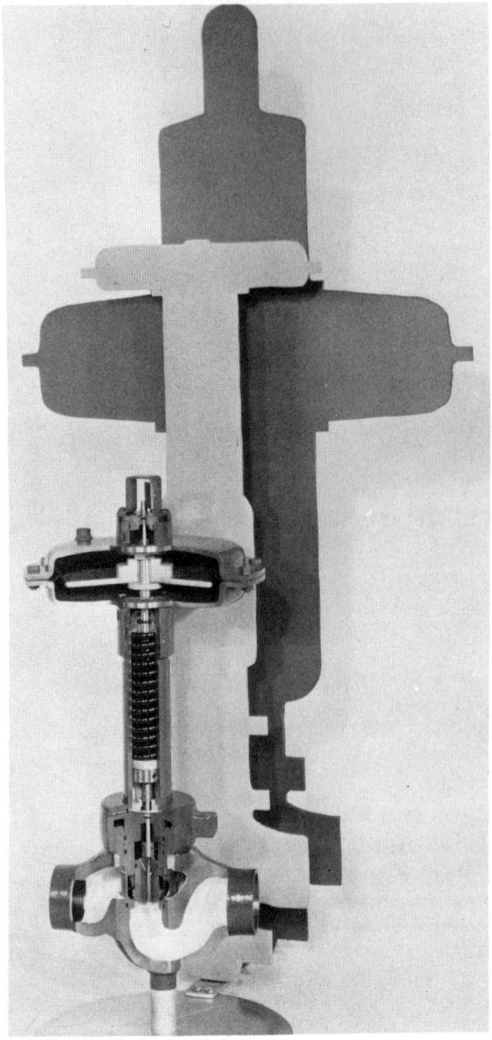

Typical dimensions and weights of the valves of Figure 55 areas follows:

Sectionalized Valve	erosive service, pressure balanced cage trim (close coupled actuator, oil field design)
	weight 72 lb., height 24 in.
1st silhouette	cage guided, balanced trim (standard open yoke actuator)
	weight 105 lb., height 36 in.
2nd silhouette	standard, non-balanced valve (open yoke with giant actuator)
	weight 250 lb., height 48 in.

Figure 55. RELATIVE SIZE OF BALANCED & NON BALANCED VALVES HAVING EQUAL C_v & ΔP OPERATING PERFORMANCE (2 in., 3000 psi WOG; flow-under; C_v 55 maximum; spring closed; 35 psi actuator air supply.).

Plug Balance

The following is an example of balancing the axial forces acting on a valve plug:

Valve & Service Conditions
2 in. valve, 1000 psi pressure drop service, flow under plug.

Physical Dimensions of Trim
2 in. seat orifice diameter, area - 3.14 in.^{2}.
3/8 in. stem diameter, area - 0.11 in.^{2}.
Seat joint width - 0.030 in./side, normal to stem axis.
Mean seat joint diameter - 2.030 in.
Mean seat joint area - 3.23 in.^{2}.

Unbalanced Design Forces
Pressure drop forces for opening: 1000 psi x 3.23= 3,230 lbs.
An air-to-close actuator would have to overcome this force plus the thrust for seating, packing friction, and spring compression.

Balanced Design Forces
Upward forces to be balanced - 3,230 lbs.
Set annular area between plug stem and cylinder to counterbalance this force +2% for downward seating thrust as the pressure drop rises:

3230 lb. + 2% of 3230 lb. = 3295 lb.

Area required:

3295 lb. ÷ 1000 lb. in.^{2} = 3.295 in.^{2}

Equivalent Diameter of Balance Cylinder Above Disc

Required Area	3.30 in.^{2}
Stem Area	0.11 in.^{2}
Cylinder Area	3.41 in.^{2}
Cylinder Diameter	2.09 in.

Note: The original unbalanced force of 3230 lbs. has been reduced to a 2% downward force of only 64 lbs.

In the open position the stem thrust from line pressure amounts to 1000 psi x 0.11 in.^{2} = 110 lbs.

It has been shown, therefore, that a force, ranging from 64 lb. (downward) to 110 lb. (upward) gives a close balance to all operating conditions. A design showing this type of plug and stem balance is illustrated in Figure 56.

Cage to Body Seals are essential in designs which shut off drop and bubble-tight. This should be a double acting seal to handle the valve pressure drop in either direction.

Guiding in cage trim valves is ideal from a design standpoint. Massive plug guiding of sufficient axial length and diameter to prevent cocking or shifting from side thrust is carried right down to the seat joint to give a near-perfect alignment. The heavy guide distributes the impact load from vibration over a greater metal area and reduces galling and guide wear. Only two parts are involved, the plug and cage. This is compared with the six parts of conventional valves: seat ring, body, bonnet, stem guide bushing, stem, and plug.

The cylinder guide for erosive service trim does not extend to the seat joint, due to the short cage length, but is located a short distance above it. Guiding is excellent. These designs are more prominent in oil field type valves.

Vent Holes in the plug allow pressure balance and, therefore, must be kept open for good valve performance. These holes should have a flow capacity that is higher than the lengthy annular clearance between the plug and the cage guide to maintain balance in the event of seal leakage, failure, or omission of the seal in field maintenance reassembly.

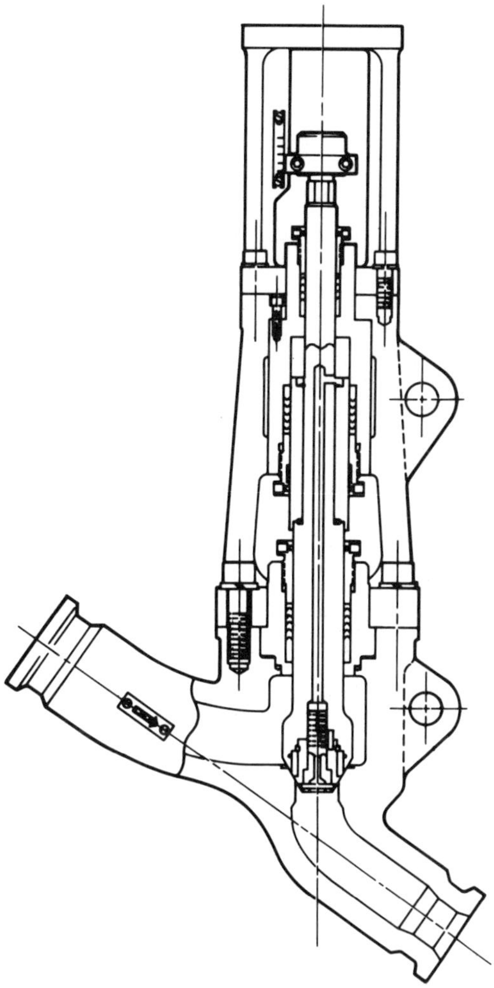

Figure 56. PLUG & STEM BALANCE DESIGN.

Thrust on the underside of the mean plug-to-seat joint area is counter-balanced by the by-piston attached to the yoke. The annular area of this piston equals the seat joint area, thus balancing both the plug and the stem thrust. The stem diameter passing through the bonnet packing equals mean plug-to-seat diameter, thereby balancing the plug to the pressure in the body cavity above the seat. When the valve is opened pressure on the underside of the plug is balanced by the yoke annular cylinder piston. Three sets of packing sense pressure drop in the line relative to atmospheric pressure.

Courtesy AMETEK, Inc., Calmec Division

Loss of Seal will not unbalance the plug. In one proof test a valve was closed by applying air pressure to the actuator until its plug touched the seat, but it had no seating load. A dial micrometer was attached to the stem and a differential pressure applied across the seat, which in an unbalanced design would have been sufficient to cause the plug to lift off the seat by 1/4 in. In this case, with the balance seal removed, the plug remained closed and only plug cylinder clearance leakage occured. No upward plug motion could be detected.

The cylindrical annular balancing space above the plug is a relatively dead area as far as flow is concerned. Flushing occurs only on valve stroking. The space is self-cleaning when the valve is in a vertical position. For entrained solids service, horizontal mounting may result in jamming, during the last 10 percent of stroke and upside-down mounting is not permitted.

Service for entrained solids are no problem. A two inch valve handling sandy water at the rate of 3600 lb. of sand mixture per hour, while stroking and standing, could not be jammed. It retained full stroke in the vertical position.

Paraffinic oils do not clog vent holes. Valves with long vent holes have been used in oil production service where the flow lines were completely blocked by paraffin and sand, yet the valves remained operable.

Special design for clogging service has included some plugs designed as cylindrical skirts with completely open venting, except for a spider at the top end for stem attachment, as shown in Figure 57.

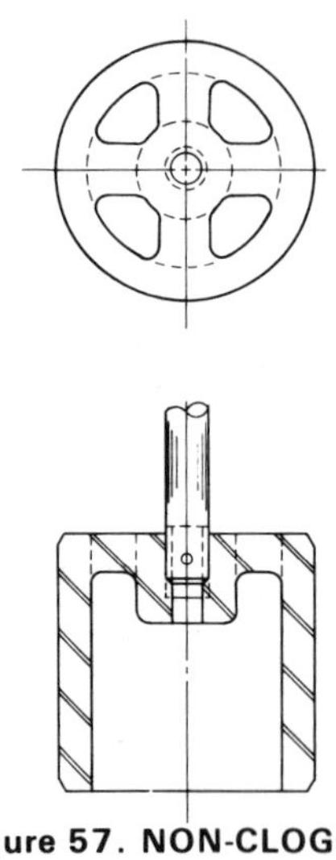

Figure 57. NON-CLOGGING, BALANCED PLUG.

Flow Capacity

Cage trim flow capacity is generally 15 to 20 percent higher than the flow capacity of regular trim. When bodies are designed specifically for cage trim, the capacities are even higher. The uniform distribution of flow around the plug by cage, reduces turbulence and offsets the fluid contraction and expansion pressure losses when passing through the cage ports.

The following example is a comparison showing the increased flow in a body designed especially for cage trim (Typical C_V values are given.).

2 in. Globe Valve (flow under)	*Plug Characteristic* Quick Opening	Linear	Equal Percentage
1) Characterized Cage 2-5/16 in. diameter orifice (non-restricting)	75 C_V	74 C_V	65 C_V
2) Standard plug and seat construction in regular body	60 C_V	55 C_V	55 C_V
Increase in flow capacity	25%	35%	19%

Greater Valve Rangeability is obtained with Cage Guided, Balanced Design. Stable throttling near the seat allows the valve to control at a degree which matches the precision of the actuator. Thus, the reliable low range of flow control is increased.

Reduced Capacity Trim change-out for dual metal seated cage designs is effected by changing only the characterized flow cage of the balanced design. A smaller set of characterized flow ports are used in the replacement cage. The number of cage ports may be reduced instead of altering their size, but at a sacrifice of balancing side thrust forces. Some manufacturers use a smaller plug, cage, seat ring and supporting members. Cage trim with characterized plugs and non-restricting cage flow ports requires a change of both plug and seat as in standard valve construction.

Quick Change Trim is found in many cage designs. The bonnet holds down the cage and the cage may hold down, or be integral with, the seat orifice. Thus, all parts may be removed through the bonnet. This applies to three-way valves as well. Screwed or welded cages are not quick change, but may be used to improve guiding for plugs having very large stems.

Cage design is so effective in stabilizing flow control that cage rings are available to screw into the body seat ring threads of both single and double ported valves.

Double Ported Cage Designs, as shown in Figure 58, are frequently used on high pressure gas and steam reducing service and also on boiler feedwater control. Their flow capacity is about 5% higher than regular trim.

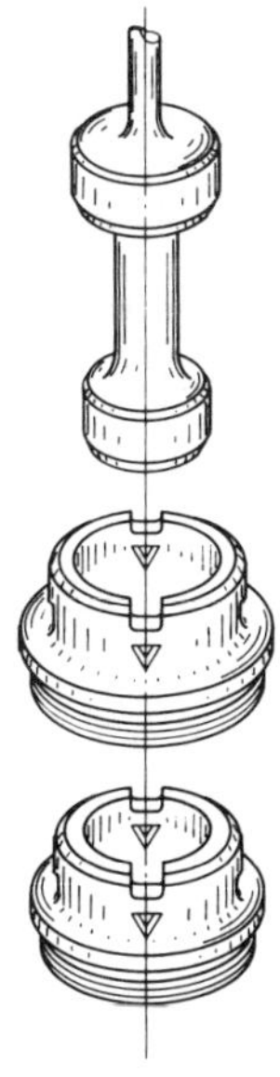

Figure 58. DOUBLE PORTED, CAGE TRIM, Screwed Type.

Copied from illustration in a Fisher Controls Company catalog.

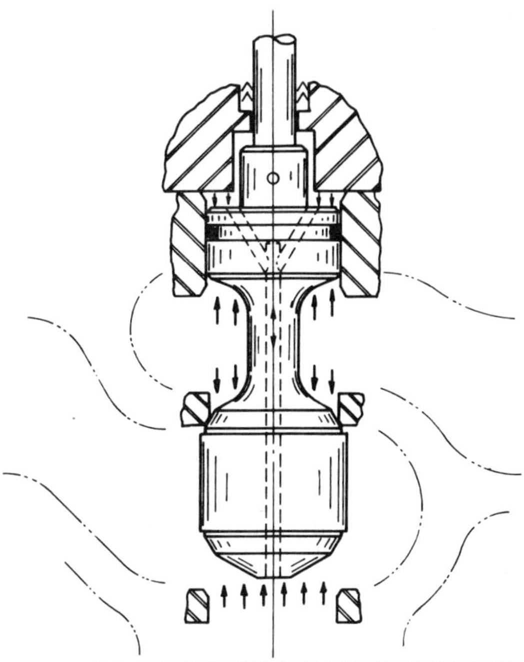

Figure 59. PRESSURE BALANCED THREE-WAY DIVERTING AND MIXING VALVE.
(2 in.; 3000 psi WOG; C_V = 55).

Pressure in the upper port is spool-balanced, while forces under the plug are equalized by the central vent hole. The stem is not balanced. Metal-to-metal seating allows tight shut-off to full 3000 psi change in pressure-in across any port.

Three-Way Valve Trim Balance

Three-way diverting and mixing valves are notoriously poor performers in high pressure shut-off or throttling service and their actuators are massive. As shown in Figure 59, balance overcomes these deficiencies for both mixing and diverting service.

By balancing, two inch valves with 1-3/4 in. metal seat joints may shut-off a full 3000 psi pressure drop across either port, with flow in any direction, and may be operated by a 50 in.^{2} diaphragm actuator that has a 35 psi air supply. Shut off with narrow joint, metal seating is drop tight (O-Ring seal used on plug balance cylinder seal).

These principles may be extended to special designs which, in addition to diverting and mixing, shut off both ports at once. These special designs, called "3-way, 3-position" valves are illustrated by Figures 60, 61, 62. Two-inch, 5000 psi, WOG valves have successfully shut off and diverted oil well flow at a well head with shut-in to 4500 psi tubing pressure. Hi-low pressure pilots also allow them to serve as safety valves.

Ordinary three-way diverting valves as shown in Figure 60, simply direct flow from one outlet to the other. In the transition period of changing outlets, fluid flows from both outlets at once. This type of valve has a full "off" position and rapidly passes through "off" in diverting the stream from process to test. One oulet must fully close before the other can be opened. Plugs float on the stem and align themselves independently in their respective seats for tight sealing. The "off" position allows one valve to perform the job formerly requiring two valves, (one for diverting and one for shut off). The stem slides through the top plug (spring seated) as the nut on the bottom of the stem lifts the bottom plug. The stem shoulder, pushing down, opens the top plug while the spring seats the bottom plug.

The valve illustrated in Figure 61 is capable of opening and closing to the differential pressure across any flow port or to a combination of differentials on all ports. Ports P_c and P_T are spool balanced while part P_I is balanced by a floating sleeve, surrounding the spring and sealed to the plugs. The stem senses no pressure. Four dynamic (moving) seals are subject to the difference between the internal and atmospheric pressures; therefore, practical high pressure designs are limited to 2 in., 3000 psi WOG, or 4 in., 1500 psi WOG. Special designs involving dual concentric stems and double diaphragms within a single case may be

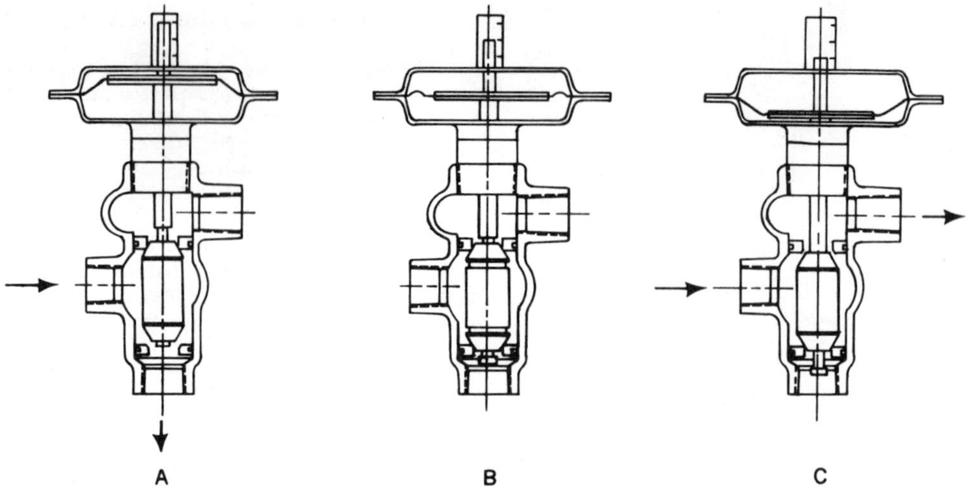

Figure 60. ORDINARY THREE WAY, THREE POSITION VALVES.

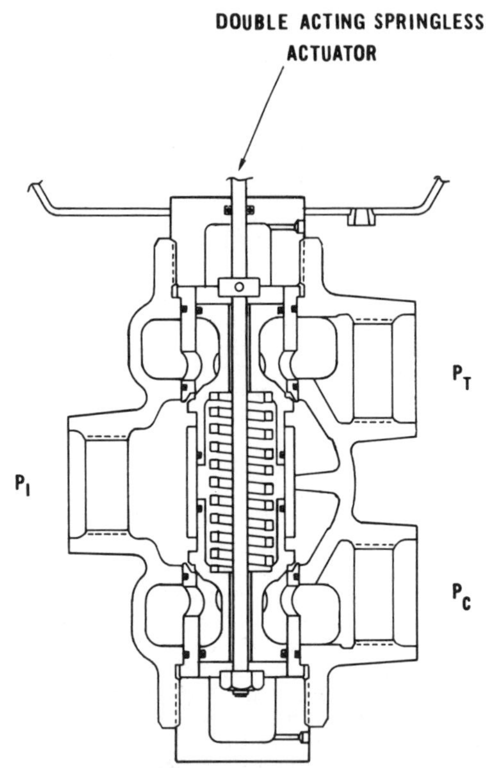

Figure 61. FULLY PRESSURE BALANCED, THREE-WAY, THREE-POSITION VALVE.

U.S. Patent No. 3,063,469.

Courtesy Rockwell Manufacturing Company

air-loaded from two sources to maintain plugs seated on the sleeve with spring compressed. This allows diverting without passing through the off position and eliminates the momentary pressure surge when changing ports. Fail-safe shut-off of all ports occurs upon relief of both loading pressures.

Application of the semi-balanced design, shown in Figure 62(a), is high-pressure, oil-production, battery manifold service at 2 in., 1500 to 5000 psi WOG. Outlets P_T and P_c connect with the test and common separator vessel manifolds, both normally operating at the same pressure. P_I is the incoming well flow line. P_T and P_c are balanced by venting P_c to the sleeve, to get behind, and tend to seat both plugs (this can be reversed to P_T, or in clean service, a shuttle valve in the stem may sense and feed-in the higher pressure). Stem balance is achieved by venting P_c to the annular shoulder on the upper stem. Three small stem seals sense internal pressure relative to atmospheric pressure. P_I will open to shut-in flow line pressures, equal to or exceeding the body pressure rating with normal actuator supply pressure. P_c will remain sealed until $(P_c - P_I) > 3000$ psi (based on spring force and slight degree of unbalance designed into the P_I side of the plug which causes the seating force to increase slightly as the shut-in flow line pressure builds up). P_T will remain closed when: $P_T = P_c$, $(P_T-P_I) > 3000$ psi; or when $P_c = 0$, $(P_T-P_I) > 1500$ psi.

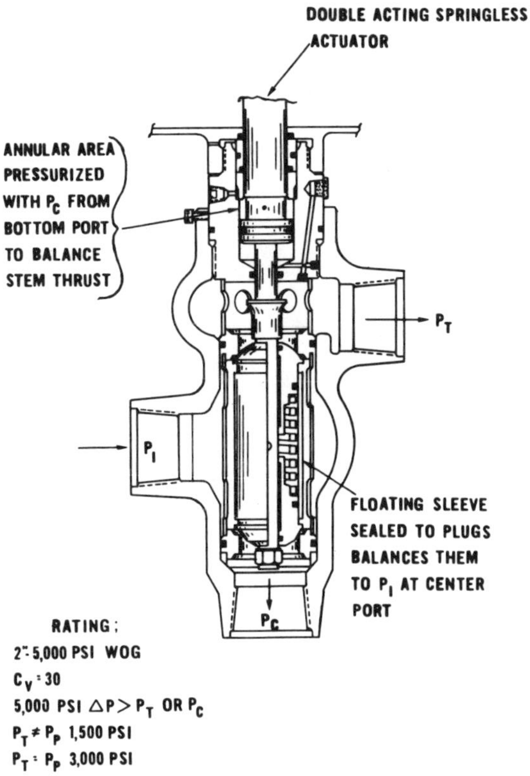

(a). "Semi-Balanced "Three-Way, Three-Position Valve.

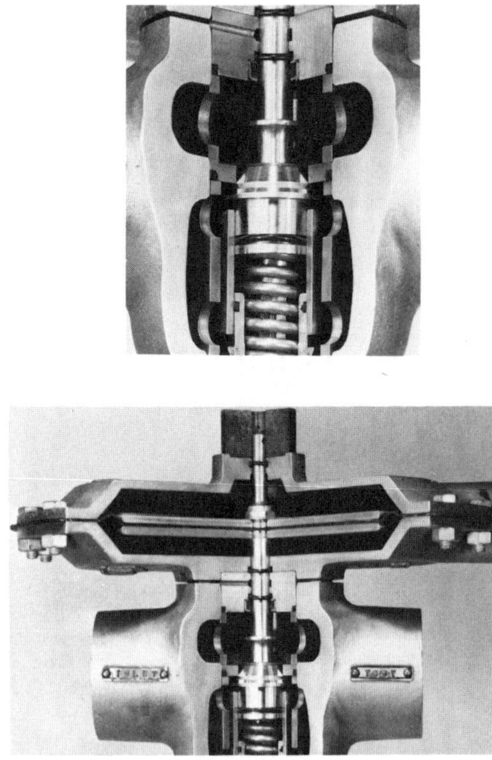

(b). "Compact", Three-Way, Three-Position Valve.

U.S. Patent Nos. 3,025,881, 2,969,811, 2,867,239

Figure 62. SEMI-BALANCED AND COMPACT, THREE-WAY, THREE-POSITION VALVES.

Courtesy Rockwell Manufacturing Company

Figure 62(b) shows a compact three-way, three-position valve that is semi-balanced (2 in.; 1500 psi WOG, Cv-20). It has balanced trim as shown in the upper illustration, and a close-coupled, double-acting springless actuator as shown in the lower view.

Large Designs And Applications

Large cage guided, balanced valve designs and applications are shown in Figures 63, 64, 65, 66, and 67. Figure 63 shows the internal and external views of a large balanced valve for boiler feed water control.

Figure 64 is a schematic of an isolation and dump valve for high pressure and large flows. It has a flow-over angle design with a separate pressure balance pilot. With the valve closed, pressure leaks past the plug, piston-to-body sleeve clearance and aids in seating the main valve. The actuator must thrust the stem downward to compress the pilot valve spring and close the pilot. Sealing is tight. To open, the actuator lifts the stem allowing the pilot to open and balance the main plug, since the pilot port has a much greater capacity than the annular plug-to-sleeve clearance area. The pilot plug shoulders against a stop in the main plug and the remaining spring-force causes these parts to remain together for the balance of the opening stroke. Unbalance, created by the plug cylinder being smaller than the seat joint, is downward. Since the main plug and pilot valve stem assembly are free to move except as loaded by the spring, the plug and stem remain rigid during opening and there is no chatter. The opening force is reduced by balancing the main valve.

In the balanced globe valve shown in Figure 65, a cage hole drains condensate out of the upper compartment, above the plug, when the valve is closed with flow-over. Thus, with the main valve and pilot valve closed, upstream pressure enters above the plug and aids seating. Seating

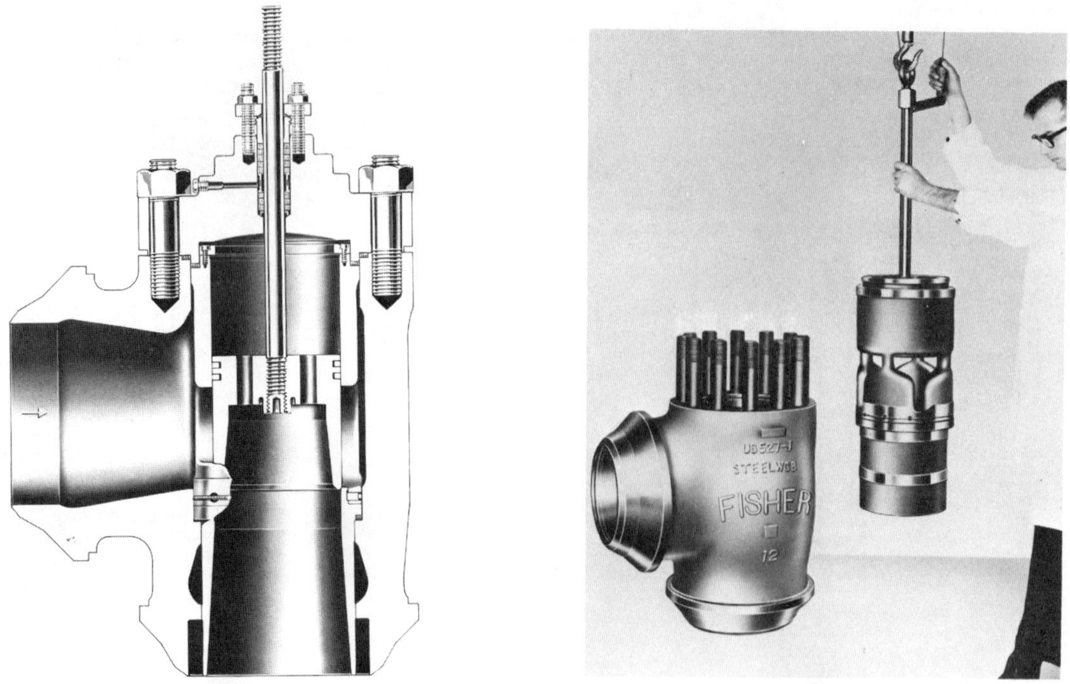

Figure 63. BALANCED VALVE FOR BOILER FEED WATER CONTROL, (12-16 in. USASI 900 Class, 2160 psi CWP).

Courtesy Fisher Controls Company

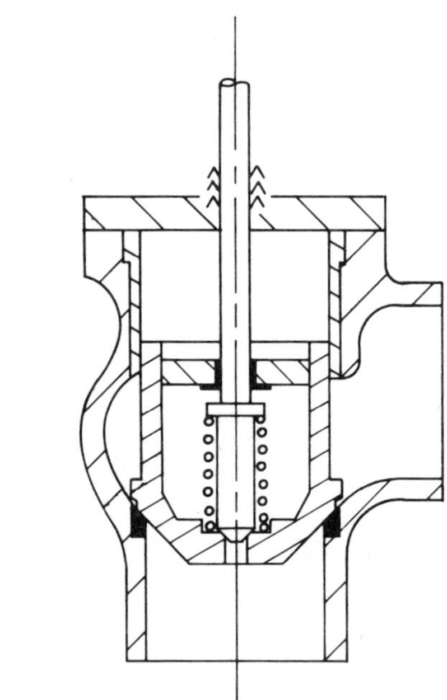

Figure 64. ISOLATION AND DUMP VALVE FOR HIGH PRESSURE AND LARGE FLOWS.

Design & functioning as described in U.S. Patent 3,428,090.

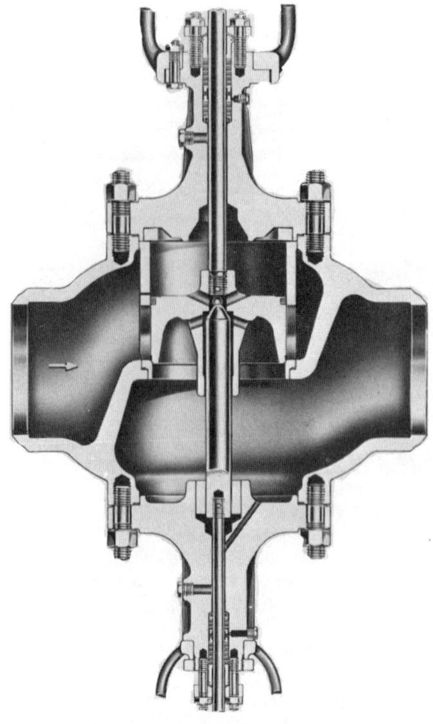

FIGURE 65. BALANCED GLOBE VALVE FOR NUCLEAR STEAM DUMP OR TURBINE BY-PASS.

Courtesy Fisher Controls Company

is tight. To open, the pilot is opened first by its actuator, thereby establishing balance across the main plug, which in turn is opened by its own actuator. Opening speed may be as fast as 350 milliseconds for a 4 in. stroke at 1000 psi pressure drops.

The sixteen-inch, balanced disc valve shown in Figures 66 and 67 is balanced to open with flow-over and, unbalanced when closed, so that the pressure drop adds to the seating force. This is accomplished by an internal pilot valve, operated by the main stem. A double-acting air piston actuator, plus a fail-safe spring closure gives rapid operation, which is regulated by a hydraulic snubber surrounding the stem inside of the nest of springs.

Summary Of Cage Balance Trim Advantages

Flow Control

1) Greatly improved throttling stability at high pressure drop.
2) Ability to close and open against any pressure-to-body rating.
3) Tighter seating
4) Greater maximum flow capacity, coupled with high pressure shut-off.
5) Increased rangeability for controlling low flows.
6) Fail-safe action.

Physical Improvements

1) No seat damage, since chattering impact is eliminated.
2) Less guide damage and noise from horizontal vibration in the plug guide.
3) Long seat life, because of reduced erosion.
4) Greatly increased packing life, because of reduced stem loads and smooth stem motion. Stems run truer with less bowing and cocking.

Actuator

1) Smaller.
2) Lighter.
3) Less expensive.
4) Faster response and better control, because of:
   a) Less dead air space;
   b) Flatter area vs. stroke curve; and
   c) Lower mass.
5) Higher spring rate, stiffer actuator, more stable control.
6) Much more actuator force may be applied to seating force. More of the actuator force may be applied to compress the spring, since very little is needed to overcome the pressure drop across the plug.
7) It is practical to spring-close large valves to high pressure drops.

Figure 66. SIXTEEN INCH, BALANCED DISC VALVE, DUMP TEST AT 1500 psi PRESSURE DROP (Designed for 1250 psi at 575°F Service).

Courtesy Rockwell Manufacturing Company

Figure 67. SIXTEEN INCH, BALANCED DISC VALVE IN A LIQUID FLOW TEST LOOP.

Courtesy Rockwell Manufacturing Company

Application

1) Simpler to apply, because of fewer restrictions, such as flow direction, air supply, orifice size, pressure drop, etc.
2) Fewer requirements for positioners.
3) A hydraulic snubber is never required.
4) Quick change trim.

SEATING, SEALING AND LEAKAGE

The three problems discussed in this section include:

Seating - The alignment and contact of the plug with the seat, including joint design and loading.

Sealing - The parameters of metal finish, joint width, and metal yielding which lead to tight sealing.

Leakage - The amount of leakage that may be expected for different sealing parameters and joint designs.

Tight sealing is becoming of greater importance to control valve users, now that improvements in designs of both valve trim and actuators allow tight shut-off. One valve may be used for both stop and throttling service at a cost saving. Diaphragm control, valve actuators with 15 to 35 psi air supplies do not develop the high seating forces used in stop valve designs with manual or automatic operation. If they did, they would be bulky and would be sluggish in their speed of stem movement for throttling action. A small amount of leakage is to be expected, because of the lower seating forces. Manufacturers normally rate their valves for maximum leakage as follows:

Double Seated Valves — $<0.1\%\ C_V$ maximum[1]

Single Seated Valves — $<0.01\%\ C_V$ maximum[1]

Development of the springless piston actuator, air loaded on both sides and using a much higher supply pressure (100-150 psi), coupled with a positioner, led to higher seating forces and tighter shut-off. Single seated valves can now be sealed, drop-tight to high pressure drops. Some cage balanced valve designs also shut off drop-tight with small diaphragm actuators.

Seating

To make a seal, a plug must first be perfectly aligned and then must fully contact the seat joint as the seating load is applied by the actuator.

If the plug strikes the seat joint slightly cocked, it will remain cocked until a higher seating load causes it to jump sideways as it slides down into the taper and slams into full joint contact. This deforms the seat, causing leakage. In time, the plug indenture will extend to form a new off center, nearly 360° seat contact. Above a certain plug cocking angle, the plug will not jump into place regardless of loading; therefore, pre-guiding and alignment of the plug before seating, is necessary.

The two principle types of mis-alignment are *Concentric* as shown in Figure 68(a) and *Axial* as shown in Figure 68(b).

Alignment

Alignment of the plug on the seat for a single ported, top and bottom guided valve involves concentric alignment of eight components and axial alignment of eleven, fit combinations; plus consideration of the operating clearances.

One can readily see the precision machining, required of the control valve manufacturer to maintain alignment of these parts. Each part must have an assembly or sliding clearance which allows a minute horizontal axial shift and a very slight cock. The flexibility of the stem is sufficient to allow the plug to move into true seat joint contact with light, initial seat-contact loading.

Seat Joint Design

Flat joints, normal to the plug axis, are used on some low pressure stop valves and soft seated control valves. It is not practical to manufacture them for high pressure service. As discussed later, tighter sealing occurs through a sliding and burnishing action of a tapered joint. Tapered joints turn the fluid gradually and are the best for high pressure drop and for erosive service. The control valve industry uses joints from 15° to 45°. Smaller angles would begin to form sticking tapers and larger angles give too much of a poppet effect, when cracked open at high pressure differentials, which would cause unstable, low-range throttling. In summary:

1) 45° seat angles find their best applications in either normally open or normally closed valves.
2) 30° seat angles are a compromise between high seating forces and streamlined flow, at low lifts, for low erosion service.

[1]Seat leakage flow is laminar rather than turbulent and the C_V formula is not applicable; therefore, this is simply a means of specifying an amount of acceptable leakage.

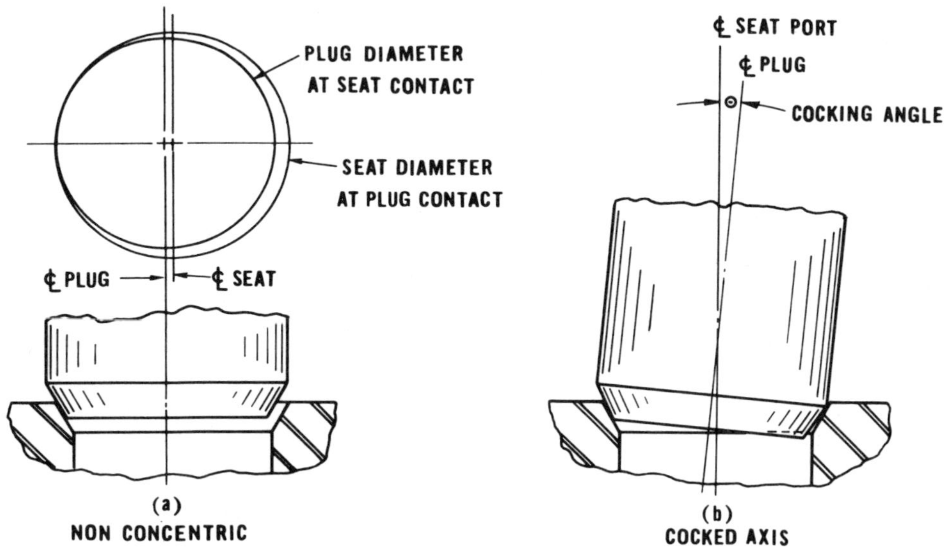

Figure 68. PLUG MISALIGNMENT WITH SEAT

3) 15° seat angles are best applied to high pressure drop, erosive service.

Seat Joint Width

Seat joint width is a balance of adding length to the flow path to increase flow resistance vs. reduction of the seating force for a given actuator seating load. A certain minimum width is essential to establish a tight seal; however, the joint must have sufficient backup strength to support the compressive seating load and must be wide enough to prevent indention of the plug. Narrow joints are much tighter than wide joints, provided they exceed the minimum width requirements.

Seat Joint Loading

Seat joint loading is usually expressed as pounds of force per linear inch of mean seat joint circumference. Loading may vary from 25-600 lb./inch as given below:

1) 25 lb./in. - Low pressure drop service; leak-tight shut-off is not required; metal-to-metal joint.
2) 50 lb./in. - Moderate pressure drop service; slight leakage expected (0.1% C_V maximum).
3) 100 lb./in. - High pressure drop service; nearly drop-tight service (0.01% C_V maximum); will seal 3000 psi pressure drop on 0.015 in. width, 30° joint of 316 SS[1].
4) 300 lb./in. - Very high pressure drop service; drop-tight (will seal 6000 psi on 0.025 in. width, 20° joint of 440-C SS, hardness 55 R_C).
5) 600 lb./in. - Extremely high pressure service.

The apparent compressive seating stresses on joints described in items 3) and 4) above, are 13,000 psi and 35,000 psi respectively, which is well below the yield point of the given trim materials. These are the stresses normal to the joint. Elastomeric and plastic yielding is occuring at the high points of each surface making a tortuous flow path for leakage restriction.

By contrast, stop-valve seat loading with hard-faced seats in steam service may be 50,000 psi, apparent stress, or four times a nominal control valve, seating load of 100 lb/linear inch for a diaphragm actuator.

To obtain the best circular seat-joint contact at low stem loading use:

1) Cage guided trim with the seat integral with the cage. Horizontal and axial alignment for seating involves

[1] 100 lbs/linear inch vertical seating force is equivalent to a force of 200 lbs/linear inch normal to a 30° joint (100 lb./sin 30°). The compressive seating stress is 200 lb/0.015 in. x 1 in. = 13,300 psi.

only two parts in sliding contract. Alignment of regular trim must be transferred from the seat to the body, to the bonnet, to the guide, to the plug, and back to the seat.

2) Cage guided trim with the seat aligned by the cage.
3) A seat that is integral with body; or
4) Extend a thin flexible seat lip above the means of seat retention as shown in Figure 29.

A radial expandable, seat ring design is illustrated by Figure 69. In this design the seating angle creates a large radial component of force which expands the seat ring against the retaining collar. The spring-out action of the non-circular ring allows a near-perfect joint contact giving drop-tight shut-off in severe thermal cycle service. This design has been successfully used on pressure drops to 4,400 psi. The large flow, entrant passage also makes the valve suitable for erosive service.

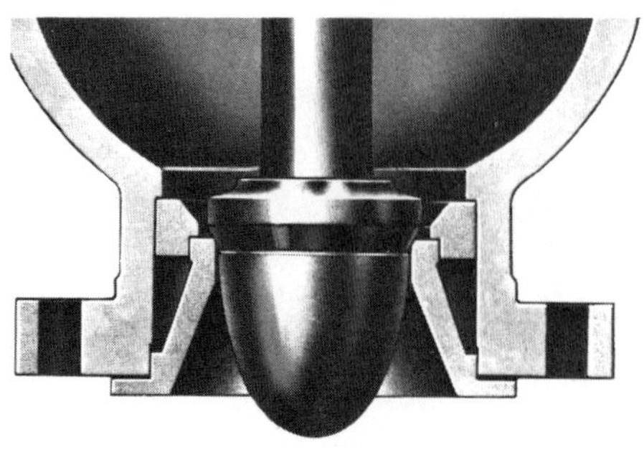

Figure 69. RADIALLY EXPANDABLE SEAT RING DESIGN.

Courtesy Conoflow Corporation

Trim Sealing

Tight sealing requires the yielding of one material into the surface "waviness" and surface roughness of the other, as illustrated by Figure 70, to block direct leakage paths; thus, making these paths long and tortuous. Compressive stresses are far below those that would yield the entire joint; therefore, the contact area is apparent. Actually, only the peaks of each surface are in contact and the concentration of force may then exceed the yield and will plastically deform the high spots on each valve closure. Additional closures require a higher seating load to achieve the same degree of tightness, until the wear particles are formed and conditions tend to stabilize.

Tapered joints provide for a sliding and burnishing action as contact is made and loading occurs. This gives a tighter initial seal than a perpendicular contact and the seal remains tighter with repeated closures.

The minimum width of a joint to seal gas to 1 x 10^{-7} cc/s/linear inch, maximum leakage, is 0.04 inch. Wider joints with the same surface finish and loading will not seal tighter. This width insures sufficient high point contact to form an adequate flow resistance path as shown by the graph of Figure 71.

Extra "super-finishing" of seat joints is unnecessary for tight sealing, because as the joint opens and closes, wear particles ball up on the surface, quickly returning it to a rougher finish. Also, some fluid contaminants tend to remain in the joint on closure and indent the surfaces. Excessive lapping generally either reduces seat tightness by increasing the actual contact area, thus, reducing the unit compressive seating stress provided by a fixed actuator seating force or it destroys the original surface geometry.

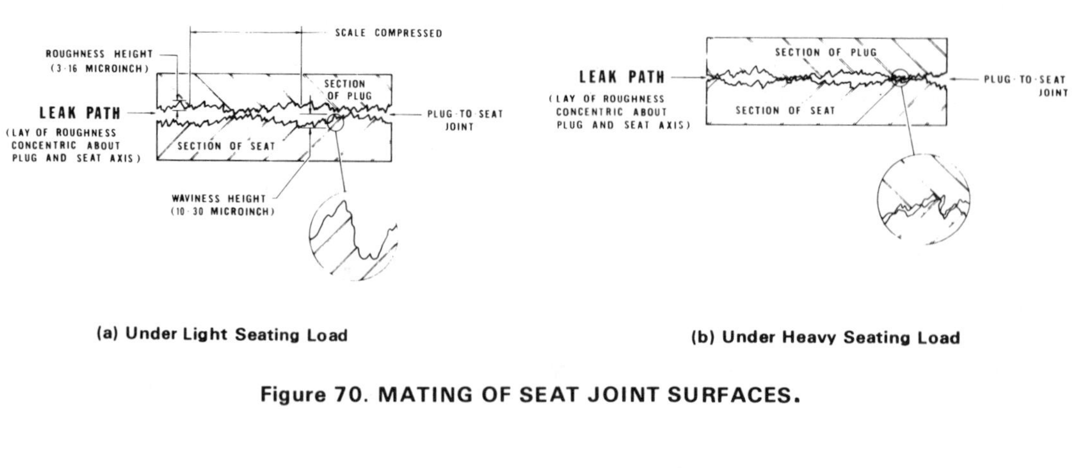

(a) Under Light Seating Load **(b) Under Heavy Seating Load**

Figure 70. MATING OF SEAT JOINT SURFACES.

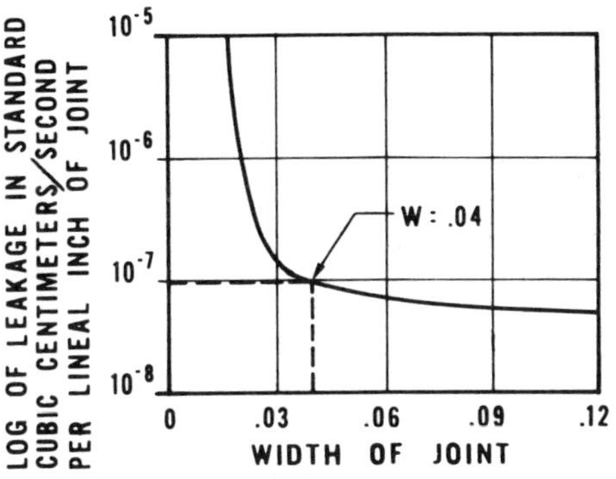

Figure 71. MINIMUM JOINT WIDTH FOR A TIGHT SEAL.

(Leakage rates for 14.7 psi ΔP helium on a flat circular joint.)

Consolidated graph taken from Reference No. 48.

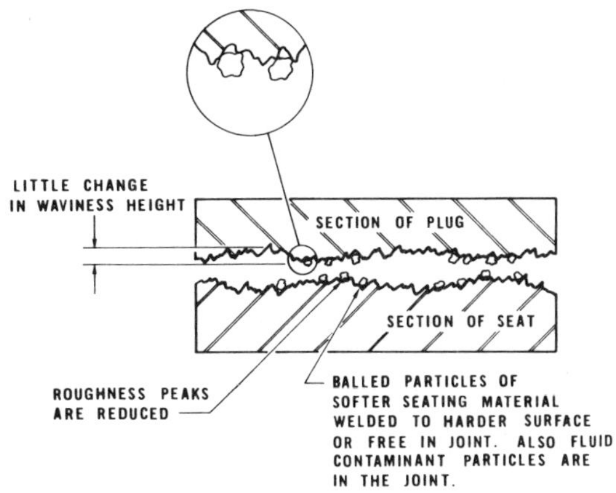

Figure 72. JOINT SURFACE AFTER REPEATED CLOSURES.

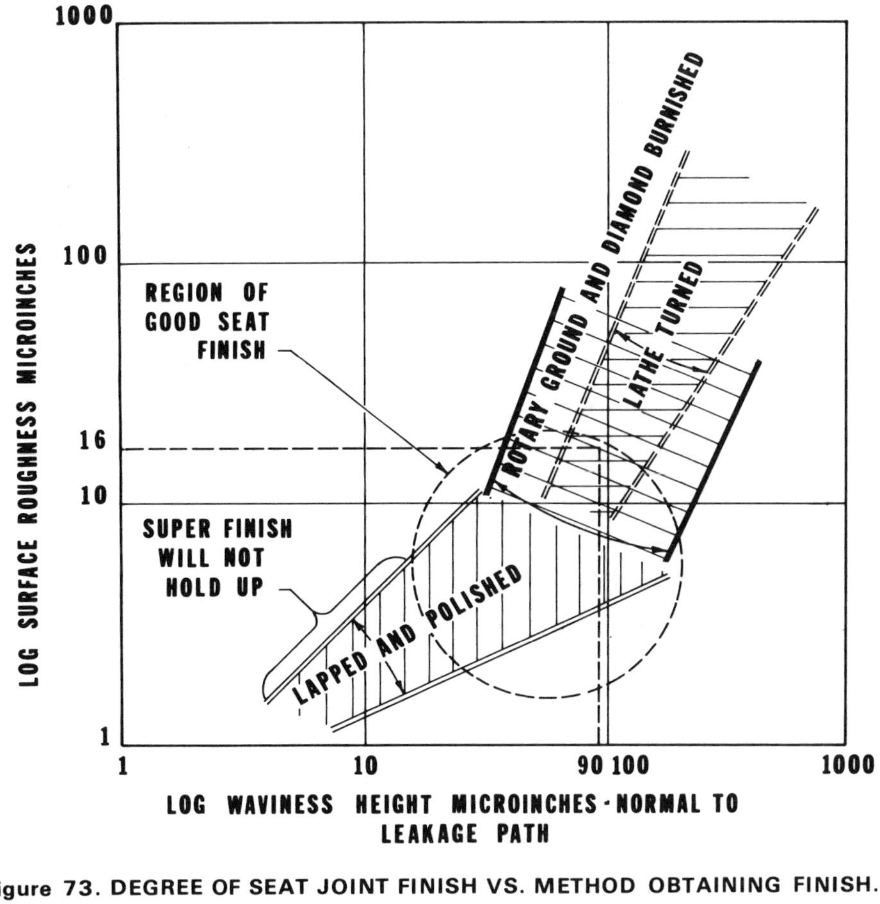

Figure 73. DEGREE OF SEAT JOINT FINISH VS. METHOD OBTAINING FINISH.

Chart presented in "Investigation of Leakage and Sealing Parameters", Paul Bauer, ITT Research Institute, Technical Report AFRPL-TF-65-153, August, 1965.

The following comparison illustrates the relative seating force to obtain same degree of tightness vs. the seat joint finishing method (concentric lay of finishing pattern). Refer to Figures 72 and 73.

Finishing Method	*Relative Seating Force*
diamond burnished (after grinding)	1.0
lathe turned	1.3
ground	2.5
lapped (excessive) (90% of apparent area of) (contact actually mated)	2.9

Trim Leakage

Definitions Relative to Trim Leakage:

Drop tight, to be meaningful, must be specified in terms of the maximum allowed number of drops per unit of time (drops/minute, cc/hour or no visible drop).

Bubble tight, to gas, should be specified as the maximum allowed number of bubbles of a given size per minute (usually 1/8 in. diameter).

Zero leakage is defined as 1×10^{-8} cc/s or about 0.3 cc/year (helium leak rate at standard conditions). Zero leakage is often specified in critical service and requires very careful joint design, material selection, finish control, and sufficient seating force on narrow joints. It is practical only for small valves, at extra cost, and may last for only a few closing cycles.

Stop valve maximum leakage rates are given in the Valve Manufacturers Standardization Society, SP-61 as:

1) Water tests at ΔP = Body CWP rating (10 cc/hour/inch of valve pipe size or about 3/drops/minute).
2) Air tests at ΔP=80 psi (0.1 SCFM/inch of valve pipe size).

Leakage Specifications, to be meaningful and allow comparison, should include test fluid, temperature, pressure, pressure drop, seating force and duration of test.

Control Valve Leakage

Properly designed control valves can achieve stop valve tightness and maintain it throughout a long service life before trim replacement; particularly with cage guided, balanced trim having elastomer plug-to-cage seals. The control valve, however, is expected to throttle and often shuts off much more frequently than stop valves. For example, some dump valves may have from 4000 to 7000 opening and closing cycles per day, handling high pressure and erosive fluids at 1000 to 4000 psi pressure drop. Few stop valves could match this performance and remain tight.

It is customary to expect a slight degree of leakage and, in general control valve service applications, this has presented few problems. Specified permissible leakage should be based on process considerations (hazards resulting from leakage after emergency shut off, etc.). Tight shut off should be specified only when necessary, because a higher quality more expensive valve is required.

Liquid Leakage is to a great extent affected by surface tension and the particular fluid wetting the joint surface, before the new fluid attempts to enter. For precise leak testing, the joint and body should be free of all traces of oil, which would preferentially wet the metal joint surfaces. Water leakage rates of an oil-free joint are higher. Oil clinging to the roughness, blocks some leakage paths, causing a higher capillary displacement pressure to establish flow of the test fluid.

Theoretical Leakage Formula

The equation relating the fluid properties, flow path geometry, and flow rate is:

$$Q_o = \frac{\pi (p_1^2 - p_2^2)\bar{h}^3}{12 \mu p_o \ln r_o / r_i} \quad \text{(See }^1\text{)}$$

where:

$\bar{h}$ = uniform channel clearance [2]
p_o = pressure at standard conditions
p_2 = exit fluid pressure
p_1 = inlet fluid pressure
Q_o = volume rate of flow at standard conditions
r_o, r_i = outside and inside radii of joint sealing area
μ = absolute viscosity

[1] Taken from Reference 50, the formula applies to a flat seat joint.

[2] This is the theoretical separation of two truly plane surfaces to give an equivalent rate of leakage caused by channels, imperfections, etc.

When the terms are rearranged, the uniform channel clearance is:

$$\bar{h} = \sqrt[3]{\frac{12 p_o \mu \ln r_o / r_i \, Q_o}{\pi (p_2^2 - p_1^2)}}$$

for gases the leakage formula becomes

$$Q_o = \left[\frac{\pi h^3 (r_o + r_i)(p_1^2 - p_2^2)}{12 \mu (r_o - r_i) p_o} \right] X$$

$$\left[1 + \frac{12.76 \, \epsilon p_o \lambda o)}{(p_1 + p_2) h} \right]$$

λo = molecular mean free path at standard conditions.
ϵ = correction factor, 0.9 for a single gas and 0.66 for a mixture.

The problem is that the leakage test gives no indication of whether the leak path is one large scratch or the sum of leakage through millions of tiny tortorous paths. Refinishing the joint may eliminate the first cause, but the condition may already be at the practical limit of seal tightness for the latter.

Type Of Gas Flow Through Seat Joint Leakage Pattern

The following is a summary of characteristics for various leak conditions.

Plug Position	*Type of Flow*	*Minimum Restricting Dimension in Leakage Path*
cracked open	nozzle flow	> 0.005 in.
seating load build up zone	turbulent channel flow	0.0005 to 0.005 in.
valve seated and leakage begins	laminar flow	0.0001 to 0.0005 in.
joint surface waviness, then roughness provides leak paths as deformation begins	transition flow (molecular and laminar)	0.000001 to 0.0001 in.
elastic and plastic yielding of joint has closed large paths	molecular flow	<0.000001 in.

When sealing to the point of achieving molecular flow, the lighter, smaller and higher velocity molecules, such as helium and hydrogen, will have the highest leakage rates.

Determination of Seat Leakage Rate

Determining the Seat Leakage Rate is not as simple as it may appear. Fluids are subject to thermal expansion and contraction during tests and air may go in or out of solution, causing volume changes in the downstream measuring system.

For accurate testing, maintain the valves and the fluids at an equal temperature and air dissolved in water at equilibrium conditions.

Helium leak detection requires a clean background with large amounts of fresh air for maximum sensitivity.

The following is a summary of the sensitivity of various test methods, given in cc/s at atmospheric conditions.

air bubbles in water (also air and soap bubbles)	1×10^{-3} to 1×10^{-4} cc/s
thermal detectors	1×10^{-4} cc/s
halogen detectors	1×10^{-5} cc/s
mass spectrometer using "sniffer" (helium leak pick up probe)	1×10^{-6} to 1×10^{-8} cc/s

Soft Seating

Resilient composition sealing materials are used to obtain bubble tight sealing with a small actuator force. Compressive seating stresses are such, that the material is elastically deformed into the surface roughness of the mating metal part, to block all leak paths. The permeability of the material, to the fluid, is the source of a very minor leakage.

Materials which are too soft, or that tend to cold-flow (creep) under load, may be stiffened with fillers such as glass. When used in thin sections, and adequately retained, the cold-flow or permanent-set problems may be eliminated.

Seals must be carefully restrained against rupture and blow-out by differential pressure. Several designs of soft-seat retention are illustrated by Figure 74. The bonding of seats to metal parts is an aid, but not a total solution, because bonds are subject to thermal shock cracking and to degradation. Sufficient pressure drop will rupture the bonding material.

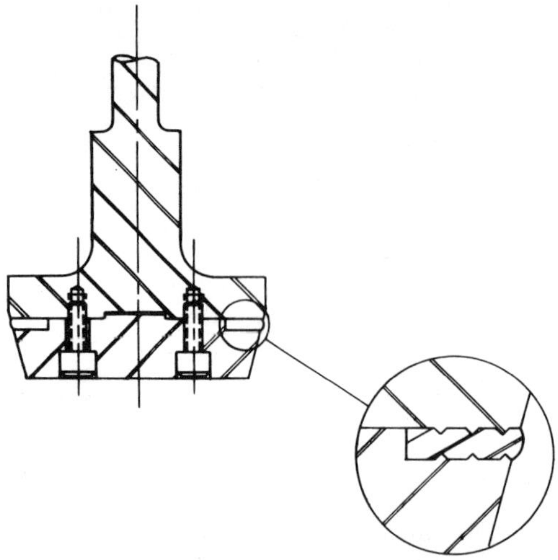

(a) Nylon seal retained in plug.

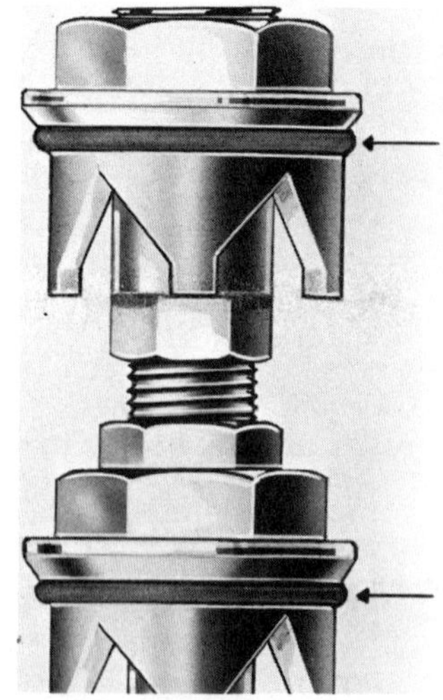

(d) Lightly restrained O-rings for low pressure drop, installed on double V-port plug.

Courtesy Fisher Controls Company

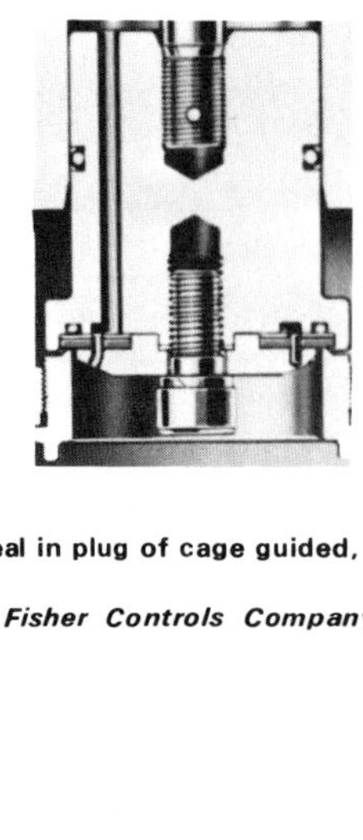

(b) TFE seal in plug of cage guided, balanced valve.

Courtesy Fisher Controls Company

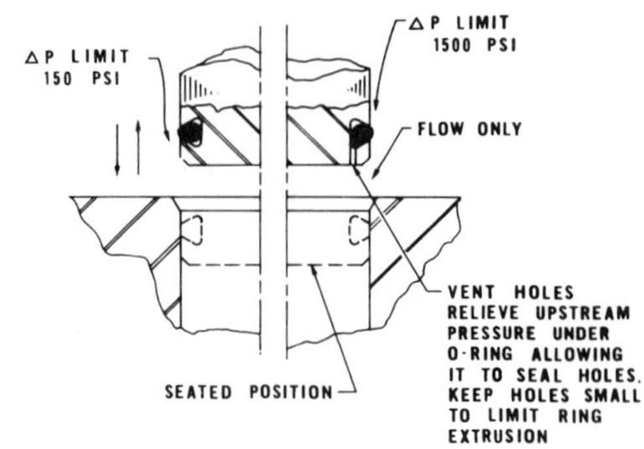

(e) O-ring in dove-tail groove for higher pressure drop. Vent holes relieve pressure under O-ring retaining it in the groove as the seat joint cracks open. Holes are small to prevent seal extrusion.

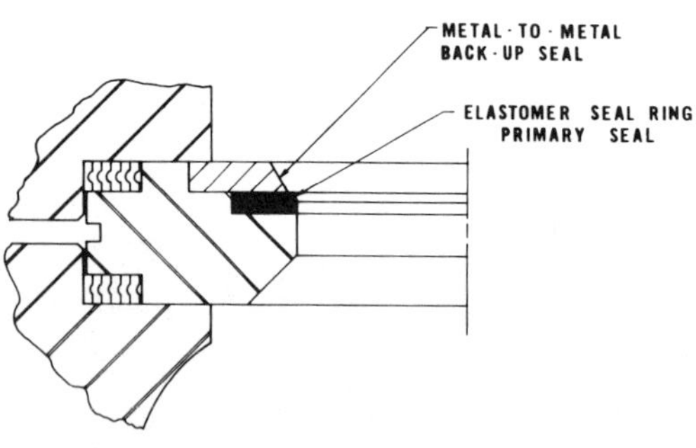

(c) Elastomer seal in seat ring of split body valve.

Figure 74. SOFT SEAT RETENTION.

A secondary advantage of soft sealing is that the seal, once compressed, is free to re-expand and to follow seat distortion as the pressure drop loading increases. To do this, the sealing material should have a rapid recovery rate upon removal of the load. This action will occur only at resilient temperatures.

Hard materials such as nylon, in thin sections carefully restrained, can handle pressure drops of several thousand psi; whereas, TFE materials are readily abraded.

Soft seals are useful for sealing where contaminants or solid material are trapped in the closed joint. Material as hard as 60D in a raised-seal beadform is capable of sealing 0.01 in. particles, bubble-tight, to 1500 psi pressure drop.

The softer the seal, the better its abrasive resistance, up to the point where it is damaged by pressure drop forces.

Large volume, high pressure blowdown systems are necessary to adequately test elastomer seals and retention means for pressure drop strength. Elastomers, under high loading, tend to act as a fluid and extrusion may occur unless the load is limited or unless a metal-to-metal stop is used. Some joint designs allow soft sealing first, followed by a metal-to-metal closure as a secondary seal in case of soft seal rupture.

The material properties to be considered in the selection of a soft seat are:

1) Fluid compatibility including, swelling, loss of hardness, permeability, degradation;
2) Hardness;
3) Permanent set;
4) Rate of recovery upon removal of load;
5) Tensile and compressive strength;
6) Distortion before rupture;
7) Modulus of elasticity.

Rarely do the physical properties given for sealing materials relate to the actual conditions of loading and strain in valve seals. There is no substitute for thermal and blow-down testing as the means to prove material, seal configuration and joint design.

SEAT ATTACHMENT AND SEALING TO BODY

Seat attachment and sealing to the body is a major consideration of valve sealing, equal in importance to joint seal, bonnet seal, and stem seal. Lack of seat-to-body sealing gives a continuous leak, often blamed on the seat-to-plug joint. In high pressure and/or steam service, leakage behind the seal will actually erode through the seat or the body bridge wall, creating an enlarging leakage path.

The plug seating load should never be applied directly to a seat-to-body compressible type gasket without a compression stop, or repeated closing will eventually loosen the seal. This also applies to the split body in which the body-joining bolt load should never be applied in this manner. A good design is for the gasket to be shouldered in a metal step, limiting the compressive loading.

Spiral wound gaskets of metal with asbestos and TFE fillers used in a metal step design require a high compression force which increases the bonnet bolting load. This is a particularly important factor where more than one gasket is required in a trim stack. The asbestos filler loses its water of crystallization above 1000° F, turning to powder; therefore, for higher temperatures, ceramic fillers may be used.

Flat metallic gaskets should not be reused if the material is work hardening. This is particularly true for stainless steel.

Split body and cage designs allow the use of a simple, non-threaded seat ring.

Various means of seat attachment are illustrated and explained in Figure 75.

a -

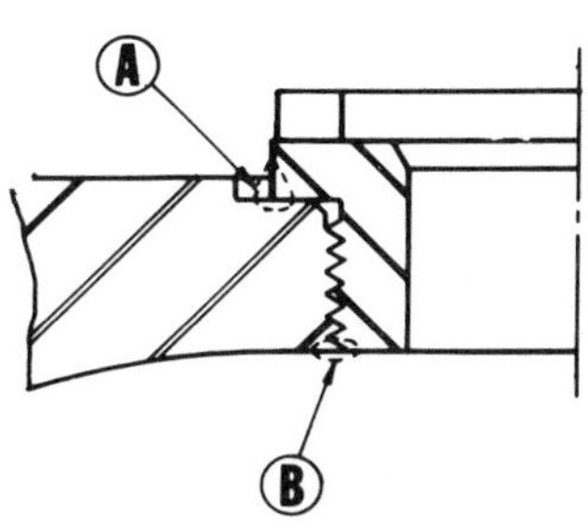

	Type and Applications	*Seal Means*
a	Screwed. Standard design up to 12 in., -150° to 450°F (seal-welded above 400° and below -50°F). Limited pressure in large sizes.	Flat metal-to-metal joint (May be seal welded at (A) or (B) for either low or high temperature service).

Figure 75. MEANS OF SEAT ATTACHMENT TO BODY.

		Type and Applications	*Seal Means*
b -	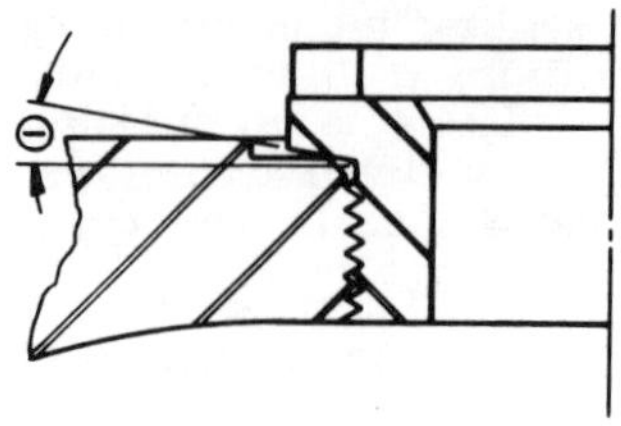	Screwed. 1/4 to 2 in., -50° to 400°F. Much lower seating torque required.	Metal-to-Metal, differential angle.
c -	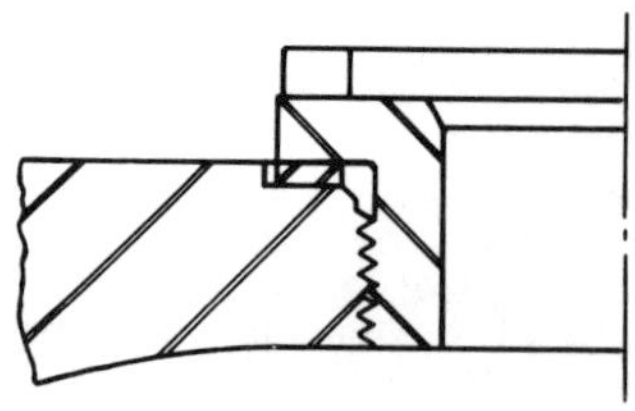	Screwed. 1/4 to 2 in., -150° to 850°F.	Soft metal, flat gasket.
d -	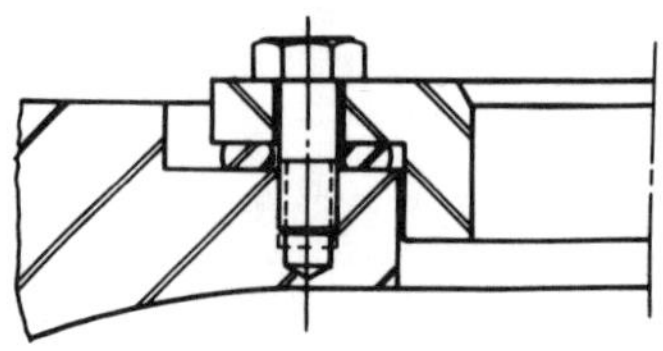	Bolted. Used in large size gas regulators to speed seat replacement, -20° to 160°F.	Flat fiber or elastomer gasket.
e -	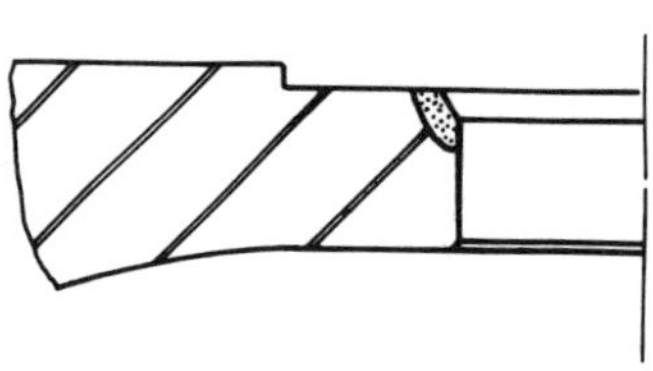	Integral.[1] All sizes and temperatures. General and slightly abrasive service.	Hard faced joint.
f	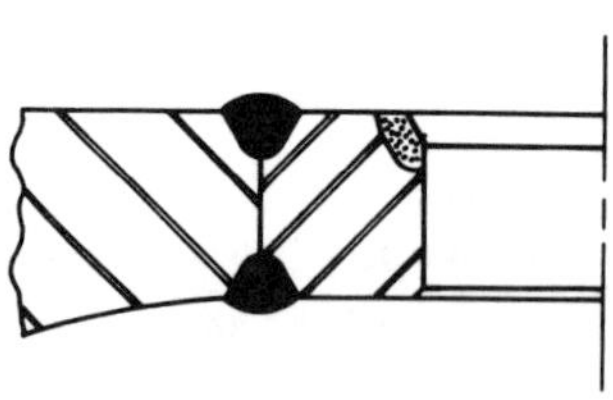	Integral welded ring.[1] Used for small restricted ports by some manufacturers.	Hard faced joint.

Figure 75. MEANS OF SEAT ATTACHMENT TO BODY (continued).

[1] NOTE: Seat to body sealing load is separate from plug seating load; improves seat-to-body seal life.

		Type and Applications	*Seal Means*
g	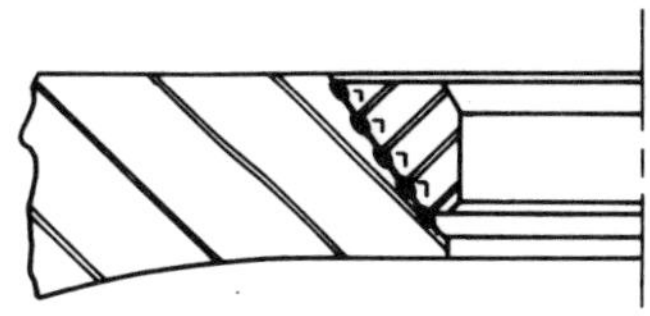	Integral. 1/4 to 2 in., all temperatures.	Projection welded. Stellite seat insert.
h	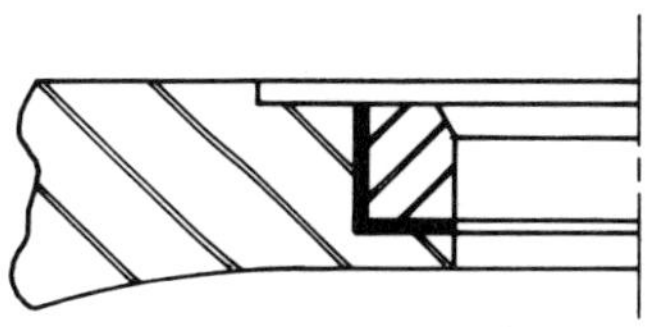	Brazed.[1] 1/8 to 1/2 in., all temperatures.	Brazed in seat bore step in body.
i	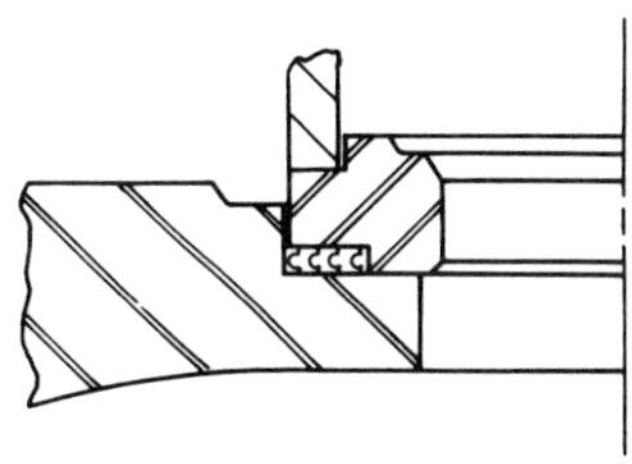	Gasketed.[1] To 1100°F.	Spiral wound gasket, metal O-ring, or flat gasket.
j	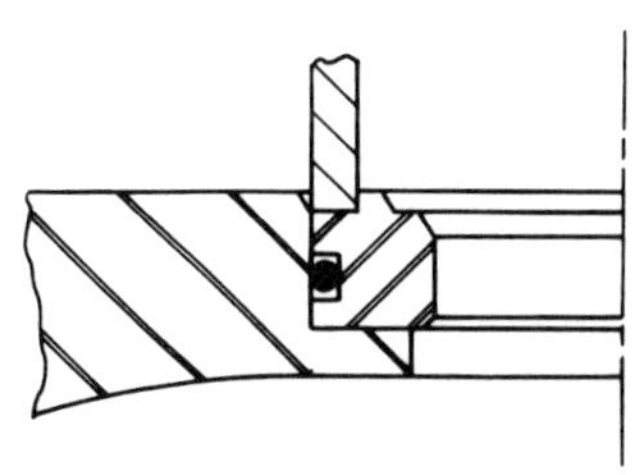	Radial.[1] Limited temperatures, -20° to 350°F.	O-Ring, elastomer.
k	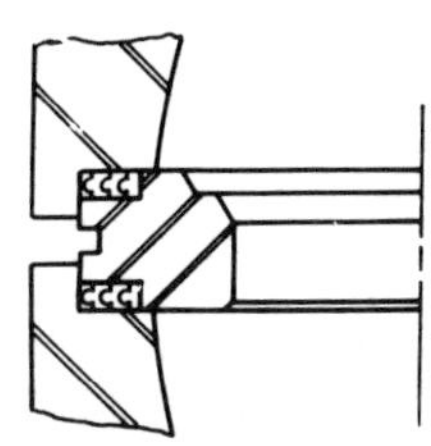	Split Body.[1] All sizes. See body rating temperatures.	Spiral wound, or flat gasket.

Figure 75. MEANS OF SEAT ATTACHMENT TO BODY (continued).

[1] NOTE: Seat to body sealing load is separate from plug seating load; improves seat-to-body seal life.

	Type and Applications	*Seal Means*
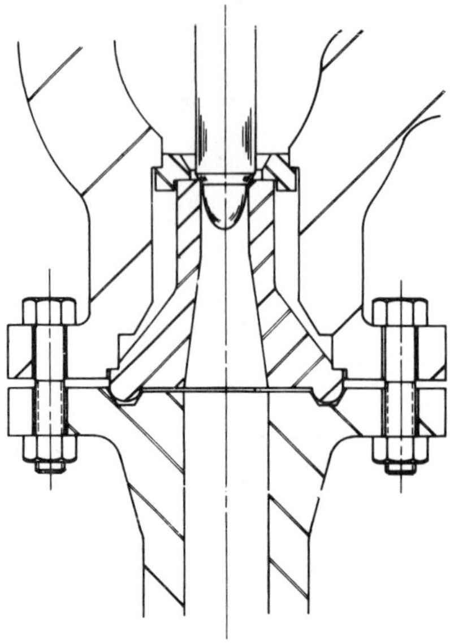	Bolted in flange.	Half-oval gasket. Shape of tail piece mates with a straight-side tapered groove of the ring-joint style, pipe flange.
	Held down by threaded ring.[1]	Any gasket (usually spiral wound metallic).

Courtesy Copes-Vulcan, Inc.

Figure 75. MEANS OF SEAT ATTACHMENT TO BODY (continued).

THREE-WAY VALVES

Diverting or Mixing Trim

Three-way valves with diverting or mixing trim have two plugs fixed to the stem and positioned so that, when one seat is closed, the other is full open, with throttling of flows through both seat orifices for in-between positions.

For mixing service, the plugs are set back-to-back in order for the flow direction of each stream to be under its plug. In diverting service, the position is reversed so flow remains under. V-port plugs are geneally furnished to reduce the amount of unbalance forces, both in regular and reversed flow directions. (See Figure 76.)

The plug characteristics are linear and the controller adjusts the valve until the proper ratio of flow is obtained. As flow rate changes in each leg of the process, so does flow resistance in that leg. Thus, it is not possible to maintain a given ratio of resistance between legs and further plug characterizaton is of no value.

Three-way valve trim is subject to thermal stress in temperature control applications. In natural gas dehydration applications, when switching dessicant towers, one flow may be 60°F while the other is 600°F. When removing LPG products from natural gas, the range may be -40° to 120°F.

Conventional three-way valves are limited in their pressure drop performance, because a high seating force is required, in opposite directions, to shut off both ports.

Cage balance trim design overcomes these difficulties and allows shut-off of

[1] NOTE: Seat to body sealing load is separate from plug seating load; improves seat-to-body seal life.

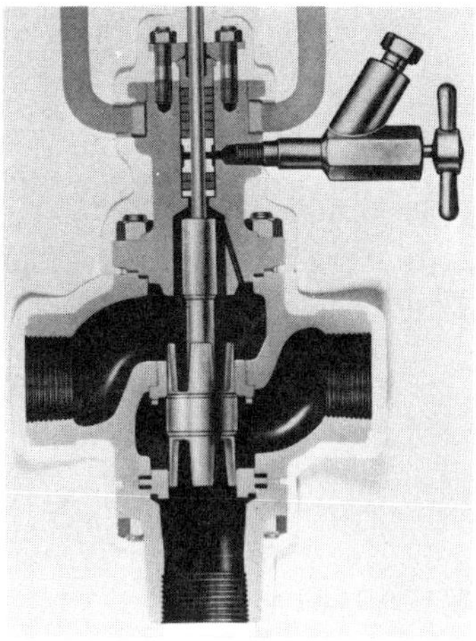

(a) Mixing service plug type.

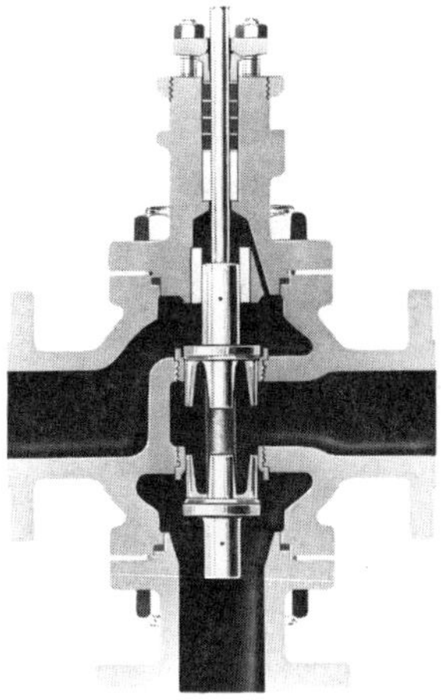

(b) Diverting service plug type.

Figure 76. THREE WAY VALVE TRIM.

Courtesy Honeywell, Industrial Division

either port, with flow in either direction, to pressure drops equal to body rating. (Refer to three-way valves in the cage guided, balanced trim section.)

Special three-way plug designs are shown in Figure 77 and 78. The valve of Figure 78 maintains a constant total flow (the sum of flow from both parts), over a broad range.

Three-Way, Three-Position Trim

Three-way, three-position trim as illustrated by Figure 79, consists of two plugs sliding on, and sealed to, the stem with a spring between them that forces the plugs to simultaneously seat on both seats, without interfering with each other. A shoulder on the stem pushes down on the top plug to open as the stem slides through the seated bottom plug, and vice versa, on opening the bottom plug while seating the top plug. Thus, one seat is closed before the other is open.

Three-way, three-position trim finds many applications in manifold service. In oil tank battery service, it either shuts-off the incoming well flow line or directs it to the production or test separator manifold. Cage balanced designs for high pressure service to 5000 psi are shown in the section on cage guided, balanced trim.

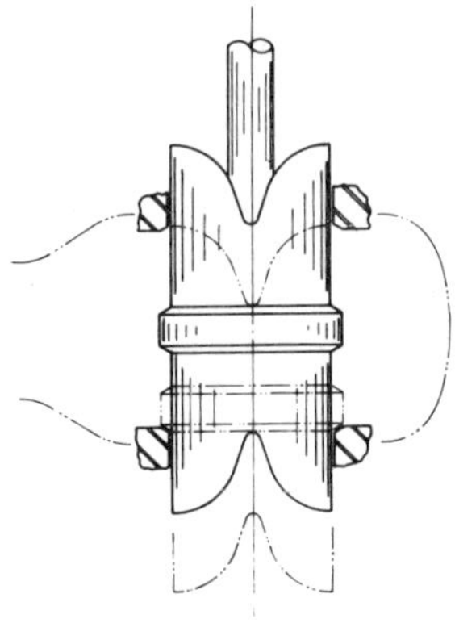

Figure 77. THREE WAY PLUG DESIGN WHICH RESTRICTS FLOW BETWEEN PORTS WHEN STROKING TO DIVERT FLUID.

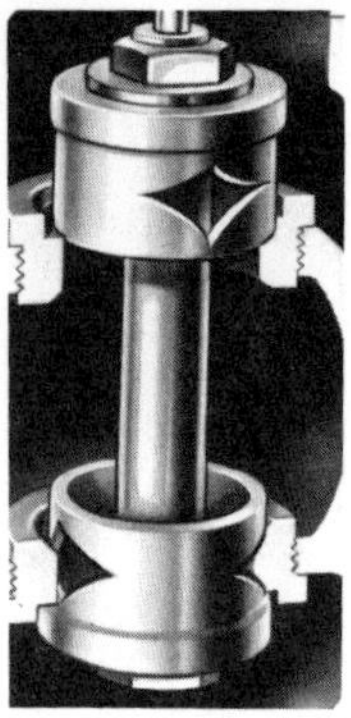

(a) Astroid plug.

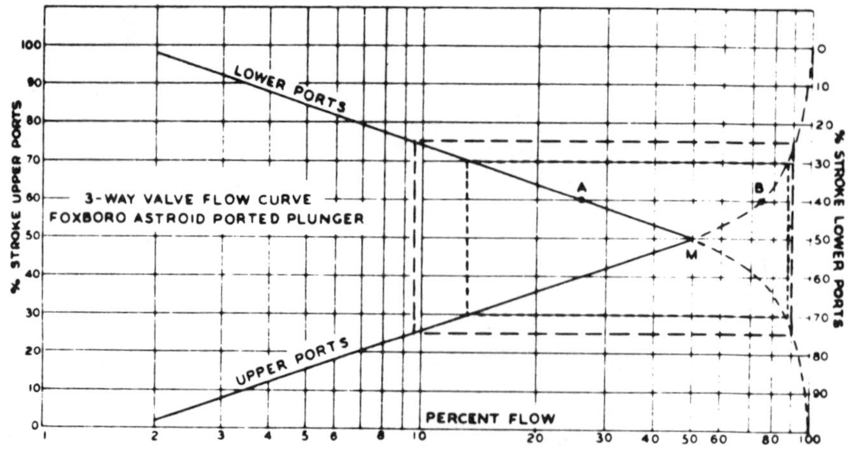

(b) Flow curves.

Figure 78. THREE WAY VALVE WITH ASTROID PORTED PLUG.

Courtesy The Foxboro Company

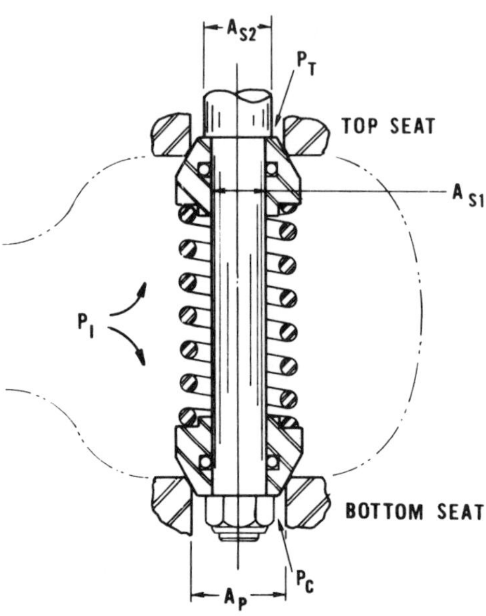

Figure 79. THREE WAY, THREE POSITION TRIM, Non-Balanced Design.

Forces Tending to Seat Plugs: $P_I\ (A_P - A_{s1})$ + spring

Forces Tending to Open Plugs:

bottom plug, $(P_c \times A_P) + P_T(A_{s2} - A_{s1})$

top plug, $P_T\ (A_p - A_{s1})$.

PACKING BOX TRIM COMPONENTS (Metal Parts Wetted by Line Fluid)

Above the top guide is the packing box (see Figure 80) in the valve bonnet neck which the stem enters through a clearance hole in the bonnet. The first part to contact the stem is the wiper ring, when spring-loaded pressure-energized chevron-type packing is used. In most cases a metal housing, with TFE and felt wiper lips, cleans the stem of dirt and abrasive particles before it slides through the packing.

If TFE V-ring seals are used, the next item above the packing box is the helical compression spring which is centered and prevented from contacting the stem by end-guiding or by the inside diameter of the packing chamber. The spring is eliminated for square packing, because it is energized by compression of the follower, rather than having an initial lip-seal as in the V-ring type.

If square compression packing is used, the wiper ring is replaced with a packing retainer ring to prevent extrusion of the packing downward into the bonnet hole clearance. This ring makes a close fit on the stem, serving as a scraper, and makes a loose fit in the packing box inside diameter. This ring is also called a "junk ring." It is usually a hardened or Stellited part so it won't gall with the stem. In some cases, a metal lantern ring may be in-

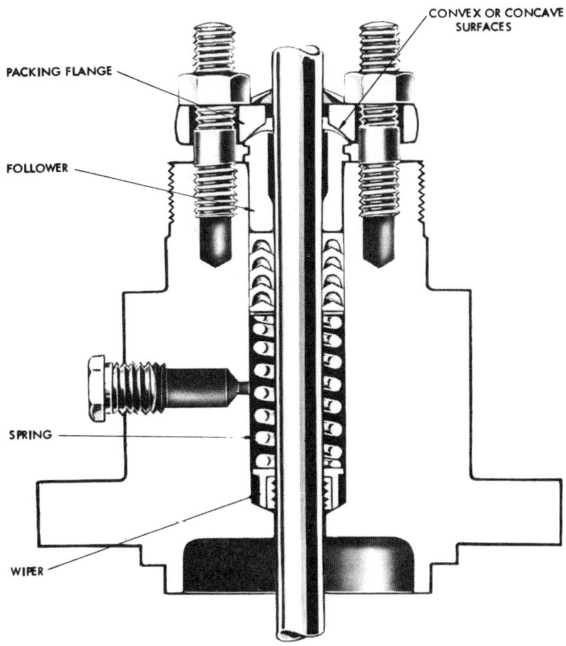

(a) Spring-Loaded, Pressure Energized Packing.

stalled between the rings of square packing in order to feed lubricant into the packing.

The packing follower, the uppermost part, is the top guide for the stem and has a narrow, close-fitting guide bearing near the bottom. The fit must be tight enough to prevent packing extrusion and "striction" or jamming of stem in the guide bearing. The follower makes a freer fit in the bonnet, letting the main stem post or cage guide center the plug in proper seat alignment. Packing intrusion does not occur in this larger clearance, because there is no relative motion.

Note, in Figure 80(b), that the follower is finished spherical on its top end so that any cocking of the packing flange, when tightening the square ring packing, cannot cock the follower guide bearing and bind the stem. The follower makes a deep engagement into the bonnet also for this reason.

Figure 81 gives the packing box materials of construction, which follow the principal trim material selection.

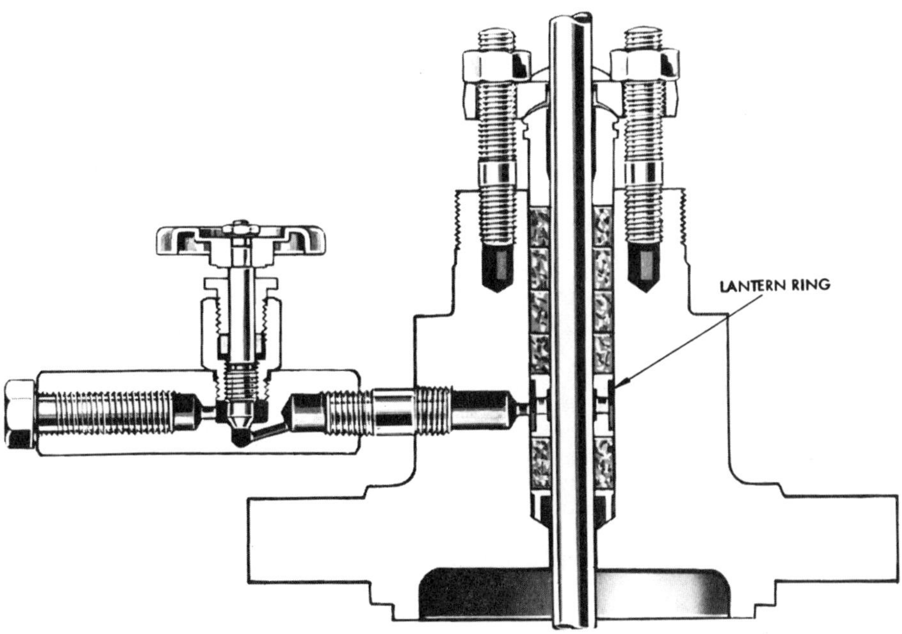

(b) Square Ring, Compression Packing Design.

Figure 80. PACKING BOXES.

Courtesy Fisher Controls Company

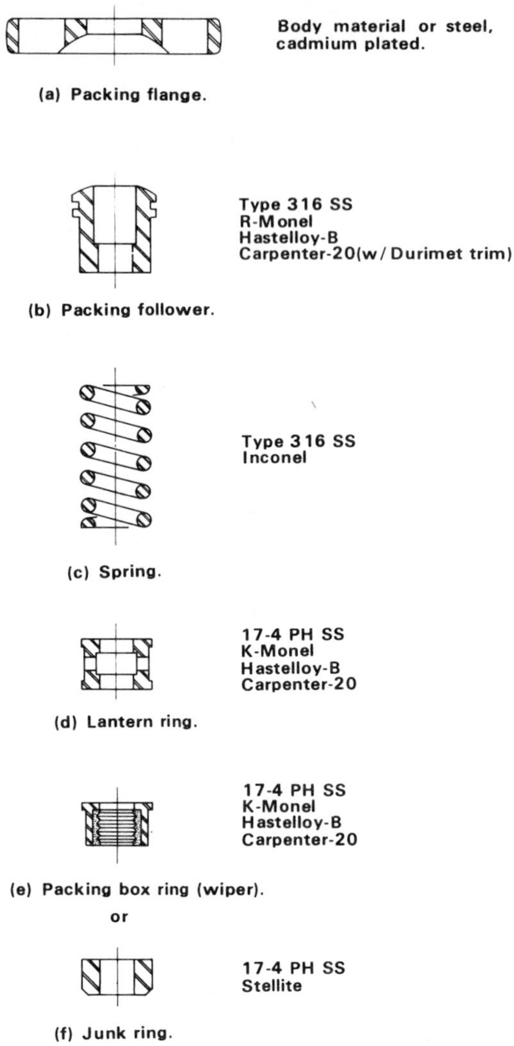

Figure 81. PACKING BOX MATERIALS OF CONSTRUCTION.

Some low pressure, liquid-level control valves (dump valves) use a rotary stem motion, placed 90° to the plug axis. Stuffing boxes seal this rotary motion with O-rings or chevron packing.

Bellows stem seals are used when no stem leakage can be tolerated or where the line fluid cannot be contained by any packing. This may be the case when the process fluid is flammable, toxic, explosive, expensive, or it rapidly destroys packing. They are also used to prevent leakage in vacuum service.

Bellows are fragile and expensive and should be avoided, whenever possible, for continuous stroking (throttling service). They are thin wall convolutions, single or multiple ply, and are subject to fatigue, cracking, and leakage if the manufacturer's rated cycle life is exceeded.

The length of the bellows usually requires an extension bonnet and a conventional packing box is added as a back-up seal. A lantern ring and lubricator valve may be added as a means of adjusting packing tightness (be sure to balance the pressure across the bellows for this test); or for providing a tell-tale leak gauge and bleed-off.

Torsional stress on the bellows from the fluid flow forces tending to rotate the plug are prevented by either:

1) a hexagonal section of the stem below the bellows which slides through a hex bushing; or by
2) a pin, extending horizontally from the stem, which slides in a vertical guide slot.

A good bellows material is A-286 stainless steel which has performed well in liquid sodium service at 50 psi and 1200 ° F for as many as 10,000 cycles. Monel is often specified and brass may be used for non-corrosive low pressure service. Regularly supplied bellows for control valve service are usually rated to 600 psi, at 750 °F.

Process fluid always surrounds the bellows tending to collapse it. Higher operating pressure may be had by counter balancing; that is, by using an inert gas pressure inside.

Two designs of bellows stem seals are shown in Figures 82 and 83.

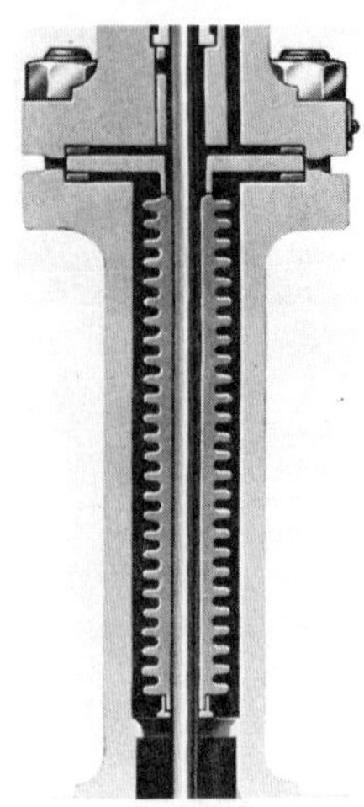

Figure 82. BELLOWS STEM SEAL.

Courtesy Honeywell, Industrial Division

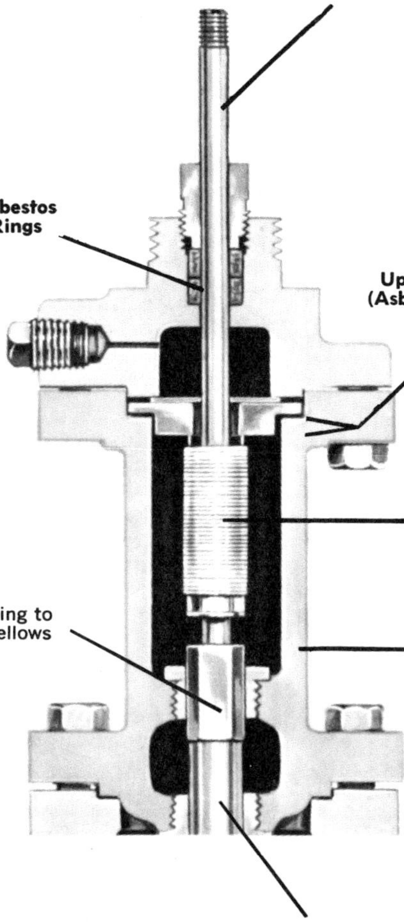

Figure 83. BELLOWS STEM SEAL WITH ANTI-ROTATION MEANS.

Courtesy Kieley & Mueller, Inc.

VALVE STEMS

The stem carries the load from the actuator to the plug, so in effect, it is a column, loaded at both ends, with horizontal restraint at each end; therefore, it will bow if overloaded. The designer's choice of stem diameter is a compromise between stiffness to prevent bowing, which causes "striction" (drag in packing follower) and results in shortened packing life, and faster bearing wear vs. the effect of line pressure thrusting on the cross sectional area of the stem. In addition, the area of the stem, contacted by the packing, is directly related to the amount of packing friction. Usually stiffness prevails and the latter considerations are counteracted by an additional actuator force requirement.

Some close-coupled, oil field designs have short stems which run very true and increase packing life. In this situation, the open-yoke, replaceable packing feature is traded for a shortened, closed barrel, which protects the stem in an oxygen environment, from corrosive elements such as salt water and acidic gas around the valve.

Cage guided, balanced trim reduces stem thrust sufficiently in high pressure service to give a noticeable increase in packing life, because of the reduction of even very minor bowing and vibration. The stem size is also reduced by using this trim.

Attachment to Plug

Stems are usually threaded and then pinned to the plug to prevent loosening. The stem may bottom in the imperfect thread runout in the plug to make a rigid connection and the pin is usually an "upset" design to anchor tightly. This design presents a maintenance problem in replacing the plug; therefore, plug-stem combinations are often replaced as a unit.

Other designs use an elastic nut to hold the plug against a shoulder on the stem. Some designs are pinned with spring pins which allows repeated plug replacement. The stem shoulders in the plug to take the underseat load, and satisfactory life may be obtained by using Monel pins. Stainless steel spring pins may break from stress corrosion cracking in tough service applications.

Pins are seal-welded in high and low temperature service.

Surface Finish and Hardness

Surface finish and hardness are important considerations for tight sealing and long packing life, concurrent with a low rate of stem wear from packing, fluid particles, and stem guide bearings. Some manufacturers get good results by finishing the portion of the stem that slides in the packing to two or three microinches. Other prefer a finish of 8 to 10 micro inches, based on the theory that the slight surface roughness will provide a small amount of fluid that will lubricate the stem-to-packing interface and they also get good results.

Although stem wipers or scraper rings are used under the lower packing ring, some line fluid abrasive particles eventually enter the stem-to-packing interface. The stem should have a hardened surface to prevent scoring and rapid packing wear and leakage. Either malcomizing or hard chrome plating are often used for highly abrasive service.

The following is a summary of stem materials and their applications (stem pin material should be the same as the stem material).

Material	*Application*
Type 316 SS	- non-abrasive service
R and K Monel	- slightly abrasive service
hard chrome plate or Malcomized[1] 316 ss.	- abrasive service
Hastelloy-B®	- with Hastelloy-B® bodies
Carpenter-20®	- with Durimet® Bodies

QUICK-CHANGE TRIM

Quick-change trim is defined as trim which may be removed and replaced by simply opening the bonnet joint; the valve body remains in the pipe line and the unscrewing of seat rings with a special clamp on lug wrench is eliminated. The seat is sealed to the body bridge wall by either a gasket or an O-ring, and a "flow cage", which holds down the seat, is in turn loaded by the bonnet. The design temperature is limited by the gasket. Two designs of quick-change trim are shown in Figures 84 and 85.

Single and double seated valves as well as three-way valves may be of quick change design. Another design enables removal of the seat supporting cage by opening the bottom cover, then the seat, plug, and stem follow. The actuator is not disturbed in this design.

The advantage of quick-change trim is the reduction in field maintenance time required to replace leaky trim or to change the trim flow capacity. Sometimes, regular trim valves must be removed from the line and taken to the maintenance shop to remove the screwed seat rings.

Some oil field designs, of one and two-inch sizes, use a threaded hammer-lug, union nut to retain the bonnet which is sealed by an O-ring; thus reducing dowtime to a minimum.

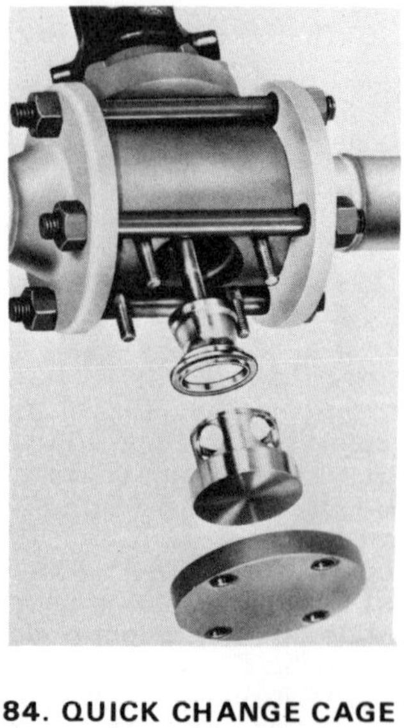

Figure 84. QUICK CHANGE CAGE TRIM.
Courtesy Fisher Controls Company

[1] A surface hardening method for 300 series stainless steel, which is similar to the nitriding of alloy steels.

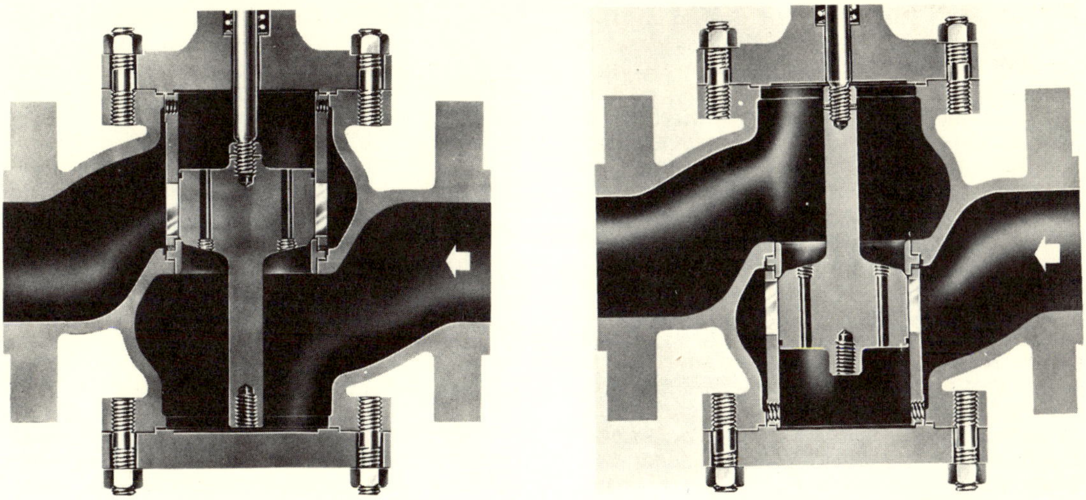

(a) Flow under plug assembly.

(b) Flow over plug assembly.

Figure 85. REVERSIBLE QUICK CHANGE TRIM.

Courtesy Kieley & Mueller, Inc.

TRIM IN NON-GLOBE TYPE CONTROL VALVES

Chapter IV primarily covers trim used in globe type valves; however, many other types of valves are frequently used and offer certain advantages over the globe style for particular applications. A full discussion of trim for these valves is not included here, but a brief discussion of the most common types is given in the following.

Ball Valves

Ball valves offer the advantages of highest flow capacity, low operating force for high pressure operation and tight shut off. The dynamic effect of high velocity flow in the solid ball design is to rotate the ball to a closed position; therefore, the actuators must counteract this effect. Floating ball designs seal tightly on the downstream seat but operating torque is from two to three times higher than for a trunnion-supported ball which seals upstream. This is due to the much smaller annular area of loading of the upstream seat. Seats in trunnion supported solid ball designs are spring-loaded and pressure-energized and account for more than one half of the operating torque; trunnion bearings plus the stem thrust bearing and the packing friction account for the remainder. This seating load is encountered in use for throttle flow and the actuators generally require positioners. Very large size valves may have pressure retractable seats to reduce the seat-wear and operating torque when throttling. The ball port may be specially shaped to improve the low-flow characteristic. Some balls are only segments of a spherical shell with a flow characterized "vee cut." They are effectively used to prevent dewatering of paper pulp mixtures. Other designs employ an eccentric-mounted, spherical surface to remain free of seat friction torque until the moment of closure.

Balls may be hard-chrome plated and polished or may be either steel or iron-coated wih a few ten thousands of an inch of TFE which remains in the valleys of the surface roughness for thousands of cycles at full pressure drop service. Breakaway time-set torque is greatly reduced by the use of both designs, because the non-metallic seal is prevented from entering and binding in the rough surface. Seat seals are usually TFE, nylon, rubber, or other elastomers and, in large valves, are encased in metal seat rings, which in turn are sealed to the body by a separate elastomer seal. Metal-to-metal seating is used in some high temperature valves. At temperature service below 600°F a viscous sealant and lubricant may be injected into the seat joint to give bubble-tight sealing. Bearings may be of either special low friction sleeve material or cylindrical roller bearings.

Trunnion-supported ball design allows "double block and bleed," that is, shut-off of both the upstream and the downstream seats simultaneously. The body cavity may be bled to check seat tightness, replace stem packing, or to prevent the mixing of fluids in the manifold if seat leakage occurs.

Butterfly Valves

Butterfly valves are prominent in control applications, because of their compactness, and their moderate torque requirements. A flat disc, pivoted in the flow stream, seats on an elastomer body liner for tight shut off. Many designs seal tightly to 175 psi WP and special seals are used for higher pressure. The linear may be either replaceable or bonded to the body wall. The valve stem axis may be either normal or canted with respect to the pipe axis. The seats may also be canted. Canted design presents less of a problem in obtaining a tight seal where the stem enters the liner, but requires a more difficult actuator mounting for the canted stem design. Stems are usually pinned to the disc and the pins must be sealed. Offset discs eliminate this problem and will turn free, once they clear the seating zone. Disc stem pins are kept high on the shaft, near the operator, to reduce torsional strain and dead band in valve operation and for the same reason dual pins must be close together so the lower pin will take its share of the load. Disc design may be modified to reduce the peak dynamic operating torque that occurs when approximately 30° to 40° open in a flow control installation. At constant pressure drop, this peak occurs between 70° to 80° (70° for a thin disc). Some designs seal at higher pressures by pressure energizing the liner against the closed disc using a separate pressure source.

Many non-metallic materials are used for liners, including the TFE's which are often energized by compressed rubber rings on their back sides. Some designs have a sealing element in the disc, thus, requiring a canted stem. Special means of seal retention allow these designs to work at much higher pressure. For high temperature service, a clearance may be allowed or a metal expansion-spring type piston ring may be placed in the disc, to give a better seal.

Digital Signal Operation

Digital signal operation of conventional valves may be had by using a device which changes the signal to either an air pressure or a mechanical loading. A unique valve design now accepts the parallel binary information directly to open or close a number of flow control orifices. Eight flow orifices, geometrically progressing in area, may be controlled by an eight-bit binary word for a 255 to 1 rangeability of flow control. Extremely fast response is assured by pilot solenoid-operated, poppet valves which are pressure balanced (20 milliseconds fully open to closed). These are located at the circumference of a flow tube. Orifices are graduated in size and are in multiples for precise flow control.

Sliding Gate Diaphragm Valves

Sliding gate diaphragm control valves employ the principle of a fixed plate with orifices and a spring-loaded sliding gate with orifices. When both sets of orifices coincide, flow is at maximum; when they are offset, the valve is closed. This provides a short-stroke, highly responsive valve. The gate and plate are self-cleaning and provide a self-lapping action with stroking. The flow characteristic is linear and much of the turbulence is reduced by the straight through flow; noise and wear on internal parts are also minimized. Thermal change does not cause leakage, because both the gate and the plate are free to expand, yet remain in close contact. Overlap of the closed seat-joint provides a long sealing path and the lapped surfaces seat tightly. The opening force is the product of the coefficient of friction and the pressure drop loading on the orifice area; thus, the operating force requirements are lower than for a non-balanced globe design of similar orifice area. The surface of the stationary orifice plate, which is in sliding contact, is hard chrome-plated, which effectively resists fluid erosion and gate wear (illustrated in Figure 9g).

In-Line Valves

In line valves are compact designs in which the flow passes through an annular space surrounding the plug. Some designs anchor the plug to the body wall with struts, and shut-off and throttling is done with a sliding seat, moved by an annular piston actuator. This presents a nearly pressure balanced design. A second version of this design incorporates a rubber plug which is compressed axially, forcing a radial expansion which throttles or seals the annular flow-way. A hydraulic cylinder in line with the plug serves as an actuator. It, in turn, is actuated by an air-operated diaphragm actuator, driving a free floating hydraulic piston to force oil into the cylinder attached to the compressible plug. These valves are frequently used in high pressure gas reduction (1000 psi down to 40 psi in one stage) and are much quieter than conventional globe style regulators. A third type of design uses a traveling, slotted tube, attached to a plug with a seat fixed to the body wall.

Pinch Valves

Pinch valves squeeze a heavy-wall, reinforced elastomer tubing section between the smooth bars of a pinch clamp connected to the actuator. Shut-off is tight and the valve cannot be jammed by a build up of solid material, because precipitates are broken away with each closing, by the flexure of the elastomer tube. It is an excellent valve for slurries.

Plug Valves

Plug valves are noted for tight shut-off and high flow capacity. Their ports may be either round, rectangular, or shaped to characterize the flow. The plugs may be solid cylindrical, split, or tapered. To seal a tapered plug, the plug must be mechanically loaded into the taper, usually a 9° included angle, in metal-to-metal joints. Smaller taper angles bind, whereas with larger angles, the plug may pop out of the taper upon standing, under a pressure drop load. Metal jointed valves usually require either some type of anti-friction plug coating or a viscous lubricant-sealant to seal tightly and maintain a reasonable operating torque. Many plug valves for corrosive service use TFE body liners or seat ring type seals, located within the plug or body wall. In cylindrical plug design, the seat rings are both spring and pressure energized, permitting an upstream seal.

Saunders Type Valves

The Saunders type valve is a straight-through construction with tight closure effected by pressing a flexible diaphragm against a weir formed in the body wall. Other designs in this category force a metal reinforced elastomer plug into contact with a shaped section of the body wall to shut off flow. Diaphragm materials may be either rubber, TFE, or Kel-F. Their temperature range is -20° to 400° F, with pressure to 150 psi. Throttling is obtained by having the actuator force a specially shaped metal shoe against the diaphragm to force it toward the weir. These valves find many applications in water filter switching and in water cooling tower service. The addition of body linings make them suitable for some chemical services.

Sleeve Valves

Sleeve valve types generally consist of a slotted section of tubing with a barrier plate between two rows of axial slots. An elastomer sleeve, overlying the slots, is actuated by line fluid and seals the sleeve against the tube. Flow is throttled by the clearance between the sleeve and the tube as the sleeve is forced outward from the tube wall by upstream pressure. Many applications for sleeve valves are found in gas pressure-reducing service.

ACKNOWLEDGMENT

This chapter is a summation of information on valve trim covered in the technical writings of the many authors listed in the bibliography and also the un-named writers of valve catalogs and bulletins. To this the author has added from his personal experience in development, production design, field testing and application of conventional control valves as well as stop, ball, plug, and gate valve types. The section on Cage Guided Balanced Trim is primarily his own effort, but much of the information in the other sections is knowledge accumulated over the years through the untiring efforts of many engineers and technicians who have worked with control valves. The author expresses his appreciation to the Rockwell Manufacturing Company for their generous assistance and permission to publish this information. Particular credit for art work is due Messers Dwight P. Bey, Melvin A. Roberts and David J. Diemert. Secretarial and stenographic credit goes to Miss Sherry Donovan and Mrs. Sandra Pumilia. Thanks is also expressed to my wife, Lucille, who expressed patience and encouragement throughout preparation of the chapter.

REGISTERED TRADE NAMES USED IN CHAPTER 4

17-4 PH (stainless steel)	ARMCO Steel Corporation
Stellite and Hastelloy	Haynes Stellite Cabot Corporation
Colmonoy	Walls Colmonoy Corp.
Nickel K/Monel S/Monel Inconel/X	International Nickel Company
Durimet	Duriron, Inc.
Malcomizing	Chapman Valve Co.

REFERENCES

STANDARDS

1. Morgan, J.P., October, 1968, "Control Valve Standards," Conference Paper No. 68-992, *Advances in Instrumentation,* Vol. 23, Part 2; Instrument Society of America, Pittsburgh.
2. *Diaphragm Control Valve Terminology,* ASME Std.-112, 1961.
3. *Standard Classification and Terminology for Power Actuated Valves,* 1963, FCI-55-1, Fluid Controls Institute, Inc.
4. Hydrostatic Testing of Steel Valves, 1961, SP61, Manufacturers Standardization Society.
5. User Engineering Specifications: (a) Engineering Specification; *Instruments and Control Systems, General,* 34B-25, 1961, American Oil Company (internal publication). (b) Instruments Standard Engineering Specification-*Control Valves, Air Diaphragm Operated,* SR121V, 1963, E. I. DuPont de Nemours & Co.

GENERAL

6. Loflin, C.A., Jr., October, 1966, "The Complete Control Valve", *Conference Paper,* No. 11.4-1-66, Instrument Society of America, Pittsburgh.
7. Beard, C.S., May, 1966, "So You Need a Control Valve," *Gas;* 42, No. 5:86.
8. Beard, C.S., 1957, *Control Valves,* 621, 812, Instruments Publishing Co., Pittsburgh.
9. Beard, C.S., 1969, *Final Control Elements (valves & actuators),* Rimbach Publishing Division, Chilton, Co., Philadelphia.

FLOW CAPACITY

10. Driskell, L.R., June, 1967, "Practical Guide to Control Valve Sizing, "Instrumentation Technology; 14: 49-58.
11. Schuder, C.B., February, 1968, "How to Size High Recovery Control Valves Correctly," *Instrumentation Technology;* 15:46.
12. Bauman, H.D., 1963, "The Introduction of a Critical Flow Factor for Valve Sizing," *ISA Transactions;* 2:107.
13. Wing, P., Jr., August, 1964, "Why Use Velocity Limits in Selecting Valves?," *ISA Journal;* 11:53 Instrument Society of America, Pittsburgh.
14. Boger, H. W., Jr., October, 1966, "The Control Valve Body - A Variable Flow Restriction," Conference Paper No. 11.4-2-66, Instrument Society of America, Pittsburgh.
15. Rockwell, R.A., April, 1954, "The Valve Flow Coefficient Cv and Its Contribution to Control Valve Sizing." Honeywell Publication SP-89.

FLOW CHARACTERISTICS

16. Lisk, R.C., 1967, "Control Valve Characteristics Made Easy," TM-19, Fisher Governor Co.
17. Schuder, C.B., March, 1967, "Control Valve Characteristics," *Instruments and Control Systems;* 40:79.
18. Lovett, O.P., Jr., September, 1963, "Valve Flow Characteristic," *ISA* Conference Paper No. 46.1.63-1; Instrument Society of America, Pittsburgh.
19. Weidmann & Rowan, 1955, "Control Valve Plug Design," ASME

EROSION

20. Morrison, R.L., March, 1966, "Control Valve for Erosive Service," *Instruments and Control Systems;* 39:99.
21. Wiedmayer, J.F., March, 1962, "How to Lessen Control Valve Erosion," *Petro/Chem. Engineer.*
22. Birdwell, Ward & Rowley, 1965, "Erosion of Various Materials by Liquir-Entrained Abrasives," ASME Paper 65-WA/PET-2.
23. Stiles, G.F., November, 1961, "Cavitation in Control Valves," Instruments & Control Systems, 34, No. 11.
24. Stiles, G.F., April, 1967, "Analyzing Cavitation in High Recovery Valves," *Instrumentation Technology;* 14:49.
25. Copson, H.R., February, 1960, "Effects of Velocity on Corrosion," *Corrosion,* 16, No. 2:86.
26. Liptak, B.G., April, 1964, "Control Valves for Slurry and Viscous Service," *Chemical Engineering,*
27. Milleville, B.J., 1961, "The Case for Higher Velocities in Feedwater Piping," Proc. of the American Power Conf., 30:475.

HIGH LOW TEMPERATURE

28. *Advanced Valve Technology for Space Craft Engines,* Report 8645-6016 RUCO, Space Technology Laboratories.
29. Fair and Barnes, September, 1967, "Cryogenic Valve Design Considerations," Conference Paper No. 34-2-CHEMPID-67, Instrument Society of America, Pittsburgh.
30. Johnson, C.M., September, 1962, *Control Valves for Cryogenic Service,* TM-10, Fisher Controls, Inc.
31. Ives, R.P., August, 1968, "Cryogenic Control Valves for Industrial Processes," *Cryogenic Engineering News,* 3, No. 8:38.

NOISE

32. Baumann, H.D., October, 1964, "Selecting Velocities of Compressible Fluids in Reducing Valves," Conference Paper No. 11.5 - 1-64, Instrument Society of America, Pittsburgh.
33. Baumann, H.D., October, 1969, "Multistep Valve Design Cuts Throttling Noise," *Instrumentation Technology;* 16:92.
34. Allen, E.E., October, 1969, "Prediction and Abatement of Control Valve Noise," Conference Paper No. 69-535; Instrument Society of America, Pittsburgh.
35. Self, R.E., October, 1968, "Why Control Velocity," Control Components Inc.

DYNAMIC FLUID FORCES

36. Brockett, G.F. and King, C.F., September, 1950, "Dynamic Force Reactions in Double-Parted Control Valves," Fisher Controls, Inc.
37. Rockwell, R.A., December, 1950, "Unbalanced Forces Reactions in Control Valve Plugs," *Instruments,* 23, No. 12:1252.
38. Lovett, O.P., September, 1961, "Control Valve Dynamics," Conference Paper No. 9-61; Instrument Society of America, Pittsburgh.
39. Stoll, H.W., September, 1961, "Selection of Control Valves Based on Process Dynamics," Conference Paper No. 16-61; Instrument Society of America, Pittsburgh.
40. Emery, J.T., "Control Valve Operating Ranges," *Instrumentation,* 20. No. 3; Honeywell, Inc.
41. Pool, E. B., September, 1963, "A New Valve for High Pressures and Temperatures," ASME Publication, 62-PET-25.

HIGH PRESSURE DROP SERVICE

42. Muller, J.T., May, 1961, "Analysis of Control Valve Problems in High Pressure Drop Service," Fourth Annual ISA Power Symposium; Instrument Society of America, Pittsburgh.
43. Teegarden, J.W., October, 1964, "High Pressure Valves," Conference Paper No. 11.5 - 2 - 64, Instrument Society of America, Pittsburgh.
44. Rosenberger, A.J., February, 1963, "Right Valves Minimize Cost," *Electric Light & Power,* 41:50.
45. Rosenberger, A.J., early 1960's, *Control Valve Engineering and Practices,* Republic Flow Meters, unpublished writings.

SEALING AND LEAKAGE

46. Baumann, H.D., August, 1966, "Should a Control Valve Leak?", *Instruments and Control Systems;* 39:111.
47. Rathbun, F. O., Jr., Report N63 18159, NASA-CR-50559.
48. *Rocket Engine Valve Poppet and Seat Design Data,* January, 1966, Report No. RPL-TDR-64-68, Rocketdyne Division of North American Aviation, Inc.
49. Heyn, W.O., 1967, "Shaft Surface Finish Is an Important Part of the Sealing System," ASME Paper 67-WA/Lub-6.
50. Bauer, P., August, 1965, *Investigation of Leakage and Sealing Parameters,* Technical Report AFRPL-TR-65-153, IIT Research Institute.
51. Pool, E.B., April, 1967, "Effect of Temperature on Sealing of Elastomers," Paper A-3, International Conference of Fluid Sealing, Cambridge, England.
52. Ives, R.P., October, 1966, "Guide to the Selection of Control Valve Actuators," Conference Paper No. 11.4-3-66; Instrument Society of America, Pittsburgh.

APPENDIX

PROPERTIES AND APPLICATIONS OF COMMON TRIM MATERIALS

Material & Composition	Temperature Range °F	Hardness as Used	Nominal Size Range	General Applications & Limitations
MILD SERVICE				*Non-erosive, non-corrosive low pressure drops, 50 psi Maximum.*
Bronze	-450° to 450°F	60 BHN	1/4 to 4 in.	Water, air, gas saturated steam.
GENERAL SERVICE				*Non-erosive, corrosive* moderate pressure drops.*
Type 316 Stainless Steel: 18% Cr, 8% Ni, 2% Mo	-40° to 750°F[1]	(14) R_c Max.	All	Steam, water, oil, and gas to 200 psi pressure drop. Cold gases to 300 psi drop.
(Most corrosion resistant of 300 series)				*For severe H_2S service specify special materials.
TOUGH SERVICE				*Higher pressure drop, erosive, slightly corrosive.*
17-4 PH Stainless Steel	-450° to 750°F	40-45R_c	All	Erosive and corrosive service (corrodes in pure water).
Type 410 Stainless Steel (plug only), 12% Cr.	-150° to 750°F[1]	37-42R_c	All	erosive service or non-corrosive service.
No. 6 Stellite® 30% Cr. 4.5% W, 3% Ni, 3% Si, 3% Fe, 2% mn, 1.5% Mo,	-450° to 1200°F	40-45R_c at 70°F 38R_c at 1200°F		
Balance Cobalt				
Cast Solid			1/4 to 2 in.	Erosive and corrosive service.
Hard Facing			1 in. and up	Slightly erosive and slightly corrosive service.

[1] Refers to seat joint. May be used at higher temperature for base trim material.

Material & Composition	Temperature Range °F	Hardness as Used	Nominal Size Range	General Applications & Limitations
No. 5 Colmonoy® Hard Facing	-150° to 1200° F	45-50R_c at 70° F	1 in. and up	Seats for erosive and corrosion resistant service.
Spray welded and fused		38R_c at 1200° F		Good impact resistance.
12% Cr, 3% B, Fe-Si-C 9% maximum, balance Ni.				Corrosion resistance similar to Inconel.
No. 6 Colmonoy® Hard Facing	-150° to 1200°	56-61 R_c at 70° F	1 in. and up	Plugs for erosive and corrosive service. Corrosion resistancc similar to Inconel.
Spray welded & Fused		(44R_c at 1200° F)		
14% Cr, 3% B, 4% Si, 5% Fe, Balance Ni (Chromium Boride crystals in nickel base)				
VERY EROSIVE SERVICE				*Abrasion, cavitation and impingement erosion.*
Type 440-C Stainless Steel	-50° to 750° F	50-60R_c	1/4 to 4 in.	Non-corrosive service.
17% Cr high carbon steel				
Hardened Tool Steel	-50° F to 450° F	60-63R_c	1/4 to 2 in.	Non-corrosive service.
Tungsten Carbide	0° to 250° F	67-72R_c	1/4 to 1-1/2 in.	Use limited by thermal shock. Brittle at low temperature. Non-corrosive service.
sintered WC with cobalt binder				

Material & Composition	Temperature Range °F	Hardness as Used	Nominal Size Range	General Applications & Limitations
VERY CORROSIVE SERVICE				*Very corrosive fluids, erosive-corrosive service*[1].
Nickel		16 R_c		Reducing solutions (hot concentrated caustic). Neutral sats.
K-Monel® 29% Cu, 1% Fe, 3% A1, Balance Ni		27-35R_c		Alkalies, salts, food products, air-free acids, reducing agents.
S-Monel® 30% Cu, 4% Mn 2% Si, Balance Ni		30-38R_c (35$_R$ at 1200° F)		Alkalies, salts, food products, air-free acids, reducing agents.
Inconel - X®	Good high temperature strength.	35-37R_c (at 1200° F)		Used for springs.
Hastelloy - B® 28% Mo, 5% Fe, 2.5% Cu, Balance Ni	1200° F	14R_c		Corrosive mineral acids, sulfuric, phosphoric wet HCl gas, oxidizing agents.
Hastelloy-C® 17% Mo, 16% Cr, 4% Fe, 2.5% Co, 4% W, Balance Ni	1200° F	23R_c		Free chlorine acid solutions of ferric or cupric salts, strong oxidizing agents.
Durimet-20® 29% Cr, 20% Ni, balance Fe				Oxidizing conditions, sulfuric acid at all concentrations.
Titanium	-320°* to 1000° F	28-35R_c (at 70° F)		Oxidizing acids, chlorides, and brines at ambient temperature.* *Do not use with liquid oxygen (may detonate).*

[1] Corrosion is a very complex process, often affected by many minor variables; therefore, specification of trim materials must be related to actual user experience in a given application. This information represents general usage areas.

CHAPTER 5

FLOW CHARACTERISTICS OF VALVES

Ralph L. Moore

E. I. DuPont de Nemours & Co.
Wilmington, Delaware

CONTENTS

INTRODUCTION

The valve plugs described in Chapter IV have been shown to be available in a variety of geometrical shapes, each with a characteristic relationship between fractional valve lift and the relative flow through the valve. This relationship is denoted as the *flow characteristic.* The proper selection of the flow characteristic is an important part of control valve application.

Flow characteristics are shown by plotting the flow through the valve at various openings versus valve stem position. Quite obviously, the flow through the valve is affected by the pressure drop across the valve or, more specifically, by the variation in pressure drop across the valve. Thus, pressure drop must be specified as a part of the flow characteristic. Valve catalog flow characteristics are shown at a constant pressure drop throughout the opening of the valve, because the manufacturer has no way of knowing the variations of pressure in systems where the valve might be applied. The characteristic at constant pressure drop is called the "inherent flow characteristic." Some of the more common flow characteristics are shown in Figure 1.

TYPES OF INHERENT FLOW CHARACTERISTICS

If the valve gain, or sensitivity, is defined as:

$$K_c = \frac{\text{change in flow}}{\text{change in lift}}$$

Then it is evident that the slope of the flow characteristic curve is the *valve gain.* Referring to Figure 1, the quick opening characteristic provides large changes in flow for small changes in lift, therefore, it usually has too high a valve gain for use in modulating control. It is limited to on-off service such as sequential operation in either batch or semi-continuous operation. For this reason the quick opening characteristic is sometimes called an "on-off" characteristic[1].

The bulk of control applications use valves with linear, equal percentage, or modified flow characteristics. The modified characteristics generally fall between the linear and equal percentage characteristics shown on Figure 1. The linear inherent flow characteristic provides a change in flow which is linear with valve lift and, thus, with the signal to the valve. The linear characteristic would seem intuitively to be the most desirable characteristic for control, since it provides constant valve gain throughout the stroke at constant pressure drop. However, inclusion of the valve into a system, with its associated piping equipment, and control loop, leads to considerations which generally make "equal percentage" the most widely applied characteristic. The equal percentage characteristic produces a change in flow, with a change in lift, that is a constant percentage of the flow before the change was made.

All valves, quite obviously, have flow characteristics and a few of the inherent characteristics[2] are shown by Figure 2.

Contrary to Murphy's Law, this figure indicates that many valve configurations have characteristics between the linear and the equal percentage types; therefore, they are conceptually useful in control application. The butterfly valve characteristic also falls within this range. The Saunder's patent valve, which has a characteristic between linear and quick opening, is also useful in control.[3]

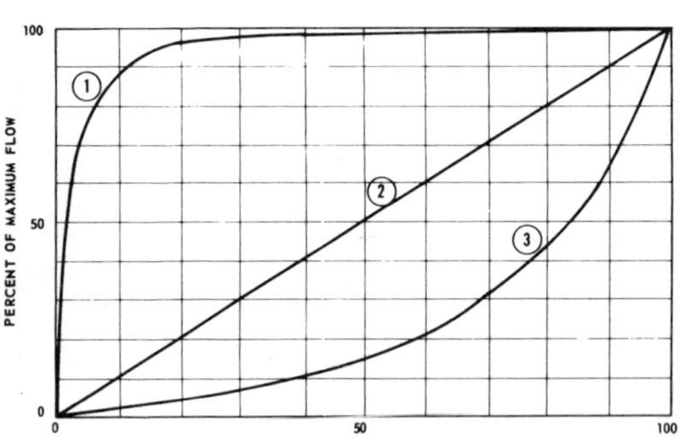

Figure 1. FLOW CHARACTERISTICS OF: 1) QUICK OPENING; 2) LINEAR; AND 3) EQUAL PERCENTAGE (rangeability = 50).

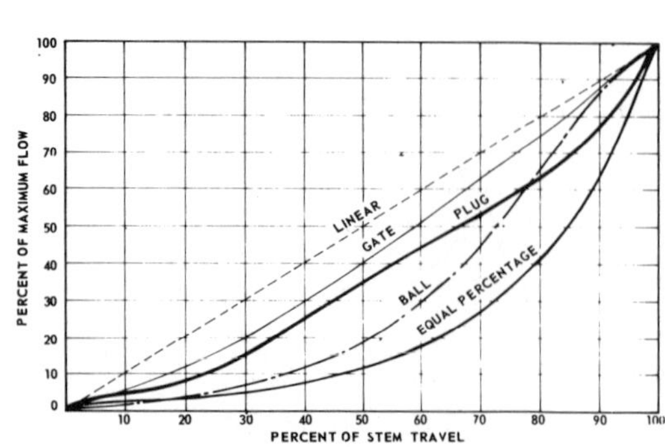

Figure 2. PERCENT OF MAXIMUM FLOW VERSUS STEM TRAVEL FOR TYPICAL VALVE TYPES.

INSTALLED FLOW CHARACTERISTICS

The inherent flow characteristic is valuable in describing the valve as supplied by the manufacturer; however, it is immediately evident that few valves actually operate at a constant pressure drop. Thus, the flow characteristic of the valve installed in the pipeline provides the basis for the proper application of the valve. This, appropriately enough, is called the "installed flow characteristic" and is unique for each valve installation; however, some generalities can be considered. Conison[6] has described the typical installation in the petrochemical industries as a centrifugal pump, supplying the pump head, the piping and the equipment in series with the valve, representing a dynamic head loss. This configuration is shown in Figure 3 where the pressure drop across the valve varies with the throughput. Kneisel[4] developed a formula for the installed flow characteristic using the simplification assumption that the pump head is constant:

$$C_o = \frac{nC_V}{\sqrt{\left(\frac{n}{m}\right)^2 + 1}} \qquad (1)$$

where, for a linear inherent characteristic:

$$m = \frac{C}{C_V} = 0.96X + 0.03 \qquad (2)$$

and for equal percentage:

$$m = \frac{C}{C_V} = (30)^{X-1} \qquad (3)$$

where

C_V = valve flow coefficient at full lift
C = valve flow coefficient at lift X.
X = valve lift, fraction of full open.
C_o = combined flow coefficient for valve and piping

$$n = \sqrt{\frac{\text{pressure drop across valve}}{\text{pressure drop across piping}}}$$

$$m = \frac{C}{C_V}$$

Equations 2 and 3 are based on a valve rangeability of 30. Successive substitutions of equations 2 and 3 into equation 1 result in the graphs of installed valve characteristics shown in Figure 4.[4] The graphs of Figure 4 show that the linear inherent characteristic takes the form of a quick opening characteristic as smaller proportions of the total-system pressure drop are taken across the control valve; therefore, the linear characteristic becomes less desirable for control under these conditions. The equal percentage inherent characteristic, on the other hand, becomes more linear with a decreasing proportion of pressure drop across the valve; therefore, it is more desirable for control. These graphs, and other similar developments in the literature[5,6,7] provide the simplest rule for flow characteristic selection: *the equal percentage characteristic is recommended if either (a) the major portion of the control system pressure drop is not available through the valve; or, (b) pressure drop is high at low flows, and vice versa.*

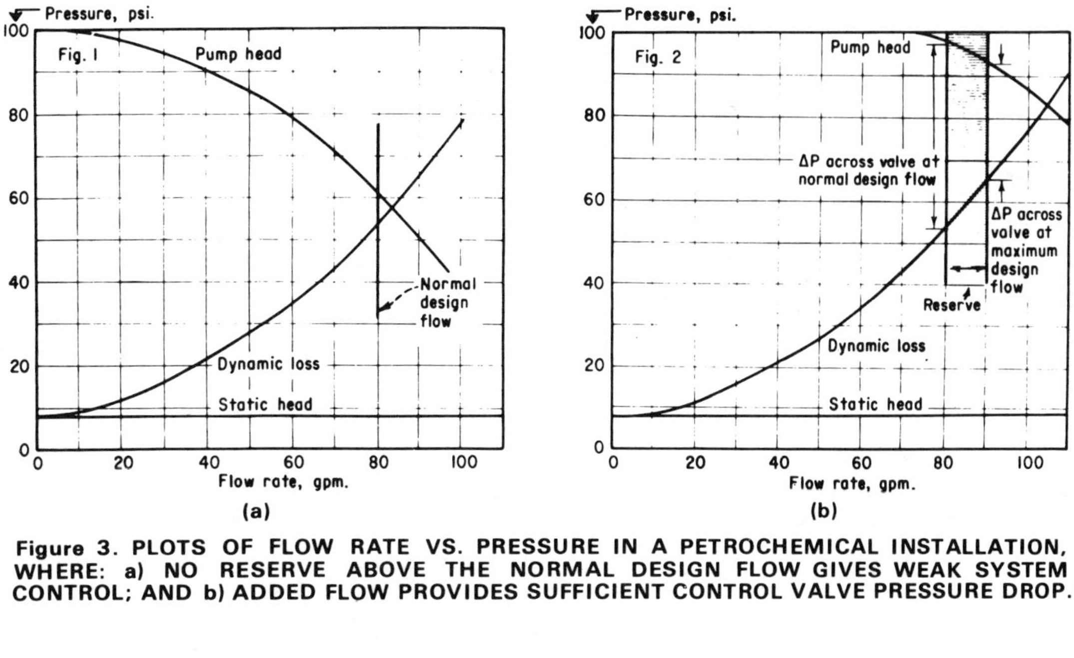

Figure 3. PLOTS OF FLOW RATE VS. PRESSURE IN A PETROCHEMICAL INSTALLATION, WHERE: a) NO RESERVE ABOVE THE NORMAL DESIGN FLOW GIVES WEAK SYSTEM CONTROL; AND b) ADDED FLOW PROVIDES SUFFICIENT CONTROL VALVE PRESSURE DROP.[6]

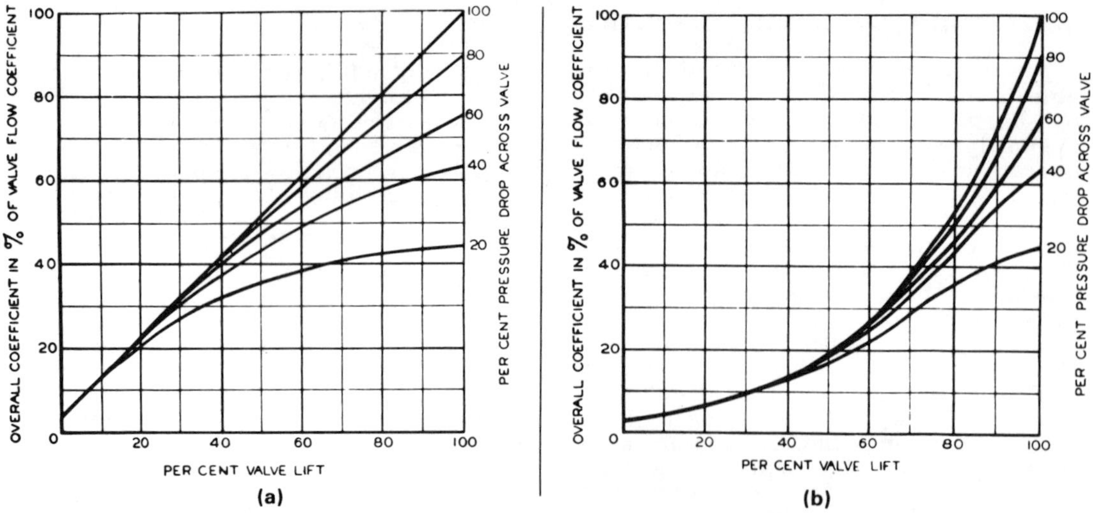

Figure 4. THE OVER-ALL FLOW COEFFICIENT IN PERCENT OF VALVE FLOW COEFFICIENT VS. PERCENT OF VALVE LIFT FOR A VALVE WITH 30 TO 1 RANGEABILITY AND a) LINEAR CHARACTERISTIC; b) EQUAL PERCENTAGE CHARACTERISTIC.[4]

Boger[5] has illustrated the effect of parallel flow paths on the inherent characteristic, which is of interest in assessing controllability under such a condition as operation with a by-pass valve open.

Consideration of the installed characteristics also emphasizes the distinction between valve rangeability and turndown. Rangeability has been described as

$$\text{rangeability} = \frac{\text{maximum controllable flow}}{\text{minimum controllable flow}}$$

while turndown is [8]

$$\text{turndown} = \frac{\text{normal maximum flow}}{\text{minimum controllable flow}}$$

Thus, turndown also decreases as the proportion of system pressure drop across the valve decreases, because less pressure drop dictates a larger valve size and a larger minimum controllable flow.

While the value of a linear installed flow characteristic in providing constant valve gain for controllability is evident, it has been pointed out that a choice of installed characteristics to provide constant open-loop gain is even more desirable. (9,10) Open-loop gain is the product of all the component gains in the control loop illustrated in Figure 5. A constant, open-loop gain results in constant attenuation of disturbances at all load conditions. Further, Buckley (11) points out that for processes with negligible process dynamics, i.e., $G_p = 1$ in Figure 5, a constant open loop gain will hold the relative stability of a system constant. A liquid flow system is given as an example. Valstar (10) and Buckley (11) suggest the hyperbolic and square root inherent flow characteristics as required to provide an approximately constant open-loop gain in some control loops. These characteristics are not commercially available in contoured valve plugs but can be obtained by characterized

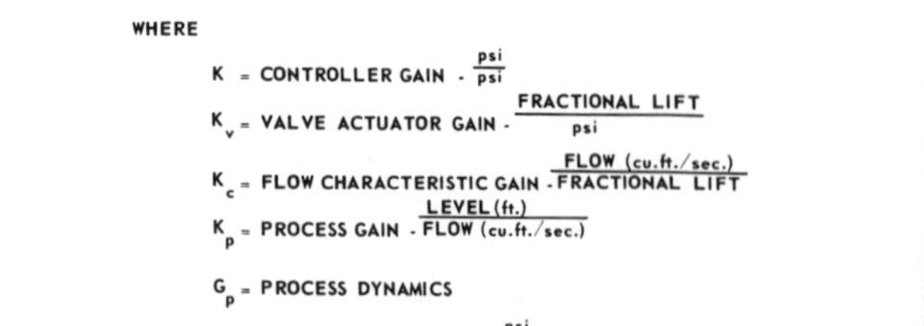

WHERE

K = CONTROLLER GAIN $\cdot \frac{\text{psi}}{\text{psi}}$

K_v = VALVE ACTUATOR GAIN $\cdot \frac{\text{FRACTIONAL LIFT}}{\text{psi}}$

K_c = FLOW CHARACTERISTIC GAIN $\cdot \frac{\text{FLOW (cu.ft./sec.)}}{\text{FRACTIONAL LIFT}}$

K_p = PROCESS GAIN $\cdot \frac{\text{LEVEL (ft.)}}{\text{FLOW (cu.ft./sec.)}}$

G_p = PROCESS DYNAMICS

K_T = TRANSMITTER GAIN $\cdot \frac{\text{psi}}{\text{LEVEL (ft.)}}$

G_T = TRANSMITTER DYNAMICS

TRANSFER FUNCTION OF REGULATOR

$$L = \frac{(K_p\, G_p)\, d}{1 + K\, K_v\, K_c\, K_p\, G_p\, K_T\, G_T}$$

Figure 5. A CONTROL LOOP TO ILLUSTRATE OPEN-LOOP GAIN.

TABLE I. OPTIMUM CHARACTERISTICS BASED ON CONSTANT PRESSURE DROPS AND ON SIGNIFICANT LINE OR PUMP DROP.

	Application	*Installed Valve Characteristic*	*Inherent Valve Characteristic* — *Constant* Δp_v	*Inherent Valve Characteristic* — *Significant Line Drop and-or Significant Pump Drop*
1.	Liquid or Gas Flow Control (negligible Process Lags)			
1.1	Orifice Measurement	$\frac{X_v}{(X_v)_{max}} = \frac{Q^2}{(Q_v)^2_{max}}$	Square Root	Linear - Equal Percentage
1.2	Linear Measurement	$\frac{X_v}{(X_v)_{max}} = \frac{Q}{(Q_v)_{max}}$	Linear	Equal Percentage - Hyperbolic
2.	Ratio Flow Control			
2.1	Open-Loop	See 1.1 and 1.2 above	See 1.1 and 1.2 above	See 1.1 and 1.2 above
2.2	Closed-Loop (Orifice or Linear Measurement)	$\frac{X_v}{(X_v)_{max}} = 1 + \frac{\text{Ln}\, Q/(Q_v)_{max}}{\text{Ln}\, \frac{(Q_v)_{max}}{(Q_v)_{min}}}$	Equal Percentage	Hyperbolic
3.	Averaging Level or Pressure Control			
3.1	Simple	$\frac{X_v}{(X_v)_{max}} = \frac{Q}{(Q_v)_{max}}$	Linear	Equal Percentage - Hyperbolic
3.2	Cascaded to Flow Control	See 1.1 and 1.2 above	See 1.1 and 1.2 above	See 1.1 and 1.2 above
4.	Mixing and Blending Control			
4.1	Measurement Exit First Tank	$\frac{X_v}{(X_v)_{max}} = \frac{Q}{(Q_v)_{max}}$	Linear	Equal Percentage/ Hyperbolic
4.2	Measurment Exit Second Tank	$\frac{X_v}{(X_v)_{max}} = 1 + \frac{\text{Ln}\, Q/(Q_v)_{max}}{\text{Ln}\, \frac{(Q_v)_{max}}{(Q_v)_{min}}}$	Equal Percentage	Hyperbolic

cam positioners. Care must be taken not to significantly alter the positioner dynamics with a characterized cam. Table I gives the optimum characteristics based on constant open-loop gain where dynamics are negligible and based on time constant ratios where dynamics are significant as developed by Buckley [11].

Brown[12] has developed a "working" flow characteristic for liquid flow control which is defined as, "the relationship between valve fractional opening and the quantity to be controlled for a constant condition of the process." A linear working characteristic thus provides a constant open loop gain. Computer calculated working characteristics [12] illustrate that, (a) different installed flow characteristics are required to maintain a constant open-loop gain in the same control loop with different disturbances, and (b) deterioration in control occurs with undersized and oversized valves. The requirement for different flow characteristics in the same loop, to compensate for different disturbances, means that no "best" flow characteristic can be selected, but that the chosen flow characteristic must be a compromise that is adequate for all anticipated conditions. Ziegler and Nichols [13] also graphically illustrate the requirement for different valve characteristics with different disturbances.

Ultimately, however, the most desirable flow characteristic is not one based on static criteria, but one which best compensates for the dynamic properties of the control loop to provide similar dynamic response at all load conditions. Thus, the relative loop stability should remain constant without controller adjustment. Consider the level control system of Figure 6. The block diagram for this process is shown as Figure 5.

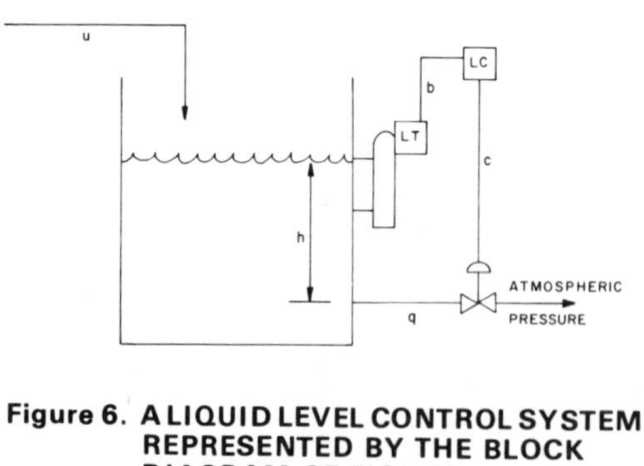

Figure 6. A LIQUID LEVEL CONTROL SYSTEM, REPRESENTED BY THE BLOCK DIAGRAM OF FIGURE 5.

The process is governed by a mass balance:

$$A\,\frac{dh}{dt} = u - q \qquad (4)$$

where

A = cross sectional area of tank, ft^2
h = height of level above drain line, ft
q = modulated flow, ft^3/s
u = disturbance flow, ft^3/s

The valve in the exit line is analogous to a variable resistance:

$$q_2 = CX\sqrt{h} \quad \text{(specific gravity = 1)} \qquad (5)$$

Letting lower case letters represent small deviations of the defined variables, and capital letters represent average values of the same variables, the linearization of equation 5 results in:

$$q = \frac{C^2X^2}{2Q}\Big|\, h + \frac{\partial q}{\partial x}\Big|\, x \qquad (6)$$

The flow coefficient, C, appears in the gain in equation 6 in order to eliminate the head, h, which varies with throughput with proportional control. Substituting equation 6 into equation 4:

$$A\frac{dh}{dt} = u - q = u - \frac{C^2X^2}{2Q}\,h - \frac{\partial q}{\partial x}\,x$$

where by comparison with Figure 4:

$$K_c = \frac{\partial q}{\partial x} \approx \frac{dq}{dx} \text{ (in differential form)}$$

then,

$$\left(\frac{2QA}{C^2X^2}\,s + 1\right) h = \frac{2Q}{C^2X^2}(u - K_c x)$$

Also by Figure 4:

$$K_p = \frac{2Q}{C^2X^2}$$

$$G_p = \frac{1}{\frac{2QA}{C^2X^2}\,s + 1}$$

and let

$$G_T = \frac{1}{0.5s + 1}$$

The closed loop response to a small change in the disturbance, u:

$$h = \frac{K_pG_p\ u}{1+\dfrac{K\ K_vK_cK_pK_T}{\left(\dfrac{2QA}{C^2X^2}s+1\right)(0.5s+1)}} \qquad (7)$$

where $K\ K_vK_cK_pK_T$ = open loop gain and $\frac{d}{dt} = s$ (the Laplacean operator).

The denominator of equation 7 is the characteristic equation which determines stability. Multiplying this:

$$\frac{QA}{C^2X^2}s^2+\left(0.5+\frac{2QA}{C^2X^2}\right)s+1+K\ K_vK_cK_pK_T$$

This second-order response equation is written in the following general form (14) to determine system stability:

$$T^2s^2+2T\zeta s+1$$

Comparing coefficients of like-order terms:

$$T=\sqrt{\frac{\dfrac{QA}{C^2X^2}}{1+K\ K_vK_cK_pK_T}}$$

$$2T\zeta=\frac{0.5+\dfrac{2QA}{C^2X^2}}{1+K\ K_vK_cK_pK_T}$$

Substituting the characteristic time, T, into the second equation, and letting $\zeta = 1$, for constant stability:

$$2\sqrt{\frac{\dfrac{QA}{C^2X^2}}{1+K\ K_vK_cK_pK_T}}=\frac{0.5+\dfrac{2QA}{C^2X^2}}{1+K\ K_vK_cK_pK_T}$$

Substituting the foregoing value for K_p, and solving this for K_c; the required installed flow characteristic for constant system stability:

$$K\ K_vK_TK_c=\frac{C^4X^4}{32AQ^2}-\frac{C^2X^2}{2Q}+\frac{A}{2} \qquad (8)$$

$$\frac{dq}{dx}=\frac{1}{K\ K_vK_T}\left[\frac{C^4X^4}{32AQ^2}-\frac{C^2X^2}{2Q}+\frac{A}{2}\right] \qquad (9)$$

Integration of equation 9 on a computer results in the installed characteristic of the valve. The gain, $K\ K_vK_T$ (which

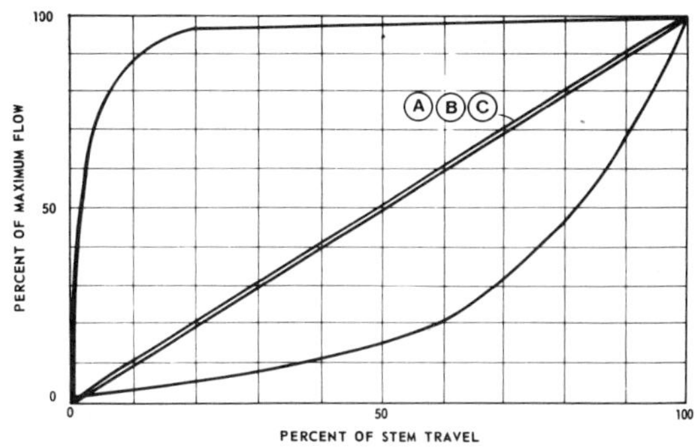

Figure 7. THE PLOTTED RESULTS OF THE INTEGRATION OF EQUATION 9, AS SHOWN IN TABLE II.

must remain constant to eliminate controller retuning), is estimated from equation 8, and values assigned to the area, A, and the coefficient, C. Values representing widely varying tank parameters and thus widely varying tank dynamics have been assigned in the following table. The results of the integration of equation 9 (values of stem lift and throughput flow) are plotted as shown on Figure 7, which is identical to Figure 1 in also showing the standard flow characteristics. Note that it would be necessary to resort to equation 1 if significant differences between the inherent and installed flow characteristics exist.

TABLE II. TABULATED SETS OF TANK LEVEL SYSTEM PARAMETERS.

Parameter	*Characteristic Curve (on Figure 7)*		
	A	B	C
h	2 ft	20 ft	2 ft
A	50 sq ft	5 sq ft	5 sq ft
q	26.8 cu ft/sec	39.9 cu ft/sec	7.9 cu ft/sec
T_p	7.4 sec	5 sec	2.5 sec
K_p	0.22	1.6	1.6
$K\ K_vK_T$	0.9 ft^{-1}	0.06 ft^{-1}	0.3 ft^{-1}

Figure 7 indicates only minor differences in the optimum valve flow characteristic for the three tabulated sets of tank level system parameters given in Table II. Further, the optimum installed characteristics are so close to being linear, that a valve with a linear characteristic would be chosen. This result conforms to the recommendations found in the literature (10,11).

The foregoing, simplified example illustrates the computational complexity in the quantitative selection of an installed valve characteristic. The complexity increases rapidly as higher order systems (such as reset in the controller) and the many possible disturbing variables are considered. Valstar [10] has dealt with this complexity by classifying the disturbances and the process dynamics, and tabulating the preferred valve flow characteristics based on these two arguments.

Quite obviously, economics dictate that most valve flow characteristics be chosen without resorting to the foregoing complex procedure, and indeed an overwhelming proportion of valve characteristics are satisfactorily chosen on the basis of the very simple rules previously listed. An even simpler rule has been suggested:[1] that is, all valves be equal percentage on the basis of (a) greater rangeability, and (b) the assumption that almost all valves have dynamic head losses in series with them which tend to convert an inherent equal percentage characteristic into a linear installed characteristic. Such a rule must be viewed as an oversimplification, however, and each application must unquestionably be given at least a cursory analysis. Ziegler and Nichols, [13] show by a series of examples that linear characteristics are superior in many common applications; therefore, they take vigorous exception to the universal use of characterized valves.

Finally, Lovett [7] has differentiated between the "theoretical" and the "actual" valve flow characteristics. The "theoretical" characteristic is described by mathematical equations and graphs and is discussed in the foregoing pages. The "actual" inherent characteristic is obtained by testing the valve after manufacture. The Fluid Controls Institute report FCI 58-2, entitled "Recommended Voluntary Standards for Measurement Procedure for Determining Control Valve Flow Capacity," contains a procedure for evaluating flow capacity, complete with test manifold design, testing procedure, and test results evaluation. This procedure suggests that valve C_V be obtained by measuring water flow rate with a nominal pressure drop across the valve (5 to 10 psi) and converting this to a flow rate at one psi pressure drop by dividing by the square root of the measured differential pressure. The actual inherent flow characteristic (relative flow vs. fractional valve lift) is obtained by repeating this procedure at various percentages of stem travel.

Figure 8 shows actual flow characteristics developed experimentally by Lovett [7] and indicates the deviation of the actual characteristic from the theoretical formulation. Lovett reports that this deviation can be as much as 40%. Thus, the precise mathematical formulation of the optimum characteristic is not only prohibitive in terms of complexity, but is also futile unless the actual characteristic of a given manufacturer's valve is known.

CONCLUSION

With the foregoing difficulties in the quantitative definition of valve flow characteristics, the characteristic is typically chosen with respect to the following guidelines:

1. The valve characteristic will have little effect if (a) controller proportional band is expected to be very narrow; or if (b) the required valve rangeability is 2 to 1 or less.

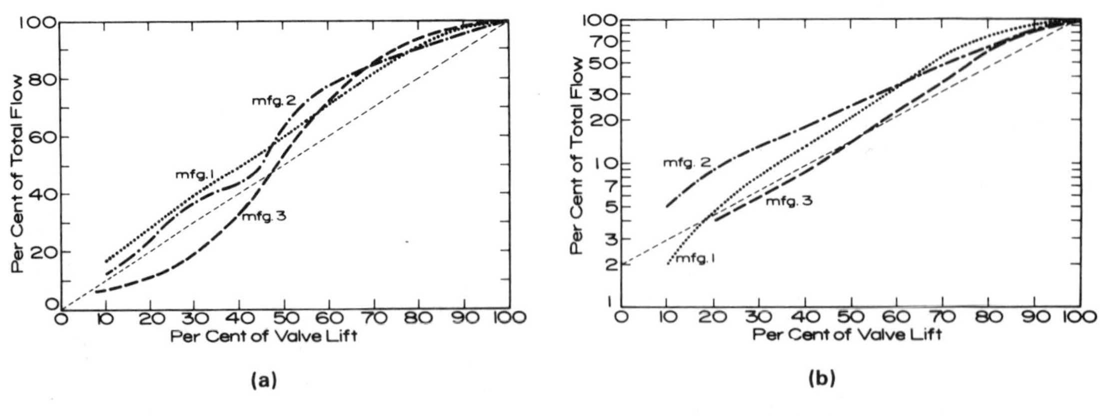

Figure 8. ACTUAL FLOW CHARACTERISTICS: a) LINEAR; b) EQUAL PERCENTAGE. [7]

2. The equal percentage flow characteristic should be used if (a) the major portion of the process system pressure drop is not available through the valve; (b) valve pressure drop is high at low flows, low at high flows; or if (c) oversizing is desirable because of limited data.
3. The linear flow characteristic should be used in flow control systems if (a) a variable head flowmeter is used; and if (b) the major portion of the process system pressure drop is available through the valve.

REFERENCES

1. Forman, E. R. and Oriolo, D.J., October, 1962, "Selection of Control Valves for Continuous Chemical Processes", Conference Paper No. 34.2.62; Instrument Society of America, Pittsburgh.
2. Ledeen, H., August, 1965, "A New Look at Valve Regulators", *Instruments and Control Systems;* 38:131-138.
3. Love, R.E., May, 1954, "Control Valves", *Instruments and Automation;* Vol. 27.
4. Kneisel, O., October, 1948, "Liquid Flow Characteristics of a Pipe Line and a Control Valve", *Petroleum Refiner;* Vol. 28.
5. Boger, H. W., November, 1966, "Flow Characteristics for Control Valve Installations", ISA Journal; 13:50-54.
6. Conison, J., December, 1958, "By Proper System Design . . . Assure Full Range Flow Control", *Chemical Engineering;* Vol. 65, No. 24.
7. Lovett, O.P., April, 1964, "Valve Flow Characteristics", *ISA Journal;* 11: 65-67.
8. Eckman, D. P., 1958, *Automatic Process Control,* p. 225, John Wiley & Sons, New York.
9. Holzbock, W. G., April 20, 1959, "How to Select Control Valve Size", *Chemical Engineering;* Vol. 66, No. 8.
10. Valstar, J. E., March, 1959, "A Fresh Look at Selecting Control Valve Characteristics", *Control Engineering;* Vol. 6.
11. Buckley, P.S., 1964, "Selection of Optimum Final Element Characteristics", *Instrumentation in the Chemical and Petroleum Industries,* Vol. 1, p. 107; Instrument Society of America, Pittsburgh.
12. Brown, G., October, 1965, "Selection and Sizing of Liquid Flow Control Valves for Normalized Working Characteristics", *Control* (British); 9:558-563.
13. Ziegler, J.G. and Nichols, N.B., January, 1949, "Valve Characteristics and Process Control", *Instruments;* 22: 75-81.
14. Eckman, D.P., 1958, *Automatic Process Control,* p. 35, John Wiley and Sons, New York.

CHAPTER 6

SIZING CONTROL VALVES

Part 1
Sizing Theory and Applications
L. R. Driskell, Blaw-Knox Chemical Plants, Inc., Pittsburgh, Pennsylvania

Part 2
Cavitation and Flashing Considerations
G. F. Stiles, Fisher Controls Company, Marshalltown, Iowa

Part 3
Noise Considerations
E. E. Allen, Fisher Controls Company, Marshalltown, Iowa

CONTENTS

PART 1 SIZING THEORY AND APPLICATIONS

(L. R. Driskell)

INTRODUCTION

The problem of sizing a control valve may be broken into several more or less distinct parts. The person sizing a valve should carefully consider the effect of these factors on valve size.

1. Flow Application Data:

   Flow rate, maximum and minimum.

   Pressure upstream and downstream (at both maximum and minimum flows).

   Temperature of the stream.

2. Fluid Data:

   Name of fluid or its generic identity.

   Fluid phase—liquid, gas, slurry, etc.

   Density (specific gravity, specific weight, molecular weight, etc.)

   Viscosity (liquids).

   Vapor pressure (liquids).

3. Piping Influence:

   Presence of reducers or other disturbances at the valve which will change the valve's normal capacity.

4. System Influence:

   Control dynamics (is oversizing unimportant?).

   Economic factors (is downstream relief valve size affected, etc.).

   Safety.

5. Style of Valve—Selection based on application:

   Capacity, order of magnitude of size.

   Rangeability.

   Corrosion or erosion resistance.

   Special requirements (tight shut-off, low noise, etc.)

6. Sizing Calculations:

   Manufacturer's sizing coefficients.

   Sizing formulas, slide rules, nomographs.

7. Judgment:

   Based on discriminating analysis of past successes and failures.

The first four items can be properly evaluated only by one who is knowledgeable about the whole process; therefore, this is usually the purchaser.

The principal responsibility of the valve manufacturer, in valve sizing, is to publish certified valve sizing coefficients. These values simply state the performance characteristics of his valves, based on standard test data and quality-control of production. In some instances a valve manufacturer may assume additional responsibilities. This usually involves sizing the valve based on data provided by the purchaser and may include the selection of valve type and materials as well. This additional engineering on the part of the vendor involves added work and liability on his part and involves some degree of risk on the part of the purchaser.

If there is divided responsibility, the application must fly without the last important ingredient—*good judgment.*

THEORY

To be a good aircraft pilot it is necessary to have the seat-of-the-pants feel of the ship. It is also important to know *why* the ship responds the way it does. For the same reasons, the art of valve sizing goes hand-in-hand with the science of fluid mechanics. It is helpful to know why valves work and what their limitations are, to properly apply the cookbook formulas.

Incompressible Fluids

A fluid flowing through control valves follows the same laws of conservation of mass and energy as expressed in the equations of fluid mechanics. First consider the flow liquids, which essentially are incompressible fluids. When any fluid flowing inside a pipe, passes through a narrower passage or restriction, it must accelerate. The energy for this acceleration must be taken from the pressure of the fluid, or the *static head.* After passing the restriction, the fluid slows down again and part of this head, or pressure, is recovered. The unrecovered part of the pressure has been converted into internal energy by friction. Figure 1 shows the pressure gradient around a valve or an orifice.

Neglecting friction and other non-ideal influences for a moment, Bernoulli's theo-

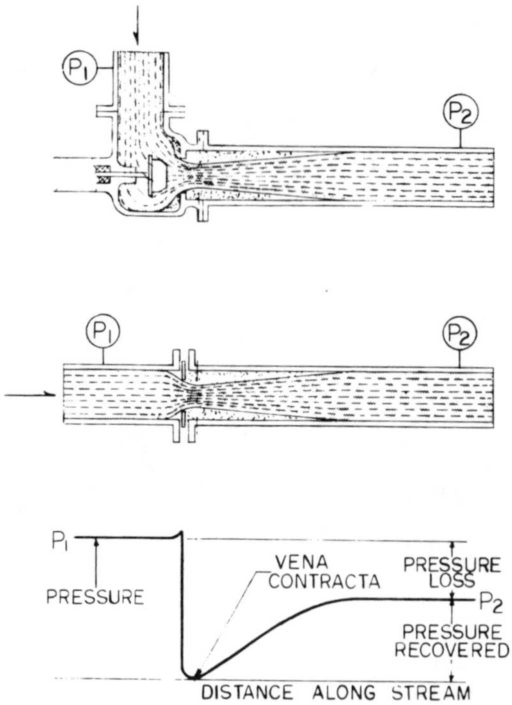

Figure 1. CONTROL VALVE AND ORIFICE PLATE INSTALLATIONS SHOWING REDUCED AREA AT VENA CONTRACTA. THE CURVE DESCRIBES THE PRESSURE GRADIENT AROUND EITHER DEVICE.

rem tells us that

$$U_2^2 - U_1^2 = 2gh \qquad (1)$$

where U_1 and U_2 are the mean axial speeds of the fluid across areas a_1 and a_2 respectively, g is the acceleration of gravity, and h is difference in head (pressure) measured in feet of the fluid.*

If area a_1 is the area of the inlet pipe, and area a_2 is the area of the stream vena contracta, as shown in Figure 1, Equation (1) can be converted to:

$$q_t = a_2 \sqrt{2g \frac{p_1 - p_2}{\gamma}} \quad \frac{1}{\sqrt{1 - \left(\frac{a_2}{a_1}\right)^2}} \qquad (2)$$

where q_t is the theoretical flow in cubic feet per second.

Since the actual flow through a restriction is always less than the indicated theoretical rate, and since the area of the vena contracta is less than the orifice area, an additional factor "C," called the discharge coefficient, must be included in the equation. This factor accounts for the contraction of the stream as well as the pressure lost to friction. The fundamental hydraulic equation then becomes:

$$q = a_2 \overline{K} \sqrt{2g \frac{p_1 - p_2}{\gamma}} \qquad (3)$$

where $\overline{K} = \dfrac{C}{\sqrt{1-(a_2/a_1)^2}}$

and a_2 is the orifice area.

It is well to be reminded at this point that the foregoing theoretical development is based on the assumption that p_2 is being measured at the vena contracta, or point of minimum stream area. If p_2 is being measured at a point farther downstream, as is customary in control valve work, another correction factor must be included.

$$F_L = \sqrt{\frac{p_1 - p_2}{p_1 - p_{vc}}} \qquad (4)$$

The relationship expressed by Equation (4) is correct only as long as the fraction of Δp_{vc} recovered remains constant. When flashing or cavitation takes place, for example, this relationship does not hold, and formulas based on p_2 must be modified.

If the fundamental hydraulic Equation (3) is converted to customary engineering units, and p_2 is being measured downstream at a point of full pressure recovery, it then becomes

$$q = \left(\frac{38.0\, a\overline{K}}{F_L}\right) \sqrt{\frac{\Delta p}{G}} \qquad (5)$$

In control valve work the value of the term in parentheses is replaced by an experimentally determined coefficient called C_v; therefore, Equation (5) becomes:

$$q = C_v \sqrt{\frac{\Delta p}{G}} \qquad (6)$$

Compressible Fluids

Liquids are nearly incompressible and, as they pass through a valve, their specific weight remains constant. Gases and vapors, on the other hand, expand as the pressure drops; therefore, their

* In this part of the discussion of theory all units are in standard absolute units, pressures in pounds per square foot, areas in square feet, etc. The working equations discussed later use customary engineering units of pounds per square inch, square inches, etc.

specific weight decreases as they pass from the valve inlet to the vena contracta within the valve body. The effective specific weight, as far as flow is concerned, is approximately mid-way between the value upstream at P_1 and the value at the vena contracta.

The most convenient way of accounting for the change of specific weight in sizing formulas for valves, as well as for orifice meters, flow nozzles, and venturis, is by applying an expansion factor "Y" to the formula for liquids. If the basic equation for liquids is converted to weight units it becomes:

$$w = 63.3\ C_V \sqrt{\Delta p\ \gamma_1} \qquad (7)$$

where w is in pounds per hour and γ_1 is in pounds per cubic foot. If the fluid is compressible, Y must be included; therefore,

$$w = 63.3\ C_V Y \sqrt{\Delta p \gamma_1} \qquad (8)$$

Theoretically, the value of Y varies with all of the following factors:

x — pressure drop ratio $\Delta p/p_1$;

m — ratio, (port area/body inlet area),

Re — Reynolds number;

k — ratio of specific heats, c_p/c_v; and the shape of the flow path.

The effect of the Reynolds number on Y is insignificant. The error due to neglecting k will not exceed 5% except for very low values of k combined with a high pressure-drop ratio. The effects of shape and area rations can be combined into a single factor which must be determined by testing.

The relationships of Y *vs.* x for several widely differing styles of valves are shown in Figure 2. It will be noticed that all of the plots are straight lines and that they terminate at a value of Y = 0.667. Although the true plots may deviate somewhat from a straight line, the test data to date indicates that the linear representation is within the tolerance of experimental error (about 2%). Termination at Y = 0.667 results from the fact that the ultimate flow through a restriction at any selected p_1 must occur when the quantity $Y\sqrt{x}$ reaches a maximum. With the linear relationship of Y to x, it can be demonstrated by differential calculus that $Y\sqrt{x}$ reaches a maximum at Y = 0.667.

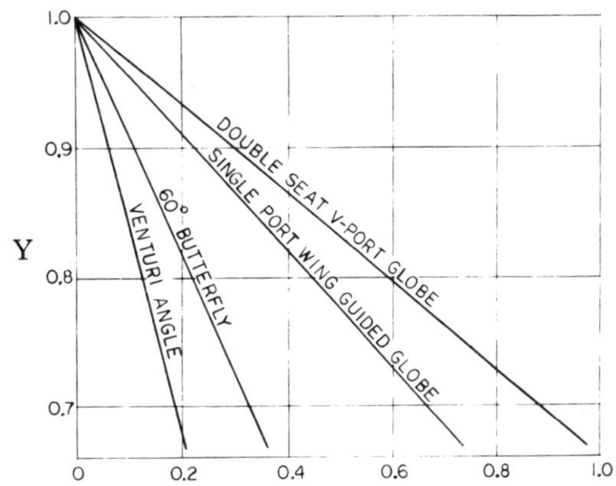

Figure 2. EXPANSION FACTOR, Y, AS A FUNCTION OF PRESSURE DROP RATIO, X, FOUR WIDELY DIFFERING STYLES OF CONTROL VALVES.

Conversely, for every restriction there is an apparent limiting or terminal value of the pressure drop ratio, which is called x_T. This means that at any fixed upstream pressure there is, for any valve, a maximum effective pressure drop. There will be no further increase in flow *if the downstream pressure is reduced beyond this limiting condition.* If the slope of Y *vs.* x or the value of x_T is determined for a particular valve, Y may be determined by:

$$Y = 1 - \frac{x}{3x_T} \qquad (9)$$

Considering the foregoing we can now arrange Equation (8) to more easily reflect the limitation, x_T:

$$w = 63.3\ C_V Y \sqrt{x p_1 \gamma_1} \qquad (10)$$

In this form, if x exceeds x_T, the value of x_T is used and Y becomes 0.667. If x is less than x_T, then Y is determined from Equation (9).

If the gas flow data is expressed in the customary volumetric units, Equation (10) can be restated as

$$q = 1360\ C_V p_1 Y \sqrt{\frac{x}{G_g T_1 Z}} \qquad (11)$$

Here, it is necessary to include the compressibility factor Z in order to provide for deviation from the universal gas law. The need for this becomes obvious when it is considered that the numerical value of Z for most real gases is usually between 0.5 and 1.5*. Flow could, therefore, be either 40% greater or 20% less than predicted, if the compressibility factor is neglected.

*See Nelson-Obert Charts in Appendix.

WORKING EQUATIONS

Liquid Formulas

$$C_V = q_f\sqrt{\frac{G_f}{\Delta p}} \quad \text{or} \quad q_f = C_V\sqrt{\frac{\Delta p}{G_f}} \qquad (6)$$

$$C_V = \frac{w}{63.3\sqrt{\Delta p\,\gamma}} \quad \text{or} \quad w = 63.3\ C_V\sqrt{\Delta p\,\gamma} \qquad (7)$$

$$C_V = \frac{w}{500\sqrt{\Delta p G_f}} \quad \text{or} \quad w = 500\ C_V\sqrt{pG_f} \qquad (8)^*$$

Limitations of Equations

High Δp - Error will not exceed 10 percent with V-port plugs or with contour plugs in flow-to-open direction when valve is wide open, if $p_2 > p_V$ for single- or double-seat globe valves. Other styles may cause serious error if $\Delta p > 0.2\ (p_2 - p_V)$. See Part 2 on cavitation and flashing for size correction.

High vapor pressure or flashing—Same effect as "High Δp" above.

Entrained gas or vapor—See the section on "Two-Phase Flow."

High viscosity—Error will not exceed 10% if $\mu = 3.7\sqrt{C_V G_f \Delta p/(\text{No. of seats})}$. For correction factors, see the section on "Low Reynolds Number."

Low Cv or low Δp—Apply the rule-of-thumb as for "high viscosity" above.

Piping effects. Error will not exceed 10 percent if the valve is between 2:1 reducers and if valve capacity is such that $C_V < 14D^2$. See the section on "Piping Effects" for the correction factor.

Example 1 *Liquid Flow*

Given:	
Fluid	Brine
Specific gravity at F.T.	1.2
Flow rate maximum	250 gpm at F.T.
Δp at maximum flow	10 psi
Flow rate minimum	40 gpm at F.T.
Δp at minimum flow	25 psi

$$C_V = q_f\sqrt{\frac{G_f}{\Delta p}} = 250\sqrt{\frac{1.2}{10}} = 87 \text{ [Equation(6)]}$$

$$C_V\ (\text{minimum} = 40)\sqrt{\frac{1.2}{25}} = 8.7$$

(Rangeability 10:1)

Gas and Vapor Formulas

$$Y = 1 - x/3x_T \quad \text{where } x \leq x_T \qquad (9)$$

$$C_V = \frac{0.0158w}{Y\sqrt{xp_1\gamma_1}} \quad \text{or} \quad w = 63.3\ C_V Y\sqrt{xp_1\gamma_1} \qquad (10)$$

$$C_V = \frac{q_g}{1360\ p_1 Y}\sqrt{\frac{G_gTZ}{x}} \quad \text{or} \quad q_g = 1360\ C_V p_1 Y\sqrt{\frac{x}{G_gT_1Z}} \qquad (11)$$

$$C_V = 0.0158\frac{w}{Y}\sqrt{\frac{v_1}{x\,p_1}} \quad \text{or} \quad w = 63.3\ C_V Y\sqrt{\frac{xp_1}{v_1}} \qquad (12)^\dagger$$

$$C_V = 0.0517\frac{w}{p_1 Y}\sqrt{\frac{T_1Z}{xM}} \quad \text{or} \quad w = 19.4\ C_V p_1 Y\sqrt{\frac{xM}{T_1Z}} \qquad (13)^\dagger$$

$$C_V = \frac{q_g}{7340\ C_V p_1 Y}\sqrt{\frac{MT_1Z}{x}} \quad \text{or} \quad q_g = 7340\ C_V p_1 Y\sqrt{\frac{x}{MT_1Z}} \qquad (14)^\dagger$$

Limitations of Equations

Mixed phase— For gas-liquid mixtures see the section on "Mixed Phase."

Ratio of specific heats— The equations are based on air (k = 1.4). The maximum error for other values of k will not exceed 10 percent. Maximum error occurs with valves of extremely high capacity, on gases with extremely low k, and with high pressure drop ratio.

Piping effects— Refer to the section on "Piping Effects" for a discussion of the effect of piping configurations at the valve. Reducers on either side of the valve will have a significant effect on high capacity styles.

Near critical pressure—Error can become significant when the pressure drop ratio is high and the vapor is near its

* Equation (8) is a variation of Equation (7) with the conversion factor for gallons per minute.

† Equations (12), (13) and (14) are variations of Equations (10) and (11), using conversion factors for customary flow units.

critical pressure. Steam begins to deviate above 1400 psia.

High Δp—Although the equations are valid for all pressure drops, extremely high outlet velocities may cause excessive noise. Pipe reducers which are normally installed at the valve outlet, can cause supersonic shock waves. To avoid this, limit the velocity at the entrance to the reducer to something less than sonic.

The minimum valve outlet diameter for gases may be determined by:

$$D > 0.0024\sqrt{\frac{qg}{p_2}\left(\frac{T_1 M}{k}\right)^{1/2}} \quad (15)$$

or

$$D > 0.047\sqrt{\frac{w}{p_2}\left(\frac{T_1}{Mk}\right)^{1/2}} \quad (16)$$

The pipe diameter for steam may be determined by:

$$D > 0.12\sqrt{\frac{w}{p_2}} \quad (17)$$

Example 2 Gas or Vapor Flow

Given:		
	Fluid	Refrigerant F-12
	Rate, w	20,000 pph
	Pressure, P_1	114.5 psia
	Pressure, P_2	54.5 psia
	Sp. vol., v_1	0.335 ft^3/lb.

If the valve selected is, for example, a single-seat, V-port, globe type, the value of x_T is about 0.75.

$$C_v = 0.0158\,\frac{w}{Y}\sqrt{\frac{v_1}{x\,p_1}} \quad \text{[Equation (12)]}$$

$x = 60/114.5 = 0.524$ [by definition]

$Y = 1 - 0.524/3\,(0.75) = 0.767$ [Equation (9)]

$$C_v = 0.0158\left(\frac{20{,}000}{0.767}\right)\sqrt{\frac{0.355}{(0.524)(114.5)}} = 31.6$$

Other Gas Equations

Other gas equations in current use are listed below along with their special symbols.

$$q = C_v C_1 p\sqrt{\frac{520}{GT_1}}\ \sin\left[\frac{3417}{C_1}\sqrt{\frac{\Delta p}{p_1}}\right]_{\text{Deg.}} \quad (18)^{(1)}$$

where the bracketed expression is limited to a value of 90°, which represents terminal flow conditions. The factor C_1 is derived by physical testing with air. It is approximately equal to $40\sqrt{x_T}$. This equation agrees with Equation (11) within the limits of measurement tolerance.

$$q = 834\,\frac{F_L C_v p_1}{\sqrt{GT_1}}\,(y - 0.148y^3) \quad (19)^{(2)}$$

where $y = \frac{1.63}{F_L}\sqrt{\frac{\Delta p}{p_1}} = 1.50$ maximum at terminal flow. This equation is based on the assumption of a fixed relationship between x_T and F_L ($x_T = 0.85F_L^2$). Valve tests by one manufacturer on air and water are said to confirm this assumption; while another manufacturer reports controverting data. The compressibility factor, Z, may be included in either of the above equations if the accuracy would be seriously affected by its omission.

$$q = 963\,C_v\sqrt{\frac{\Delta p\,(p_1 + p_2)}{GT_1}} \quad (20)^{(3)}$$

where Δp and p_2 are limited to terminal conditions of $p_2 \geq 0.5\,p_1$. Flow predicted by this formula is almost always more than actual. In order for the error not to exceed 10%, the pressure drop ratio must not exceed the following limits:

If X_T= 0.77	0.70	0.60	0.50	0.40	0.30	0.20
Then x < 1.00	0.40	0.28	0.20	0.14	0.10	0.065

Slide Rules and Nomographs

Many control valve manufacturers provide slide rules and nomographs to be used for valve sizing. When properly designed and used, the results of these tools are as accurate as direct calculations. They, of course, have their advantages and disadvantages. Generally these methods are faster, because all constants in the equations are incorporated into the scales and require no additional settings. The tools must be used frequently in order to avoid re-reading the procedural instructions. The units of measurement used must conform to those employed by the slide rule or nomograph. The chances of error for the various methods depends upon the calculator. Usually no record is preserved when these devices are used; therefore, checking is difficult.

LOW REYNOLDS NUMBER

The nature of flow in pipes depends upon the combination of four variables (diameter, viscosity, density, velocity) arranged to produce a dimensionless number called the Reynolds number. The value of this number determines whether the flow is laminar or turbulent, and it also affects the coefficient of discharge of an orifice.

The flow coefficient, C_v, is determined at a reasonably high Reynolds number with the fluid in turbulent flow. The C_v changes very little as the Reynolds number is increased. However, as the Reynolds number is decreased, the effective Cv becomes smaller and a correction factor must be applied. At extremely low Reynolds numbers the flow becomes laminar rather than turbulent, and the flow rate is proportional to Δp rather than $\sqrt{\Delta p}$. This change in flow mechanism is provided for by the correction factor, F_R.

For a particular valve, the Reynolds number decreases as viscosity increases and as differential pressure decreases. To calculate the Reynolds number directly, however, would involve making certain physical measurements of the valve. Making these measurements is usually impractical. For example, one would need to know the hydraulic radius, which is defined as the cross-sectional area of the flowing fluid, divided by the wetted perimeter of the channel. The hydraulic radius could only be found indirectly by conducting flow experiments in the laminar region.

As a practical solution, we may shorten the rigorous method of calculation by collapsing the theoretical equations into working formulas and making use of dimensionless factors. Such a procedure is described in the following section.

Correction Factor F_R

For a low Reynolds number, resulting from high viscosity or extremely low C_v or Δp, use the appropriate formula to find θ, which is a function of R_e^{-1}. With this value of θ find F_R from Figure 3. The factor F_R is used to modify C_v as used in Equations (6), (7), and (8).

To find valve size: (Curve 1)
$$\theta = \frac{\mu F_d}{\sqrt{\Delta p\, G_f\, (C_v)_u}} \quad (21)$$

To find flow rate: (Curve 3)
$$\theta = \frac{\mu F_d}{\sqrt{\Delta p\, G_f\, C_v}} \quad (22)$$

To find Δp: (Curve 2)
$$\theta = \frac{\mu F_d \sqrt{C_v}}{q_f G_f} \quad (23)$$

$$C_v = 0.07 \left(\frac{q_f \mu}{\Delta p}\right)^{2/3} \quad (25)$$

$(C_v)_u$ = Uncorrected C_v required, calculated by Equation (6), (7), or (8) without the factor, F_R.

F_d = Experimentally determined factor related to hydraulic radius. See Table 1.

Limitation of equations

Equation (25)— When F_d is unknown, this approximate formula may be used to determine the order of magnitude of the required valve C_v. Best results are achieved when C_v is greater than 20 and when flow is definitely laminar.

Non-Newtonian Liquids— All gases and most simple, low-viscosity liquids are Newtonian; that is, the shear rate is directly proportional to the shear stress. Unfortunately, most high viscosity liquids such as polymers, slurries, sols, gels, colloids, etc., are non-Newtonian. For these fluids the *effective* viscosity must be used in the equations.

Tolerance — The values of F_d, given in Table I, are based on limited test data taken on small-size valves with contoured and V-port plugs. It is expected that the results will be conservative for butterfly or ball-type valves which should yield about 15% more flow than calculated.

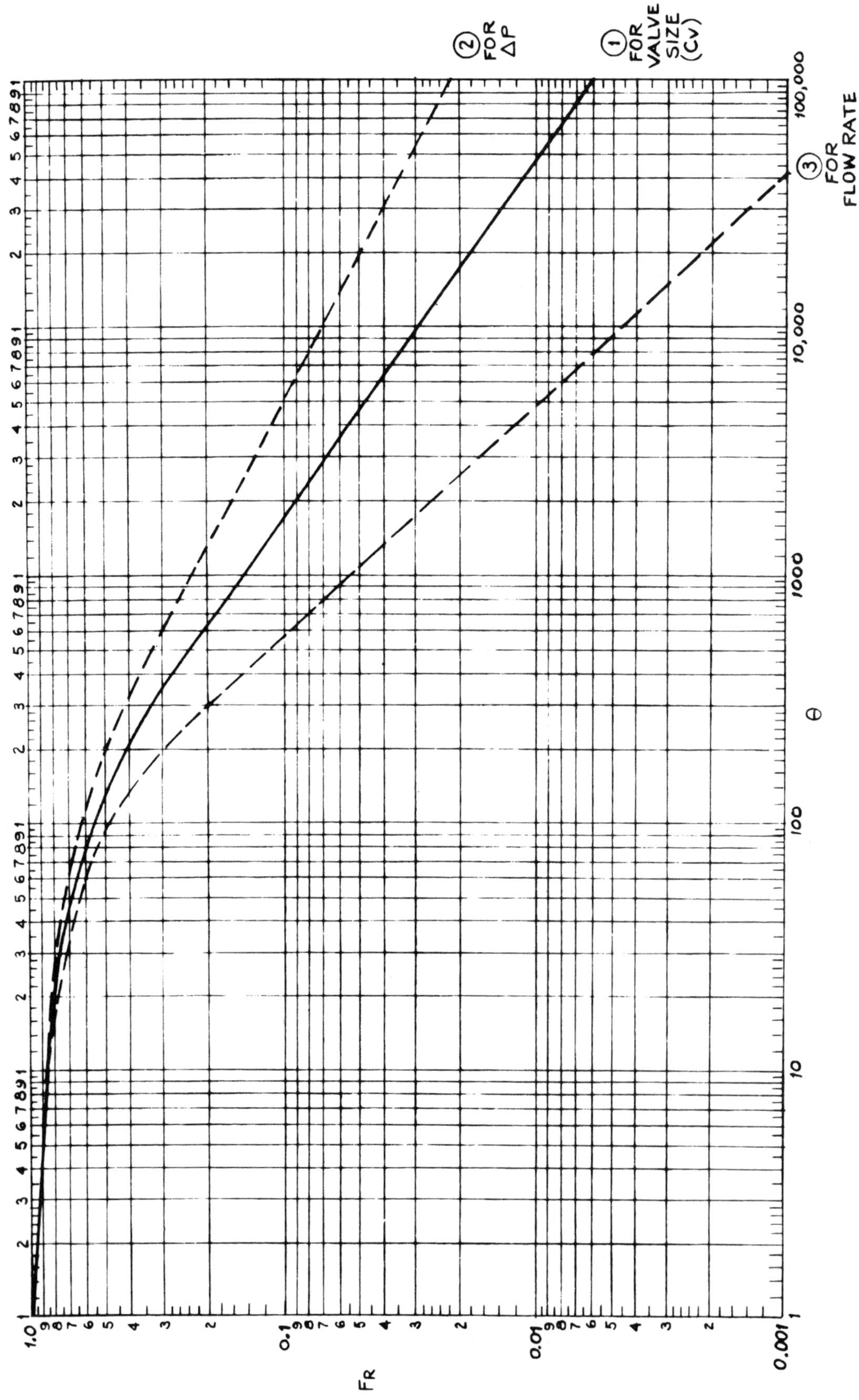

Figure 3. THE VALVE RATED C_V MUST BE MULTIPLIED BY THE FACTOR F_R WHEN THE STREAM REYNOLDS NUMBER IS LOW. HIGH VISCOSITY, LOW Δp AND SMALL C_V ALL CONTRIBUTE TO LOWERING OF REYNOLDS CRITERION.

Table I. HYDRAULIC RADIUS CORRECTION

Use *all* applicable factors	*Fd*
Single port	1.0
Double port (including butterfly)	1.4
Size, 1/2 inch	1.7
1 inch	1.1
Over 1 inch	1.0
Style, Hollow plug V-port	0.7
Contour plug T & B Guided (flow-to-open)	1.0

MIXED PHASE FLUIDS

Liquid / Gas

In this portion of the "Mixed Phase" section, the sizing procedure described applies only to homogeneous mixtures of liquids and non-condensible gases. The equations are based on the premise that the pressures within the valve body are always high enough to prevent cavitation or flashing of the liquid. If vaporization occurs, choked flow may result, and the allowable pressure drop will determine the capacity limit. (Refer to Part 2 of this chapter on cavitation and flashing.) There are no known experimental data for flash-

Example 3 *High Viscosity Liquids*

Given:	
Fluid	Newtonian liquid
Specific gravity, G_f	0.90
Pressure, p_1	85 psia
Pressure, p_2	65 psia
Temperature, T_1	70° F
Flow rate, q_f	50 gpm
Viscosity, μ	20,000 cp

For rough approximation of size to aid in selection of valve type, use Equation (25):

$$Cv \approx 0.07\left(\frac{q_f\,\mu}{\Delta p}\right)^{2/3}$$

$$Cv \approx 0.07\left(\frac{50\,(20{,}000)}{20}\right)^{2/3} = 95$$

Select single-seat contour plug. From Table I, Fd = (1.0) (1.0) (1.0) = 1.0

Uncorrected C_v, $(C_v)_u = q_f\sqrt{\frac{G_f}{\Delta p}}$ [Equation (6)]

$$(C_v)_u = 50\sqrt{\frac{0.90}{85\text{-}65}} = 10.6$$

$$\theta = \frac{\mu F_d}{\sqrt{\Delta p\, G_f\, (C_v)_u}}$$ [Equation (21)]

$$\theta = \frac{20{,}000\,(1)}{\sqrt{20\,(0.9)\,10.6}} = 1450$$

$F_R = 0.11$ from Curve 1 (Figure 3)

$$C_v = \frac{(Cv)u}{F_R} = \frac{(10.6)}{0.11} = 95.4$$

ing, mixed-phase fluids.

When a liquid/gas mixture passes through a valve orifice, the liquid's specific weight remains constant. The gas, on the other hand, expands; therefore, its effective specific weight is reduced. The factor that expresses this effect is the expansion factor, Y, which has been described previously.

It will be seen from equation (12) that the weight rate of flow of a gas is proportional to $Y\sqrt{\frac{x}{v_1}} = \sqrt{\frac{x}{v_e}}$, where v_e is the effective specific volume. It follows that $v_e = v_1/Y^2$. In other words, the weight rate of flow of a compressible fluid with a specific volume of v_1 is the same as the weight rate of flow of an incompressible fluid with a specific volume of v_e when other conditons are the same.

To calculate valve C_v for liquid/gas mixtures it is only necessary to use Equation (7), replacing γ_1 with the effective value γ_e, hence:

$$C_v = \frac{w}{63.3\sqrt{\Delta p \gamma_e}} \qquad w = 63.3\, C_v \sqrt{\Delta p \gamma_e} \qquad (26)$$

$$\gamma_e = \frac{w}{V_{ge} + V_f} \qquad \text{lb/ft}^3 \qquad (27)$$

$$V_{ge} = w_g v_e = \frac{10.7\, T_1 w_g}{M p_1\, Y^2} \qquad \text{ft}^3\text{/hr.} \qquad (28)$$

$$V_f = \frac{w_f}{\gamma_f} \qquad \text{ft}^3\text{/hr.} \qquad (29)$$

Limitations Of The Above Formulas

1. These formulas agree with experimental data gathered on air and water mixtures in turbulent flow through small size conventional globe valves, in horizontal lines. For similar situations the error should not exceed approximately 10%.
2. The performance of valves on mixed phase flow in vertical lines has not been evaulated experimentally. If the fluid is homogeneous the formulas are theoretically valid.
3. Stratified streams, where the gas and liquid are not thoroughly mixed, are not predictable.
4. Vaporization will occur within the valve body if the pressure falls below the vapor pressure of the liquid. This can result in choked flow at some limiting value of the pressure drop ratio, x. Experimental data is unavailable; however, it is known that as the gas volume fraction of the stream approaches zero, the limit of x approaches $F_L^2\,(1-F_F p_v/p_1)$; where F_L is the body-loss coefficient, F_F is the flashing coefficient, and p_v is the liquid vapor pressure. (Refer to Part 2 of this chapter on cavitation and flashing.) As the gas volume fraction approaches one, the limit of x approaches x_T.

Liquid / Vapor

This part of the "Mixed Phase" section is devoted to a discussion of streams composed of a homogeneous mixture of a liquid and its own vapor. At the present state of the technology it is impossible to predict valve capacity accurately under these conditions.

A large amount of the experimental results are puzzling, to say the least. Some of the data appears to imply that the stream density actually increases as it passes through the valve orifice. This indicates that the calculated value of the stream fraction which is in the vapor state does not coincide with the true value; that is, thermodynamic equilibrium is not attained.

The difficulty in predicting the valve flow capacity on liquid/vapor streams can be appreciated if the following points are considered:

1. The mechanism of flow through a valve orifice is not well understood even when the vapor and liquid are in equilibrium.
2. Prediction of the exact vapor fraction at the valve inlet is more of a problem than it would appear, since the fluid may exist in a metastable state.
3. Extremely small changes in the weight fraction of the stream that is vaporized result in large changes in the fluid density.

The formula recommended for liquid/vapor applications is Equation (26); that is

$$C_v = \frac{w}{63.3\sqrt{\Delta p \gamma_e}} \qquad w = 63.3\, C_v \sqrt{\Delta p \gamma_e} \qquad (26)$$

Example 4 Mixed Phase Liquid/Gas

Given:	Air Flow, w_g	460 lb/hr.
	Water Flow, w_f	20,000 lb/hr.
	Pressure, p_1	100 psia
	Δp	36 psi
	Temperature	60°F

Select: Valve type - double seat V-port with $x_T = 0.75$

Expansion factor, $Y = 1 - \frac{\Delta p/p1}{3x_T} = 1 - \frac{36/100}{3\,(0.75)} = 0.84$ [Equation (9)]

Effective gas volume, $V_{ge} = \frac{10.7\ T_1 w_g}{M\ p_1\ Y^2}$ [Equation (28)]

$$V_{ge} = \frac{(10.7)\ (520)\ (460)}{(29)\ (100)\ (0.84)^2} = 1250 \text{ cfh}$$

Liquid volume, $V_f = w_f/\gamma f$ [Equation (29)]

$V_f = 20{,}000/62.3 = 321$ cfh

Total effective volume, $V_e = V_{ge} + V_f = 1571$ cfh [Definition]

Total weight flow, $w = w_g + w_f = 20{,}460$ lb/hr. [Definition]

Effective sp. wt., $\gamma e = w/V_e = 13.1$ lb/ft.^{3} [Equation (27)]

$$C_v = \frac{w}{63.3\sqrt{\Delta p \gamma e}} = \frac{20{,}460}{63.3\ \sqrt{(36)\ (13.1)}}$$ [Equation (7)]

$C_v = 14.7$

Example 5 Mixed Phase Liquid/Vapor

Given:	Water, w_f	50,000 pph
	Steam, w_g	1,700 pph
	Pressure, p_1	100 psia
	Pressure, p_2	40 psia

Select: Double ported globe, $x_T = 0.75$

Expansion factor, $Y = 1 - x/3x_T = 1 - 0.6/3\ (0.75) = 0.73$ [Equation (9)]
Vapor volume correction factor, $Y^{-2} = 1.88$ [Given]
Sp. Vol. of vapor, $v_{g_1} = 4.43$ ft^3/lb. [Steam Table]
Sp. Vol. of liquid, $v_{f_1} = 0.0177$ ft^3/lb. [Steam Table]
Liquid Volume, $V_f = w_f v_{f_1} = (50{,}000)\ (0.0177) = 885$ cfh [Definition]
Effective Vapor Vol., $V_{ge} = w_g v_{g_1} Y^{-2} =$
$(1700)\ (4.43)\ (1.88) = 14{,}160$ cfh [Var. Equation (28)]
Total effective volume, $V_e = V_{ge} + V_f =$
$(885 + 14{,}160 = 15{,}045$ cfh [Definition]
Effective sp. wt., $\gamma e = (w_f - w_g)/V_e =$
$(50{,}000 + 1700)/15{,}045 = 3.44$ lb/ft^3 [Equation (27)]

$$C_v = \frac{w}{63.3\sqrt{\Delta p \gamma e}} = \frac{51{,}700}{63.3\ \sqrt{(60)\ (3.44)}}$$ [Equation (26)]

$C_v = 57$

The limitations of Equation (26) for liquid/vapor applications include:

1. The errors due to inaccurate determination of true specific weight at the valve inlet far exceed the limitation of the formula.
2. The presence of a pipe reducer at the valve inlet will have a significant effect on the vapor fraction, as well as having the usual effect on inlet pressure.
3. Flashing, which occurs at high pressure-drop ratios, can cause choked-flow to limit the effective pressure differential; however, the sharp cut-off exhibited by cavitating liquid flow is not observed. Choked-flow tends to occur only at high pressure-drop ratios and with an extremely low vapor fraction at the valve inlet. As the vapor fraction increases, the limit of x approaches x_T.

PIPING EFFECTS

The sizing coefficients for control valves are determined from tests with the valve body mounted in a straight run of pipe which is the same diameter as the valve body. If the piping configuration is different from that used in evaluating the coefficients, the valve capacity will be different. The magnitude of this effect depends principally on the type of pipe fittings involved, how near they are to the valve and the ratio of the valve orifice area to the pipe area. This situation can be compared to the effect of piping on a metering orifice, where the same factors affect the relationship of flow to differential pressure.

Piping effects are most pronounced with "high capacity" valves such as butterfly, Saunders patent, Y-pattern, ball, or plug types, all of which have high area ratios, m.

Block valves, if they are full area gate or ball type, do not disturb the flow significantly, but if they are preceded by either a reducer or an elbow, the effect of these fittings will be transmitted. Full-area valves whose openings are not round, such as rectangular port plug cocks, may have an additional effect and cause a re-

Example 6 Valve with Reducers—Liquid

Given:		
	Fluid	Water
	Specific Gravity, G_f	1.0
	Flow Rate, q_f	360 gpm
	Line Size, D	6 in.
	Δp	9 psi

$$F_P C_v = q_f \sqrt{\frac{G_f}{\Delta p}} \qquad \text{[var. Equation (6)]}$$

$$F_P C_v = \frac{360}{3} = 120$$

A preselected style has a C_d = 18, hence a 3-inch valve with a C_v = 162 seems reasonable. The 3 in. x 6 in. reducers being used are assumed to have resistance coefficients such that $\Sigma K = 1.5(1\text{-}d^2/D^2)^2 = 0.84$ [Abrupt transition]

$$F_P = \frac{1}{\sqrt{1 + \frac{\Sigma K C_d^2}{890}}} = \frac{1}{\sqrt{1 + \frac{(0.84)(18)^2}{890}}} \qquad \text{[Equation (30)]}$$

$$F_P = 0.87$$

$$F_P C_v = (0.87)(162) = 141$$

At the required flow, the valve will be operating at 85% of capacity, which is satisfactory.

duction in valve capacity greater than indicated by the calculated pressure loss. This is especially true when the cocks are preceded by reducers.

An elbow, when located close to a valve inlet, can cause significant effects in certain valves. If a reducer precedes the elbow the flow may be affected in an unpredictable manner. Even with valves of moderate capacity ($C_V/d^2 < 15$) the flow may be reduced by 15 percent.

Concentric reducers adjacent to a valve, although they do not skew the velocity profile, cause a reduction in over-all flow capacity. The primary reason for this is the change in area ratio, m. Since line pressure will be measured in a pipe that has a larger cross sectional area than that of the valve body, the velocity of approach, F, has been changed, and F is a component of the valve flow coefficient C_V. In addition, there are losses due to wall friction and losses due to the fluid friction caused by the contraction and expansion at the reducers.

Several methods have been proposed to correct for the effect of reducers. One method is straightforward; it simply depends upon actual valve testing to produce new coefficients for the combination of valve plus reducers. A second method, based on theory, relies on calculation to produce correction factors. The advantage of the second method is that it can be applied generally to all types of valves, for reducers with any specified taper, and even for sets of reducers of different sizes.

The formulas for arriving at the correction factors to be applied when valves are installed with reducers are listed

---

Example 7 Valve with Reducers—Gas

Given:

	Fluid	Sulfur Dioxide Gas
	Molecular Weight	64
	Upstream Pressure	15 psig= 29.7 psia
	Downstream Pressure	5 psig
	Temperature	150°F
	Flow Rate	200,000 lb/hr.
	Line Size	18 inches

$$(F_P C_V) = 0.0157 \frac{w}{p_1 Y} \sqrt{\frac{T_1 Z}{xM}} \qquad \text{[Equation (13)]}$$

$$(F_P C_V) = 0.0517 \frac{(200{,}000)}{29.7\,Y} \sqrt{\frac{(150+460)}{\left(\frac{10}{29.7}\right)(64)}}$$

$$YF_P C_V = 1860$$

Select a 60° opening butterfly valve. Since Y and FP are both less than one we will try a 14-in. size with $C_V = 3330$, and $x_T = 0.38$. The 14 in. x 18 in. reducers being used have resistance coefficients $K_1 = 0.077$ and $K_2 = 0.153$. The Bernoulli coefficient $K_B = 1\text{-}(d/D)^4$ or 0.63.

$$F_P = \frac{1}{\sqrt{1 + \frac{\Sigma K C_d^2}{890}}} = \frac{1}{\sqrt{1 + \frac{(0.23)(17)^2}{890}}} = 0.965 \qquad \text{[Equation (30)]}$$

$$x_{TP} = \frac{x_T}{F_P^2}\left[1 + \frac{C_d^2 x_T (K_1 + K_{B1})}{1000}\right]^{-1} = \frac{0.38}{0.93}\left[1 + \frac{(17)^2(0.38)(0.71)}{1000}\right]^{-1} = 0.38 \qquad \text{[Equation (33)]}$$

$$Y = 1 - \frac{x}{3\,x_{TP}} = 1 - \frac{0.34}{3(0.38)} = 0.70 \qquad \text{[Equation (9)]}$$

therefore

$$C_V = \frac{1860}{F_P Y} = \frac{1860}{(0.965)(0.70)} = 2750$$

The size selected is satisfactory.

Note: F_P and x_{TP} may be taken from Table 2.

TABLE II

CALCULATED VALUES OF F_P AND x_{TP} FOR VALVES INSTALLED BETWEEN SHORT PIPE REDUCERS ASSUMING TWO REDUCERS OF THE SAME SIZE WITH ABRUPT CHANGE IN AREA.

C_d	10					
X_T	.40	.50	.60	.70	.80	
d/D	X_{TP}					F_P
.80	.40	.49	.59	.68	.78	.99
.75	.40	.50	.59	.68	.78	.99
.67	.40	.50	.59	.69	.78	.98
.60	.41	.51	.60	.70	.79	.97
.50	.41	.51	.61	.70	.79	.96
.40	.42	.52	.62	.71	.80	.95
.33	.43	.53	.63	.72	.82	.94
.25	.44	.54	.64	.73	.83	.93

C_d	15					
X_T	.40	.50	.60	.70	.80	
d/D	X_{TP}					F_P
.80	.40	.49	.58	.67	.75	.98
.75	.40	.49	.58	.67	.75	.97
.67	.41	.50	.59	.67	.75	.95
.60	.42	.52	.60	.69	.78	.93
.50	.44	.53	.62	.71	.79	.91
.40	.45	.55	.64	.73	.82	.89
.33	.46	.56	.66	.75	.83	.88
.25	.47	.57	.67	.76	.85	.87

C_d	20				
X_T	.40	.50	.60	.70	
d/D	X_{TP}				F_P
.80	.39	.48	.56	.64	.96
.75	.40	.49	.57	.65	.94
.67	.41	.50	.58	.66	.92
.60	.44	.53	.61	.69	.89
.50	.46	.55	.64	.72	.85
.40	.49	.58	.67	.75	.82
.33	.50	.60	.69	.78	.81
.25	.52	.62	.71	.79	.79

C_d	25				
X_T	.20	.30	.40	.50	
d/D	X_{TP}				F_P
.80	.21	.30	.39	.48	.94
.75	.22	.31	.40	.48	.91
.67	.23	.33	.42	.50	.88
.60	.25	.36	.45	.54	.84
.50	.28	.39	.49	.58	.79
.40	.30	.42	.53	.62	.76
.33	.31	.43	.54	.64	.74
.25	.33	.45	.57	.66	.72

C_d	30			
X_T	.15	.20	.25	
d/D	X_{TP}			F_P
.80	.16	.21	.26	.92
.75	.17	.22	.27	.88
.67	.19	.24	.29	.84
.60	.21	.27	.33	.79
.50	.23	.30	.36	.74
.40	.26	.33	.40	.70
.33	.28	.35	.42	.68
.25	.29	.37	.44	.66

below. In the case of liquids, two factors are pertinent: F_P which corrects the valve coefficient C_v, and F_{LP} which supplants both coefficients F_P and F_L in the choked flow condition.

$$F_P = \frac{1}{\sqrt{\dfrac{1+\Sigma K C_d^2}{890}}} \qquad (30)$$

Where $C_d = C_v/d^2$, and $\Sigma K = K_1 + K_{B1} + K_2 - K_{B2}$ (the coefficients of the inlet and outlet reducers respectively). The Bernoulli coefficients K_{B1} and K_{B2} may be omitted if the inlet and outlet reducers are the same size, since one term is positive and the other negative. If the reducers are of different size, or if there is only one reducer, they must be included. As a simplifying assumption, K may be taken as $1.5\left(1-\frac{d^2}{D^2}\right)^2$ if the reducers are the same size. This value is based on abrupt transition and hence is conservative.

The loss (recovery) factor of a valve is F_L. This factor is important when either flashing or cavitation exists. When a valve is installed between reducers, the F_L factor for the combination may be called $(F_L)_P$. Its value is calculated from the intrinsic loss factor (F_L) of the valve as follows:

$$(F_L)_P = \frac{1}{F_P\sqrt{\dfrac{1}{F_L^2}+\dfrac{C_d^2\,(K_1+K_{B1})}{890}}} \qquad (31)$$

Since the factors $F_P F_L$ are always used together in the choked flow equation, a simpler equation results if Equation (31) is rearranged for a single factor, $F_{LP}=(F_L)_P F_P$:

$$F_{LP} = \left[\frac{1}{F_L^2} + \frac{C_d^2\,(K_1+K_{B1})}{890}\right]^{-1/2} \qquad (32)$$

In the case of compressible fluids, such as gas or vapor, the expansion factor Y and its determiner x_T are affected by reducers. The corrected value of x_T becomes:

$$x_{TP} = \frac{x_T}{F_P^2}\left[1 + \frac{x_T C_d^2\,(K_1+K_{B1})}{1000}\right]^{-1} \qquad (33)$$

See Table II for calculated values.

REFERENCES

1. Buresh, J.F. and Schuder, C.B., October, 1963, "The Development of a Universal Gas Sizing Equation for Control Valves," Conference Paper No. 46.2.63; Instrument Society of America.
2. Boger, H.W., July, 1968, "Recent Trends in Sizing Control Valves," *Instruments and Control Systems*, 41:117-121.
3. *Recommended Voluntary Standard Formulas for Sizing Control Valves*, 1962, FCI-62-1, Fluid Controls Institute.

BIBLIOGRAPHY

1. Driskell, L.R., June 1967, "Practical Guide to Control Valve Sizing", *Instrumentation Technology*, 14: 47-58.
2. *Fluid Meters, Their Theory and Application*, 1959, The American Society of Mechanical Engineers.
3. Cunningham, R.G., July 1951, "Orifice Meters with Supercritical Compressible Flow", *Transactions* ASME.
4. Driskell, L.R., July 1969, "New Approach to Control Valve Sizing", *Hydrocarbon Processing*.
5. Spink, L.K., 1967, *Principles and Practice of Flow Meter Engineering*, Ninth Edition, Plimpton Press.
6. Nelson, L.C. and Obert, E.F., "Generalized PVT Properties of Gases", *Transactions* ASME, 76:1057.
7. Stiles, G.F., January 1964, "Liquid Viscosity Effects on Control Valve Sizing", Paper presented at Texas A&M Nineteenth Annual Symposium on Instrumentation for the Process Industries.
8. Baumann, H.D., September 1965, "Why Limit Outlet Velocities in Reducing Valves—" *Instruments & Control Systems*, 38:135-139.

NOMENCLATURE

a — Area of orifice or valve opening, $in.^2$

C — Coefficient of discharge, dimensionless. Includes effect of jet contraction and Reynolds number, mach number (gas at high velocities), turbulence.

C_d — Relative capacity (at rated C_v) $C_d = C_v/d^2$.

c_p — Specific heat at constant pressure, Btu/lb. °F.

c_v — Specific heat at constant volume, Btu/lb. °F.

C_v — Valve coefficient, $38\ a\overline{K}/F_L$, gpm/$\sqrt{psi}$.

d — Valve inlet diameter, inches.

D — Pipe diameter, inches.

F — Velocity of approach factor $= \dfrac{1}{\sqrt{1-m^2}}$.

F_d — Experimentally determined factor relating valve C_v to an equivalent diameter for Reynolds number. (See Table I)

F_L — Pressure loss factor, to convert valve coefficient to choked flow conditions when capacity is governed by the apparent pressure at the valve vena contracta. When the valve is not choked.

$$F_L = \sqrt{(p_1 - p_2)/(P_1 - P_{vc})}$$

F_P — Correction factor for piping around valve (e.g. reducers) $F_P C_v$ = effective C_v for valve/fitting assembly.

$(F_L)_P$ — Pressure loss factor F_L modified for valve/fitting assembly.

F_{LP} — Combined pressure loss and piping geometry factors for valve/fitting assembly.

F_R — Correction factor for Reynolds number, where $F_R C_v$ = effective C_v.

g — Acceleration due to gravity (taken as 32.17 $ft./s^2$).

G_f — Specific gravity of liquids relative to water at 60° F.

G_g — Specific gravity of gas relative to air with both at standard temperature and pressure.

h — Effective differential head, ft. of fluid.

$\overline{K}$ — Flow coefficient = CF, dimensionless.

ΣK — Sum total of effective velocity head coefficients where $K(V^2/2g) = h$.

K_B — Bernoulli coefficient $= 1-(d/D)^4$.

K_1 — Resistance coefficient for inlet fitting.

K_2 — Resistance coefficient for outlet fitting.

k — Ratio of specific heats of a gas $= c_p/c_v$, dimensionless.

M — Molecular weight.

m	Ratio of areas.
p	Absolute static pressure, psia.
q	Volume rate of flow, q_f = gpm, q_g = scfh.
R	Gas constant = 144 pv/T.
Re	Reynolds number.
T	Absolute temperature, degrees Rankine.
U	Average velocity, fps.
v	Specific volume ($1/\gamma$), ft^3/lb.
V	Volume flow rate, ft^3/hr.
w	Weight rate of flow, lb/hr.
x	Ratio of differential pressure to absolute inlet static pressure, $x = (p_1 - p_2) / p_1$
x_T	Terminal or ultimate value of x, used to establish expansion factor, Y.
x_{TP}	Value of x_T for valve/fitting assembly.
Y	Expansion factor. Ratio of flow coefficient for a gas to that for a liquid at the same Reynolds number (includes radial as well as longitudinal expansion effects).
Z	Compressibility factor = 144 pv/RT (See Nelson-Obert Compressibility Charts, pp. 272-276).
γ (gamma)	Specific weight (1/v), lb/ft^3.
Δ (delta)	Difference, (eg., $\Delta p = p_1 - p_2$).
μ (mu)	Viscosity, centipoise.
θ (theta)	Index for viscosity correction.

SUBSCRIPTS

1	Upstream
2	Downstream
e	Effective value
f	Liquid
g	Gas
t	Theoretical
T	Terminal or ultimate value
u	Uncorrected value
vc	Vena contracta (point of minimum jet stream area)

PART 2 CAVITATION AND FLASHING CONSIDERATIONS

(G. F. Stiles)

INTRODUCTION

Two phenomena often encountered in liquid flow, which are completely absent in gas or vapor flow, are cavitation and flashing. These phenomena are of significant interest in any comprehensive discussion of control valves since their occurrence will affect the valve sizing procedures, may introduce noise and vibration, and also may limit the life expectancy of the valve components and the immediate downstream piping. While there are definite similarities between cavitation and flashing there are also important differences so that it will be necessary to distinguish between them at the outset of this discussion.

CAVITATION AND FLASHING DEFINED

Cavitation is a two-stage phenomenon, the first stage of which is the formation of voids or cavities within the liquid system. The second stage is the collapse or implosion of these cavities back into an all-liquid state. Some have defined cavitation as merely the formation of cavities but this seems to be an impractical definition from a control valve standpoint, because most of the effects and evidences of cavitation are related directly to the collapse, rather than the formation, of the cavities.

Investigators have theoretically identified and defined two types of cavitation; gaseous and vaporous. Although a rigorous analysis of the difference between these is beyond the scope of this treatment of the subject, it is now understood that both types require the presence of some nucleating agent for their inception. These very tiny nuclei, which will contain either vapor or dissolved gas, or perhaps both, will enlarge into finite cavities within the liquid. Even though it is agreed that nuclei are required, the explanation of what they are or how they work is not completely understood. In one popular theory, dust particles of microscopic dimension are believed to act as hosts for the gas or vapor. Most industrial liquids contain various solid contaminants so that the nuclei necessary for cavitation inception are almost universally present.

Based on their physical structures, ideal liquids would have tensile or fracture strengths of at least the order of tens of thousands of pounds per square inch. By contrast, the first stage of cavitation is observed in most industrial liquids when the local pressure drops to the vapor pressure of the liquid. Several investigators have shown that the inception of cavitation may be delayed significantly in test liquids which have been carefully degassed or held under high pressures for sufficient time for gas and vapor to be completely dissolved.

If the cavitation process were to be halted before the completion of the second stage, so that vapor persists downstream of the region where bubble collapse normally takes place, the process is known as flashing. Obviously, since flashing is directly related to the first stage of cavitation the statements regarding the inception of cavitation also apply to flashing.

CONTROL VALVE ANALOGY

In order to visualize how these phenomena occur in control valves, consider a liquid flowing in a piping system in which a restriction such as a concentric orifice has been placed. In this case the orifice may be considered analogous to a control valve at some fixed opening. The upper part of Figure 4 illustrates the system and, in the lower part, the pressures and velocities are plotted for points along the flow stream.

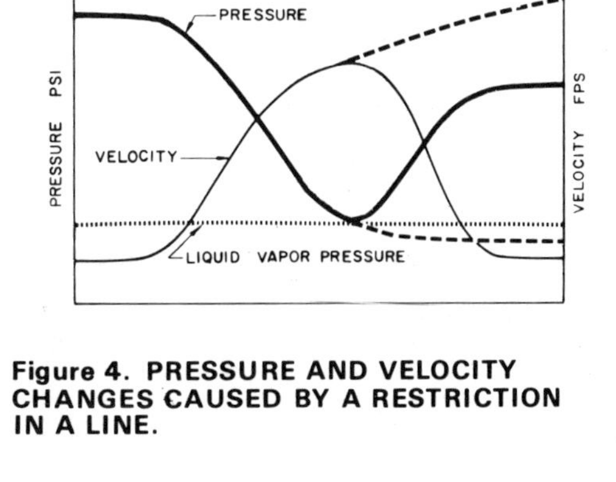

Figure 4. PRESSURE AND VELOCITY CHANGES CAUSED BY A RESTRICTION IN A LINE.

At point p'_1, upstream of the orifice, assume the liquid stream to have some static pressure above its vapor pressure. Also assume some flow rate so that the stream velocity may be determined. The total energy level of the stream (ignoring any elevation factor) may then be obtained by summing the contributions of the pressure and velocity written in some dimensionally homogeneous form, feet of head, for example.

The *First Law of Thermodynamics* states that, if no energy is lost through the boundaries of the system in the form of work or heat, the total energy level will remain constant at every point along a given datum plane in the system.

As the fluid stream approaches the restriction in the line its cross-sectional area must decrease in order to pass through the orifice. The velocity is inversely proportional to the stream area and, therefore, must increase. Since the sum of the pressure and velocity heads will remain approximately equal, an energy interchange must take place with the pressure head losing what the velocity head gains.

Immediately downstream of the orifice the stream will reach its minimum cross section and thus its maximum velocity and minimum pressure. This point is called the vena contracta (p_{vc}). If the velocity is increased sufficiently, the pressure will fall to the vapor pressure, thus permitting the formation of voids in the stream, which is the first stage of cavitation.

Downstream from the vena contracta, fluid friction causes the stream to decelerate with resultant increases in both stream cross section and pressure. This reversal of the energy interchange between the velocity and pressure heads is called "pressure recovery" and, as we shall see later, plays an important role in valve sizing. Vapor bubbles, formed as a result of reducing the pressure at the vena contracta to the vapor pressure, cannot exist at the increased pressure downstream and are forced to collapse or implode back into the liquid state. When this has occurred, the cavitation process is complete.

If the pressure in the downstream piping system were being maintained at a level equal to or less than the inlet vapor pressure, the fluid proceeding downstream would have an increasing percentage of vapor, the velocity of the stream would continue to increase and the end result would be flashing rather than cavitation.

REQUIREMENTS FOR OCCURRENCE

As cavitation and flashing have been defined, the following restrictive statements may be made regarding their occurrence in control valves.

Cavitation

1. The fluid at both the inlet and outlet must be in an all-liquid condition. That is, there can be no vapor present in the piping upstream or immediately downstream of the valve.
2. The liquid must be in a sub-cooled state at the inlet. Obviously if the fluid were all liquid, but in a saturated state, at the inlet any pressure drop across the valve would result in residual vapor present downstream.
3. The valve outlet pressure must be either at or above the vapor pressure of the liquid. Conceivably, it would be possible for cavitation to exist if a saturated, but all-liquid, condition were present at the downstream location.

Flashing

1. The fluid at the inlet must be in an all-liquid condition while some vapor must be present at the valve outlet. Obviously, if vapor were present at the inlet, any pressure differential taken across the valve would result in the formation of additional vapor. This restriction, requiring absence of vapor at the inlet, is imposed primarily because the valve sizing procedures are considerably more complicated when the inlet stream contains vapor. Valve sizing procedures, which will be discussed later, will be limited to the all-liquid inlet condition.
2. The fluid at the inlet may be in either a saturated or a subcooled condition.
3. The valve outlet pressure must be either at or below the vapor pressure of the liquid.

CAVITATION EVIDENCES

The occurrence of cavitation in a control valve is always accompanied by one or more of several characteristic evidences which include noise, vibration, physical damage and a decrease in efficiency. The numbers manifested, and their magnitudes, are dependent upon the proximity of the instantaneous operating point to the point of incipient cavitation. All of the effects of cavitation can be traced to the implosion of the vapor bubbles, with the exception of the decrease in efficiency which is traceable to their formation.

Noise

In a control valve the evidence of incipient cavitation is usually a hissing sound, sometimes intermittent, emanating

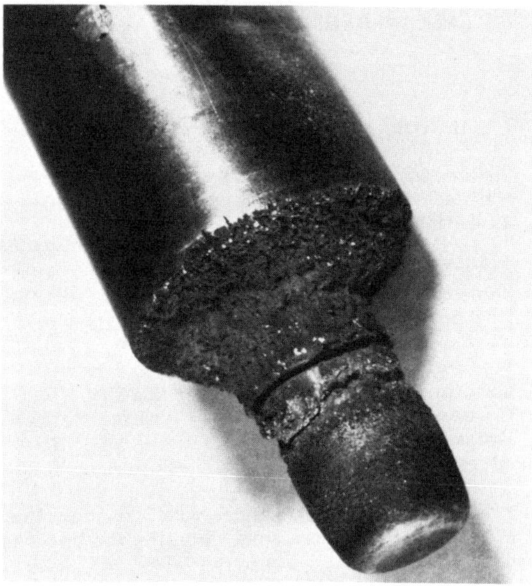

Figure 5. EXAMPLE OF CAVITATION DAMAGE TO A VALVE PLUG.
Courtesy Fisher Controls Company

from the downstream side of the valve body. In its very early stages, however, this sound may be masked by the normal flow noises. As the intensity of the cavitation increases, due to increasing pressure differentials, the sound level also increases. Fully developed cavitation in a control valve produces a rattling sound which gives the listener the impression that gravel is passing through it.

Vibration

The physical process which produces noise also results in vibration. How serious a problem this will be, will depend upon several factors, including the mass of the system, how well the system components are anchored, and whether valve-mounted instruments are vibration-sensitive. Field reports on a variety of control valve types indicate that, in severe cavitating service, these vibrations can reach serious proportions.

Material Damage

Probably the most dramatic effect of cavitation is the material damage which it inflicts. Under severe cavitating conditions, extremely hard control valve components have failed in a matter of a few hours time. An example of this is shown in Figure 5 which pictures a valve plug for a conventional globe style body that had been subjected to severe cavitating conditions on water service in a power generating station. In this example, the plug tip material was solid Stellite No. 6, a material which has been shown to be relatively high in resistance to cavitation damage, while the beveled seating area was 18-8 stainless steel overlaid with Stellite No. 6.

Figure 6 pictures two identical valve plugs machined from Alloy 1100-F (commercially pure) aluminum. The plug on the left is in the untested condition whereas the plug on the right has been subjected to cavitating cold water for a ten-minute period with the valve inlet pressure maintained at 2000 psig. The pressure drop across the valve was 2000 psi. In this case the very soft material was selected to illustrate that substantial metal loss can occur from the effects of cavitation even though exposure times are very short.

Numerous examples could be shown where cavitation has eroded holes completely through valve body or downstream piping walls, has destroyed valve plug seating surfaces and flow contours, and has ruined the guiding surfaces of post guided plugs.

In general it may be stated that resistance to cavitation damage increases with increasing hardness. Numerous investigators have studied this problem and have made attempts to correlate damage resistance on the basis of various material properties including tensile strength, modulus of elasticity and strain energy in addition to hardness. Up to the present

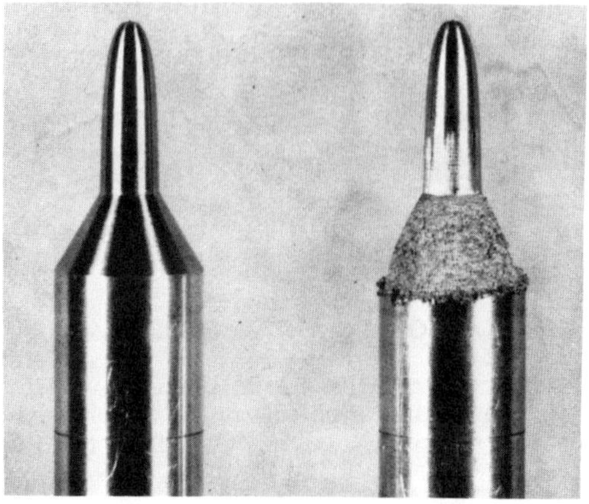

Figure 6. BEFORE AND AFTER APPEARANCE OF A SMALL VALVE USED IN CAVITATION DAMAGE TESTS.
Courtesy Fisher Controls Company

time no single property has been found which will yield good correlations for all experimental data. Other considerations, especially corrosion resistance, often preclude the use of the hardest available material in a particular application.

In the relatively hard materials from which control valve components are normally fabricated, cavitation damage is uniquely characterized by a very rough cinder-like appearance which makes it easily distinguishable from most other types of flow damage that are usually quite smooth in appearance.

Materials have been evaluated by a variety of methods for their resistances to cavitation damage. One method which has been used quite extensively employs a magnetostrictive* device to induce rapid oscillations of a test specimen supported in some test fluid. The oscillating motion of the specimen produces alternate high and low pressures in the fluid, forcing it to vaporize and reliquify at the driving frequency. Cavitation thus produced, damages the test specimen. Various test materials may be evaluated by comparing the relative weight losses for a given test time.

Another method that has been employed to duplicate actual control valve operating conditions, involves machining small valve plugs, as illustrated in Figure 6, from the desired test materials and subjecting them to identical flowing cavitating conditions. The following table of relative resistances of several common valve body and trim materials was developed from tests of this type. A sample of Type 316 stainless steel, a popular body and trim material, was tested until it had received a certain amount of damage. The time required for this damage was noted. Subsequently, samples of other materials were tested until they had achieved approximately the same magnitude of metal loss. Values of an "Index of Resistance to Cavitation Damage" were determined by dividing the times required to produce damage in the various materials by the time required to produce equivalent damage in the Type 316 stainless steel sample (See Table III).

TABLE III
INDEX OF RESISTANCE TO CAVITATION DAMAGE

Material	*Hours Tested*	*Index*
Stellite No. 6 over Type 316	120	20*
Die Steel (Latrobe BR-4FM)	120	20*
17-4 PHSS Hardened to 45 RH-C	12	2*
Type 316 SS Electrolized	8	1.33
Type 316 SS	6	1.00
Cast Iron (A126 Class C)	4.5	0.75
Chrome Moly Steel (C5)	4.0	0.67
Nodular Iron (A395)	2.5	0.42
Carbon Steel (WCB)	2.25	0.38
Carbon Steel (AISI C1213)	1.0	0.17
Brass (ASTM B16)	0.5	0.08
Aluminum (110-F)	(2-min.)	0.006

*Indicates that the damage was less than that of Type 316 stainless steel when the the test was discontinued.

CAVITATION DAMAGE MECHANISM

Several theories have been advanced relative to the mechanism by which cavitation damage occurs. One fact that is known, however, is that damage always takes place in close proximity to the imploding bubbles.

Earlier investigators have reported pressures as high as 100,000 psi in collapsing cavities. One theory suggests that a high pressure shock wave emanates from each and when these shock waves originate near a solid boundary, a highly concentrated, but minutely small blow is struck. Repeated strikes on any given increment of surface tend to fatigue it until eventually the endurance limit is reached and a tiny bit of metal is removed. Bubbles that implode at a sufficient distance from a solid surface are believed to be incapable of physical damage, because their energy is absorbed by the flowing liquid.

Credibility is given to the above theory by the fact that the harder materials can be subjected to known cavitating conditions for appreciable periods of time without exhibiting visible damage. Conversely, very soft materials such as commercially pure aluminum, from which the valve plugs in Figure 6 were machined, will exhibit microscopically visible individual indentations after exposure for a few seconds. In the case of a hard material, the stress imposed by the shock wave is

*Editor's Note: Magnetostriction is a term which literally implies magnetic contraction, but is generally understood to also include a number of closely allied phenomena relating to certain types of material under magnetic influence.

less than the yield stress; therefore, some finite period of time is required to fatigue the surface metal. For the soft material, the yield stress is exceeded by the implosion stress and the result is immediate damage.

Another theory, while subscribing to the high pressure shock wave as the basic damage mechanism, holds that some portion should be attributed to a chemical attack. This theory suggests that when an increment of surface is struck it undergoes a localized rise in temperature due to absorption of energy. Such localized rise in temperature may be sufficient to permit a chemical reaction between the flowing fluid and the surface material, forming a film which is broken up and washed away by subsequent exposure.

Although there may be some disagreement regarding the actual damage mechanism, one fact is clear—*there is no engineering material, available today, which can indefinitely resist the stresses imposed by severe cavitating conditions.*

FLASHING EFFECTS

When a fluid is flashing in a control valve, there are two major problems which must be considered: physical damage and decreased efficiency. It will be recalled that these were also two of the problems associated with cavitation.

Material Damage

Physical damage associated with the flashing process has a very smooth appearance in contrast to the cinder-like resemblance of cavitation damage. The appearance is often compared to a fine sandblasted surface, although in some material the surface may be even smoother. Actually, the damage mechanism probably closely resembles sandblasting. In sandblasting the grit is borne along by some compressible media, usually steam or air, and achieves a velocity approaching that of the carrier. These high velocity particles impact on the surface causing deformation and ultimately removal of the surface material.

In a flashing liquid, the volume of vapor is frequently greater than the volume of liquid so that the liquid droplets tend to achieve the high velocity of the vapor. Their impact upon a surface will have the same tendency to deform and remove material as will solid particles, although perhaps not to the same extent.

Physical damage to control valves passing flashing liquids is usually confined to the downstream section and it is not uncommon for the piping immediately downstream to be eroded. If the liquid is near saturation at the valve inlet, so that flashing begins in the upstream section of the body, the valve plug's seating surface and flow contours may also be affected.

Selection of proper materials will usually solve most flashing damage problems. The practice of specifying chromium-bearing steel alloys such as WC6, WC9 and C5 for valve body materials has significantly reduced the incidence of the flashing damage often referred to as "body washout". A good example of this is the now common specification of C5 body material for control valves on heater drain service in power generating stations.

Decreased Efficiency - Cavitation and Flashing

In one respect the cavitation and flashing phenomena produce the same type of result. The occurrence of either cavitation or flashing within a control valve will result in a decreased ability of the valve to convert pressure drop across it into mass flow rate. This problem has long been recognized for flashing, but was often completely overlooked for cavitation until recent years.

For non-vaporizing flow, the relationship between flow rate, pressure drop, and liquid specific gravity at any given valve opening is usually expressed in a very simple formula which is often referred to as the "Basic Liquid Sizing Equation." This equation, which can be derived easily using the Bernoulli Theorem and the Continuity Equation in conjunction with some simplifying assumptions is:

$$q = C_V \sqrt{\Delta p / G_f} \qquad (6)$$

where

q = liquid flow rate in gpm
Δp = pressure differential across valve in psi
G_f = liquid specific gravity (water at 60° F = 1.0)
C_V = liquid flow coefficient

From this it may be observed that the flow rate is proportional to the square root of the pressure drop and that the constant of this proportionality is the liquid flow coefficient, C_V.

Most control valve manufacturers determine C_V in accordance with a standard procedure which, if adhered to, permits coefficients produced in the test facilities of one manufacturer to be compared directly with those of another manufacturer. Among other stipulations, the procedure requires test valve pressure drops to be within a 5-10 psi range and the downstream tap pressure to have a minimum value of 2 psig. Using water with a specific gravity of 1.0, flow rates are measured at each of three pressure drops within the above range. These flow rates are then divided by the square root of the appropriate pressure drop to yield C_V values. These experimental values are subsequently averaged to produce a result that is indicative of the capacity of the test valve at the test opening. Under such conditions the observed C_V for any type of control valve should be within the expected limits of experimental accuracy.

Increasing the pressure drop beyond the above range, while holding a constant inlet pressure, will ultimately result in a point where the calculated C_V values will begin to decrease. This apparent decrease in the flow coefficient has been shown to be an indication of cavitation within the valve, with the point of initial departure from the proportional relationship between q and $\sqrt{\Delta p}$ indicating incipient cavitation of the main flow stream. It is entirely possible that some extraneous cavitation due to localized high velocity regions, caused by sharp protrusions in the flow stream, may occur at slightly lower pressure drops.

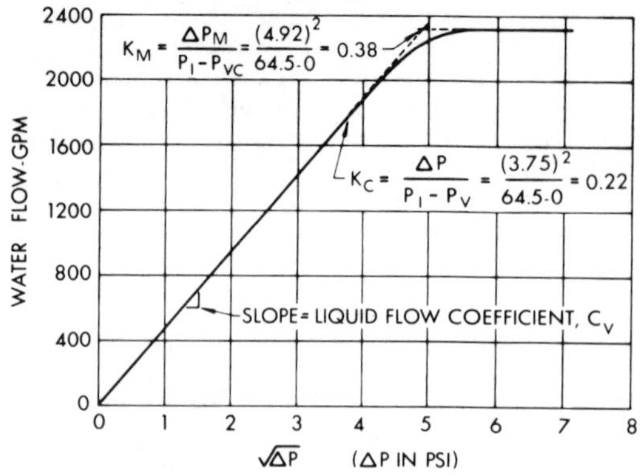

Figure 7. TYPICAL TEST CURVE SHOWING K_c AND K_m POINTS.

CAVITATION INDEX

A dimensionless ratio, experimentally determined from a plot of q versus $\sqrt{\Delta p}$ at fixed values of inlet pressure and valve opening, is used to describe the point of initial departure from a proportional relationship. This ratio is called the *Cavitation Index,* K_c, and is defined as follows:

$$K_c = \frac{p_1 - p_2}{p_1 - p_v} = \frac{\Delta p}{p_1 - p_v} \qquad (34)$$

where

p_1 = valve inlet pressure in psia
p_2 = valve outlet pressure in psia
p_v = vapor pressure of liquid at valve inlet in psia (will also be the pressure at the vena contracta)
Δp = pressure differential across body at point of deviation in psi

A typical plot of q versus $\sqrt{\Delta p}$, showing the K_c point, is shown as Figure 7.

Inspection of the defining relationship for K_c indicates that it represents the fraction of the total available drop from the inlet pressure to vapor pressure which may be taken as drop across the valve before cavitation begins. For a variety of control valves, K_c values have been observed in the approximate range of 0.15 to 0.80. The reasons for this variation will become apparent in a later discussion.

The vapor bubbles formed at the vena contracta have a much lower specific weight than the liquid at the same pressure and temperature. Therefore, as bubbles begin to form, the average specific weight at the vena contracta starts to decrease and the flow begins to be choked.

After cavitation has begun, further decreases in the valve outlet pressure (i.e., increased pressure drops) result in increased vaporization, increased cavitation intensity and further decreases in the apparent liquid flow coefficient. Referring again to Figure 7 it is noted that, with sufficient pressure drop, the flow becomes fully choked, so that increasing the pressure drop results in no increase in flow rate. Increasing the pressure drop after choked flow has been reached will result in increased amounts of cavitation damage until the valve outlet pressure is decreased to the value that will permit flashing.

Whether choked flow is the result of the specific volume increasing faster than the velocity at the vena contracta, or is the result of reaching the sonic velocity of the liquid-vapor mixture, is not known. It is obvious, however, that the results are in agreement with sonic flow criteria where pressure impulses cannot move upstream against the sonic barrier and thus further changes in downstream pressure cannot affect the upstream flow pattern.

VALVE RECOVERY COEFFICIENT

An additional experimental coefficient is determined by flow test to approximate the point above which no increase in flow rate is achieved for an increase in pressure drop. This coefficient is called the *Valve Recovery Coefficient* because its value is a measure of the amount of pressure that is recovered between the vena contracta and the valve outlet. It is designated by the symbol, K_m, and is defined as follows:

$$K_m = \frac{\Delta p_m}{\Delta p_{vc}} = \frac{\Delta p_m}{p_1 - p_{vc}} \qquad (35)$$

where:

Δp_m = pressure drop across the valve body required to produce choked flow in psi

Δp_{vc} = pressure drop between valve inlet and vena contracta at choked flow in psi

p_1 = valve inlet pressure in psia

p_{vc} = vena contracta pressure at choked flow in psia

On an experimental plot of q versus $\sqrt{\Delta p}$, such as that shown in Figure 7, K_m is determined at the intersection of the straight lines: 1) where q is proportional to $\sqrt{\Delta p}$ (i.e., where C_v is constant); and 2) where q is constant. It will be obvious from an inspection of the curve in Figure 7 that the point at which K_m is calculated, while adequately describing the ultimate choked flow rate, does introduce a slight error in the pressure drop. Analyses of a great many plots of this type have not yielded a mathematical expression to describe the shape of the curve between the K_c point and the point of initial choked flow. Values of K_m ranging from 0.20 to 0.95 have been observed in a laboratory using cold water as the test fluid. In laboratory determinations of K_m values, the inlet pressures will usually be 50 psig or greater. Since the vapor pressure of cold water is approximately 0.2 to 0.4 psia and since the vena contracta pressure will be shown to be less than vapor pressure, the contribution of the vena contracta pressure in the computation of K_m values is negligible.

From the definition, it is clear that K_m represents the fraction of the difference between inlet pressure and the choked flow vena contracta pressure that may be taken as pressure drop across the valve, before fully choked flow occurs. An understanding of the physical significance of the valve recovery coefficient and reasons for the wide range of observed values may be gained by consulting Figure 8, in which the pressure profiles through two valves

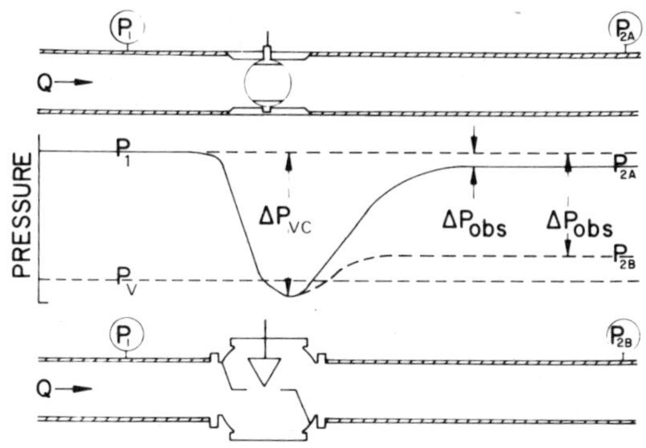

Figure 8. PRESSURE RECOVERY IN TWO STYLES OF CONTROL VALVES.

with significantly different internal geometries have been plotted. A ball valve in the full-open position is shown at the top of the illustration with a conventional globe style valve at the bottom.

The ball valve is representative of a class of devices which have come to be known as "high recovery valves" whereas the globe valve represents devices at the opposite end of the scale or "low recovery valves". Assume for both of the valves that:

1. inlet pressures are equal;
2. the flowing fluid is the same, and is at the same termperature, so that the vapor pressures are equal; and
3. overall pressure drops have been imposed which are just sufficient to produce choked flow.

It will be noted from the pressure plot of Figure 8, that the pressure at the vena contracta of each valve has fallen to the same value. This appears logical in view of the above assumptions. Note also that the vena contracta pressure is shown to be less than the vapor pressure. It will be recalled that cavitation began when the pressure at the vena contracta reached the vapor pressure. Since more flow is realized at the choked flow point than at the incipient cavitation point, it is logical that the pressure at the vena contracta must be lower than the vapor pressure.

Pressure recovery begins after the fluid has passed the vena contracta. In the case of the ball valve, which is represented by the solid line, the pressure recovery is substantial so that only a small overall drop across the valve is observed ($\Delta p_{obs} = p_1 - p_{2a}$). On the other hand the pressure recovery for the globe valve is minimal, resulting in a significantly greater overall pressure drop (Δp_{obs} =

$p_1 - p_{2b}$). If the C_V's of the two valves were equal, then equal flow rates would have been observed even though the overall pressure drops, as measured from valve inlet to outlet, were substantially different. Stated differently, the pressure drop that is effective in producing flow through any valve is the drop that exists from the inlet to the vena contracta. This is true whether the overall drop is less than, equal to, or greater than the drop required to produce choked flow.

Why do some valves exhibit very high pressure recovery while the recovery in others is very low? The answer to this question is found in the valve's internal geometry. It will be noted that the flow passage through the ball valve is relatively free of obstructions that would introduce turbulence into the flow stream. Compare this with the tortuous path that the fluid must follow in traversing the globe valve. It is axiomatic to state that anything which introduces turbulence into the flow stream will reduce pressure recovery. Pressure recovery in a well-designed venturi tube may be as much as 99%.

In addition to the conventional ball valve, other examples of high recovery control valves include butterfly valves, segmental ball valves with "V" and "U" shaped openings, and certain flow-closed angle valves. Most globe valves and flow-opened angle valves fall into the low recovery classification. Obviously, examples can be found spanning the complete range of pressure recovery. Generally, control valves that exhibit high recovery in the full-open position lose at least some of their recovery efficiency at decreased openings, while valves with low recovery at full opening, tend to remain that way throughout their travel. Table IV lists representative full-open K_m values for a number of control valve types; however, for specific sizes and types it is recommended that the literature published by the appropriate manufacturer be consulted.

TABLE IV.
REPRESENTATIVE FULL-OPEN K_M VALUES

Body Type	K_m
Globe: single port, flow opens	0.70 - 0.80
Globe: double port	0.70 - 0.80
Angle: flow closes	
Venturi outlet liner	0.20 - 0.25
Standard seat ring	0.50 - 0.60
Angle: flow opens	
Maximum orifice	0.70
Minimum orifice	0.90
Ball Valve:	
V-Notch	0.40
Conventional	0.30
Butterfly valve:	
60° open	0.55
90° open	0.30

Referring again to the pressure plot in Figure 8 it may be seen that the ratio of the overall drop, Δp_{obs}, to the drop from the inlet to the vena contracta, Δp_{vc}, is a measure of pressure recovery. This observation agrees with the definition of the valve recovery coefficient, K_m, as previously stated. Note that the drop, Δp_{obs}, is what would be obtained as the difference between readings on pressure gauges installed upstream and downstream of the body. The pressure drops used for Figure 7, where the method for experimentally determining K_m values is shown, were obtained in just such a manner.

By rearranging the defining K_m equation, an expression may be obtained for the maximum effective overall pressure drop across the valve body:

$$\Delta p_m = K_m\ (p_1 - p_{vc}) \qquad (35a)$$

This equation, however, still cannot be solved for those cases where p_{vc} is something other than approximately zero, even though K_m values can be obtained from the manufacturers' literature.

THEORETICAL CRITICAL PRESSURE RATIO

The problem, therefore, is to find a way to predict, for any fluid at any vapor pressure, what the vena contracta pressure will be at the lowest overall pressure drop across the valve, which will produce choked flow. In order to do this on a theoretical basis, it is necessary to make an assumption concerning the thermodynamic process that the fluid undergoes between the valve inlet and the vena contracta. The end points of the process from valve inlet to valve outlet will have equal enthalpies, providing that the fluid is all liquid and that the flow areas are equal at both locations. In order for choked flow to exist, however, some vapor must be present at the vena contracta. Also, the fluid velocity is obviously much greater there than at the inlet; therefore, the assumption of an isenthalpic process is impractical.

A constant entropy, or istentropic, process, requires that no fluid friction exist; however, in a flowing fluid stream this is impossible. Thermodynamic texts will agree, however, that most of the losses in flowing fluid streams which undergo contraction and subsequent expansion occur during the expansion. The frictional losses, therefore, will be much lower during the contraction to the vena contracta and this fact is used to justify the assumption of a constant entropy process to the vena contracta.

In addition to the constant entropy assumption, assume that the fluid at the valve inlet is in an all-liquid, but saturated condition. If increasing flow is induced through the valve by successive reductions in the pressure at the vena contracta, the quality of the resulting mixture may be computed for each value of vena contracta pressure by:

$$X = \frac{S_{vc} - S_f}{S_{fg}} \qquad (36)$$

where

X = mixture quality
S_{vc} = S_1 = entropy at inlet and vena contracta in BTU/lb-°F
S_f = saturated liquid entropy at vena contracta pressure in BTU/lb-°F
S_{fg} = difference between saturated vapor and liquid entropies at vena contracta pressure in BTU/lb-°F

With the quality of the mixture known, rearrangement of the above relationship with substitution of specific volume for entropy, may be used to compute the vena contracta specific volume at each value of assumed vena contracta pressure.

$$\upsilon_{vc} = \upsilon_f + X \upsilon_{fg} \qquad (37)$$

where

υ_{vc} = vena contracta specific volume in ft^3/lb
υ_f = saturated liquid specific volume at vena contracta pressure in ft^3/lb
υ_{fg} = difference between saturated vapor and liquid specific volumes at vena contracta pressure in ft^3/lb

A curve may be plotted from this data, defining the relationship between the pressure and specific volume of the fluid at the vena contracta for the assumed saturated inlet pressure to the valve. A typical curve for such a plot is shown by Figure 9.

An expression may be derived for the theoretical mass rate of flow through restrictions in control devices. One such expression is:

$$W/A = 1/\upsilon \sqrt{2g \int \upsilon_{vc}\, dp_{vc}} \qquad (38)$$

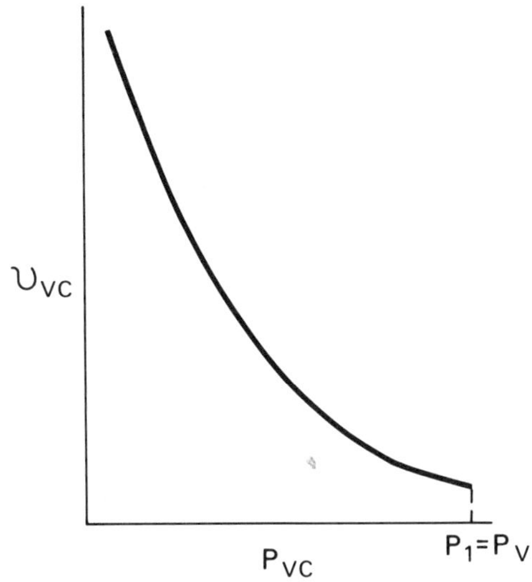

Figure 9. TYPICAL PLOT OF RELATIONSHIP BETWEEN PRESSURE AND SPECIFIC VOLUME AT VENA CONTRACTA.

where

W/A = theoretical mass rate of flow per unit of flow area in $lb/s\text{-}ft^2$
g = acceleration of gravity in ft/s^2
υ = vena contracta specific volume at upper limit of integration in ft^3/lb
υ_{vc} = vena contracta specific volume in ft^3/lb
dp_{vc} = differential of vena contracta pressure in psfa (lb/ft^2 abs.)

Since the mathematical relationship between vena contracta specific volume and pressure is not known, the integral in the above equation must be obtained graphically. Values obtained for this integral then permit the computation of W/A at a variety of vena contracta pressures for which values may then be plotted. Such plots have a steep slope as the vena contracta pressure is reduced from the saturated liquid condition. The slope of the curve begins to fall off, however, as further reductions are made until eventually a zero slope will be observed indicating a theoretical choked flow condition. It will be recalled that such a condition was also observed experimentally (see Figure 7). Computing mass flow rates at vena contracta pressures lower than that, at which

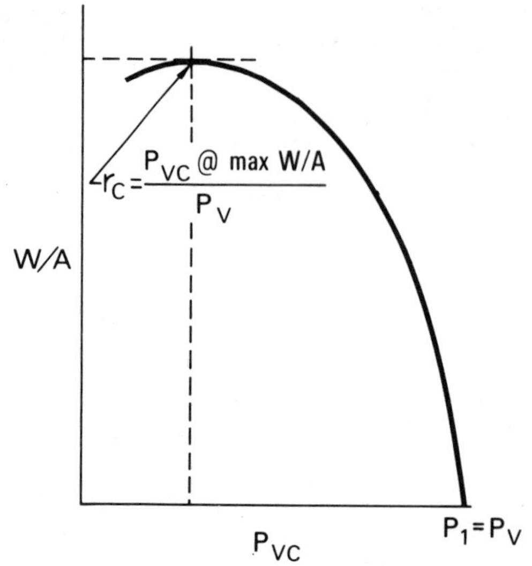

Figure 10. TYPICAL PLOT OF RELATIONSHIP BETWEEN W/A AND VENA CONTRACTA PRESSURE.

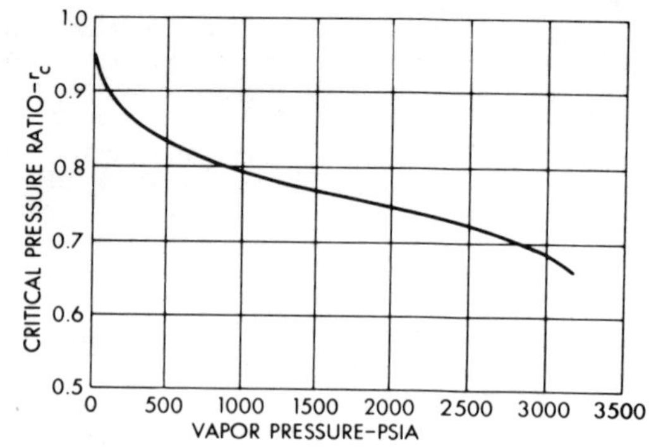

Figure 11. CRITICAL PRESSURE RATIOS FOR WATER.

choked flow is indicated, show apparent reductions in flow rate. In actual practice, however, the flow rate is observed to reach a choked condition and to maintain this rate despite further reductions in downstream valve pressure. Figure 10 illustrates the typical shape of a plot of W/A versus p_{vc}. The choked flow rates have been experimentally observed to exceed the theoretical values computed using the techniques outlined here.

The ratio of p_{vc} at maximum W/A to the p_v which was applicable to the entire plot in Figure 10 is defined as the *theoretical critical pressure ratio* and is given the symbol, r_c.

$$r_c = \frac{p_{vc}}{p_v} \tag{39}$$

By rearranging the above equation, it may be seen that the vena contracta pressure required to produce choked flow will be obtained as the product of the theoretical critical pressure ratio and the vapor pressure.

$$p_{vc} = r_c p_v \tag{39a}$$

The relationship of r_c to p_v over the range of values from zero to the thermodynamic critical pressure, p_c, is shown for water in Figure 11. In order to predict the choked flow vena contracta pressure, the plot is entered at the appropriate p_v, the corresponding r_c read on the ordinate and the product of the entering and exiting values obtained.

While compiling such curves for a variety of fluids including two refrigerants, a hydrocarbon and a single-element liquid in addition to water, it was noted that a marked similarity existed in the shape of the curves, as well as in the range of r_c values. With little sacrifice in accuracy, these may be combined into a single curve as shown by Figure 12, where the abscissa is p_v/p_c rather than p_v. Critical pressures, p_c, for several commonly encountered fluids are given in Table V.

TABLE V. CRITICAL PRESSURE OF VARIOUS FLUIDS, PSIA

Fluid	Critical pressure, psia
ammonia	1636
argon	705.6
butane	550.4
carbon dioxide	1071.6
carbon monoxide	507.5
chlorine	1118.7
Dowtherm A	465
ethane	708
ethylene	735
fluorine	808.5
helium	33.2
hydrogen	188.2
hydrogen chloride	1198
isobutane	529.2
isobutylene	580
methane	673.3
nitrogen	492.4
nitrous oxide	1047.6
oxygen	736.5
phosgene	823.2
propane	617.4
propylene	670.3
Refrigerant 11	635
Refrigerant 12	596.9
Refrigerant 22	716
water	3206.2

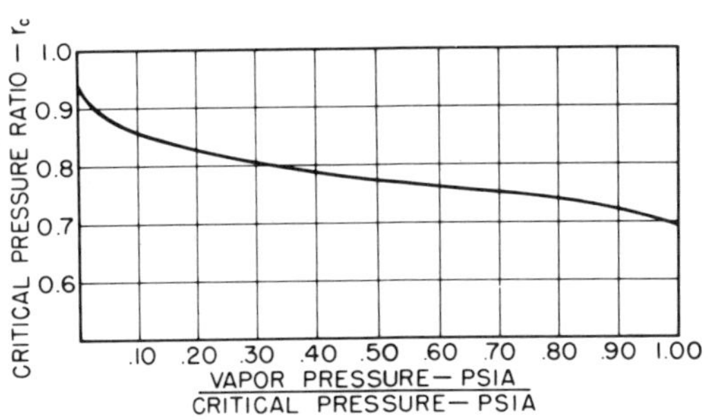

Figure 12. UNIVERSAL CRITICAL PRESSURE RATIO CURVE.

CHOKED FLOW EQUATION

Having shown that the choked flow vena contracta pressure may be approximated by the product of r_c and p_v, this product may be substituted for p_{vc} in Equation 35a so that it will then read:

$$\Delta p_m = K_m \ (p_1 - r_c p_v) \qquad (40)$$

This is referred to as the *Choked Flow Equation* and is intended for use in both cavitation and flashing situations. For the special case where the liquid is saturated at the valve inlet, p_v may be substituted for p_1 and then factored out, leaving the relationship:

$$\Delta p_m = K_m p_v (1 - r_c) \qquad (41)$$

The sole purpose of either form of the choked flow equation is to determine the maximum effective overall pressure drop across a control valve. Note from the terms in the equation that, with K_m, it takes into account the type of valve body being used and that, with both p_v and r_c, the thermodynamic properties of the flowing liquid are considered.

In normal valve sizing operations where the pressure conditions for the valves being sized have already been fixed, and where it cannot positively be determined by inspection, that cavitation of flashing will not occur, the choked flow equation should always be solved for Δp_m. The value obtained must then be compared with the actual Δp that the valve will see in service. The smaller of the two values will then be used in the liquid sizing equation to ascertain the required liquid flow coefficient, C_v.

$$C_v = \frac{q}{\sqrt{\Delta p / G}} \qquad (6a)$$

It may be shown, referring again to Figure 7, that if the calculated Δp_m is less than the actual service Δp, the flow rate will be in the choked region. On the other hand, if the calculated Δp_m is greater than the actual Δp, it simply indicates that the valve will be operating in the zone where q is proportional to $\sqrt{\Delta p}$.

Ignoring Δp_m when it is the applicable sizing drop and using, instead, the actual Δp in the sizing equation will result in a calculated C_v that is smaller than what is actually required. Thus, the result may be an undersized valve with all the attendant problems.

When Δp_m is the applicable sizing drop as a result of a cavitating situation, the valve user is thus alerted that severe cavitation damage problems may be encountered so that preventive or protective measures can be taken. The question often arises regarding the severity of cavitation damage at the drop defined by K_m, as compared with the K_c point. In general, it has been found that permitting drops up to the K_m point will not seriously affect the life expectancy of the valve. Obviously there is no magic line of demarcation with serious damage on one side and none on the other. There is growing evidence, however, that in some high recovery valves, significant damage will be experienced at the K_m point.

When process systems are still in the design stage and valve types and locations are being selected, the choked flow equation can be of significant assistance. If, for example, it is determined that a certain valve will exhibit choked flow, a different location, such as at a lower elevation or at the beginning rather than the end of a long line, may permit a higher inlet pressure that would eliminate the problem. It may also be possible to select a type of valve with a higher K_m that would permit more pressure drop.

Experimental Verification Tests

Over a period of several years the choked flow equation has been successfully used to size a wide variety of valves, on a wide variety of fluids and over a wide inlet condition ranging from saturated to substantially subcooled. No reports of incorrect sizing have been received from users; however, until recently no controlled laboratory tests have been conducted specifically for the purpose of substantiating the equation.

Refrigerant-12 (dichlorodifluoromethane-CCl_2F_2) was chosen as the test fluid because it fulfilled several important criteria. The fact that it is most commonly used as a refrigerant was of no importance in its selection. It was desired that the tests be performed with values of vapor pressure covering as much of the range from 0 to p_c as possible. To accomplish this without an excessively expensive test installation, required that the critical pressure and temperature, p_c and T_c respectively, be moderate. These values for R-12 are 596.9 psia and 233.6 °F respectively. By contrast, comparative values for water, which would otherwise be an ideal fluid, are 3206.2 psia and 705.4 °F respectively. Another requirement was that reasonably complete thermodynamic property data be available, and in this regard, the published information was more than adequate. Since the tests were to be conducted in an indoor laboratory the fluid had to have a very low toxicity and again, R-12 was satisfactory.

Prior to installation in the actual test loop, the entire test section, which included the test valve, upstream and downstream pressure control valves, upstream and downstream pressure taps, and an inlet temperature thermometer well, was installed in a laboratory water test facility. Here the test valve liquid flow coefficient, C_v, and valve recovery coefficient, K_m, were carefully determined in accordance with practices outlined earlier in this chapter.

After installation of the test section into the test loop, the system was evacuated to an absolute pressure of 0.01 in. Hg or less, before charging the system with R-12. The precaution avoided contaminating the R-12 with atmospheric air which would alter the fluid properties. In addition to the test valve section and appropriate flow, pressure, and temperature instrumentation, the closed-loop test system incorporated a water-cooled condenser, a pump, and a steam heater.

Flow versus $\sqrt{\Delta p}$ curves similar to that shown as Figure 7 were acquired at numerous vapor pressure values. For each value of vapor pressure, data runs were made at several values of p_1/p_v. The choked flow for each of the above runs was compared with the flow predicted using the basic liquid sizing equation in which the applicable pressure drop had been calculated from the choked flow equation. The combination of these equations would appear as follows:

$$q\ pred = C_v \sqrt{\Delta p_m/G} \tag{6}$$

and (36)

$$= C_v \sqrt{K_m\ (p_1 - r_c,\ p_v)/G}$$

In Figure 13, the percent errors in predicted flow rates are plotted as a function of the variable, $(p_1 - p_v)/p_v$. The numerator is a measure of the amount of subcooling and this has been divided by the vapor pressure to make the variable dimensionless. It may be seen that at values of the above variable, greater than approximately 0.8, no errors were noted. However, at decreasing values, increasing errors were observed which approached 20% when the test fluid was saturated. Significantly, the observed choked flow rates were greater than predicted flow rates, so that the use of the choked flow equation to predict the appropriate sizing drop would result in an oversized, rather than an undersized installation. The data shown in Figure 13 is shown only for descriptive purposes and is not intended to be used for actual sizing problems.

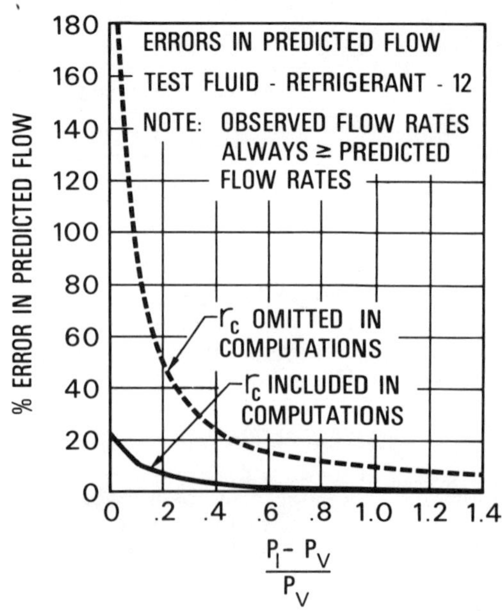

Figure 13. RESULTS FROM R-12 REFRIGERANT TESTS.

To show the necessity for the inclusion of r_c in the choked flow equation, choked flow sizing drops were calculated, omitting r_c from the equation, and flow rates were predicted using these drops. Percent errors in predicted flow were then plotted (Figure 13). Examination of these results will show that, even with substantial subcooling, errors of 7% were observed. Obviously, when p_v is very small in relation to p_1, this error will disappear. Near saturated inlet conditions, the errors become very great and theoretically would be infinite at saturation.

Although use of the choked flow equation does result in some sizing errors near saturated conditions, its use is recommended because it is the most accurate method presently available.

NOMENCLATURE

C_v	Liquid flow coefficient.
G_f	Liquid specific gravity (water at 60°F - 1.00).
g	Acceleration of gravity, in ft/s^2.
K_c	Cavitation Index.
K_m	Valve Recovery Coefficient.
p_1	Valve inlet pressure, in psia.
p_2	Valve outlet pressure, in psia.
p_c	Thermodynamic critical pressure, in psia.
p_v	Vapor pressure of liquid at valve inlet, in psia.
p_{vc}	Vena contracta pressure, in psia.
dp_{vc}	Differential of vena contracta pressure, in psfa (lb./ft^2 abs).
Δp	Pressure differential, in psia.
Δp_m	Pressure differential across the valve body required to produce choked flow, in psi.
Δp_{vc}	Pressure differential between valve inlet & vena contracta, in psi.
q	Liquid flow rate, in gpm.
r_c	Theoretical Critical Pressure Ratio.
S_f	Saturated liquid entropy, in BTU/lb-°F.
S_1	Entropy at valve inlet, in BTU/lb-°F.
S_{fg}	Difference between saturated vapor & liquid entropies, in BTU/lb-°F.
S_{vc}	Entropy at vena contracta, in BTU/lb-°F.
v_f	Saturated liquid specific volume, in ft^3/lb.
v_{fg}	Difference between saturated vapor & liquid specific volumes, in ft^3/lb.
v_{vc}	Vena contracta specific volume, in ft^3/lb.
W/A	Theoretical mass rate of flow per unit of flow area, in lb/sec-ft^2.
X	Mixture quality.

BIBLIOGRAPHY

1. Eisenberg, Phillip, February, 1963, "Cavitation," *International Science And Technology.*
2. Knapp, Robert T., 1958, "Cavitation and Nuclei," *Transactions Of ASME.*
3. Wheeler, W.H., 1959, "Identation of Metals by Cavitation," ASME Paper No. 59-Hyd-15.
4. Stiles, G.F., November, 1961, "Cavitation in Control Valves," *Instruments And Control Systems,* Vol. 34, No. 11.
5. Allen, W.F., Jr., 1951, "Flow of a Flashing Mixture of Water and Steam Through Pipes and Valves," *Transactions of ASME.*

SUPPLEMENTARY NOTES ON CAVITATION AND FLASHING IN VALVES*

Cavitating flow in valves is much more common than may generally be realized. Transient cavitation is commonly heard as a "hiss" or a whistle when an on-off valve is either opened or closed, whether in a high pressure power plant or at the kitchen sink. Liquid control valves may give satisfactory service for months or even years, while operating in a moderately cavitating mode with the characteristic hissing or roaring sounds which accompany the process. Yet, in extreme cases cavitation may destroy valve trim or produce catastrophic damage to adjacent piping in a matter of hours. Between these extremes, there lies a broad gulf of uncertainty which challenges the skill and experience of valve designers and application engineers who must select valves for difficult service.

The key point is that cavitation may vary broadly in intensity. In many applications, the cost or complexity of the steps required to *eliminate* cavitation would be prohibitive; therefore, the objective often is to either limit the intensity of cavitation to an "acceptable" level or to confine the cavitation to a region where its effects can be tolerated.

There can be no absolute definition of an acceptable intensity of cavitation, because the criteria for acceptability are certainly dependent upon valve application requirements. For instance, limitations on environmental noise levels might dictate that cavitation be limited to an intensity much lower than that which would cause rapid damage to valve trim.

Research on cavitation in valves has produced a great amount of data but few confirmable conclusions. Unfortunately, due to variations in the criteria used to evaluate valve tests, results produced by various researchers are difficult to compare. The purpose of this supplement is to illustrate, at least qualitatively, a correlation between some of the important phenomena observed in cavitating flow tests.

Noise and Vibration

Tests have been conducted in a water flow loop using sensitive accelerometers to monitor valve vibration under various conditions of flow, line pressure, and differential pressure. Electronic analysis equipment permitted frequency and amplitude relationships to be identified. Simultaneous notation of audible noises permitted a qualitative correlation of the instrument-recorded data with the observations which might be noted in other valves under field conditions.

Results are best understood if a *system cavitation parameter*, K_{sc}, is introduced as a measure of the tendency of system conditions to cause cavitation in a valve:

$$K_{sc} = \frac{\Delta p}{p_1 - p_v} \tag{42}$$

where

K_{sc} - System cavitation parameter.
Δp - Pressure drop across valve, psi.
P1 - Absolute upstream pressure, paia.
Pv - Absolute vapor pressure of fluid flowing, psia.

In typical tests at low values of K_{sc} (low flows and pressure drops), the instrumentation detects no vibrations different from the background levels measured with no flow. In this range, there is no flow-induced noise near the valve. As K_{sc} is increased, a "lower threshold" value is reached where small amplitude, very high frequency vibrations are detected (generally at frequencies above the audible range). With a slight increase in K_{sc} above this threshold value, the instrumentation detects vibrations at somewhat lower frequencies and an audible hissing sound can be heard near an uninsulated valve. It is considered that this range corresponds closely with the "critical cavitation index" as discussed in Reference (1) and numerous other papers.

With further increase in K_{sc}, above the lower threshold value, measurements reveal vibrations of progressively lower frequencies, but these frequencies remain in the acoustic range and amplitudes remain relatively small until an "upper threshold" value of K_{sc} is attained. Between the lower and upper thresholds, audible noise increases in intensity and becomes lower in frequency, producing a roaring sound as cavitation gradually becomes more intense.

The upper threshold value of K_{sc} is characterized by the onset of random vibration patterns in the low frequency range (e.g., well below 100 Hz). Since

*Earl A. Bake, Rockwell Manufacturing Co., Pittsburgh, Pennsylvania.

*Editor's Note: Continuing investigation and development is at present in progress by several different companies on the phenomena and characteristics of cavitation. Some of the findings of Rockwell Manufacturing Company are given here as a supplement to the preceding section by G.F. Stiles.

piping systems often have natural frequencies in this range, measured amplitudes are influenced by the piping configuration; nevertheless, visible mechanical vibration is common at K_{sc} values above the upper threshold, and these may take on dangerous proportions when a valve is poorly supported. The audible noise produced in this range is the characteristic sound of bulk cavitation; often compared to rocks and gravel passing through the valve.

The numerical values of K_{sc} at the lower and upper thresholds vary broadly among valve types and are outside the scope of this supplement. Both thresholds occur at lower K_{sc} values with high-recovery valve types than with low-recovery valves.

Cavitation Effects on Flow

The preceding section on Cavitation and Flashing in Control Valves notes that a reduction of the calculated liquid flow coefficient of a valve, C_v, will be observed if pressure drop is increased beyond a certain limit at a constant upstream pressure. If the pressure drop is increased further, while holding upstream pressure constant, a second point is reached where flow no longer increases with increasing pressure drop. The cavitation index, K_c, defines the first point where reduction of C_v can be observed from experimental data. The recovery coefficient K_m approximates the second point where flow becomes choked.

K_c and K_m can be considered as two values of the system cavitation parameter K_{sc}, as defined in this supplement, characterizing two levels of cavitation intensity in a specific valve. Flow data were logged along with noise and vibration data in the aforementioned tests in the attempt to correlate K_c and K_m with the cavitation thresholds discussed above. The following observations are typical of observations made in testing a number of different types of valves:

1. The lower threshold of cavitation, as determined by vibration measurements or audible noise, was at a lower value of the system cavitation parameter, K_{sc}, than that characterized by the cavitation index, K_c. Practically, this means that there was always detectible cavitation at pressure drops below that where an effect on the valve flow coefficient could be discerned.
2. At conditions where an effect on the valve flow coefficient could just be detected ($K_{sc} = K_c$), the intensity of cavitation, as observed from audible noise and vibration measurements, was quite variable from valve to valve, ranging from mild to moderately severe. Hence, it appeared that the cavitation index, K_c, determined from flow data does not provide a positive indication of the intensity of cavitation which will be produced by a valve in service.
3. The upper threshold of cavitation, as determined from noise and vibration tests, correlated well with the recovery coefficient, K_m, in most tests. Generally, operation at system cavitation parameters exceeding K_m, produced visible vibration and severe "rocks and gravel" noise. An exception was noted in cases of nearly-closed valves (e.g., butterfly valve 10° open) where data revealed choked-flow without excessive noise or low-frequency vibration.

In summary, the recovery coefficient of a valve, K_m, as defined in the preceding section has been found to be a good indicator of conditions producing severe or bulk cavitation, but the cavitation coefficient, K_c, does not correlate well with other experimental methods of establishing the point of inception of cavitation in a valve. Methods based on noise and vibration indicate that cavitation begins at lower pressure drops than would be indicated by K_c; the discrepancy is variable, but quite significant in some cases. While the intensity of the cavitation which is produced at system cavitation parameters below K_c is often mild or moderate, it might, nevertheless, prove to be troublesome in certain applications. Specifically, environmental noise might be unacceptable or damage could be produced in long-term service of valves which do not employ cavitation-resistant trim. Therefore, the cavitation index, K_c, is useful in valve sizing and control analysis work, but caution should be exercised in applying this index to any decision regarding whether or not a valve application will limit cavitation to an "acceptable" level.

REFERENCES

1. Tullis, J. Paul and Skinner, M.M., November 1968, "Reducing Cavitation in Valves," *Journal of the Hydraulics Division*, ASCE, HY6, Proc. Paper, pp 1475-1488.
2. Tullis, J. Paul and Marschner, B.W., January 1968, "Review of Cavitation Research on Valves," *Journal of the Hydraulics Division*, ASCE, HY1, Proc. Paper, pp 1-16.

PART 3 CONTROL VALVE NOISE

(E. E. Allen)

INTRODUCTION

The enactment of antinoise laws by local, state, and federal governments has provided the primary impetus for much of the attention being given to the problem of industrial noise. Excessive noise levels are inherent to many fluid process and transmission systems. Control valves are often a major source of these noises. This section is devoted to: a brief review of noise terminology; identification of the sources of control valve noise; presentation of an empirical equation for prediction of valve noise; and discussion of the methods of noise control. The information presented herein is based on the results of an extensive research study, conducted by the writer's company, on noise generated by flow through control valves and adjacent piping.

NOISE TERMINOLOGY

Noise is commonly defined as unwanted or annoying sound. Sound is the auditory sensation evoked by oscillatory pressure fluctuations propagated in the ambient atmosphere. Socio-political behavior of the average person in response to sound environment is the accepted criterion for determining what is noise. Thus, dependent upon the audience, a specific sound could be considered either acceptable or annoying.

Because of its vague definition, neither an absolute method of measurement nor specific dimensions of noise have been devised. For lack of a better method, noise is frequently described or specified by the physical characteristics of sound. The definitive properties of sound are the magnitude of sound pressure and the frequency of pressure fluctuation, as illustrated by Figure 14.

Loudness of a sound is determined primarily by the amplitude of sound pressure (p_s). Sound pressure measurements are normally root mean square values of sound pressure expressed in microbars. Because the range of sound pressure of interest in noise measurements is about 10^8 to 1, it is customary to deal with sound pressure level instead of sound pressure. Sound pressure level is a logarithmic function of the relative amplitude of sound pressure and is expressed mathematically as:

Sound Pressure Level (SPL) =

$$20 \log_{10} \frac{Ps}{0.0002 \ \mu \ \text{bars}} \text{ decibels (dB)} \tag{43}$$

The arbitrarily selected reference sound pressure of 0.0002 microbars is approximately the sound pressure required at 1000 Hz to produce the faintest sound that the average young person with normal hearing can detect. The characteristic of the SPL scale is such that each change of 6 decibels in level represents a change in the amplitude of sound pressure by a factor of two.

The apparent loudness of a sound varies not only with amplitude of sound pressure, but also as a function of frequency. The human ear responds to sounds having a frequency between 20 - 18,000 Hertz. The normal ear is most sensitive to pressure fluctuations in the neighborhood of 3000 Hz; therefore, the degree of annoyance created by a specific sound is a function of both sound pressure and frequency.

Figure 15 shows a plot of sound pressure level *vs.* frequency. On this plot are

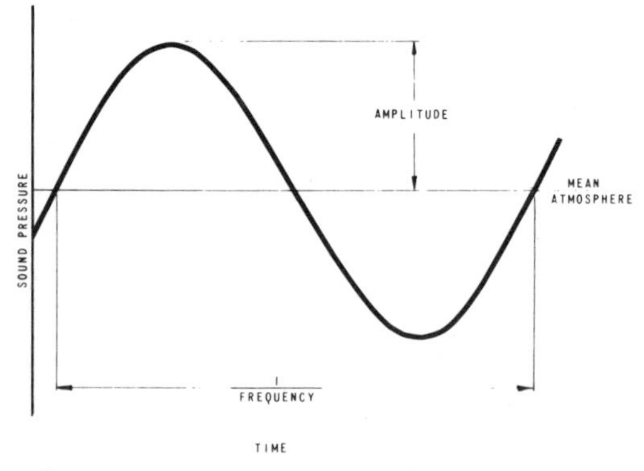

Figure 14. DEFINITIVE PROPERTIES OF SOUND.

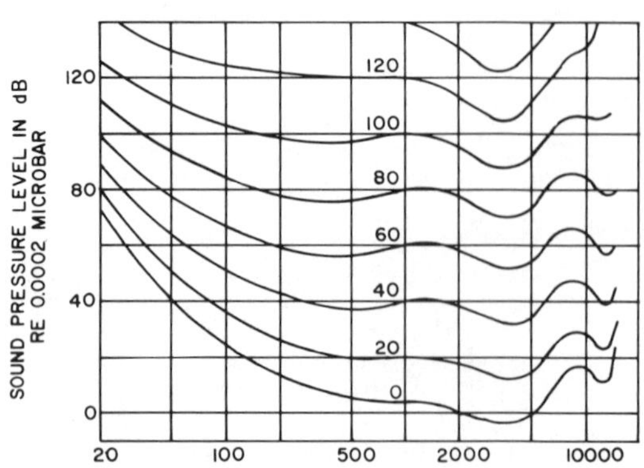

Figure 15. EQUAL LOUDNESS CURVE.

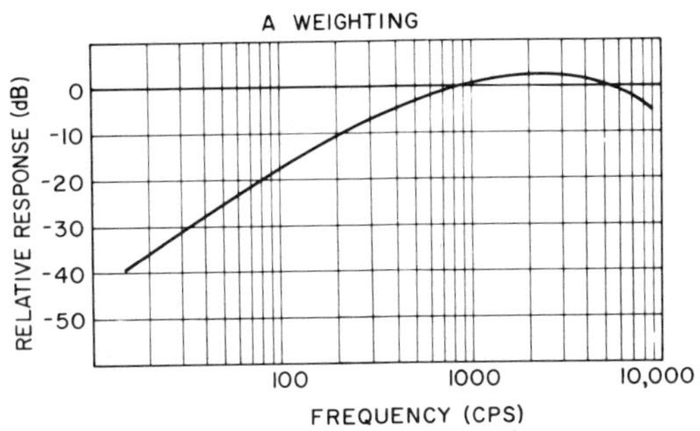

Figure 16. 'A-WEIGHTING' CURVE.

TABLE VI. APPROXIMATE SOUND LEVEL OF FAMILIAR SOUNDS.

Sound	*Sound Level Decibels*
Pneumatic rock drill	130
Jet takeoff (at 200 ft.)	120
Boiler factory	110
Electric furnace (area)	100
Heavy street traffic	90
Tabulation machine room	80
Vacuum cleaner (at 10 ft.)	70
Conversation	60
Quiet residence	50
Electric clock	20

equal-loudness contours for the average ear. Each contour gives combinations of sound pressure levels and frequencies that seem to be equally loud; that is, have the same loudness level. For example, a sound at 30 Hz and a sound pressure level of 100 dB is identical in loudness level to a sound at 1000 Hz and a sound pressure level of 80 dB.

Sound pressure level measurements are often weighted to adjust the frequency response. Weighting that attenuates the lower frequencies to approximate the equal loudness contours of the human ear is called "*A-weighting.*" Figure 16 shows SPL correction as a function of frequency for A-weighted octave band analysis.

Sound intensity (I) is a measure of acoustic energy and is defined as the acoustic sound power transmitted through a unit area, perpendicular to a specified direction. The common unit of measurement for sound intensity is watts per square centimeter. Sound intensity for a plane wave is given by the relation:

$$I = \frac{ps^2}{\rho c} \tag{44}$$

where the product of the density (ρ) and sonic velocity (c) of the transmitting medium represents the characteristic impedance.

For measurement and comparison of sound intensities, it is more convenient to deal with *sound intensity levels,* than with absolute values of sound intensity. sound intensity level is commonly defined by the following relation:

Sound Intensity Level =

$$10 \log_{10} \frac{I}{10^{-16}} \quad \text{decibels} \tag{45}$$

The reference sound intensity is arbitrarily selected as 10^{-16} watts per square centimeter. This is approximately the minimum intensity audible to the average human ear at 1000 cycles per second for standard air. The characteristic of the sound intensity level scale is such that each change of 3 dB in level represents a change in sound intensity by a factor of two.

Sound Measurement

A sound-level meter is the most common means for measuring sounds and noises. Meters are available with built-in weighting networks and contiguous frequency band analyzers. Table VI presents the approximate overall sound levels of some familiar sound environments.

SOURCES OF VALVE NOISE

The major sources of control valve noise are:

1. mechanical vibration of control valve components;
2. cavitation; and
3. aerodynamic noise.

Vibration of valve components is a result of random pressure fluctuations within the valve body and/or fluid impingement upon the movable or flexible parts. The most prevalent source of noise resulting from mechanical vibration is the lateral movement of the valve plug relative to the guide surfaces. Sound produced by this type of vibration will normally have a frequency less than 1500 Hertz and is often described as a metallic rattling. The physical damage incurred by the valve plug and associated guide surfaces, as shown in Figure 17, is generally of more concern than the noise emitted.

Figure 17. FAILURE OF PLUG AND GUIDE SURFACE, LOW FREQUENCY.
Courtesy Fisher Controls Company

A second source of mechanical vibration noise is a valve component resonating at its natural frequency. Resonant vibration produces a single pitched tone, normally having a frequency between 3000 and 7000 Hertz. This type of vibration produces high levels of stress that may ultimately produce fatigue failure of the vibrating part. An example of fatigue resulting from resonant vibration is shown in Figure 18. Valve components susceptible to natural frequency vibration are contoured valve plugs with hollow skirts and flexible members such as the metal seal ring of a ball valve.

Noise that is a by-product of the vibration of valve components is usually of secondary concern and may even be beneficial since it warns that conditions exist that could produce valve failure. Noise resulting from mechanical vibration has for the most part been eliminated by improved valve design and is generally considered a structural problem rather than a noise-control problem.

Cavitation, a phenomenon that occurs in liquid flow, is another source of noise associated with control valves. This noise is caused by the implosion of vapor bubbles that are formed in the cavitation process. Noise produced by cavitation has a broad frequency range and is frequently described as ranging from a hissing sound to a rattling sound similar to that which would be anticipated if gravel were in the fluid stream. Cavitation may produce severe damage to the solid boundary surfaces that confine the cavitating fluid; however, noise produced by cavitation is usually of secondary concern. Flow conditions that will produce cavitation can accurately be predicted. Thus, valve noise resulting from cavitation can be eliminated by the application of appropriate limits to the service conditions.

The third and major source of control valve noise is due to the Reynolds stresses or shear forces that are a property of turbulent flow. Because of the relative velocities, high intensity levels of noise resulting from turbulent flow are more common to valves handling gas than to those controlling liquids. Noise resulting from turbulent flow of a gas is called aerodynamic noise. Aerodynamic noise can be classified as a non-periodic or random noise with the predominant frequencies occurring between 2000 and 8000 Hertz. The significance of noise having this frequency range is apparant in Figure 16.

Sources of turbulence in gas transmission lines are obstructions in the flow path, rapid expansion or deceleration of high velocity gas, and sharp turns or bends of the fluid stream. Specific areas that are inherently noisy are headers, pressure regulators, line size expansions, sharp bends and orifice meters.

Figure 18. FATIGUE FAILURE OF PLUG.
Courtesy Fisher Controls Company

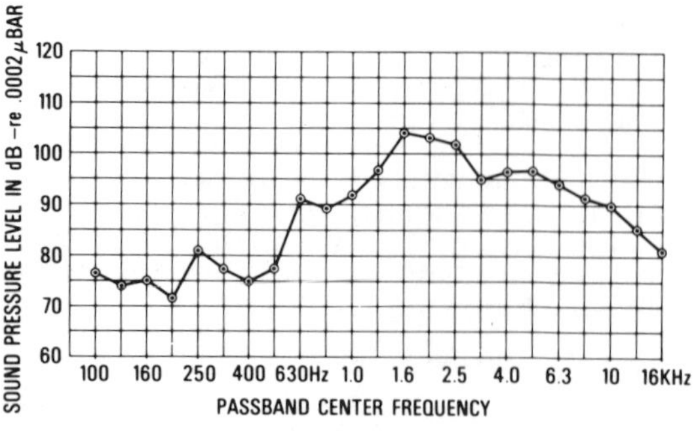

Figure 19. TYPICAL SPECTRUM OF CONTROL VALVE NOISE.

General Theory of Aerodynamic Noise

Classical theories on the generation of aerodynamic noise by flowing gas streams ascribe that the associated noise level is largely a function of the turbulent level of the fluid stream. Turbulence level is defined as the ratio of the rms velocity fluctuation to that of the mean velocity of the fluid stream at the same point.

Analytical Approach

Most of the theories, to date, on aerodynamic noise have been developed with special reference to the noise of free jets as opposed to a gas stream restrained by solid walls. A standard reference on this topic is Lighthill's paper (1).

Several hypotheses describing the acoustic power generated by gas flow through a control valve have been advanced on the postulates: (a) Lighthill's theory is applicable; and (b) the turbulent level of a flow stream can be accurately defined analytically. The validity of either or both postulates is very questionable, because:

(a) Lighthill's theory is based on a free jet and would not appear applicable to the vastly more complex flow field produced by high subsonic and sonic gas streams contained within the solid boundaries of a control valve.

(b) Very little, either qualitatively or quantitatively, is known concerning the turbulence level of a bounded fluid stream.

Empirical Advantage

Development of an analytical model which accurately describes the precise mechanism of aerodynamic noise generation by control valves is very unlikely, in view of the complexity of the problem and the lack of knowledge concerning absolute values of the relevant parameters. It would seem obvious, therefore, that an accurate technique for predicting valve noise must be based on empiricism. Such an approach requires the noise characteristic for each valve style and adjacent piping configuration to be established by actual test. Obviously, it is neither practical nor feasible to determine noise levels for all valve applications by laboratory tests. Dimensional analysis, a very useful and powerful tool in formulating problems which defy analytical solution and must be solved experimentally, would be the logical tool for correlation and extrapolation of the test data.

Advantage of Dimensional Analysis

In fluid mechanics, the relevant parameters in any physical situation may be combined into independent, dimensionless groups that characterize the flow. Two flow streams are considered similar, if they are geometrically similar and if the dimensionless groups have constant values for both flows. This technique of expressing a physical relationship between quantities by utilizing dimensionless ratios of relevant parameters is the field of mathematics known as "dimensional analysis."

Dimensionless groups (or as they are called, numbers) may be obtained directly either from the differential equations that describe the flow field or by simply combining the relevant variables into as many independent dimensionless numbers as possible. An advantage of the latter system is that one does not need to know the governing equations or the laws for complex problems; however, one must identify all the variables.

The variables to be considered for a dimensional analysis approach to the problem of predicting aerodynamic noise generated by compressible fluid flow through control valves are p_s, Δp, V, ρ, D, μ, c, C_p, C_v and λ. The independent dimensionless numbers formed from the above list are:

1. Prandtl Number, ($P_r = C_p\ \mu/\lambda$);
2. Reynolds Number, ($Re = \rho VD/\mu$);
3. mach number, ($M = V/c$);
4. specific heat ratio, ($k = C_p/C_v$); and
5. ratio of sound pressure to differential pressure across valve, ($p_s/\Delta p$).

Basic Noise Relationship

Correlation of over 250,000 data points obtained in laboratory tests substantiates a theory proposed by C. B. Schuder, Director of Research at Fisher Controls Co., which purports that M and k are the significant, independent, dimensionless ratios effecting sound pressure. (2)

In addition to the vast amount of empirical data supporting Schuder's theory, the following brief arguments are offered.

1. The Prandtl Number, P_r, is nearly the same for all gases and can be considered insignificant in the analysis of processes where heat transfer does not play an important role. Generation of aerodynamic noise is considered an adiabatic process and thus P_r can be dropped from consideration.

2. Reynolds Number, Re, is a measure of the ratio of inertia forces to viscous forces. Thus, in the study of compressible fluids, the role of Reynolds Number as an important parameter is limited to areas concerning low velocities or the boundary layer. For control valve noise levels of concern, the principal noise generation mechanism is located in the area of high subsonic or supersonic velocity, thus the Reynolds Number would not be considered a relevant parameter.
3. From the preceding argument it is obvious that the mach number, M, should be considered the most important factor.
4. Not enough experimental data has been obtained to accurately define the influence that the specific heat ratio, k, has on noise generation; however, it has a relatively weak influence on the flow stream characteristic.

The theory briefed above led to the development of a basic relationship of the form

$$p_s^2 \propto C_g \, (\Delta p)^2 \, [f(M, k)] \qquad (46)$$

where C_g (critical flow coefficient) is used as a scaling factor to relate valve size, and f(M, k) is a complex function that must be determined experimentally for each valve style, size, opening, and adjacent piping configuration.

Conclusions

Examination of relationship (46) for its most practical application as a tool for predicting valve noise produced the following conclusions:

1. In view of argument (4) presented above, and the fact that k remains relatively constant for most gases; at this "state-of-the-art," k can justifiably be dropped from the complex f(M, k).
2. Test data shows f(M) can be presented as $f(\Delta p/p_1)$, providing a substantial gain in simplicity without any significant change in final results.
3. Acceptance of a prediction accuracy of ± 5 dB would accommodate a 10 to 1 reduction in graphical presentation of information required for noise prediction of greater accuracy.

NOISE PREDICTION TECHNIQUE

Graphical solution of the following equation, based on relationship (46) and the arguments above, provides a very expedient and accurate technique for predicting ambient noise levels resulting from the flow of compressible fluids through control valves.

$$SPL = SPL_{\Delta p} + \Delta SPL_{C_g} + \Delta SPL_{\Delta p/p_1} + \Delta SPL_K \qquad (47)$$

where

SPL = overall noise level in dB at a predetermined point (48 in. downstream of the valve outlet and 29 in. from the pipe surface).

$SPL_{\Delta p}$ = base SPL in dB, determined as a function of Δp.

ΔSPL_{C_g} = correction in dB for C_g.

$\Delta SPL_{\Delta p/p_1}$ = correction in dB for valve style and pressure ratio.

ΔSPL_K = correction in dB for acoustical treatment; i.e., heavy wall pipe, insulation, inline silencers, etc.

The following example serves as a guide for using the noise prediction technique. Information presented in Figures 20 through 31 facilitates the prediction technique.

Example 8. Noise Prediction Technique

Given:

4 in., 300 lb ANSI, Fisher Design ES (a cage style globe body) installed in an 8 in. Sch 40 pipe line.

p_1	= 615 psia	
Δp	= 348 psi	
C_g	= 4000	[calculated required]
$SPL_{\Delta p}$	= 103.5 dB	[Figure 20]
ΔSPL_{C_g}	= - 2 dB	[Figure 21]
$\Delta SPL_{\Delta p/p_1}$	= + 9.5 dB	[Figure 23]
ΔSPL_K	= 0 dB	[Table VII]
SPL	= [103.5 + (-2) + 9.5 + 0] dB	
SPL	= 111 dB	

As a possible alternate valve selected to minimize noise, we will repeat the above problem using an 8 in. x 4 in. Design EW (a cage style globe body) with Whisper Trim™. Note: the only change in our calculation will be the $\Delta SPL_{\Delta p/p_1}$ value.

$\Delta SPL_{\Delta p/p_1}$	= -7.5	[Figure 22]
SPL	= [103.5 + (-2) + (-7.5) + 0] dB	[Equation (47)]
SPL	= 94 dB	

If we were to use Schedule 80 pipe from TABLE VII, ΔSPL_K = - 9 dB

SPL	= (94 - 9) dB
SPL	= 85 dB

TABLE VII. CORRECTION TO BASE SPL IN dB FOR PIPE WALL ATTENUATION.

Nominal Pipe Size	*Schedule Number*							
(In.)	30	40	80	120	160	STD	XS	XXS
2	--	0	-6	--	-12	0	-6	-16
4	--	0	-7	-10	-13	0	-7	-16
6	--	0	-8	-12	-15	0	-8	-18
8	+3	0	-9	-14	-18	0	-9	-16
10	+3	0	-9	-14	-19	0	-6	--
12	+4	0	-10	-16	-20	0	-5	--

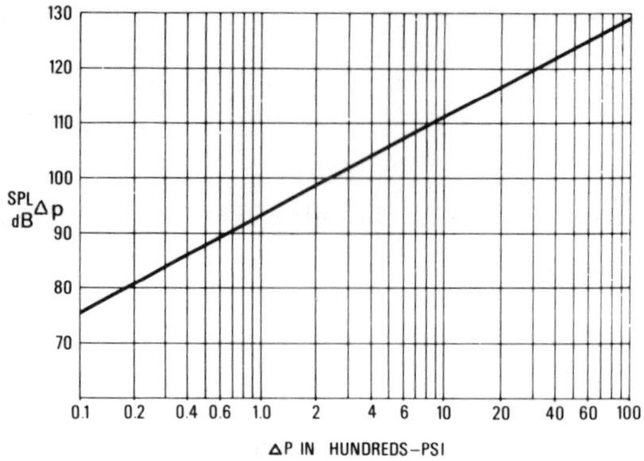

Figure 20. BASE SPL, ALL VALVE STYLES.

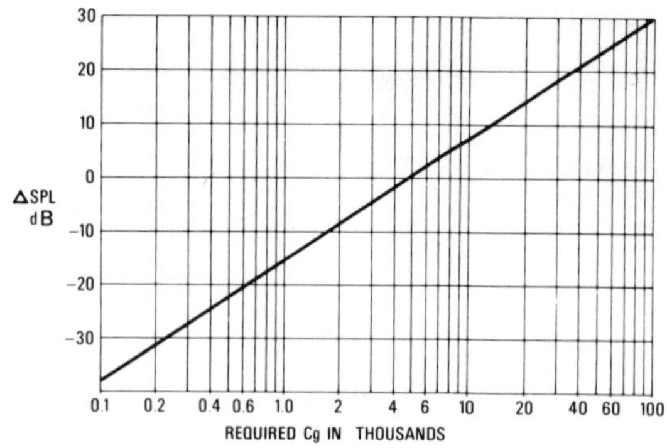

Figure 21. SPLcg CORRECTION, ALL VALVE STYLES.

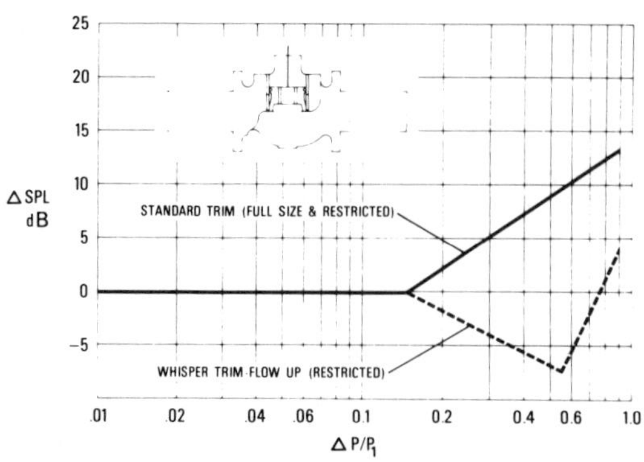

Figure 22. CAGE STYLE GLOBE VALVES, LINE SIZE EQUALS BODY SIZE (4-12 in.)

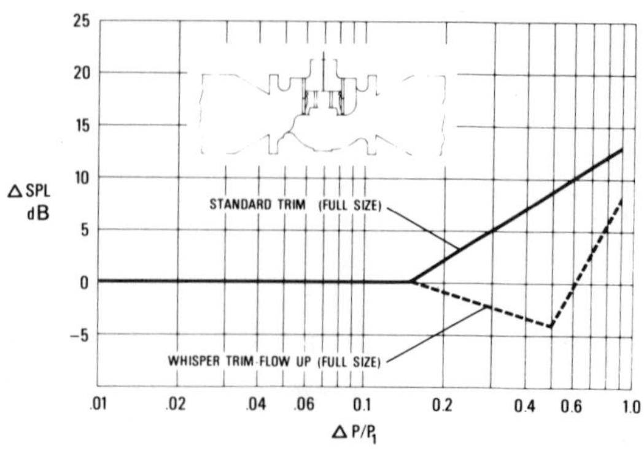

Figure 23. CAGE STYLE GLOBE VALVES SWAGED, (3 - 6 in.).

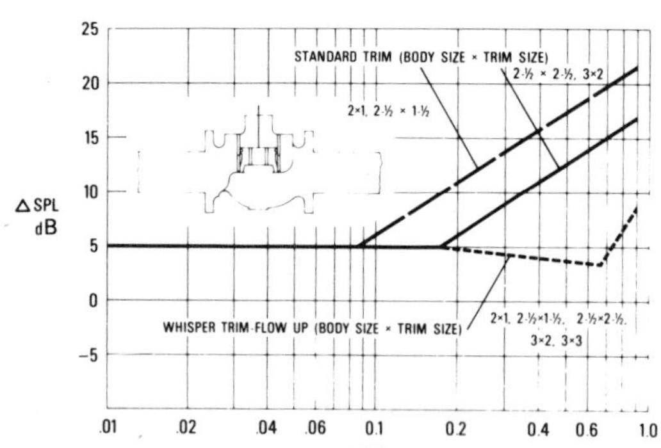

Figure 24. CAGE STYLE GLOBE VALVES, LINE SIZE EQUALS BODY SIZE (2-3in.).

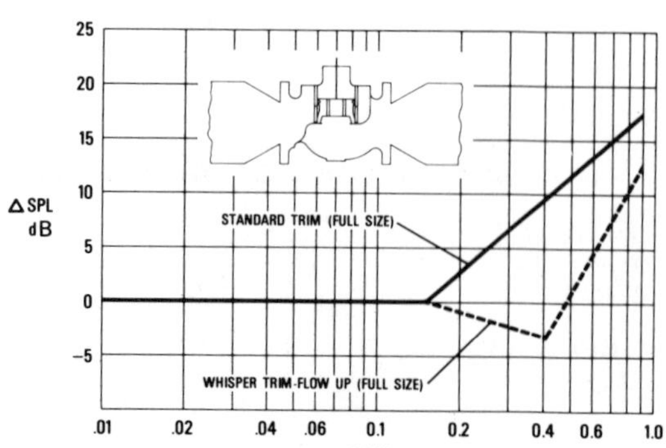

Figure 25. CAGE STYLE GLOBE VALVES, SWAGED (2 in.).

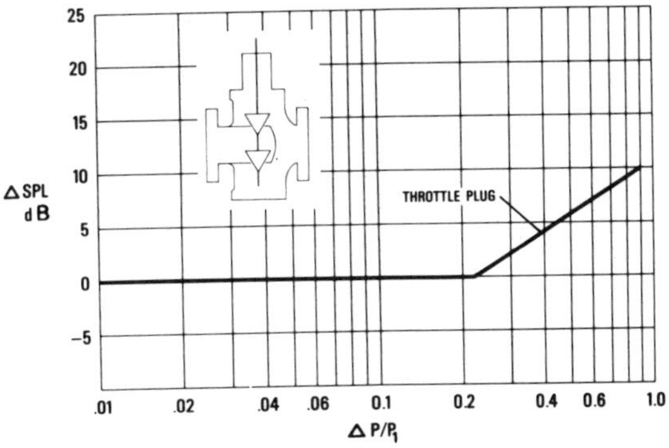

Figure 26. TOP AND BOTTOM GUIDED GLOBE VALVES, ALL LINE TO BODY SIZE RATIOS (All Valve Sizes).

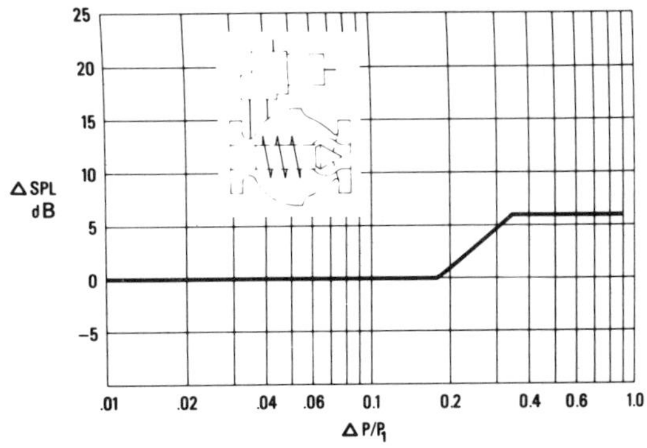

Figure 27. SLIDING SLEEVE REGULATORS, ALL LINE TO BODY SIZE RATIOS (All Valve Sizes).

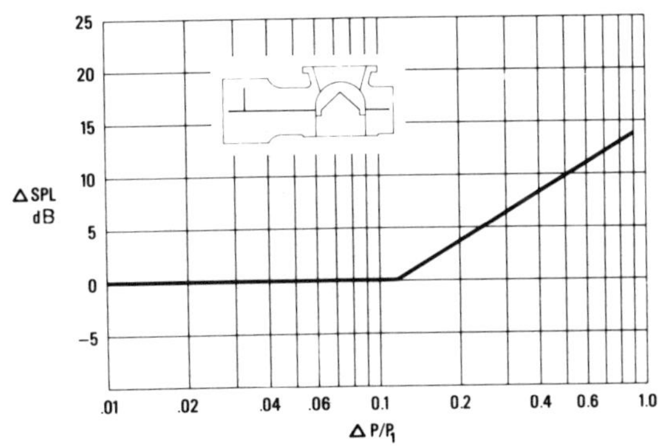

Figure 28. V-NOTCH BALL VALVES, ALL LINE TO BODY SIZE RATIOS (All Sizes).

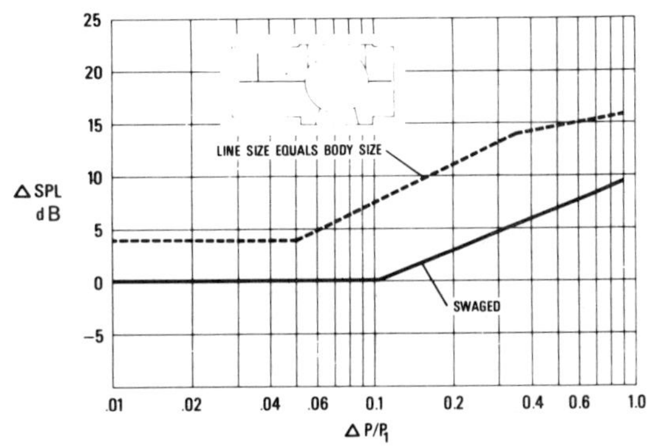

Figure 29. STANDARD BALL VALVES, ALL LINE TO BODY SIZE RATIOS (All Valve Sizes).

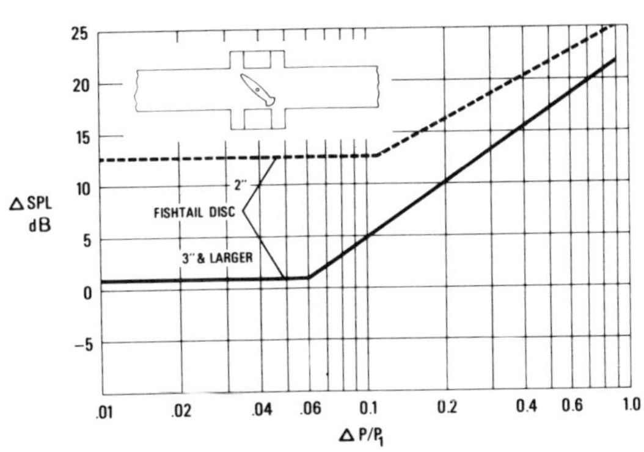

Figure 30. BUTTERFLY VALVES, LINE SIZE EQUALS BODY SIZE (All Sizes).

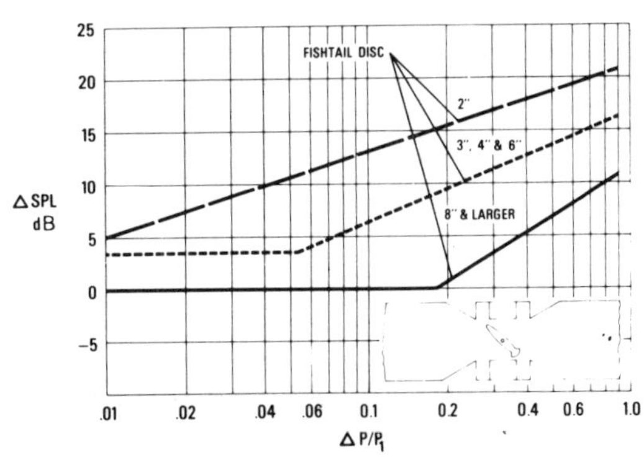

Figure 31. BUTTERFLY VALVES, SWAGED (All Sizes).

NOISE CONTROL

Noise control employs either one or both of the following basic approaches:

1. Source Treatment—Prevention or attenuation of the acoustic power at the source.

2. Path Treatment—Reduction of noise transmitted from a source to a receiver.

Attentuation of acoustic power at the source is the most desirable approach to noise control, if it is economically and physically feasible. The turbulent intensity of a fluid stream, and consequently the acoustic intensity level generated, is very sensitive to the geometrical shape of the fluid path. One would anticipate; therefore, that it may be possible to design a valve that would generate acoustic power at a level substantially lower than that of conventional valves.

One design uses cage type trim that utilizes multiple orifices of special shape, size, and spacing to minimize noise resulting from gas flow through globe style valves. The use of "whisper" trim, in conjunction with a properly selected body size, can result in noise levels 20 dB lower than those levels obtained with a valve selected only on the basis of capacity and control requirements, as shown in Figure 32. (Editors Note: Several additional types of low noise valves are described in Chapter 3).

A second approach to noise control is that of path treatment. Sound is transmitted via longitudinal waves through the elastic medium or media that separate the source from the receiver. The speed and efficiency of sound transmission is dependent on the properties of the medium through which it is propagated. Path treatment consists of regulating the impedance of the transmission path to reduce the acoustic energy that is communicated to the receiver.

The fluid stream is an excellent noise transmission path. When critical flow exists (fluid velocity at the vena contracta is at least at the sonic level), the vena contracta acts as a barrier to the propagation of sound upstream via the fluid. At sub-critical flow, however, valve noise can be propagated in the upstream direction almost as efficiently as it is downstream. The impedance to the transmission of noise upstream at sub-critical flow is primarily a function of valve geometry. The valve geometry that provides a direct line of light thru the valve (i.e., ball valves and butterfly valves), offers little resistance to noise propagation. Globe style valves provide approximately 10 dB attenuation. In any path treatment approach to control valve noise abatement, consideration must be given to the amplitude of noise radiated by both the upstream and downstream piping.

The dissipation of acoustic energy, by acoustical absorbent materials, is one of the most effective methods of path treatment. The acoustical material should be located in the flow stream either at, or immediately downstream from the noise source, whenever possible. This approach to control valve noise abatement is accommodated by in-line silencers. In-line silencers effectively attentuate the noise level transmitted to the solid boundaries and minimize the propagation of noise within the fluid stream. Figure 33 is a cross-sectional view of a typical in-line silencer. Where there are high mass flow rates and/or high pressure ratios across the valve, in-line silencers are often the most realistic and economic approach to noise control. The usage of in-line silencers can provide 30 dB or more attenuation in ambient noise levels.

Noise that cannot be eliminated within the boundaries of the flow stream must be eliminated by external treatment or isolation. This approach to the abatement of control valve noise includes the use of heavy walled piping, acoutical insulation of the exposed solid boundaries of the fluid stream, and insulated boxes, rooms and buildings, to isolate the noise source.

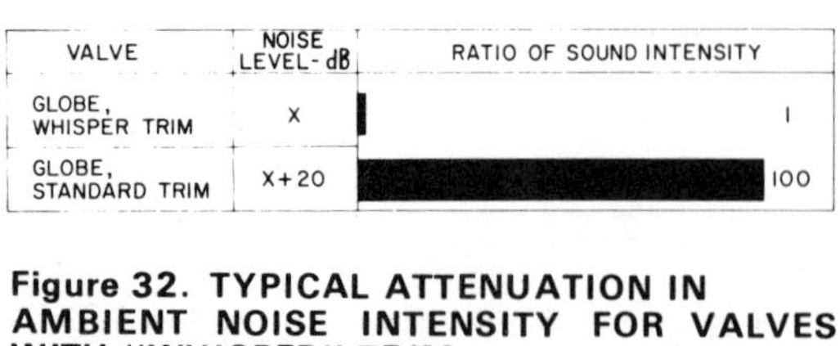

Figure 32. TYPICAL ATTENUATION IN AMBIENT NOISE INTENSITY FOR VALVES WITH "WHISPER" TRIM.

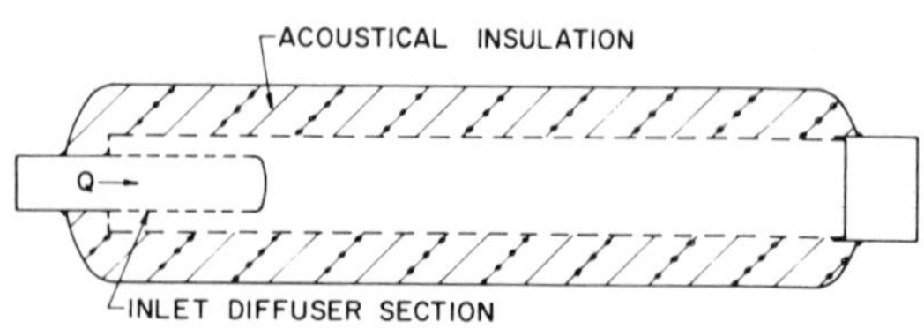

Figure 33. TYPICAL INLINE SILENCER.

The amplitude of the ambient noise level is proportional to the amplitude of vibration exhibited by the emanating surface; therefore, the solid boundaries (pipes, headers, regulator bodies, etc.) of the fluid stream can be a significant impedance to the transmission of noise. It has been shown that the radial vibration of pipe, resulting from pressure pulsations of the contained fluid, is inversely proportional to the wall thickness. Results of limited laboratory tests to determine the attenuation of ambient noise level as a function of pipe wall thickness on a 4-inch installation are shown in Table VII.

Acoustical insulation of the exposed solid boundaries of the fluid stream is an effective means of noise abatement for localized areas. Test results indicate that ambient noise levels can be attenuated as much as 10 dB per inch of thickness of the insulation. The effectiveness of path treatment such as the use of heavy wall pipe and acoustical insulation terminates at the point that this treatment is discontinued.

CONCLUSION

Because of the numerous sources of noise in the fluid process and the transmission systems, noise abatement is often a complex problem. The ideal approach to noise abatement is to design for quiet operation. Frequently, the need for costly noise abatement programs can be avoided if noise control is given proper consideration when the system is designed.

Prior to 1969, the approach to abatement of control valve noise had primarily been a pragmatic function. Noise control programs can now be formulated with a high degree of confidence, because of 1) improved noise prediction techniques; 2) the availability of fluid control elements that are designed to minimize noise; and 3) the increased knowledge concerning methods of noise abatement.

NOMENCLATURE

c	Velocity of sound.
C_g	Gas sizing coefficient.
C_p	Specific heat at constant pressure.
C_v	Specific heat at constant volume.
I	Sound intensity.
I_o	Reference sound intensity.
IL	Sound intensity level.
k	Specific heat ratio, C_p/C_v .
M	Mach number, V/c.
p	Pressure, pais.
p_1	Upstream pressure, psia.
Δp	Pressure differential.
p_o	Reference sound pressure.
p_r	Prandtl Number, $C_p\ \mu/\lambda$.
p_s	Sound Pressure.
Re	Reynolds Number, ρ VD/μ .
SPL	Sound pressure level.
V	Velocity.
λ	Thermal conductivity.
μ	Absolute viscosity.
ρ	Density.

REFERENCES

1. Lighthill, M.J., 1952 (Part 1), 1954 (Part 2), "On Sound Generated Aerodynamically,"
2. Schuder, C.B., "Control Valve Noise —Prediction and Abatement", *Chemical Engineering,* October 19, 1970.

SUPPLEMENTARY METHOD: CONTROL VALVE NOISE LEVEL ESTIMATION*

The following is a mathematical method that can be used to estimate the sound pressure level (in dB) existing three feet from any throttling control valve working under choked-flow conditions.

SPL =

$$10 \log_{10} (10^{11} \eta p_1 p_2 C_v F_L) - TL + S_g \quad (48)$$

where

$$TL = 17 \log_{10} \left[m\, n \frac{1}{\sqrt{\left(p_1/p_2 - \frac{1}{1 - 0.47 F_L^2}\right)} C_v F_L} \right] + 35$$

and where

C_v	actual required flow coefficient
F_L	critical flow factor of control valve at valve lift (a ratio)
m	weight of pipe wall, lb/ft^2 (see Table X)
n	number of apparent frequency producing orifices (see Table IX)
p_1	inlet pressure, psia
p_2	outlet pressure, psia
S_g	gas property correction factor, dB (see Table VIII)
SPL	sound pressure level, dB (ref. 2 x 10^{-4} microbar)
TL	transmission loss through pipe wall, dB
η	acoustical efficiency (a ratio—see Figure 34)

Note: SPL values refer to the "A-weighted" frequency scale. The accuracy of the equation, as shown here, is limited to pressure ratios p1/p2 above 2:1.

TABLE VIII. GAS PROPERTY FACTOR, S_g.

	(dB)
saturated steam	-2
superheated steam	-3
natural gas	-1
hydrogen	-10
oxygen	+0.5
ammonia	-2
air	0
acetylene	-1
carbon dioxide	+1
carbon monoxide	0
helium	-6.5
methane	-1
nitrogen	0
propane	+1
ethylene	-1
ethane	-1

TABLE IX. n FACTOR FOR SEVERAL VALVE TYPES.

Valve Type	n *Factor*
Control ball valve	1.0
Camflex Single port globe valve Angle valve Butterfly valve	1.4
Double port valve 4 port cage valve	2.2

TABLE X. WEIGHT OF PIPE WALL.

Pipe Size (inches)	"m" *Schedule 40* (lb/Ft^2)	"m" *Schedule 80* (lb/Ft^2)	"m" *Schedule 160* (lb/Ft^2)
1	5.4	7.3	10.2
2	6.3	8.9	14
3	8.8	12.2	17.8
4	9.7	13.7	21.6
6	11.4	17.6	29.2
8	13.1	20.4	37.0
10	14.9	24.1	46.0
12	16.5	28.0	53.4

REFERENCE

1. Baumann, H.D., 1970 "On the Prediction of Aerodynamically Created Sound Pressure Level of Control Valves", ASME Paper WA70/FE-28.

*H. D. Baumann, Masoneilan International, Inc.

Editors Note: Control valve noise is a subject under continuing investigation. In the preceding section E. E. Allen has presented an empirical method for the prediction of control valve noise. This supplement is a mathematical method, presented by H. Baumann, for estimating aerodynamically-created sound pressure levels.

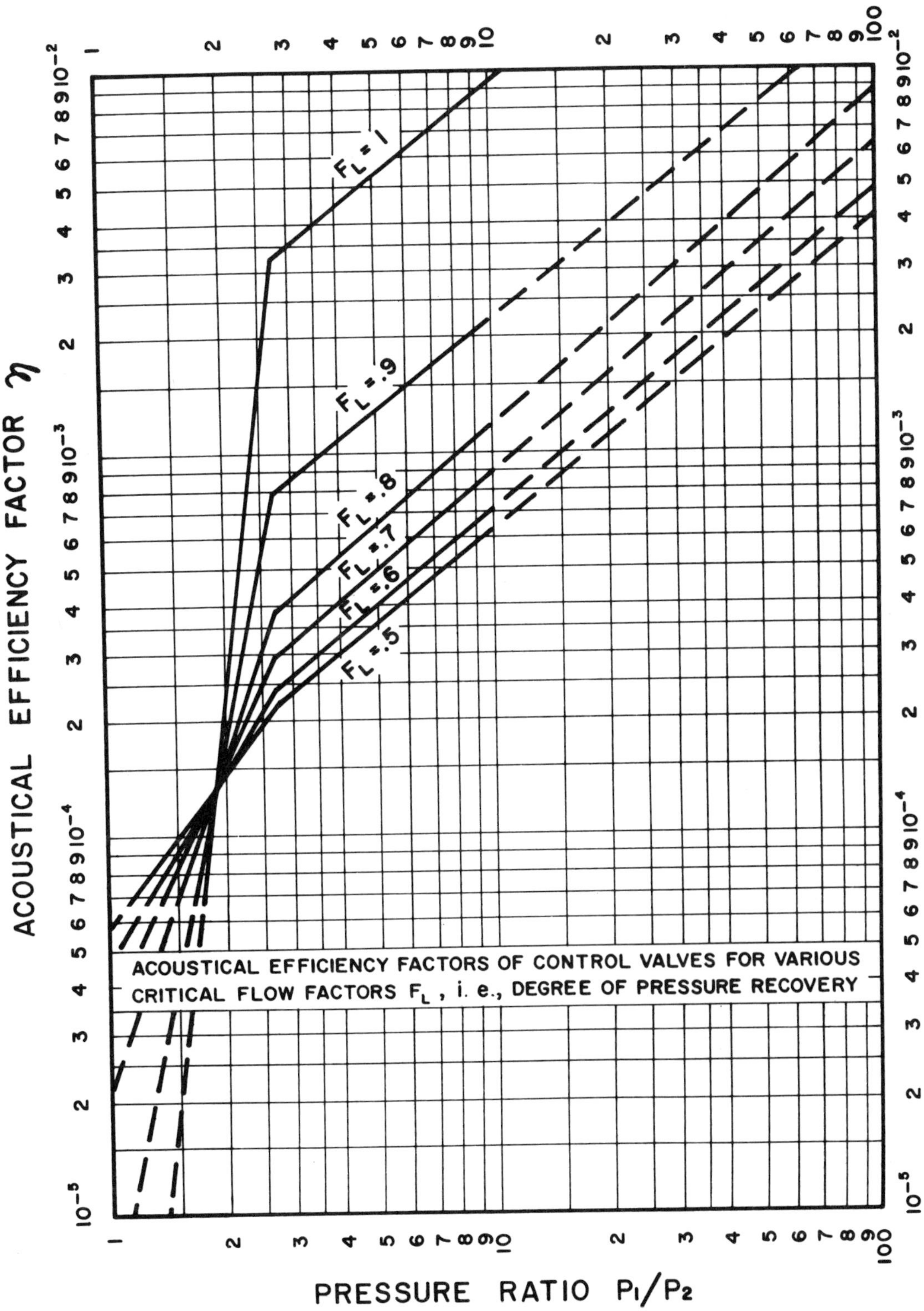

Figure 34. ACOUSTICAL EFFICIENCY FACTORS OF CONTROL VALVES FOR VARIOUS CRITICAL FLOW FACTORS, F_L (i.e., Degree of Pressure Recovery).

CHAPTER 7

CONTROL VALVE ACTUATORS

J. O'Connor, Jr.

Kieley & Mueller, Inc.
Middletown, New York

CONTENTS

INTRODUCTION

There is a wide variety of valve types used in control systems as has been discussed in great detail in prior chapters. Each valve type (i.e., globe valve, butterfly valve, or ball valve) imposes varying problems with respect to actuation.

For example, a standard globe control valve, which is a device to introduce a pressure loss within a system, generates a force that is determined by its ability to induce a pressure loss. This generated force is essentially the pressure drop, multiplied by the port diameter of the control valve. Further, this force is generated in a linear fashion, that is, in the direction the valve stem moves.

On the other hand, a butterfly valve generates angular forces within itself. The forces or torques induced by pressure drop within a butterfly valve are not simply the product of an area times a pressure drop, but are greatly affected by the ability of the butterfly valve to behave as a lifting body.

The control valve body assembly is selected by the control or instrumentation engineer, based on the variety of service conditions under which the control valve must operate and compared with the wide variety of control valve body types available. In the same fashion, the control engineer has at his disposal, many actuators, among which there is a great deal of diversity, with each type having some particular advantage for application in a particular condition.

This chapter discusses the general types of actuators currently in widest use in the operation of control valves. These actuators may be classified into six general types:

1. Pneumatically operated, diaphragm actuators;
2. Piston (cylinder) actuators;
3. Electro-hydraulic actuators;
4. High-performance servo actuator;
5. Electro-mechanical actuators; and
6. Manual handwheel actuators.

In actual practice today, it is quite probable that more than 90% of all control valve actuators in use are the pneumatically operated piston or diaphragm type; therefore, this chapter deals most extensively with them.

SELECTION PARAMETERS

Before considering any actuators, it is well to consider the parameters under which an actuator is to be selected. It is perhaps even more important to consider the features of the actuators to be placed in a control loop.

In any control loop, we have of course, the process which is being controlled and which is capable of being, within the measured variable, in error. To that extent, the first component of a control loop is, therefore, the error sensing device; that is, the instrument which measures the controlled variable and compares it to an acceptable value. If an error between the measured variable and the acceptable value is determined, the instrument automatically delivers an error-correcting signal to the control valve.

This error-correcting signal can take many forms. It can be a low-power electric signal, a hydraulic signal, or a pneumatic signal. These error-correcting signals may be of sufficient magnitude to be used within the control valve as a source of power to control the process, particularly if they are hydraulic or pneumatic signals. The error correcting signal (which may also be power) is then transmitted to the final control element, the device which causes an energy conversion to occur within the system by means of pressure loss within the control valve. It is the actuator that moves the control valve to that position which will change the energy input to the process so that it may return to an acceptable value, as determined by the error measuring device.

The actuator is, therefore, that part of the final control element which, in response to the error correcting signal from the error sensing device, moves the control valve. It moves the control valve either in a linear direction as in a globe valve, or in a rotary direction as in butterfly and ball valves.

Among other things, the actuator has to be compatible with the error-correcting signal. If, for example, the error-correcting signal from the instrument is hydraulic, the actuator must be compatible with the hydraulic fluid.

The actuator must be capable of generating enough force from the error-correcting signal to overcome the maximum forces that may be generated by the pressure drop phenomena within the valve body, whether the forces are the linear pressure times area relationship, as in a globe valve, or are varying forces such as the torque in a butterfly valve.

Lastly, the actuator must be compatible with the environment in which it will operate. This environmental compatibility may, for example, require explosion-proof housings on electric or electro-hydraulic actuators in service in oil refineries or chemical plants where the hazard of electrically caused fires or explosions is high. The environmental compatibility may require preparation by special corrosion-resistant or acid-resistant paint, particularly

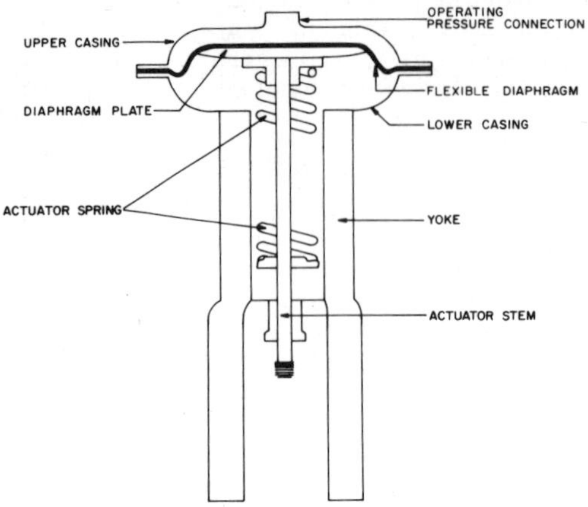

Figure 1. A DIRECT-ACTING DIAPHRAGM ACTUATOR.

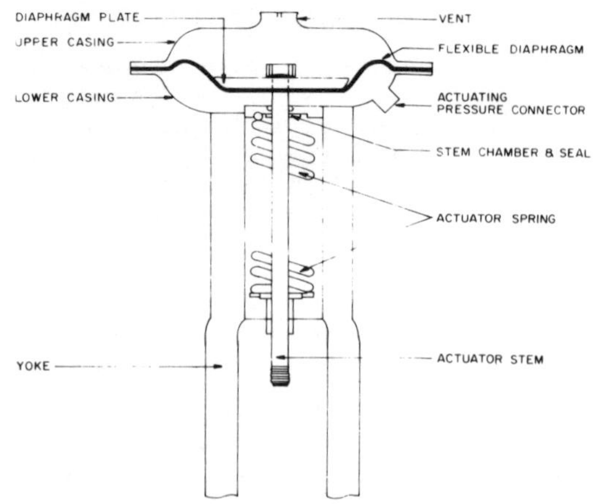

Figure 2. A REVERSE-ACTING DIAPHRAGM ACTUATOR.

where the environmental conditions might expose the actuator to acid vapors, or other corrosive agents.

TYPES OF ACTUATORS

Diaphragm Actuators

Diaphragm actuators usually take the form of a flexible diaphragm, placed between two stampings or cast "casings," and at least one section of the two resultant chambers is made pressure tight. A diaphragm actuator generally has a "range" spring opposing the force generated within the actuator. The control system air signal is connected to the pressure tight chamber, and an increase or decrease in the air pressure results in a force which is used to overcome the pressure-drop forces within the valve body, the hysteresis forces within a control valve body (stuffing box), and the force of the actuator's range spring.

A diaphragm actuator, in which the pressure-tight chamber is above the diaphragm, and in which an increasing pressure within that chamber will result in a downward motion, is generally described as being a direct-acting diaphragm actuator.

By using a seal around the actuator stem, the lower chamber can be made pressure-tight and the control system air signal may be connected to it. By venting the upper chamber to atmosphere and by changing the spring mounting arrangement so that upward stem motion will compress it, an increase in signal pressure in the lower chamber will result in an upward motion of the stem. This type of diaphragm actuator is ordinarily called a reverse-acting actuator. Cross sections of each type of diaphragm actuator, together with their essential nomenclature are shown in Figures 1 and 2.

Diaphragm actuators, whether direct or reverse acting, are not necessarily directly related to valve size in terms of pipeline inches. Each diaphragm actuator has to be calculated to make sure that it is powerful enough to overcome the forces within the valve body assembly during service, because the sole purpose of the actuator is to move the valve in response to the error-correcting signal. All major manufacturers of control valves catalog a fairly wide range of actuator sizes and generally specify them in relation to the square inch area of the diaphragm; however, it is quite possible that a very large diaphragm actuator may be required for a small valve, if the service conditions were such that a high pressure drop was being induced across a small line-size valve. Conversely, in a large valve with low pressure drop, it is entirely possible that a small actuator can be used.

Consider the sketch of a single seated direct acting valve body assembly, shown schematically in Figure 3. Assuming an inlet pressure (P_1) of 40 psig, an outlet pressure (P_2) of 10 psig, and a port area of 12.56 sq. inches (nominally a 4" valve), the force generated along the axis of the valve stem, tending to open the valve, would be P_1-P_2 x port area (30 psi x 12.56 in^2), or 376.8 pounds.

We would expect that this valve would

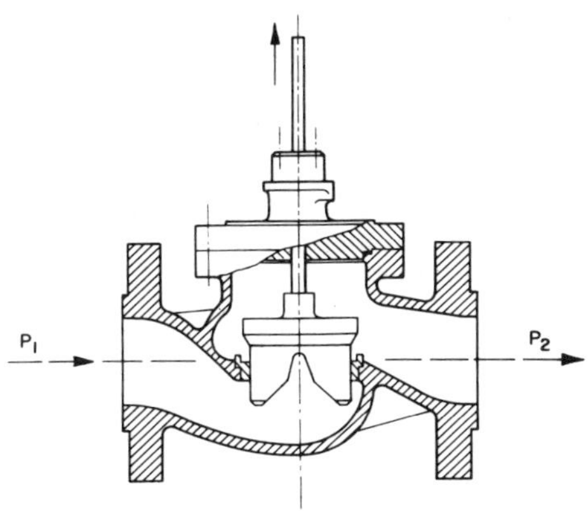

Figure 3. PLUG FORCE DEVELOPMENT.

begin to move downward at the 3 psig signal; assuming that the instrument signal is 3-15 psig, which is a conventional signal from a controlling instrument, and assuming further a direct acting response, (that is, increased pressure on the actuator results in downward motion of the actuator), air-to-close, and proportional service. We could, therefore, assuming a constant stem force upward over the full length travel of the valve, divide the 3 psig instrument signal into the total force of 376.8 pounds, and determine that an actuator having 125 sq. inches in area would be sufficient to begin to move the valve. However, we would also then have to consider that at the conclusion of the stroke of the valve, we would have 15 psig (instrument pressure) x 125 sq. inches, or 1875 lb. force on the valve stem. Therefore, we have an excess force at the seat, to the extent of 1875 pounds, less 376.8 pounds, or a net surplus of force on the valve seat of 1499 pounds. It will therefore, be necessary for us to absorb this additional or surplus force by the use of the range spring, if we're fortunate enough to find a manufacturer who has a 125 square inch diaphragm. Assuming that the travel of the valve is one inch, we would have to put into the valve, a spring that would have a compression rate of 1499 pounds per inch. The additional pressure that may be required for seating force has not been accounted for here.

It should be noted that the above is really an ideal situation, assuming that a diaphragm of 125 square inches area could be located and that spring rates could be chosen to fit a given situation. In actual practice, this is not usually the case and most often, some compromise has to be made with spring range and diaphragm area in order to come as close as possible to the instrument signal and power output. Most manufacturers of control valves carry a sufficiently wide range of diaphragm actuators and springs so the normal instrument signal and power ranges (3 to 15 psig, 3 to 27 psig, and 6 to 30 psig) can be accommodated by 1/2 psig within the given range.

In concluding the discussion of diaphragm actuators, reference is made to Figure 4 which shows the reverse-acting actuator (a direct acting actuator would leave the same components) of one manufacturer, with all of the parts called out. It will be noted that there are really only two parts that can be greatly affected by ambient conditions. These two parts are

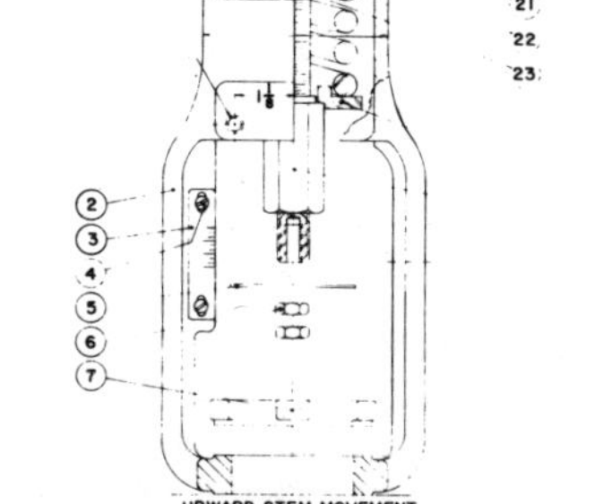

NO	PART NAME
1	NAMEPLATE
2	YOKE
3	SCALE, INDICATOR
4	SCREW, IND. SCALE
5	INDICATOR
6	LOCKNUT, STEM
7	LOCKNUT, YOKE
8	LOCKNUT
9	LOCKWASHER
10	DIAPHRAGM PLATE
11	ASS'Y, UPPER CASING
12	DIAPHRAGM
13	BOLT, CLAMPING
14	NUT, CLAMPING
15	ASS'Y, LOWER CASING
16	SCREW, CAP, MOUNTING
17	GASKET
18	WASHER, DIAPHRAGM
19	O-RING, BUSH/YOKE
20	O-RING, BUSH/STEM
21	BUSHING, O-RING
22	SPRING
23	STEM

Figure 4. DIAPHRAGM ACTUATOR ASSEMBLIES, COMPONENTS, AND NOMENCLATURE.

the diaphragm in either the direct or reverse acting motor and the stem seal in the reverse acting motor. Most valve manufacturers use neoprene diaphragms reinforced with fiberglass, cotton, or nylon for normal ambient conditions. If the ambient conditions exceed 200° F, higher temperature materials such as silicones, Viton®, or polyacrylics with Dacron or fiberglass fabric may be used. Occasionally a hydraulic fluid or other source of pressure may be used in the diaphragm actuator; it may be necessary to either back or overlay a diaphragm with a thin flexible piece of some fluid resistant material such as teflon or Kel-F®. With respect to the material used for the stem seals (which are ordinarily O rings) or the diaphragms, most major control valve manufacturers can offer special solutions to given ambient or corrosion problems that may exist.

Diaphragm actuators are, as we have indicated, the most widely used type in proportional control today. The chief advantages are a) their adaptability to a wide variety of valve sizes and circumstances, because of the wide range of sizes available through most manufacturers; and b) the diaphragm actuator is still the least expensive proportioning actuator available on the market.

Despite its advantages of wide adaptability and low cost, however, there are a number of instances in which diaphragm actuators cannot be applied satisfactorily. One of the inherent disadvantages of a diaphragm actuator is the fact that it must use large diaphragm areas, which in turn require large casings. As these casings become larger, the pressure which they are able to contain effectively, becomes less. One manufacturer, for example, has a 300 square inch actuator, but the maximum pressure that can be withstood within the diaphragm chamber itself is 30 psig. Although this is sufficient to generate a 9,000-pound force, the outside diameter of the diaphragm casing is approximately 27 inches; therefore, where small valves with large unbalanced or generated forces are involved, the entire assembly tends to become top heavy, imposing other stress problems on the valve and operator connection.

Piston Actuators

Another type of pneumatic or hydraulic actuator, commonly available from most valve manufacturers, is the piston actuator (also called cylinder actuator). Although the piston actuator for proportional control generally requires the use of valve positioners, which are described in Chapter 8, these cylinder actuators do offer some advantages that are not available in diaphragm actuators.

The cylinder is usually either cast or wrought iron designed to withstand substantially higher pressures than the cast or stamped casings in the diaphragm actuator. Because of its higher pressure rating; therefore, it is capable of delivering more force for smaller diameters. The cylinder actuator can be a very effective means of coping with high pressure drops, particularly when used in conjunction with a valve positioner which is capable of delivering pressures to the actuator, often up to the limit of the plant air rated pressure.

Cylinder actuators may be spring opposed although, in general, the cylinder actuators are usually used in proportioning or positioning control valves in conjunction with a valve positioner, by use of air pressure alone.

Figure 5 shows a cylinder actuator with a valve positioner, and air-spring mounted. The instrument signal is fed into the valve positioner at the top of the cylinder and through the feedback spring that is placed below the valve positioner. By the pressure on the piston, the positioner correlates the valve position with the instrument signal and begins to either load or unload air to the top of the piston to move the valve to the required position.

Beneath the piston, a constant air pressure (called air spring) is maintained by an air regulator. If the piston is pushed down, the airset relieves the pressure; if it is required that the piston move upward, the airset increases its pressure to maintain a constant pressure below the piston.

Figure 5 also shows the same type of cylinder which is hooked up so that, instead of being direct-acting (that is, an increasing instrument signal results in the downward motion of the piston), an increase in instrument signal will result in an upward motion of the operator stem by use of a reverse acting positioner. This particular hook-up also demonstrates another feature; that it is possible for the valve stem to go down in the event of air failure. If the air fails, a check valve in the supply line closes, the three-way valve opens and permits air pressure from an accumulator to go directly into the cylinder chamber, pushing it down. In the meantime, air pressure that may have remained below the chamber from the air spring regulator, will have been bled out by the downward motion of the piston.

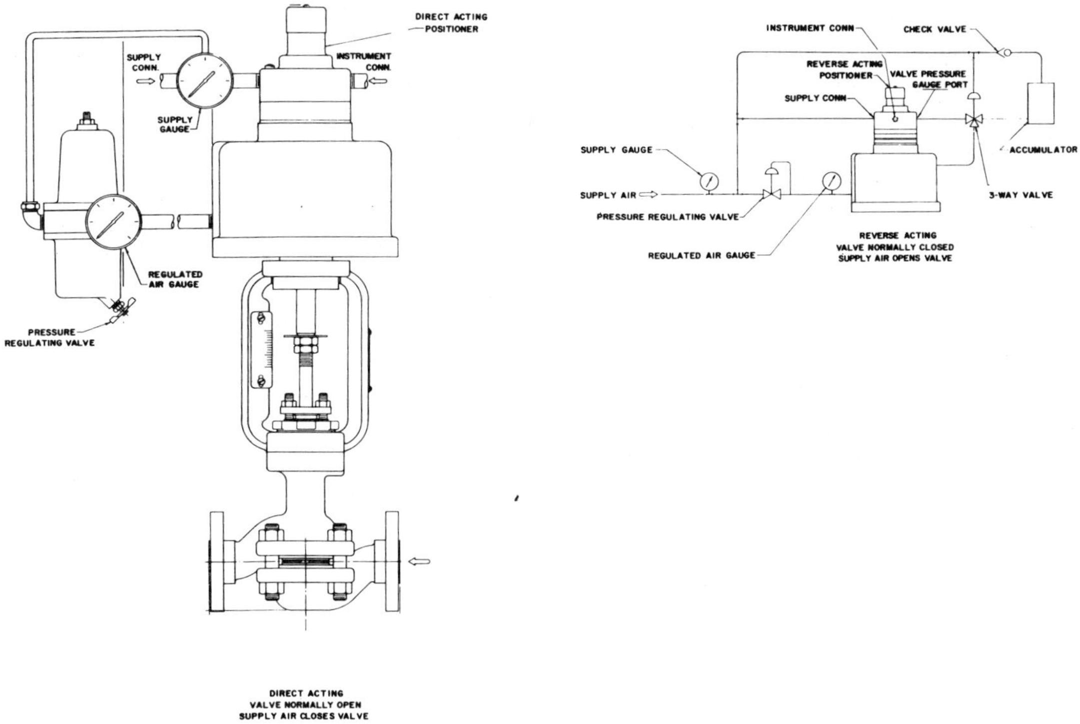

Figure 5. A DOME ACTUATOR CONTROL VALVE.

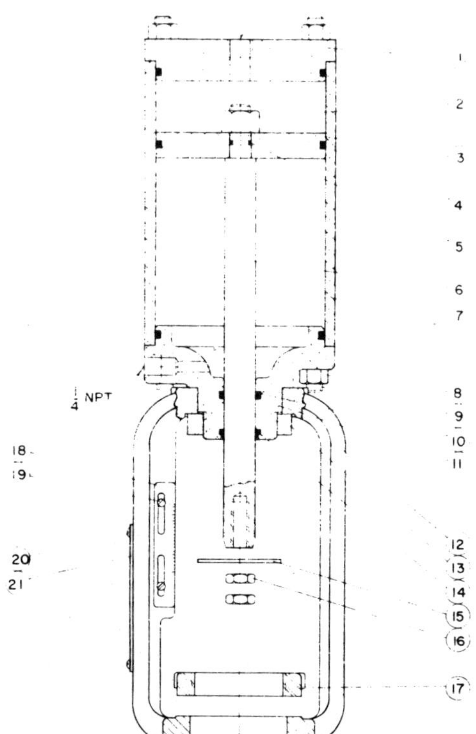

NO.	PART NAME
1	CLOSURE, BLIND END
2	NUT
3	PISTON
4	"O" RING, CLOSURE/PISTON HSG.
5	"O" RING, ROD TO PISTON
6	ROD, PISTON
7	HOUSING
8	NUT, TIE ROD
9	ROD, TIE
10	"O" RING, CLOSURE TO ROD
11	CLOSURE & INTEGRAL BUSHING
12	YOKE
13	RING, LOCK
14	RING, WIPER
15	INDICATOR, TRAVEL
16	NUT, LOWER STEM JAM
17	NUT, YOKE LOCK
18	SCALE, TRAVEL
19	SCREW, SCALE MTG
20	SCREW, NAMEPLATE MTG
21	NAMEPLATE

Figure 6. A DOUBLE-ACTING CYLINDER ACTUATOR.

Figure 6 shows a double-acting cylinder, without any accessory instrumentation or control devices. This actuator, which is intended essentially for an on-off service, can supply very swift open-shut action when used with a three-way solenoid air pilot valve and is useful in batching operations or cyclic work. This actuator can be made from either tubing or centrifugal castings and can be used with extremely high operating pressures for high thrust service.

Figure 7 shows a cylinder actuator used on a conventional diaphragm motor yoke with a spring opposing its motion. It can be used in the same manner as a conventional diaphragm motor, but can offer longer strokes of travel for a given yoke and spring combination. It can also be used with a positioner, as shown in the inset, and does not require the addition of the "air spring" and the accompanying piping.

Electro-hydraulic Actuators

There has been an increasing acceptance of electric controlling instruments in the instrumentation and controls industry over the past several years. These instruments, instead of producing a pneumatic error-correcting signal, yield a variety of low-power electric signals which can be either transduced or transposed at a pneumatic control valve in the field, as described in Chapter 8; or when combined with an electro-hydraulic valve actuator, can make a single control loop without any addition-

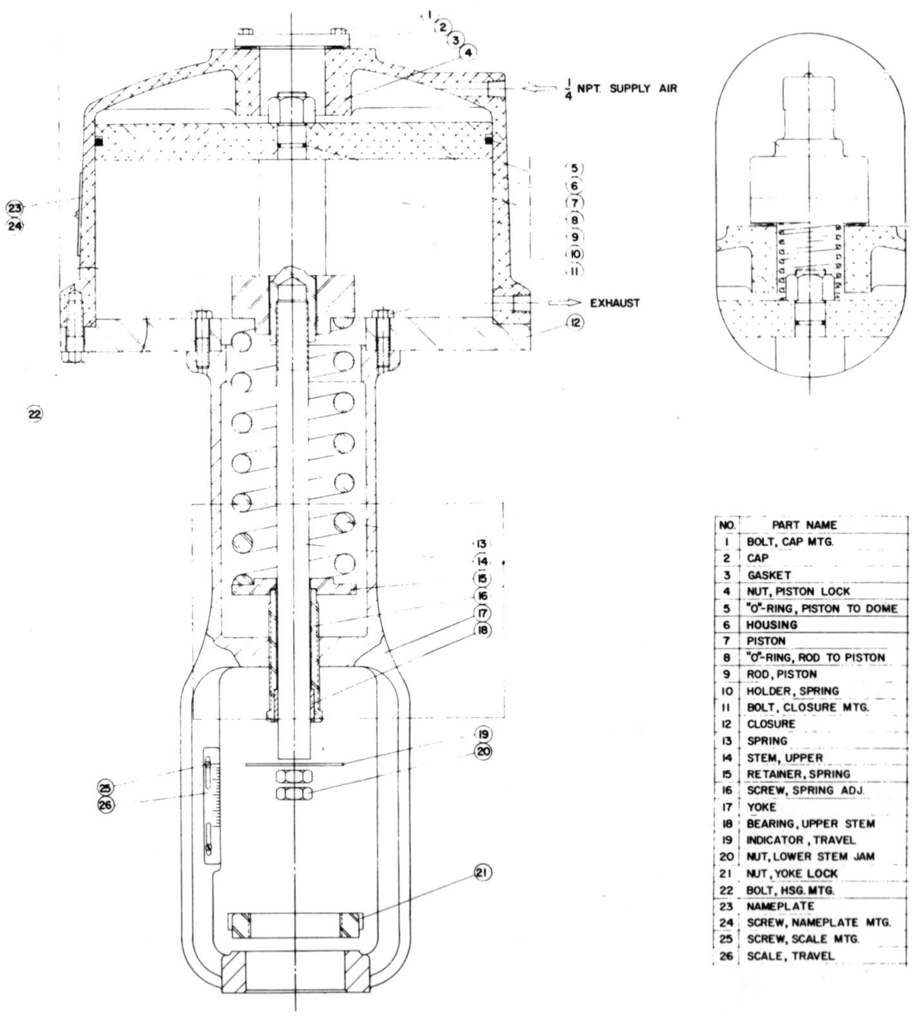

NO.	PART NAME
1	BOLT, CAP MTG.
2	CAP
3	GASKET
4	NUT, PISTON LOCK
5	"O"-RING, PISTON TO DOME
6	HOUSING
7	PISTON
8	"O"-RING, ROD TO PISTON
9	ROD, PISTON
10	HOLDER, SPRING
11	BOLT, CLOSURE MTG.
12	CLOSURE
13	SPRING
14	STEM, UPPER
15	RETAINER, SPRING
16	SCREW, SPRING ADJ.
17	YOKE
18	BEARING, UPPER STEM
19	INDICATOR, TRAVEL
20	NUT, LOWER STEM JAM
21	NUT, YOKE LOCK
22	BOLT, HSG. MTG.
23	NAMEPLATE
24	SCREW, NAMEPLATE MTG.
25	SCREW, SCALE MTG.
26	SCALE, TRAVEL

Figure 7. A DIRECT ACTING CYLINDER ACTUATOR ASSEMBLY.

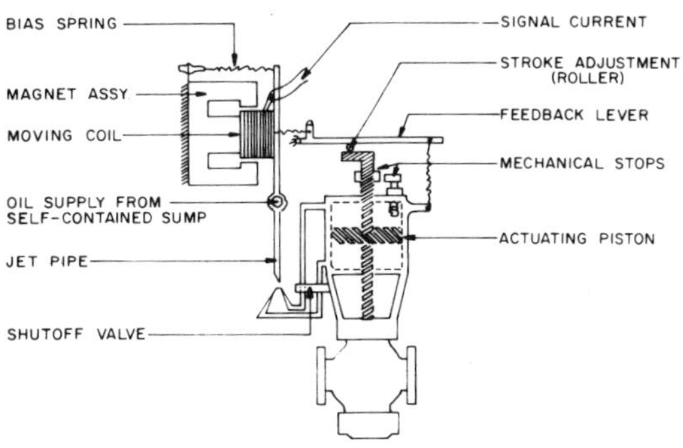

Figure 8. AN ELECTRO-HYDRAULIC VALVE ACTUATOR.

al requirement for a pneumatic power source. They do, however, require a separate electric power source.

These low power electrical signals are generally 24 to 65 volts d.c., and are of the order of 1 to 5, 4 to 20, or 10 to 50 milliamperes.

The current signal in one representative design, as illustrated by Figure 8, is fed into a moving coil. As the signal current varies, the coil moves correspondingly within the field of the permanent magnet. The coil in this particular design is also attached to a pivoted nozzle, through which a high pressure hydraulic oil flows. As the signal varies the position of the moving coil through the pivot, the nozzle injects fluid at a high pressure into one or the other of two cylinder pressure pickups. which either increases or decreases the hydraulic pressure on either side of the piston, causing the valve to move. The forces acting on the nozzle are balanced when the valve reaches a new position; therefore, the nozzle re-centers between the cylinder pressure pickups and there is no further valve movement.

Other electro-hydraulic actuators on the market are similar to the one described here, except that they differ in the way in which the hydraulic pressure is picked up by the cylinders. At least one other manufacturer uses a conventional flapper nozzle design of the coil to change the pressures on either side of the piston.

Electro-hydraulic actuators, in general, offer the advantage that they can be placed remotely from an instrument, where there may be no other auxiliary pressures, such as a pneumatic pressure, to operate a valve. They have not been widely received in the chemical and petroleum processing plants, probably because they continue to be expensive in relation to a diaphragm actuated control valve with a transducer, and because the essence of the design requires a constant source of pressure, which means a constant use of electric power to pump the hydraulic power fluid. Further, their operating speeds are sometimes lower than can be obtained with a diaphragm actuator, and even at this writing, their maximum stem thrusts are somewhat lower than can be obtained with large diaphragm actuators or high pressure cylinder actuators.

High Performance Servo-Actuators

A variation in the electro-hydraulic actuator has been developed by a number of companies, for service where extremely high pressures and pressure drops within a valve are encountered and which may be accompanied by high shut-off or opening speed requirements, extremely swift valve stem movement requirements, or extremely high frequency response characteristics.

Figure 9 shows the essentials of the operation of this high performance type servo-actuator. The valve stem motion, through a potentiometer or linear differential transformer, is compared with the instrument signal within a servo amplifier. A signal is then sent to a high performance servo-valve which either loads or unloads high-pressure hydraulic fluid to one side or the other of a high pressure cylinder actuator that is directly affixed to the valve stem. The excess hydraulic fluid is returned to a reservoir.

Ordinarily, these high performance servo-operators use hydraulic operating pressures that range from 1000-3000 psig. With these pressure and flow rates, extremely high valve stem speed can be obtained, extremely fast response to changing instrument signal conditions can be obtained,

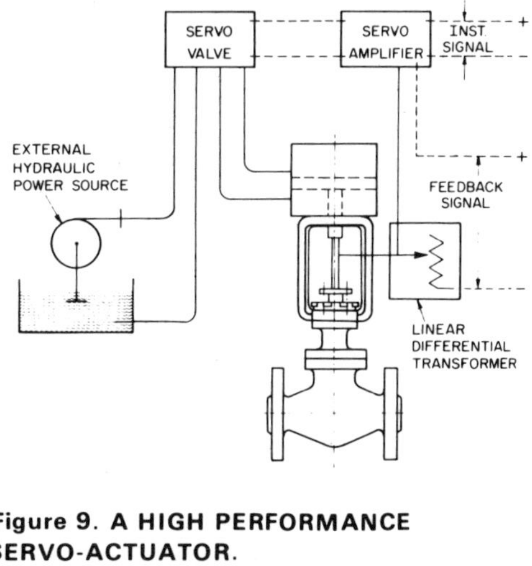

Figure 9. A HIGH PERFORMANCE SERVO-ACTUATOR.

and extremely high pressures or pressure drops within the main valve can be accommodated.

Their chief disadvantage is that they do require a very powerful external hydraulic source. Of course, this source can be used to supply a number of valves; however, their operating costs and installation costs are such that they have found a limited market, chiefly in the aerospace field.

Electro-mechanical Actuator

Another type of actuator which has been developed for proportional control in recent years is the electro-mechanical valve actuator. One of the principle disadvantages of the electro-hydraulic actuator, which is the constantly running motor and pumps, is not present in the electro-mechanical design. An electro-mechanical valve actuator is composed of a motorized gear train and screw assembly which drives the valve stem, an electronic positioner, and an electronic feedback device. A typical example is shown schematically in Figure 10. The varying input signal, whose magnitude corresponds to the required position of the inner valve stem, is fed into

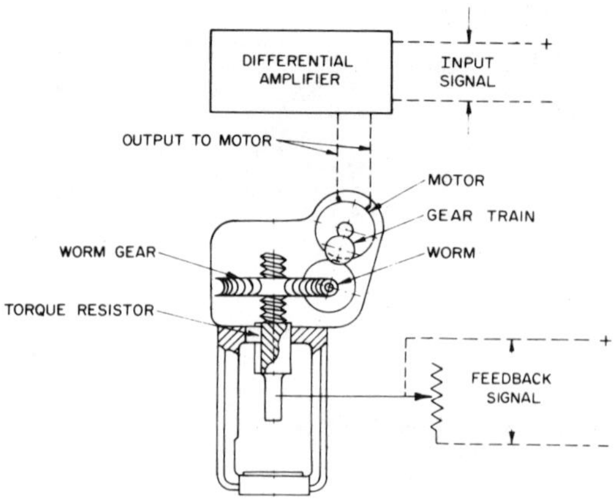

Figure 10. AN ELECTRO-MECHANICAL VALVE ACTUATOR.

the positioner (usually a differential amplifier) and produces a voltage to actuate the motorized gear train and screw. The resultant movement of the stem and the take-off attached to it and to a potentiometer or linear differential transformer, produces a voltage that increases with

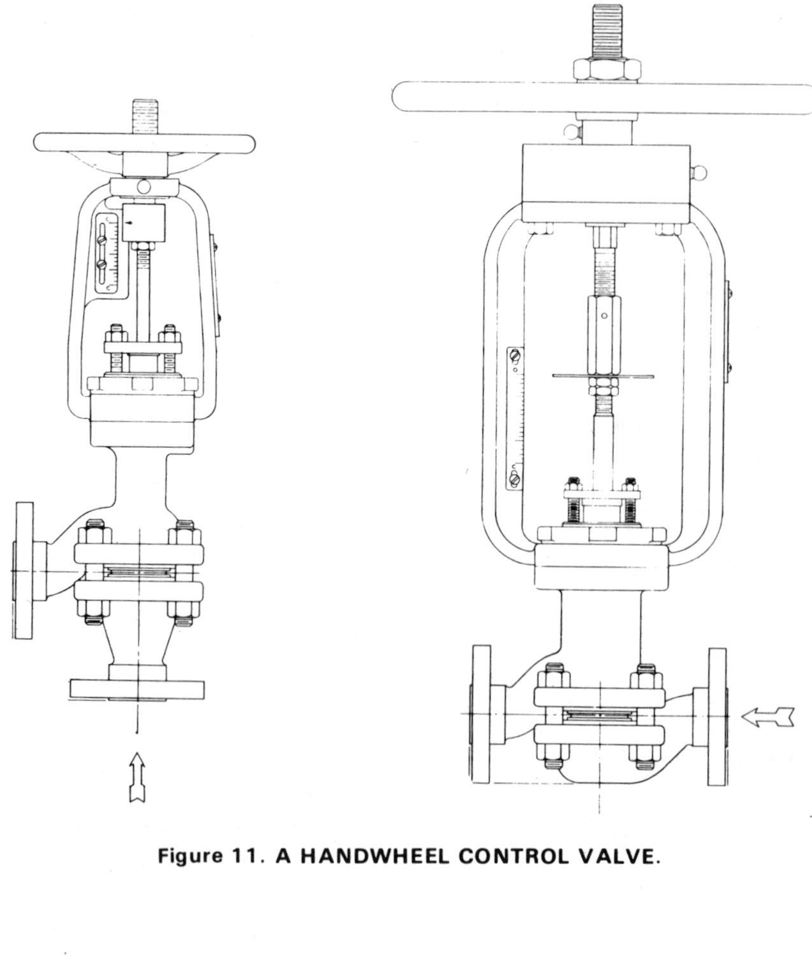

Figure 11. A HANDWHEEL CONTROL VALVE.

stroke, and is sent back into the positioner. When the input signal voltages and feedback voltages are equal, the output voltage to the motor goes to zero and the motor stops with the valve stem at the required position. Conversely, when the voltages are not equal, the motor is run in the direction to make them equal.

The electro-mechanical valve actuator has essentially the same advantages as the electro-hydraulic actuator with respect to field use. It is capable of being used over long distances with only inconsequential signal transmission delays. It is also immune from the pneumatic system problem of freeze-up in extremely cold ambient conditions. Electro-mechanical actuators are still, however, generally more expensive, although more efficient, than electro-hydraulic units.

Manual Actuators

Manual actuators are not, strictly speaking, proportioning actuators; however, they are briefly described in this section because they are used in control systems.

One type, as shown in Figure 11, is a straight manual actuator, with no power supplied. This really makes the control valve body assembly nothing more than a hand valve. There are occasions, however, when flows must be passed through valves precisely, even though not in conjunction with an automatic control system. Control valve body assemblies with manual actuators can serve this purpose where the flow characteristic is known. These valves can have their position read directly from the valve travel indicator, or can be equipped with accurate auxiliary indicators for even more precise work.

Two other types of handwheel operators are the top-mounted handwheel actuator, shown in Figure 12, and the side-mounted handwheel operator shown in Figure 13. Both of these operators are actually designed to override the control signal that is positioning the valve and to give operating personnel the option to override the control system, if warranted. They can also be used to facilitate the start-up of a control system, for example, to preposition a valve to a given flow. They can also be used as devices to shut off the valve, eliminating the need in some systems for costly bypass valve arrangements.

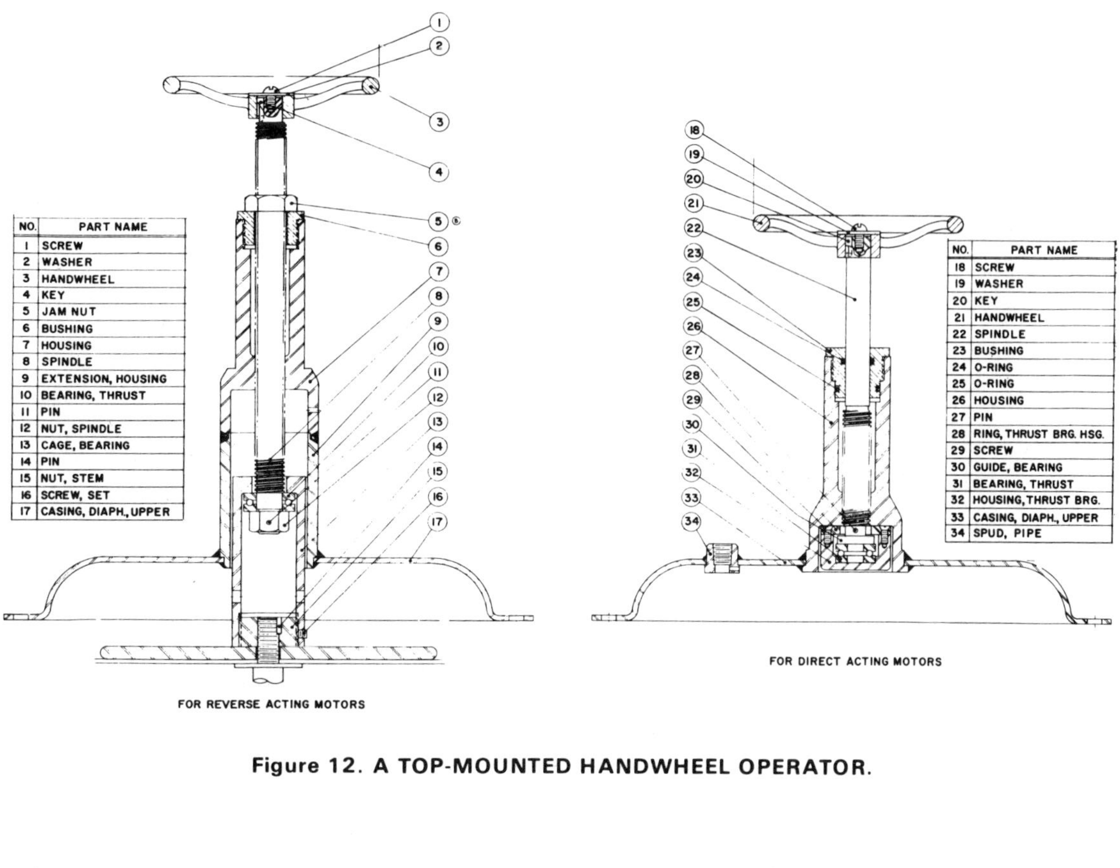

Figure 12. A TOP-MOUNTED HANDWHEEL OPERATOR.

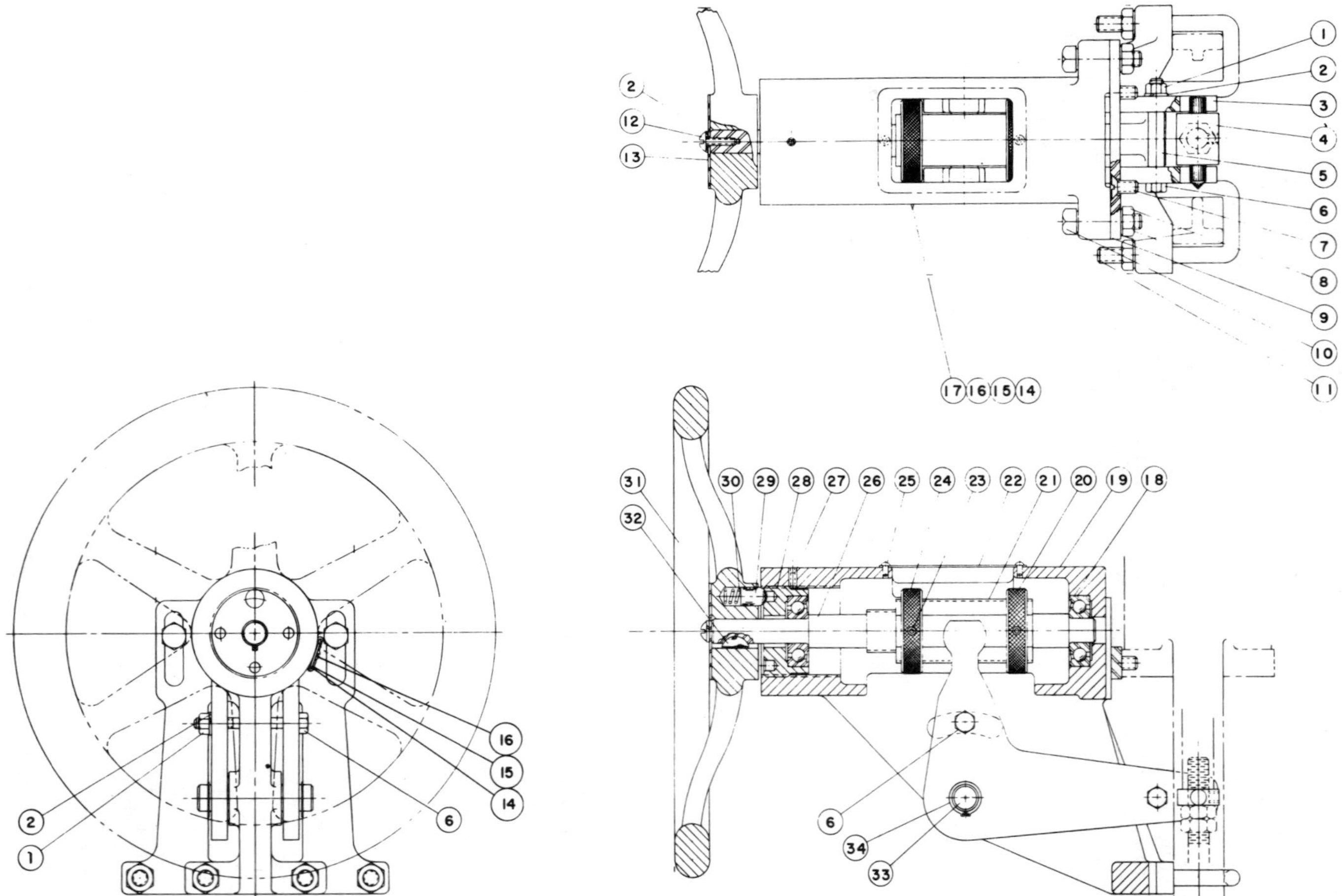

Figure 13. A SIDE-MOUNTED HANDWHEEL OPERATOR.

NO.	NAME
1	NUT, HEX.
2	LOCKWASHER
3	ARM, ROCKER
4	TAKE OFF, VALVE STEM
5	SPACER, ARM
6	BOLT, HEX. HD.
7	SCREW, MACH. FL. HD.
8	NUT, HEX.
9	PLATE, YOKE ADAPTOR
10	BOLT, HEX. HD.
11	U-BOLT
12	SCREW, MACH. R. H.
13	WASHER
14	SCALE, TRAVEL
15	CLIP, NEUTRAL POSITION
16	SCREW, SELF TAP.
17	INDICATOR, NEUTRAL POSITION
18	BEARING
19	HOUSING, HANDJACK
20	COLLAR, FORWARD THRUST
21	BLOCK, TRAVEL
22	COVER PLATE
23	SCREW, SET, ALLEN HD.
24	COLLAR, AFTER THRUST
25	SCREW, SELF TAP.
26	SHAFT, HANDWHEEL
27	SCREW, SET, ALLEN HD.
28	BUSHING, BEARING
29	DETENT
30	SPRING
31	HANDWHEEL
32	KEY, WOODRUFF
33	SHAFT, ROCKER ARM
34	RING, SNAP

CHAPTER 8

POSITIONERS AND OTHER DIAPHRAGM CONTROL VALVE ACCESSORIES

Edward L. Vollmer, III
Honeywell, Inc.
Fort Washington, Pennsylvania

CONTENTS

POSITIONERS

This section describes some of the most common types of pneumatic control valve accessories. Since the positioner is often considered to be the most important of them, it is covered first.

A pneumatic valve positioner is a device which precisely positions, by the use of air, the moving part (or parts) of a pneumatically operated valve in accordance with a pneumatic signal. A typical positioner is shown in Figure 1.

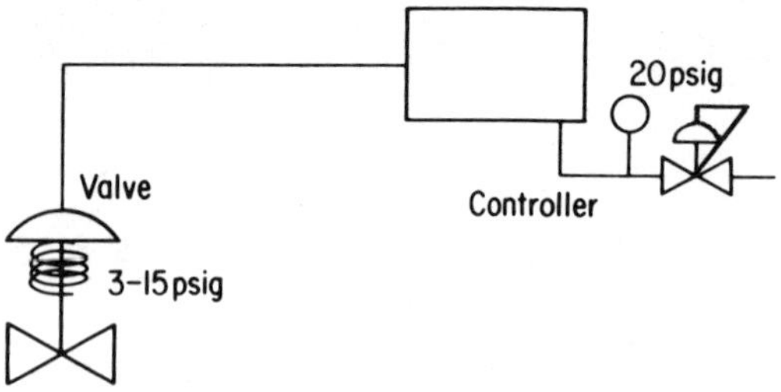

Figure 2. SINGLE-SEATED CONTROL VALVE WITHOUT POSITIONER, AIR-TO-CLOSE.

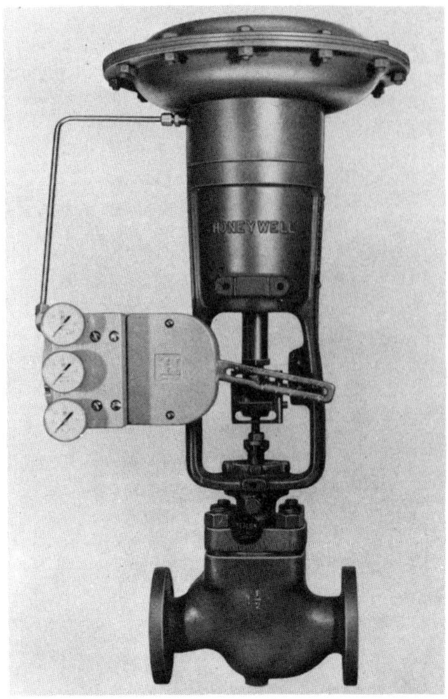

Figure 1. SIDE MOUNTED VALVE POSITIONER ON DIAPHRAGM CONTROL VALVE.

Courtesy Honeywell, Inc.

Principles of Operation

Figure 2 shows schematically a valve that is operated directly by a controller. A typical controller output varies from a minimum of 0 psig to a maximum of 20 psig. The most common valve spring operates from 3-15 psig. Fifteen psig is required to compress the valve spring. The air pressure above 15 psig is used to close the valve against the upstream pressure.

Figure 3 shows a valve with a positioner, in addition to the controller. The signal from the controller in this case goes to the positioner instead of directly to the control valve. The positioner compares the signal with the valve stem position. If the stem is incorrectly positioned, the positioner either adds or exhausts air from the valve actuator until the correct valve stem position is obtained.

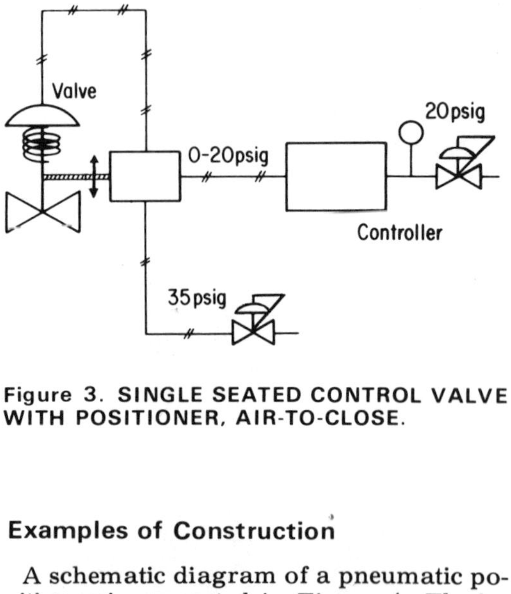

Figure 3. SINGLE SEATED CONTROL VALVE WITH POSITIONER, AIR-TO-CLOSE.

Examples of Construction

A schematic diagram of a pneumatic positioner is presented in Figure 4. The instrument control signal is applied to the signal diaphragm. An increasing signal will drive the diaphragm and pilot stem to the right. The pilot stem will then open the supply flapper admitting supply pressure into the valve chamber. The exhaust flapper remains closed when the pilot stem is deflected to the right. Since the valve chamber is connected to the valve actuator, whatever pressure exists in the valve chamber will also exist in the actuator.

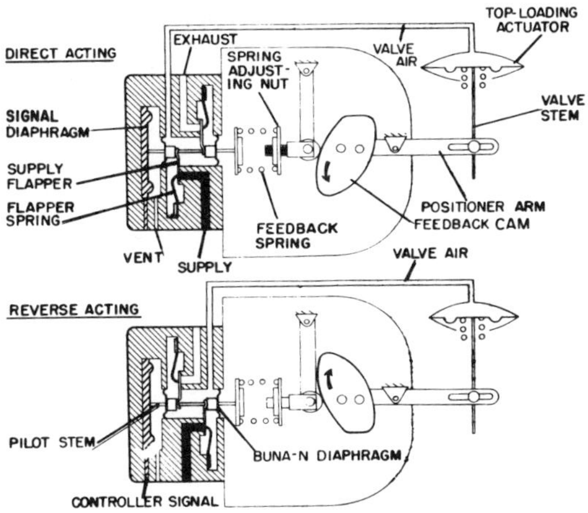

Figure 4. SCHEMATIC DIAGRAM OF PNEUMATIC VALVE POSITIONERS.

In the direct acting type, an increasing instrument signal increases the pressure in the valve chamber and in the valve actuator. This increased pressure on the actuator drives the valve stem downward and rotates the positioner arm clockwise. This rotation compresses the feedback spring through the cam. When the valve stem reaches the position called for by the controller, the compression on the feedback spring will give a force equal to the force caused by the controller signal acting on the signal diaphragm. When this "force balance" exists, the pilot stem moves back to the left and both flappers are closed.

If the controller signal is decreased, the force exerted by the signal diaphragm will also decrease and the force from the feedback spring will push the pilot stem to the left, opening the exhaust flapper. This causes a decrease in valve chamber pressure and allows the valve stem to move upward until a new force balance is established.

The positioner illustrated in Figure 5 employs the principle of force balance operation by spring loading the bellows unit (from the actual motion of the valve stem) to the same force as that exerted by the control-instrument air pressure.

Either a rise or a fall of the control-instrument air pressure will move the bellows and operate the air pilot. Supply air is either admitted to the diaphragm-actuator chamber or exhausted past the upper pilot stem, to the atmosphere. As the valve moves, the balance-spring load is changed. This action brings the bellows and pilot back to their original positions,

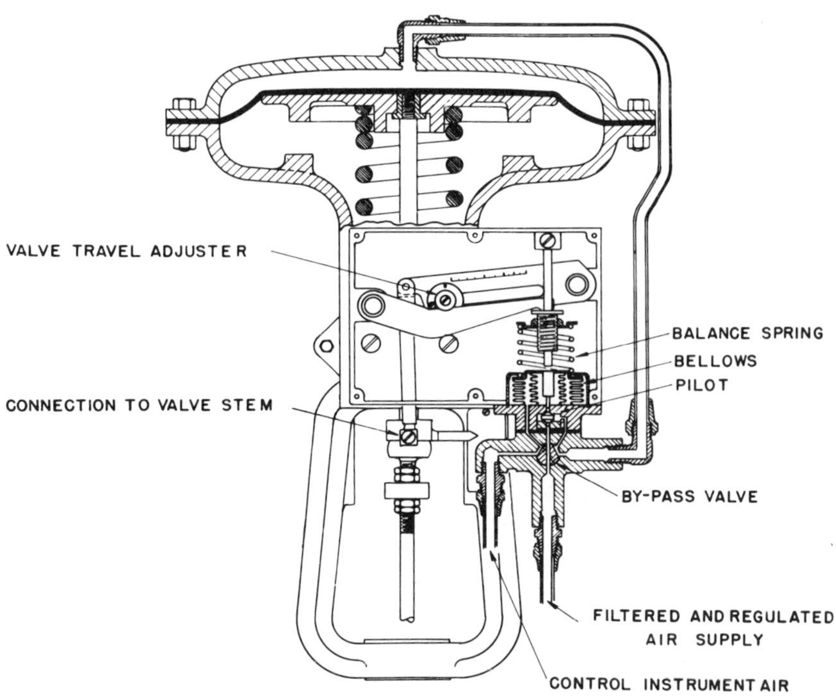

Figure 5. SIDE MOUNTED VALVE POSITIONER.

Courtesy Moore Products Company

thus restoring the balance between the spring loading and the control-instrument pressure.

Bellows characteristics do not affect the calibration, because the bellows remain in virtually the same position at all points of balance. The arrangement of the levers compensates for angularity errors. A linear relationship exists between the control-instrument pressure and the valve stem position.

The bellows and air pilot are a self-contained unit, without elements that would introduce friction and cause errors. Any friction in the lever bearings is on the valve side; therefore, it does not influence the performance of the positioner.

Applications

Some of the common applications of positioners are listed below, along with pertinent discussion. These applications should be tempered by the limitations given under the subheading entitled, "Limitations of Positioners."

To Overcome Friction

A positioning accuracy of 0.001 in. with a friction of 10 per cent to 15 per cent of full diaphragm actuator power is possible with some positioners.

To Increase the Shut-Off Rating of Single-Seated Valves

A common application is the use of a positioner, with a supply pressure of 35 psig, on a single seated valve. If a 3-15 psig spring is used in the valve, the first 15 psig from the positioner output compresses the valve spring. The remaining 20 psig (35 psi minus 15) is available to close the valve against the upstream line pressure. This is considerably more air pressure than is available from the controller; therefore, the close-off rating of the valve against line pressure is increased accordingly.

The supply pressure to the positioner is generally limited by the maximum allowable air pressure to the control valve actuator. A further gain in control valve shut-off rating can be obtained by reducing the initial compression in the control valve spring. For example, a 3 to 15 psig spring range can be adjusted to a 1 to 13 psig range. If 35 psig is the positioner supply pressure, then only 13 psig is needed to compress the valve spring and the remaining 22 psig is available to close the valve against the up-stream pressure. A further increase in valve shut-off rating can be obtained by using a 3-9 psig spring in the control valve. Here, the extra pressure available for closeoff is 26 psig (35 minus 9). As long as the positioner input spring is 3 to 15 psig and sufficient positioner supply pressure is available, the valve will operate precisely over a 3-15 psig range regardless of the control valve spring range.

Figure 6 shows an air-to-open valve with a 6 to 30 psig spring. Here the heavier spring and high initial compression provides greater close off capability than that which could be obtained with a 3-15 psi valve spring. The control valve's shut-off rating is limited by the initial compression in the valve spring, because the positioner can only remove the air pressure to the control valve actuator to close the valve.

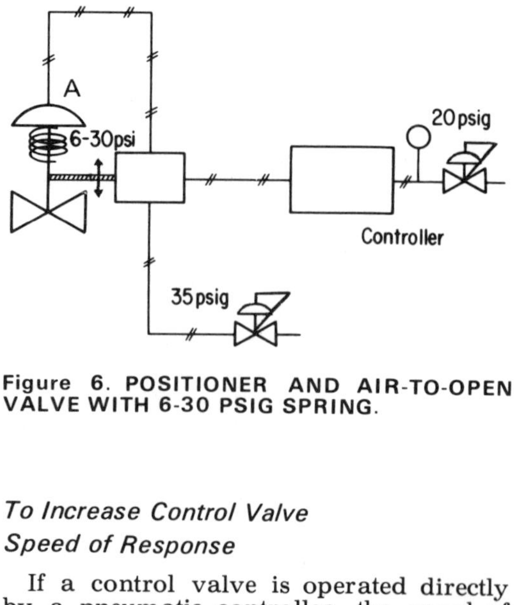

Figure 6. POSITIONER AND AIR-TO-OPEN VALVE WITH 6-30 PSIG SPRING.

To Increase Control Valve Speed of Response

If a control valve is operated directly by a pneumatic controller, the speed of operation of the valve depends upon: 1. the length and size of the connecting tubing; 2. The actuator volume; and 3. the capacity of the pilot relay in the controller.

If a positioner is used, the controller signal goes to the positioner instead of to the control valve actuator. Since the input to the positioner requires approximately one cubic inch or less of air, large quantities of air need not flow between the controller and the positioner. This, in itself, can increase the speed of valve response. Also, the pilot relay in the positioner generally has a larger supply port than the controller; therefore, the air from the positioner to the valve actuator moves faster and in greater quantity to achieve greater speed of valve response.

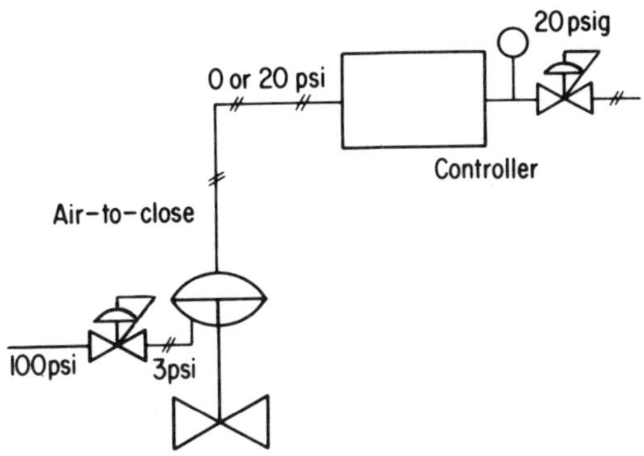

Figure 7. SPRINGLESS ACTUATOR CONTROL VALVE WITHOUT POSITIONER, FOR ON-OFF SERVICE.

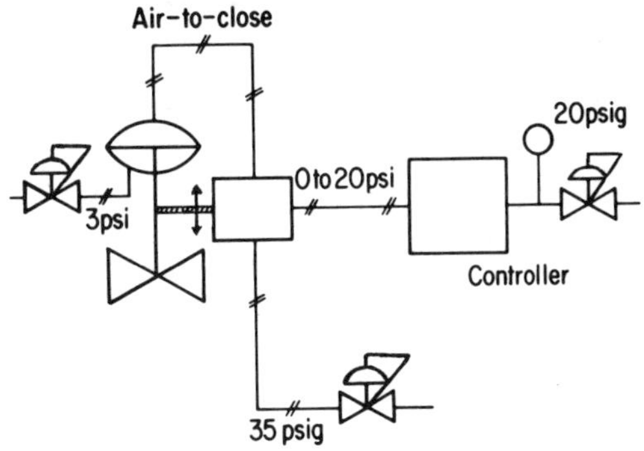

Figure 8. SPRINGLESS ACTUATOR CONTROL VALVE WITH POSITIONER, FOR THROTTLING SERVICE.

To Operate Springless Actuators

Figure 7 shows a controller operating a pneumatic control valve with a springless actuator. The springless actuator is the diaphragm type with the lower chamber supplied with a constant loading pressure of 3 psig air. This configuration is generally used only with two-position control since any air pressure above 3 psig on top of the diaphragm tends to close the valve, and any pressure less than 3 psig above the diaphragm permits it to open. Its chief advantage is that it provides greater shut-off forces to close the valve, since there is no spring to compress.

Figure 8 shows a pneumatic control valve with a springless actuator and a positioner. The controller is of the throttling type, with an output varying from approximately 0 to 20 psig. The positioner will apply the air pressure (up to 35 psig) necessary to correctly position the valve stem in accordance with the controller signal. The chief advantage of this arrangement is a higher control valve close-off rating against the upstream pressure, since there is no spring to compress. The chief disadvantage is a decrease in stability, due to the absence of a valve actuator spring. It also fails to provide reliable fail-safe performance.

Split Range Operation

It is sometimes desirable to operate a pneumatic control valve over only part of the controller output range. This can be accomplished by either changing or adjusting the input spring of the positioner. A common arrangement is to have one valve and positioner operate over 3 to 9 psig of the controller output, while another valve and positioner operates over 9 to 15 psig of the controller output. In the above example, it is only necessary to modify the positioner springs. The control valve springs may be standard 3 to 15 or 6 to 30 psig springs.

To Reverse Action

A positioner whose output air pressure increases as the input signal increases is called a direct-acting positioner. A positioner whose output air pressure decreases as the input signal increases is called a reverse-acting positioner. Positioners are available, whose action can easily be changed in the field.

The use of a reverse-acting positioner is not common and is not necessarily required with a reverse-acting control valve actuator. One application, however, is for heating and cooling service with two normally closed valves having a split-range operation.

To Change Control Valve Flow Characteristics

Most positioners are linear. That is, they change the valve stem position linearly with respect to controller output pressure. However, some positioners are provided with a means (generally a cam) to change this relationship to effect different control valve flow characteristics. Figure 9 shows this situation graphically. The combined results may not offer an improvement

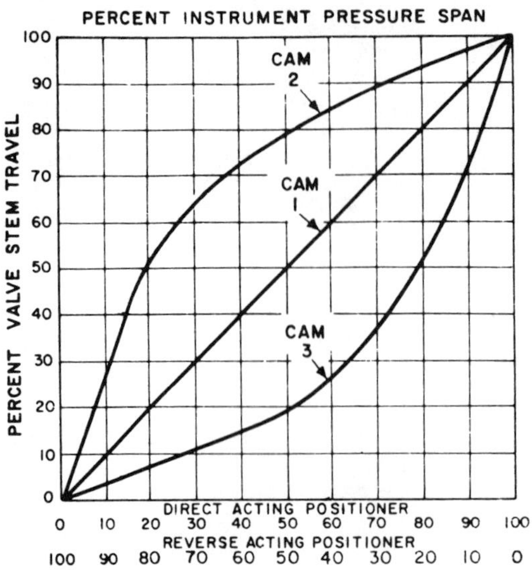

Figure 9. CAM CHARACTERISTICS.

in certain cases, because the valve time constant as well as the gain are affected by the cam shape.

Wide Proportional Band

When the valve must respond to very small changes in air pressure (less than 0.25 psi), a valve positioner is often recommended. This is the situation when the proportional band is wide.

Nature of Flow Medium

If the process fluid tends to gum or cake on valve parts and increase the friction between moving parts, a valve positioner is often helpful to provide the additional force required to overcome the friction.

High Pressure Flow Medium

On high pressure applications, the valve stem packing may tighten considerably resulting in higher friction on the valve stem which increases hysteresis and deadband. A valve positioner can sometimes be used to advantage under these circumstances. (See References 1 and 2)

Accessories For The Positioner

An integral pneumatic switch is sometimes supplied with a positioner and is used to bypass the positioner. The bypass may not be recommended where the valve will not operate without the positioner.

Pressure gages are often supplied to indicate the controller output pressure, positioner output pressure, and supply pressure to the positioner.

Limitations Of The Positioner

The above applications are traditional; however, recent research indicates that the use of a positioner may be detrimental to the quality of control in fast processes such as liquid pressure and flow processes. Where required (for example, higher air pressures to close the valve, or to speed the operation of the valve) pressure or volume may be used instead of positioners.

For slow processes, such as most thermal systems, liquid-level, and some large volume gas pressure processes, the positioner will improve the quality of control even though it appears not to be needed. (For further information see References 1 and 2).

Other Types of Positioners

Top-Mounted

The top-mounted valve positioner as shown in Figure 10, operates on the pneumatic force-balance principle. Control-instrument air pressure exerts an upward force on the larger (center) diaphragm of the positioner, because of the difference in the diaphragm areas. This force, plus the upward force of the range spring, is balanced by the downward force of the suppression spring. At balance, a small

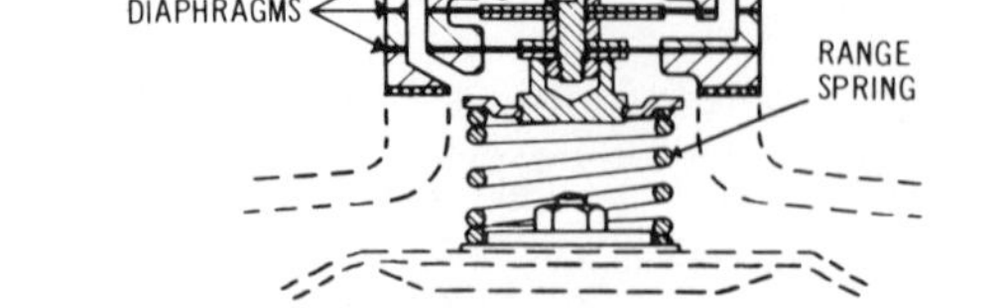

Figure 10. TOP MOUNTED VALVE POSITIONER.

Courtesy Moore Products Company

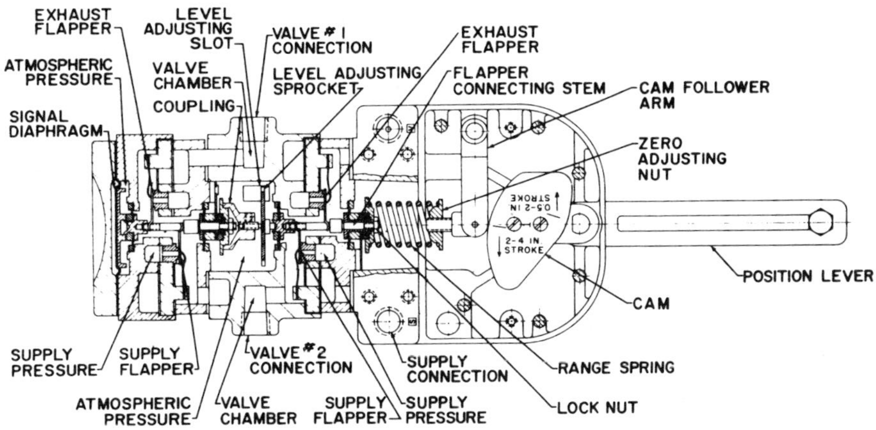

Figure 11. DOUBLE ACTING VALVE POSITIONER.

pilot flow (approximately 0.15 scfm) provides instant response to the slightest change in the control-instrument air pressure.

An increase in control-instrument air pressure temporarily raises the diaphragm assembly, closes the exhaust, and opens the pilot valve. The increase in valve pressure moves the motor diaphragm downward, reducing the force on the range spring. When the valve stem is in the position required by the control-instrument air pressure, the forces are again in balance.

Double-Acting Positioner

The double-acting pneumatic positioner is a force-balance instrument which positions a control valve stem in accordance with a pneumatic signal received from either a controller or a manual loading station. Figure 11 shows a direct-acting positioner set up for use on a piston actuator. With this configuration, an increasing signal will drive the piston stem downward.

In operation, the instrument signal is applied to the signal diaphragm. An increasing signal will drive the diaphragm and flapper connecting stem to the right. The flapper connecting stem then opens the supply flapper on "valve #2," admitting supply pressure into the upper chamber. The exhaust flapper on "valve #2" remains closed when the flapper connecting stem is deflected to the right. On "valve #1" the opposite action occurs, exhausting the lower chamber. An increasing instrument signal increases the pressure in the upper chamber of the piston and decreases pressure in the lower chamber. This pressure differential on the piston drives the valve stem downward and rotates the positioner lever clockwise. Clockwise rotation of the positioner lever compresses the range spring through the cam. When the valve stem reaches the position called for by the controller, the compression on the range spring creates a force equal to the force caused by the controller signal acting on the signal diaphragm. When this *force balance* exists, the flapper connecting stem moves back to the left and all flappers are closed.

If the controller signal is decreased, the force exerted by the signal diaphragm will also decrease and the force from the range spring will push the flapper connecting stem to the left, opening the exhaust flapper on "valve #1" and closing the supply flapper on "valve #2." This causes an upward force on the piston and allows the valve stem to move upward until a new force-balance is established.

Electropneumatic Positioner

The Electropneumatic Positioner, shown in Figure 12, is a force-balance device which positions a control valve in ac-

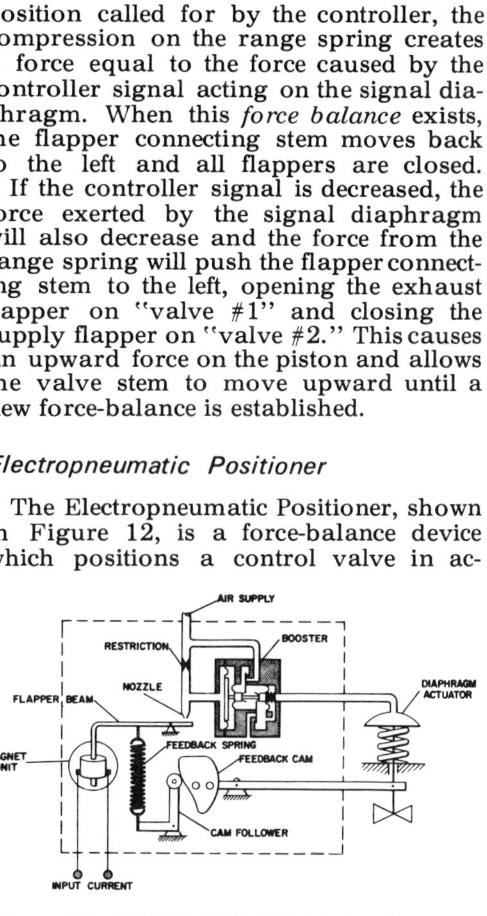

Figure 12. ELECTROPNEUMATIC VALVE POSITIONER.

Figure 13. CONTROL VALVE WITH DIGITAL PNEUMATIC POSITIONER.

Courtesy Honeywell, Inc.

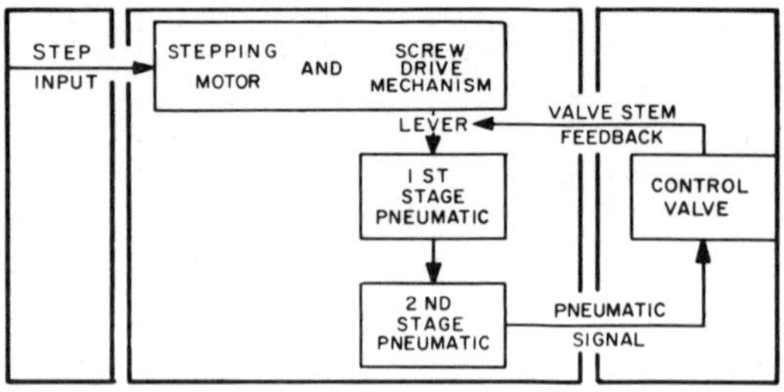

Figure 14. BLOCK DIAGRAM OF DIGITAL PNEUMATIC VALVE POSITIONER CONTROL LOOP.

cordance with the current signal from a controller. As shown in the figure, the positioner supplies air to the control valve until the force from the feedback spring balances the force of the electromagnetic coil.

Digital Positioner

The digital positioner is a single-acting, two stage, force balance instrument. It receives a digital electrical signal and positions a pneumatic control valve. Figure 13 is a photograph of a digital pneumatic positioner and Figure 14 is a block diagram of the digital positioner in a control loop.

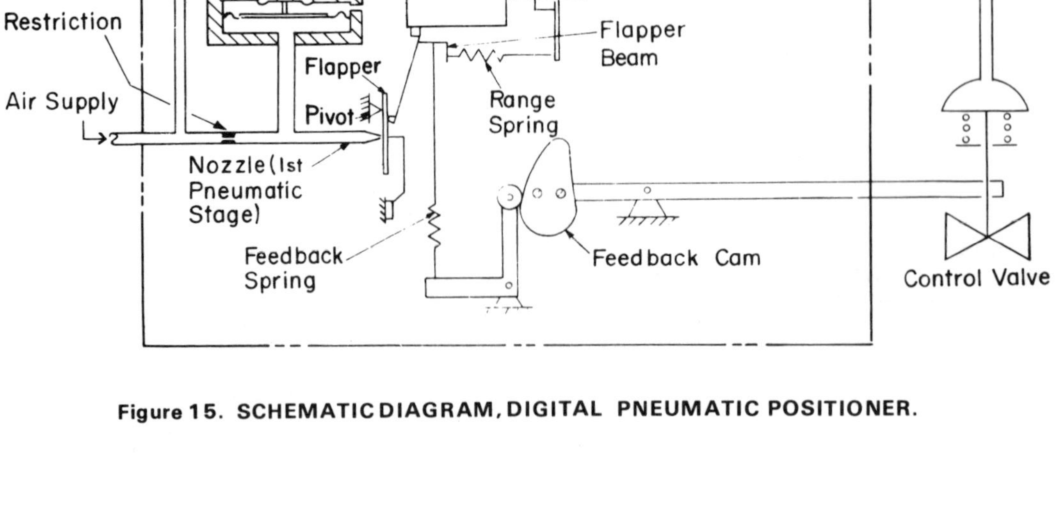

Figure 15. SCHEMATIC DIAGRAM, DIGITAL PNEUMATIC POSITIONER.

Figure 15 is a schematic illustration of a digital positioner. The input signal goes to a dc stepping motor which rotates in discrete steps. As the threaded motor shaft rotates, a traverse nut assembly travels on it varying the tension of a range spring. If this tension decreases, it causes a clockwise rotation of the flapper beam and a counter-clockwise rotation of the flapper allowing more air to flow out of the nozzle. This effect reduces the back pressure applied to the diaphragm of the second stage pilot, causing a reduction of output pressure applied to the control valve diaphragm. Reduced pressure on the control valve diaphragm will cause the valve stem to move up and the feedback cam to rotate counter-clockwise. Action of the feedback cam and feedback spring rotates the flapper beam in a counter-clockwise direction and the flapper in a clockwise direction. This action decreases the amount of air flowing out of the nozzle. When the moment on the flapper beam caused by the range spring, equals that of the feedback spring, the valve has reached the new position called-for by the signal. The valve will remain in this position until the signal is changed.

PNEUMATIC VOLUME AND PRESSURE BOOSTERS (See Fig. 16)

Figure 16. VOLUME AND/OR PRESSURE BOOSTER.
Courtesy Kendall, Inc.

Volume Boosters

A volume booster can be used to increase the speed of operation of the diaphragm control valve. A typical arrangement is shown in Figure 17.

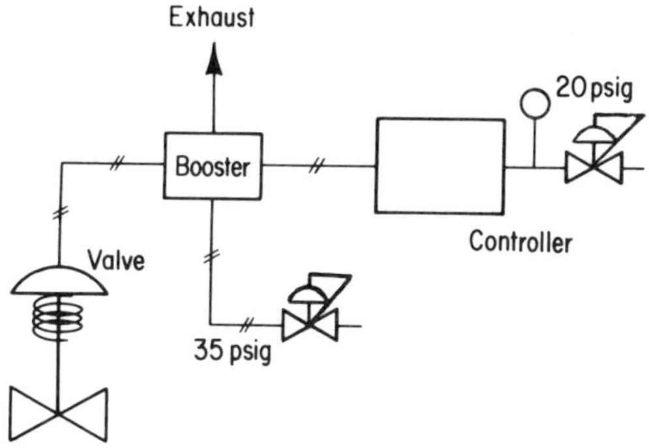

Figure 17. VOLUME BOOSTER IN CONTROL VALVE LOOP.

Figure 17 shows a volume booster being used in conjunction with a controller and a pneumatic valve. The controller applies its output signal to the booster instead of to the control valve. Only about one cubic inch of air is required to position the pilot in the booster, thus, the volume of air moved through the connecting tubing is small. The air that operates the valve comes through the pilot in the booster and since this pilot has a large capacity (35 scfm), the stroking time for the control valve is substantially reduced.

If a pressure reducing valve is used for the supply pressure to the booster, it should be of the high-capacity type so as not to limit the output capacity of the booster.

Pressure Boosters

Pressure boosters are generally also volume boosters; however, their main function is to increase the pressure from the controller to above 20 psig in certain valve applications.

Figure 18 shows a single-seated, air-to-open valve with a 6-30 psi spring. The heavy spring may be used to close the valve against a high upstream pressure; however, the normal controller output (20 psi maximum) cannot open the valve. A 2:1 booster makes the system operational.

The booster may be preferred over a positioner because of its lower cost. Also, the booster does not close a loop around the valve, a fact which may enhance stability in fast-response systems.

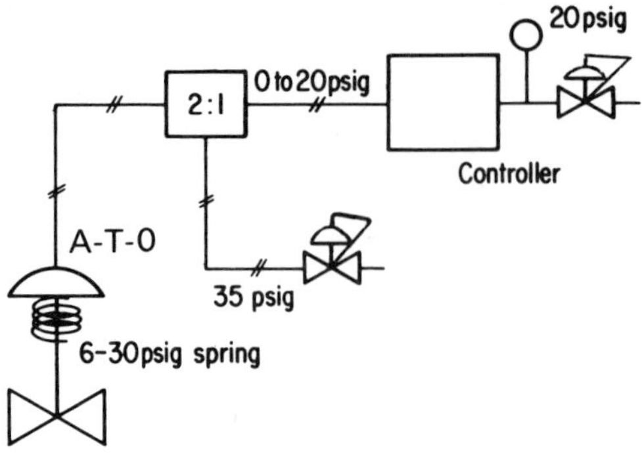

Figure 18. PRESSURE BOOSTER IN CONTROL VALVE LOOP.

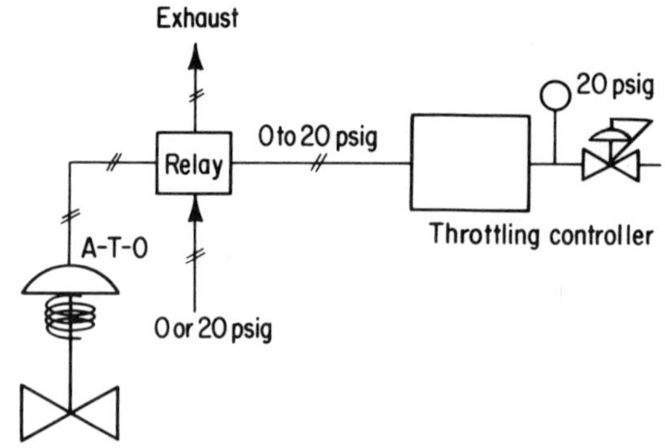

Figure 20. THROTTLING SERVICE, TWO-POSITION PNEUMATIC RELAY FOR EMERGENCY SHUT-DOWN.

ON-OFF AIR RELAYS

An on-off air relay may be used in some applications where the maximum output (20 psi) of the controller is insufficient to operate a diaphragm control valve. Figure 19 shows an on-off pneumatic controller operating an air relay. The relay in turn applies 35 psig or 0 psig to the diaphragm control valve to open or close the valve. The 35 psi pressure allows the single-seated diaphragm control valve to close against higher line pressure than would be possible with the 20 psig output from the controller.

The pneumatic relay may be used with a throttling controller where the relay is used in emergency service, as shown in Figure 20. In this application the relay is normally positioned so that the exhaust port is closed and the path through the relay from controller to the diaphragm control valve is kept open. With proper piping, an emergency signal (0 or 20 psi) can be applied to the relay to close off the controller signal and to open the exhaust port. With the relay exhaust port connected to the control valve, the air-to-open valve is rapidly closed by the valve spring.

SOLENOID VALVES

An electric solenoid valve, as shown in Figure 21, can be used to advantage in combination with diaphragm control valves in some applications.

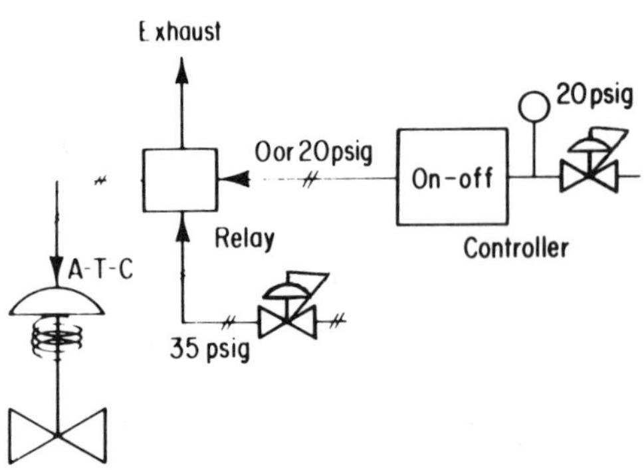

Figure 19. ON-OFF PNEUMATIC RELAY IN CONTROL VALVE LOOP.

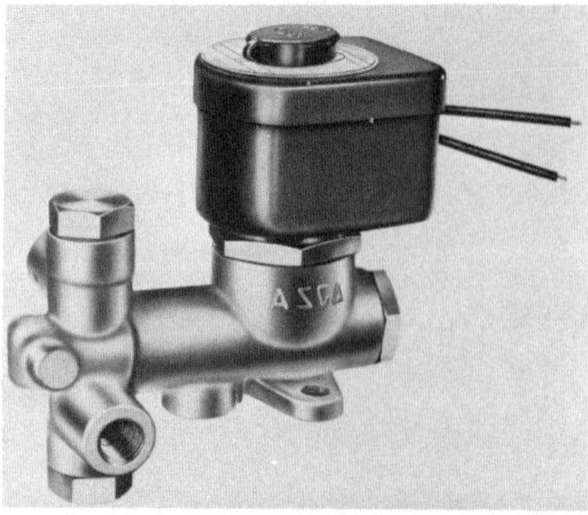

Figure 21. THREE-WAY ELECTRIC SOLENOID VALVE.
Courtesy Automatic Switch Company

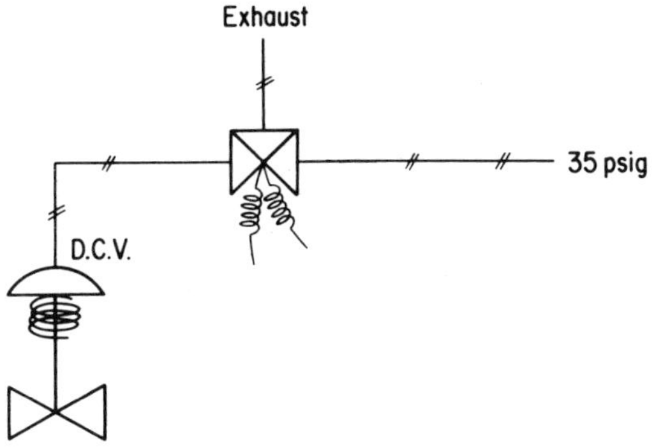

Figure 22. THREE-WAY SOLENOID VALVE USED TO ACTUATE A DIAPHRAGM CONTROL VALVE FOR ON-OFF SERVICE.

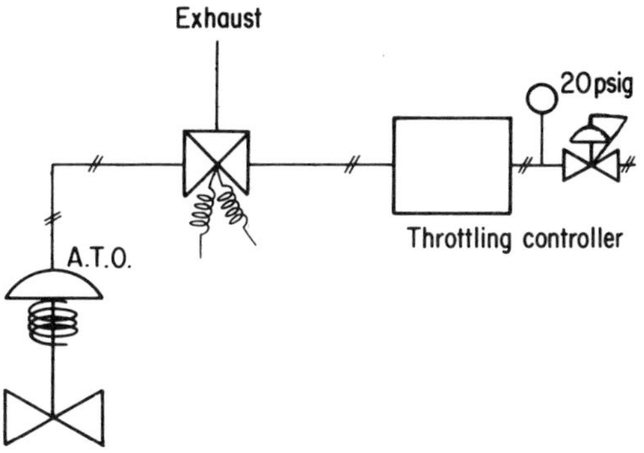

Figure 23. THREE-WAY SOLENOID VALVE IN A DIAPHRAGM CONTROL VALVE LOOP FOR THROTTLING SERVICE.

Figure 22 illustrates a three-way solenoid valve supplying or exhausting air from a diaphragm control valve to achieve on-off operation. Depending on the size, this arrangement may be considerably less expensive and faster responding than an all electric valve.

The three-way solenoid valve may be used with a pneumatic throttling controller where the solenoid is used in emergency service, as illustrated by Figure 23. In this application the solenoid three-way valve is normally positioned so that the exhaust port is closed and the path between the controller and diaphragm control valve is kept open. With proper piping an emergency signal, in the form of electric current through the solenoid, can cause the controller output to be blocked and the exhaust port to be opened. With the solenoid valve exhaust port connected to the diaphragm control valve, the air-to-open valve is rapidly closed by the valve spring.

POSITION INDICATING SWITCHES

For remote, two-position indication of valve stem position, either plunger or roller-actuated switches (see examples in Figure 24) can be mounted directly on the actuator yoke. This indication is limited to two-position and is usually accomplished by having the switch operate red and green indicating lights.

AIR LOCKS

This device is used for applications that require a control valve to hold its position in the event that the plant air supply pressure falls below a given level.

Figure 24. VALVE POSITION INDICATOR SWITCHES.
Courtesy Honeywell, Inc.

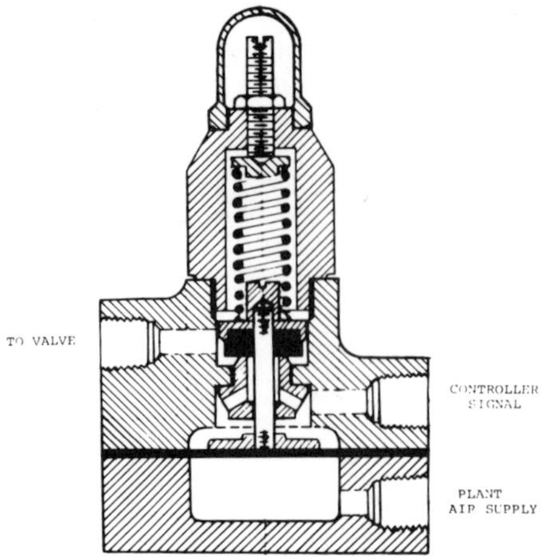

Figure 25. AIR LOCK.

One type is shown in Figure 25. The plant air supply is fed into a chamber sealed by a spring-opposed diaphragm. In the event that the plant air decreases to a predetermined lower limit, the spring closes the connection to the actuator and locks the existing controller-signal pressure in the line connecting the valve operator.

MOTION TRANSMITTERS

The pneumatic motion transmitter (Figure 26) is a device for continuous indication of valve stem position. It is generally yoke mounted and attached to the valve stem through a linkage. Through this linkage, it senses valve stem position and transmits a proportional 3-15 psig signal to a remotely located receiver.

Devices which transmit the valve stem position electrically are also available.

ELECTROPNEUMATIC TRANSDUCERS

Electropneumatic transducers convert the electrical output signals from electric controllers into pneumatic signals which may be used to operate diaphragm control valves. They may be either current-to-pressure (I/P) transducers, or voltage-to-pressure (E/P) transducers. A typical (I/P) transducer is schematically shown in Figure 27.

The input current is applied to the magnet unit coil. The combined action of the input current and the magnetic field of the permanent magnets creates a force which is applied to the beam through the strap connecting it to the magnet unit. The beam, which is pivoted at one end, is deflected by this force. In operation, an increase in input current pivots the magnet coil

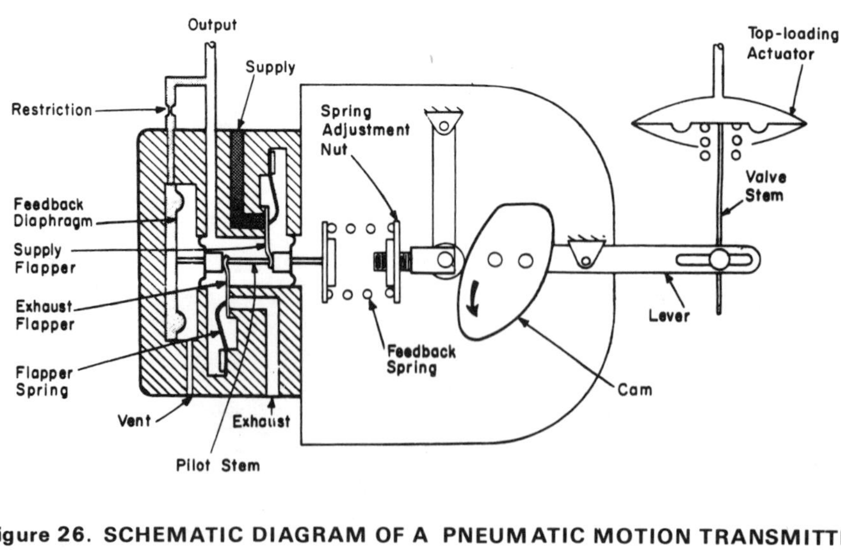

Figure 26. SCHEMATIC DIAGRAM OF A PNEUMATIC MOTION TRANSMITTER.

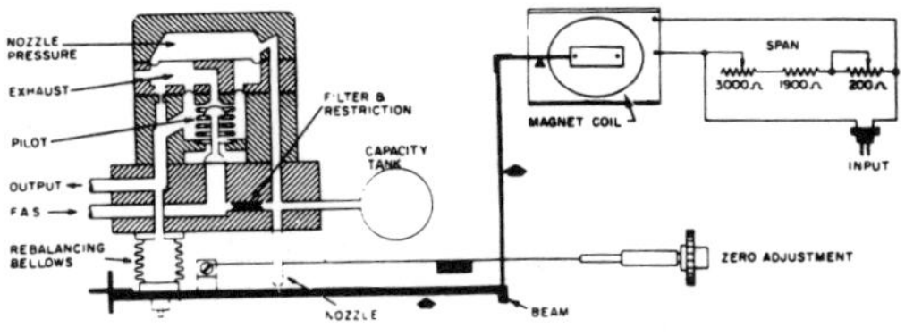

Figure 27. SCHEMATIC DIAGRAM OF AN ELECTROPNEUMATIC (I/P) TRANSDUCER.

clockwise pulling up the strap and moving the beam closer to the nozzle. This increases the nozzle pressure transmitted to the pilot valve and, therefore, increases the output pressure from the pilot valve to both the output line and the rebalancing bellows. At the bellows, the increased output pressure is changed into a balancing force which is applied to the beam to counteract the unbalanced force applied by the magnet unit, thus rebalancing the beam. Zero adjustment is obtained by applying a zeroing force to the beam through the zero spring.

REFERENCES

1. Lloyd, S.G., May, 1969, *Guidelines for the Application of Valve Positioners,* Paper No. TM-23, Fisher Controls, Inc., Marshalltown, Iowa.
2. Lloyd, S.G., October, 1968, *Guidelines for the Use of Valve Positioners and Booster Amplifiers,* Paper No. 68-922, Instrument Society of America, Pittsburgh.

CHAPTER 9

SELECTION OF VALVE MATERIALS OF CONSTRUCTION

C. F. King
E. L. Dixson
D. J. Gaffney

Fisher Controls Company
Marshalltown, Iowa

CONTENTS

INTRODUCTION

Material of construction is of the utmost importance in control valve selection since control valves are called upon to handle all types of fluids, from clean dry air to corrosive chemicals at temperatures ranging from near absolute zero to well above 1000° F., and pressures from near a perfect vacuum to 50,000 psig or higher. There are many materials that can be used and some services require the use of exotic alloys and metals to withstand corrosive fluids. However, since the majority of applications are relatively non-corrosive and at reasonable pressures and temperatures, cast iron and carbon steel are the most common valve body materials.

Most control valve materials can be placed in two categories:

1. The pressure containment materials for the valve body, bonnet, bottom flange and bolting.
2. The valve trim materials for the valve plug, seat ring, cage, valve stem, guide bushings, and packing box parts.

Table I. VALVE BODY (PRESSURE CONTAINMENT) MATERIALS

General Classification	*Applicable Material Specification*	
	Forgings	*Castings*
Cast Iron		ASTM A126
Ductile Iron		ASTM A395
Carbon Steel	ASTM A105 Grade I ASTM A105 Grade II	ASTM A215 Grade WCB
Carbon Steel (Low Temp)	ASTM A350 Grade LF1 & LF2	ASTM A352 Grade LCB
Carbon Moly	ASTM A182 Grade F1	ASTM A217 Grade WC1
1-1/4 Cr - 1/2 Mo	ASTM A182 Grade F11	ASTM A217 Grade WC6
2-1/4 Cr - 1 Mo	ASTM A182 Grade F22	ASTM A217 Grade WC9
5 Cr - 1/2 Mo	ASTM A182 Grade F5a	ASTM A217 Grade C5
9 Cr - 1 Mo	ASTM A182 Grade F9	ASTM A217 Grade C12
Type 304	ASTM A182 Grade F304	ASTM A351 Grade CF8
Type 347	ASTM A182 Grade F347	ASTM A351 Grade CF8C
Type 316	ASTM A182 Grade F316	ASTM A351 Grade CF8M
3-1/2 Ni	ASTM A350 Grade LF3	ASTM A 352 Grade LC3
Aluminum	ASTM B247	ASTM B26
Bronze		ASTM B61 ASTM B62
Inconel	ASTM B166	
Monel		ASTM A296-M35
Hastelloy B		ASTM A296-N12M
Hastelloy C		ASTM A494 (Ni-Mo-Cr)
Titanium	ASTM B381	ASTM B367
Nickel		ASTM A296-CZ-100
Alloy 20		ASTM A296-CN-7M

MATERIALS FOR PRESSURE CONTAINMENT PARTS

The selection of materials for the pressure containment parts is primarily dependent upon the following conditions of the fluid media: pressure, temperature, corrosion properties, and erosion properties.

Sometimes a compromise must be made in material selection. For example, a material with good erosion resistance may not be satisfactory because of poor corrosion properties when handling a particular fluid.

Pressure and Temperature Requirements

Pressure and temperature ratings for the pressure containment parts have been established for the more common materials by the American National Standards Institute (ANSI). Other materials may be used, but are generally confined to the materials and allowable stresses listed in Section VIII of the American Society of Mechanical Engineers, *Boiler and Pressure Vessel Code.*

The recommended materials for specified physical and chemical requirements are given in standards established by the American Society for Testing and Materials (ASTM). It is considered good practice, especially for the pressure containment parts, to specify materials by the proper ASTM designation. Table I lists the more popular pressure containment materials and their respective ASTM specification designations.

Control valves are used in services with a wide variety of temperatures. Table II lists valve body (pressure containment) materials with their respective upper and lower temperature limits.

Since most materials have a reduction in allowable working stress at elevated temperatures, the pressure-temperature ratings must also be considered in choice of materials. These are tabulated for cast iron (Table III), ductile iron (Table IV), steel and steel alloys (Table V), and bronze (Table VI).

Control valve body bolting materials are selected on considerations of strength and temperature. Table VII indicates materials used for stuffing box, bonnet, and bottom flange bolting for various temperature ranges.

Table II. VALVE BODY MATERIAL TEMPERATURE LIMITS DEGREES F.

Material	*Lower*	*Upper*
Cast Iron	-20	410
Ductile Iron	-20	650
*Carbon Steel (Grade WCB)	-20	1000
Carbon Steel (Grade LCB)	-50	650
Carbon Moly (Grade WC1)	-20	850
1-1/4 Cr - 1/2 Mo (Grade WC6)	-20	1000
2-1/4 Cr - 1 Mo (Grade WC9)	-20	1050
5 Cr - 1/2 Mo (Grade C5)	-20	1100
9 Cr - 1 Mo (Grade C12)	-20	1100
Type 304 (Grade CF8)	-425	1500
Type 347 (Grade CF8C)	-425	1500
Type 316 (Grade CF8M)	-425	1500
3-1/2 Ni (Grade LC3)	-150	650
Aluminum	-325	400
Bronze	-325	550
Inconel	-325	1200
Monel	-325	900
Hastelloy B	-325	700
Hastelloy C	-325	1000
Titanium		600
Nickel	-325	500
Alloy 20	-50	300

*The carbon phase of carbon steel may be converted to graphite upon long exposure to temperatures above 775° F. (Check applicable codes for maximum temperature rating of various materials.)

Table III. PRESSURE-TEMPERATURE RATINGS OF CAST IRON PIPE FLANGES AND FLANGED FITTINGS*

	125 lb ASTM A126			250 lb ASTM A126		
Temperature Degrees F.	*Class A Sizes 1-12*	*Class B Sizes 14-24*	*Class B Sizes 30-48*	*Class A Sizes 1-12*	*Class B Sizes 14-24*	*Class B Sizes 30-48*
-20 to 150	175	150	150	400	300	300
200	165	135	115	370	280	250
225	155	130	100	355	270	225
250	150	125	85	340	260	200
275	145	120	65	325	250	175
300	140	110	50	310	240	150
325	130	105		295	230	125
350 [1]	125	100		280	220	100
375				265	210	
400 [2]				250	200	

All pressure ratings given in PSIG. Allowable pressures may be interpolated between temperatures.

[1] 353 F (max) to reflect temperature of saturated steam at 125 psig.
[2] 406 F (max) to reflect temperature of saturated steam at 250 psig.

Note: Class A Cast Iron - 21,000 psi min. tensile strength
Class B Cast Iron - 31,000 psi min. tensile strength

LIMITATIONS: 250 lb. - when used for liquid service the tabulated pressure-temperature ratings in sizes 14 in. and larger are applicable to 250 lb. flanges only and not to 250 lb. fittings. (Consult manufacturer for ratings of valves 14" and larger).

* Extracted from ANSI Standard Cast Iron Pipe Flanges and Flanged Fittings (ANSI B16.1-1967), with the permission of the publisher, the American Society of Mechanical Engineers, United Engineering Center, 345 East 47th Street, New York, N.Y. 10017.

Table IV. PRESSURE-TEMPERATURE RATINGS FOR NODULAR IRON COMPONENTS MANUFACTURED IN COMPLIANCE WITH ANSI B16.5. †

	Type of Flange		
Service Temperature Degrees F	*150 lb. Steel*	*300 lb. Steel*	*600 lb. Steel*
-20 to 100	220	575	1000
150	203	568	1000
200	191	560	1000
250	180	550	1000
300	167	542	1000
350	155	540	1000
40	143	532	1000
450	131	520	1000
500	120	500	1000
550	120	500	1000
550	112	470	950
600	104	445	888
650	96	410	825

Cast Nodular Iron Must Conform To ASTM A395.
Pressure ratings in this table are 80% of those in ANSI B16.5 for carbon steel.

† Extracted in part from American National Standards Institute "Standard Code for Pressure Piping, Power Piping," ANSI B31.1.0-1967, with the permission of the publisher, The American Society of Mechanical Engineers, United Engineering Center, 345 East 47th Street, New York, N.Y. 10017.

Table V. PRESSURE-TEMPERATURE RATINGS FOR STEEL AND STEEL ALLOYS*

The pressure and temperature ratings given below apply to all flange facings used on steel valve bodies. Screwed steel fittings have the same rating as shown under 600 pound. All pressures are in pounds per square inch gauge (psig)

Service Temperature Degrees F.	150 POUND									300 POUND								
	WCB	WCI	WC6	WC9	C5	C12	304	347	316	WCB	WCI	WC6	WC9	C5	C12	304	347	316
-20 to 100	275									720						615		720
150	255									710						585		710
200	240									700						550		700
250	225									690						520		690
300	210									680						495		680
350	195									675						470		675
400	180									665						450		665
450	165									650						430		650
500	150									625						410		625
550	140									590						395		590
600	130									555						380		555
650	120									515						370		515
700	110									470	480			485		355		495
750	100									425	445			450		340		470
800	92									365	410			415		330		450
850	[82]	82								[300]	370			385		320		425
875	[75]	[75]								[260]	[355]			365		315		415
900	[70]	[70]								[225]	[335]			350		310		400
925	[60]	[60]								[190]	[320]			335		305		390
950	[55]	[55]								[155]	[300]			315		305		380
975	[50]	[50]								[120]	[280]			300		300		370
1000	[40]	[40]								[85]	[215]	265	265	250	290	300		355
1025	—	—	—	—	—	—	—	—	—	—	—	230	235	215	240	295		345
1050	—	—	—	—	—	—	—	—	—	—	—	190	200	190	190	290		335
1075	—	—	—	—	—	—	—	—	—	—	—	165	170	145	150	275		325
1100	—	—	—	—	—	—	—	—	—	—	—	135	145	115	115	255		310
1125	—	—	—	—	—	—	—	—	—	—	—	110	125	95	95	225		300
1150	—	—	—	—	—	—	—	—	—	—	—	85	105	75	75	195	260	290
1175	—	—	—	—	—	—	—	—	—	—	—	65	85	65	65	175	215	260
1200	—	—	—	—	—	—	—	—	—	—	—	40	70	50	50	155	170	235
1225	—	—	—	—	—	—	—	—	—	—	—	—	—	—	—	135	140	205
1250	—	—	—	—	—	—	—	—	—	—	—	—	—	—	—	110	115	180
1275	—	—	—	—	—	—	—	—	—	—	—	—	—	—	—	100	95	160
1300	—	—	—	—	—	—	—	—	—	—	—	—	—	—	—	85	75	135
1325	—	—	—	—	—	—	—	—	—	—	—	—	—	—	—	75	65	115
1350	—	—	—	—	—	—	—	—	—	—	—	—	—	—	—	60	50	95
1375	—	—	—	—	—	—	—	—	—	—	—	—	—	—	—	55	45	80
1400	—	—	—	—	—	—	—	—	—	—	—	—	—	—	—	50	40	70
1425	—	—	—	—	—	—	—	—	—	—	—	—	—	—	—	40	35	60
1450	—	—	—	—	—	—	—	—	—	—	—	—	—	—	—	35	30	50
1475	—	—	—	—	—	—	—	—	—	—	—	—	—	—	—	30	30	45
1500	—	—	—	—	—	—	—	—	—	—	—	—	—	—	—	25	25	35

* Extracted from ANSI Standard Steel Pipe Flanges and Flanged Fittings (ANSI B16.5 - 1968) with the permission of the publisher, the American Society of Mechanical Engineers, United Engineering Center, 345 East 47th Street, New York, New York, 10017.

Note: Figures set in brackets must be used in accordance with the Introductory Notes, Paragraph 2.3, of ANSI B16.5-1968. (Applicable codes warn of possible graphitization upon prolonged exposure to temperatures corresponding to bracketed figures.)
Abbreviations used in the column headings may be identified by referring to Table 2.

... PRESSURE-TEMPERATURE RATINGS FOR STEEL AND STEEL ALLOYS*

The pressure and temperature ratings given below apply to all flange facings used on steel valve bodies. Screwed steel fittings have the same rating as shown under 600 pound. All pressures are in pounds per square inch gauge (psig)

Service Temperature Degrees F.	600 POUND									900 POUND								
	WCB	WCI	WC6	WC9	C5	C12	304	347	316	WCB	WCI	WC6	WC9	C5	C12	304	347	316
-20 to 100	1440						1235		1440	2160						1850		2160
150	1420						1165		1420	2130						1750		2130
200	1400						1095		1400	2100						1645		2100
250	1380						1040		1380	2070						1565		2070
300	1365						985		1365	2050						1480		2050
350	1350						945		1350	2025						1415		2025
400	1330						900		1330	2000						1350		2000
450	1305						860		1305	1955						1290		1955
500	1250						825		1250	1875						1235		1875
550	1180						795		1180	1775						1190		1775
600	1110						765		1110	1660						1145		1660
650	1030						735		1030	1550						1105		1550
700	940	960			965		710		985	1410	1440			1450		1065		1480
750	850	890			900		685		940	1275	1330			1350		1025		1410
800	730	815			835		660		895	1100	1225			1250		985		1345
850	[600	745			765		640		850	[900	1115			1150		960		1275
875	525	[710			735		630		825	785	[1060			1100		945		1240
900	445	670			700		620		805	670	1010			1050		930		1205
925	375	635			665		615		780	565	955			1000		920		1175
950	310	600			635		610		760	465	900			950		915		1140
975	240	555			600		605		735	360	835			900		905		1105
1000	170]	430]	535	535	500	585	600		715	255]	645]	800	800	750	875	900		1070
1025	—	—	[455	465	430	480	595		690	—	—	[685	700	645	720	890		1035
1050	—	—	375	400	355	375	585		670	—	—	565	595	535	565	875		1000
1075	—	—	325	[345	290	300	550		645	—	—	490	[515	435	455	825		970
1100	—	—	275	290	225	225	515		625	—	—	410	430	340	340	770		935
1125	—	—	225	245	190	190	455		600	—	—	335	370	285	285	680		900
1150	—	—	170	205	150	150	395	520	585	—	—	255	310	225	225	590	780	875
1175	—	—	125	170	125	125	350	430	525	—	—	190	255	190	190	525	650	785
1200	—	—	80]	135]	105	105	310	345	465	—	—	125]	205]	155	155	465	515	700
1225	—	—	—	—	—	—	265	285	415	—	—	—	—	—	—	400	425	620
1250	—	—	—	—	—	—	225	225	365	—	—	—	—	—	—	335	340	545
1275	—	—	—	—	—	—	195	190	320	—	—	—	—	—	—	295	285	480
1300	—	—	—	—	—	—	170	150	275	—	—	—	—	—	—	250	225	410
1325	—	—	—	—	—	—	145	125	230	—	—	—	—	—	—	220	190	345
1350	—	—	—	—	—	—	125	105	185	—	—	—	—	—	—	185	155	280
1375	—	—	—	—	—	—	110	95	160	—	—	—	—	—	—	165	140	240
1400	—	—	—	—	—	—	95	80	135	—	—	—	—	—	—	145	125	205
1425	—	—	—	—	—	—	80	70	120	—	—	—	—	—	—	125	110	180
1450	—	—	—	—	—	—	70	60	105	—	—	—	—	—	—	105	95	155
1475	—	—	—	—	—	—	60	55	85	—	—	—	—	—	—	90	85	130
1500	—	—	—	—	—	—	50	50	70	—	—	—	—	—	—	75	75	105

*Extracted from ANSI Standard Steel Pipe Flanges and Flanged Fittings (ANSI B16.5 - 1968) with the permission of the publisher, the American Society of Mechanical Engineers, United Engineering Center, 345 East 47th Street, New York, New York, 10017.

Note: Figures set in brackets must be used in accordance with the Introductory Notes, Paragraph 2.3, of ANSI B16.5-1968. (Applicable codes warn of possible graphitization upon prolonged exposure to temperatures corresponding to bracketed figures.) Abbreviations used in the column headings may be identified by referring to Table 2.

Table V Cont'd. PRESSURE-TEMPERATURE RATINGS FOR STEEL AND STEEL ALLOYS*

The pressure and temperature ratings given below apply to all flange facings used on steel valve bodies. Screwed steel fittings have the same rating as shown under 600 pound. All pressures are in pounds per square inch gauge (psig)

Service Temperature Degrees F.	1500 POUND									2500 POUND								
	WCB	WCI	WC6	WC9	C5	C12	304	347	316	WCB	WCI	WC6	WC9	C5	C12	304	347	316
-20 to 100	3600						3085		3600	6000						5145		6000
150	3550						2915		3550	5915						4855		5915
200	3500						2740		3500	5830						4565		5830
250	3450						2605		3450	5750						4340		5750
300	3415						2470		3415	5690						4115		5690
350	3375						2360		3375	5625						3930		5625
400	3330						2245		3330	5550						3745		5550
450	3255						2150		3255	5430						3585		5430
500	3125						2055		3125	5210						3430		5210
550	2955						1985		2955	4925						3305		4925
600	2770						1910		2770	4620						3180		4620
650	2580						1845		2580	4300						3070		4300
700	2350	2400			2415		1775		2465	3920	4000			4025		2960		4110
750	2125	2220			2250		1710		2355	3550	3700			3745		2850		3920
800	1830	2040			2080		1645		2240	3050	3400			3470		2745		3730
850	[1500	1860			1915		1595		2125	[2500	3100			3190		2660		3540
875	1305	[1770			1830		1570		2070	2180	[2950			3055		2620		3445
900	1115	1680			1750		1545		2010	1855	2800			2915		2580		3350
925	945	1590			1665		1535		1955	1570	2650			2775		2560		3260
950	770	1500			1585		1525		1900	1285	2500			2640		2540		3165
975	600	1395			1500		1510		1840	1000	2320			2500		2520		3070
1000	430]	1070]	1335	1335	1250	1455	1500		1785	715]	1785]	2230	2230	2085	2430	2500		2975
1025	—	—	[1140	1165	1070	1200	1485		1725	—	—	[1900	1945	1785	2000	2470		2880
1050	—	—	945	995	890	945	1455		1670	—	—	1570	1655	1485	1570	2430		2785
1075	—	—	815	[855	730	755	1370		1615	—	—	1355	[1430	1215	1255	2285		2690
1100	—	—	685	720	565	565	1285		1555	—	—	1145	1200	945	945	2145		2595
1125	—	—	555	615	470	470	1135		1500	—	—	930	1030	785	785	1895		2500
1150	—	—	430	515	375	375	985	1305	1455	—	—	715	855	630	630	1645	2170	2430
1175	—	—	315	430	315	315	880	1080	1310	—	—	530	715	530	530	1465	1800	2185
1200	—	—	205]	345]	255	255	770	855	1165	—	—	345]	570]	430	430	1285	1430	1945
1225	—	—	—	—	—	—	665	710	1035	—	—	—	—	—	—	1105	1185	1730
1250	—	—	—	—	—	—	555	565	910	—	—	—	—	—	—	930	945	1515
1275	—	—	—	—	—	—	490	470	795	—	—	—	—	—	—	815	785	1330
1300	—	—	—	—	—	—	420	375	685	—	—	—	—	—	—	700	630	1145
1325	—	—	—	—	—	—	365	315	575	—	—	—	—	—	—	605	530	955
1350	—	—	—	—	—	—	310	255	465	—	—	—	—	—	—	515	430	770
1375	—	—	—	—	—	—	275	230	405	—	—	—	—	—	—	455	385	670
1400	—	—	—	—	—	—	240	205	345	—	—	—	—	—	—	400	345	570
1425	—	—	—	—	—	—	205	180	300	—	—	—	—	—	—	345	300	500
1450	—	—	—	—	—	—	170	155	255	—	—	—	—	—	—	285	255	430
1475	—	—	—	—	—	—	150	140	215	—	—	—	—	—	—	250	235	355
1500	—	—	—	—	—	—	130	130	170	—	—	—	—	—	—	215	215	285

* Extracted from ANSI Standard Steel Pipe Flanges and Flanged Fittings (ANSI B16.5 - 1968) with the permission of the publisher, the American Society of Mechanical Engineers, United Engineering Center, 345 East 47th Street, New York, New York, 10017.

Note: Figures set in brackets must be used in accordance with the Introductory Notes, Paragraph 2.3, of ANSI B16.5-1968. (Applicable codes warn of possible graphitization upon prolonged exposure to temperatures corresponding to bracketed figures.)

Abbreviations used in the column headings may be identified by referring to Table 2.

Table VI. PRESSURE-TEMPERATURE RATINGS BRONZE FLANGES AND FLANGED FITTINGS*
150 and 300 lb.

	Material	*Pressure psi (Max)*	*Temperature Deg. F (Max)*
150 lb	ASTM B62	150 225	406 150
	ASTM B61	150 225	500 150
300 lb	ASTM B62	300 500	422 150
	ASTM B61	300 500	550 150

*Extracted from American Standard Bronze Flanges and Flanged Fittings, 150 and 300 Lb (ANSI B16.24 - 1962) with the permission of the Publisher, The American Society of Mechanical Engineers, United Engineering Center, 345 East 47th Street, New York 17, N.Y. 10017.

Table VII. VALVE BODY BOLTING MATERIALS

Body Materials	*Temperature Degrees F.*	*Material Specifications* *Bolts*	*Nuts*
Cast Iron	-20 to 410	ASTM A307 Gr B	ASTM A307 Gr B
Ductile Iron	-20 to 650	ASTM A193 Gr B7	ASTM A194 Gr 2H
Carbon Steel (Grade WCB)	-20 to 800	ASTM A193 Gr B7	ASTM A194 Gr 2H
Carbon Steel (Grade LCB)	-50 to 650	ASTM A193 Gr B7	ASTM A194 Gr 2H
Carbon Moly (Grade WC1)	-20 to 850	ASTM A193 Gr B7	ASTM A194 Gr 2H
1-1/4 Cr - 1/2 Mo (Grade WC6)	-20 to 1000	ASTM A193 Gr B7	ASTM A194 Gr 2H
2-1/4 Cr - 1 Mo (Grade WC9)	-20 to 1000 1000 to 1050	ASTM A193 Gr B7 ASTM A193 Gr B16	ASTM A194 Gr 2H ASTM A194 Gr 4
5 Cr - 1/2 Mo (Grade C5)	-20 to 1000 1000 to 1100	ASTM A193 Gr B7 ASTM A193 Gr B16	ASTM A194 Gr 2H ASTM A194 Gr 4
9 Cr - 1 Mo (Grade C12)	-20 to 1000 1000 to 1100	ASTM A193 Gr B7 ASTM A193 Gr B16	ASTM A194 Gr 2H ASTM A194 Gr 4
Type 304 (Grade CF8)	-425 to 100 100 to 1500	ASTM A320 Gr B8 *ASTM A193 Gr B8	ASTM A194 Gr 8 *ASTM A194 Gr 8
Type 347 (Grade CF8C)	-425 to 100 100 to 1500	ASTM A320 Cr B8 *ASTM A193 Gr B8	ASTM A194 Gr 8 *ASTM A194 Gr 8
Type 316 (Grade CF8M)	-425 to 100 100 to 1500	ASTM A320 Gr B8 *ASTM A193 Gr B8M	ASTM A194 Gr 8 *ASTM A194 Gr 8M
3-1/2 Ni (Grade LC3)	-150 to -50	ASTM A320 Gr L7	ASTM A194 Gr 4 With Charpy Test or 8M
	-50 to 650	ASTM A193 Gr B7	ASTM A194 Gr 2H

*Alloy steel bolting, A193 Gr B7 bolts and A194 Gr 2H nuts can be used at moderate temperatures depending on the permissible differential expansion.

Corrosion Resistance

The corrosion properties of the fluid media are important factors in material selection. Table VIII is intended to give only a general indication of how a limited selection of materials will react when in contact with certain fluids. The recommendations cannot be absolute because concentration, temperature, contaminants, pressure and other conditions may alter the suitability of a particular material. For example, Table VIII shows chlorine gas as 'A' for carbon steel and 'C' for titanium. While this is true for dry gas, the opposite is true if the gas is wet. It is therefore emphasized that this Table should be used only as a guide. References 1-4 provide additional reading on the subject of corrosion.

Erosion Resistance

Erosion is damage caused by the impingement of high velocity particles on the material surfaces. Entrained sand, slurries, catalyst fines, and flashing liquids are sometimes associated with this type of wear.

The selection of materials for the pressure containment parts is rather limited and the use of sacrificial liners at the areas of impingement have been successful. Figure 1 illustrates an angle type control valve and Figure 2 illustrates a ball type control valve with this type of construction. Experience has shown that longer life can be expected from a flashing liquid application by using materials alloyed with chromium, such as ASTM A217 Grade C5 or ASTM A351 Grade CF8M.

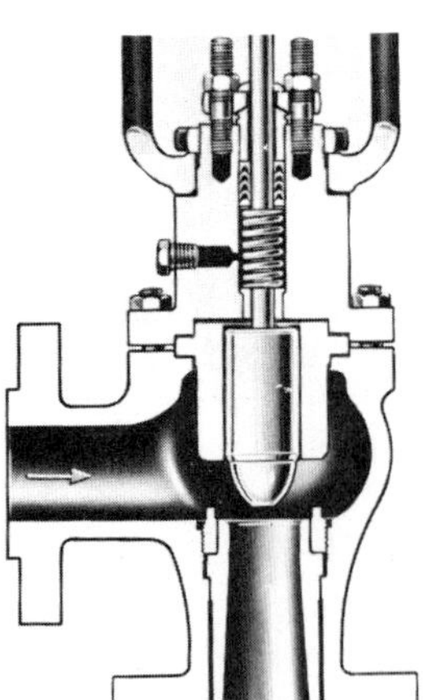

Figure 1. ANGLE VALVE.

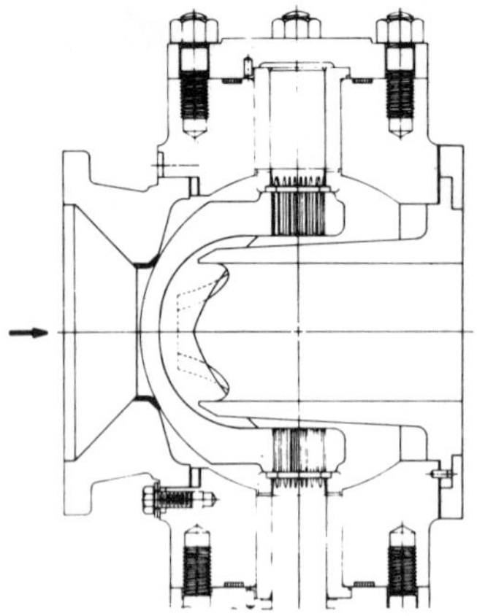

Figure 2. BALL VALVE.

MATERIALS FOR TRIM PARTS

The control valve trim parts which include valve plug, seat ring, cage, valve stem, guide bushings, and packing box parts have a wider lattitude in the selection of materials as they do not comprise the pressure containment parts of the control valve assembly and, therefore, do not come under the jurisdiction of various codes. The selection of these materials is generally predicated by the factors of corrosion, erosion, wear and galling, and temperature.

Corrosion Resistance

Table VIII may be used as a general guide for the corrosion resistance of trim materials. A typical valve plug failure caused by corrosion is illustrated in Figure 3.

Figure . VALVE PLUG FAILURE CAUSED BY CORROSION.

Table CORROSION SELECTION

	Carbon Steel	Cast Iron	302 or 304 Stainless Steel	316 Stainless Steel	Bronze	Monel	Hastelloy "B"	Hastelloy "C"	Durimet 20	Titanium	Alloy 6 (co-cr)	416 Stainless Steel	440 C Hard Stainless Steel	17-4 Ph Hard Stainless Steel
Acetaldehyde	A	A	A	A	A	A	I.L.	A	A	I.L.	I.L.	A	A	A
Acetic Acid, Air Free	C	C	B	B	B	B	A	A	A	A	A	C	C	B
Acetic Acid, Aerated	C	C	A	A	A	A	A	A	A	A	A	C	C	B
Acetic Acid Vapors	C	C	A	A	B	B	I.L.	A	B	A	A	C	C	B
Acetone	A	A	A	A	A	A	A	A	A	A	A	A	A	A
Acetylene	A	A	A	A	I.L.	A	A	A	A	I.L.	A	A	A	A
Alcohols	A	A	A	A	A	A	A	A	A	A	A	A	A	A
Aluminum Sulfate	C	C	A	A	B	B	A	A	A	A	I.L.	C	C	I.L.
Ammonia	A	A	A	A	C	A	A	A	A	A	A	A	A	I.L.
Ammonium Chloride	C	C	B	B	B	B	A	A	A	A	B	C	C	I.L.
Ammonium Nitrate	A	C	A	A	C	C	A	A	A	A	A	C	B	I.L.
Ammonium Phosphate (Mono-Basic)	C	C	A	A	B	B	A	A	B	A	A	B	B	I.L.
Ammonium Sulfate	C	C	B	A	B	A	A	A	A	A	A	C	C	I.L.
Ammonium Sulfite	C	C	A	A	C	C	I.L.	A	A	A	A	B	B	I.L.
Aniline	C	C	A	A	C	B	A	A	A	A	A	C	C	I.L.
Asphalt	A	A	A	A	A	A	A	A	A	I.L.	A	A	A	A
Beer	B	B	A	A	B	A	A	A	A	A	A	B	B	A
Benzene (Benzol)	A	A	A	A	A	A	A	A	A	A	A	A	A	A

KEY:
A - Can be or is being successfully used.
B - Proceed with caution.
C - Should not be used.
I.L. - Information lacking.

Table VIII cont'd. CORROSION SELECTION

	Carbon Steel	Cast Iron	302 or 304 Stainless Steel	316 Stainless Steel	Bronze	Monel	Hastelloy "B"	Hastelloy "C"	Durimet 20	Titanium	Alloy 6 (co-cr)	416 Stainless Steel	440 C Hard Stainless Steel	17-4 Ph Hard Stainless Steel
Benzoic Acid	C	C	A	A	A	A	I.L.	A	A	A	I.L.	A	A	A
Boric Acid	C	C	A	A	A	A	A	A	A	A	A	B	B	I.L.
Butane	A	A	A	A	A	A	A	A	A	I.L.	A	A	A	A
Calcium Chloride (Alkaline)	B	B	C	B	C	A	A	A	A	A	I.L.	C	C	I.L.
Calcium Hypochlorite	C	C	B	B	B	B	C	A	A	A	I.L.	C	C	I.L.
Carbolic Acid	B	B	A	A	A	A	A	A	A	A	A	I.L.	I.L.	I.L.
Carbon Dioxide, Dry	A	A	A	A	A	A	A	A	A	A	A	A	A	A
Carbon Dioxide, Wet	C	C	A	A	B	A	A	A	A	A	A	A	A	A
Carbon Disulfide	A	A	A	A	C	B	A	A	A	A	A	B	B	I.L.
Carbon Tetrachloride	B	B	B	B	A	A	B	A	A	A	I.L.	C	A	I.L.
Carbonic Acid	C	C	B	B	B	A	A	A	A	I.L.	I.L.	A	A	A
Chlorine, Gas	A	A	B	B	A	A	A	A	A	C	A	C	C	C
Chlorine, Liquid	C	C	C	C	B	C	C	A	B	C	B	C	C	C
Chromic Acid	C	C	C	B	C	A	C	A	C	A	B	C	C	C
Citric Acid	I.L.	C	B	A	A	B	A	A	A	A	I.L.	B	B	B
Coke Oven Gas	A	A	A	A	B	B	A	A	A	A	A	A	A	A
Copper Sulfate	C	C	B	B	B	C	I.L.	A	A	A	I.L.	A	A	A
Cottonseed Oil	A	A	A	A	A	A	A	A	A	A	A	A	A	A
Creosote	A	A	A	A	C	A	A	A	A	I.L.	A	A	A	A
Ethane	A	A	A	A	A	A	A	A	A	A	A	A	A	A
Ether	B	B	A	A	A	A	A	A	A	A	A	A	A	A
Ethyl Chloride	C	C	A	A	A	A	A	A	A	A	A	B	B	I.L.
Ethylene	A	A	A	A	A	A	A	A	A	A	A	A	A	A
Ethylene Glycol	A	A	A	A	A	A	I.L.	I.L.	A	I.L.	A	A	A	A

KEY:
A - Can be or is being successfully used.
B - Proceed with caution.
C - Should not be used.
I.L. - Information lacking.

Table VIII cont'd. CORROSION SELECTION

	Carbon Steel	Cast Iron	302 or 304 Stainless Steel	316 Stainless Steel	Bronze	Monel	Hastelloy "B"	Hastelloy "C"	Durimet 20	Titanium	Alloy 6 (co-cr)	416 Stainless Steel	440 C Hard Stainless Steel	17-4 Ph Hard Stainless Steel
Ferric Chloride	C	C	C	C	C	C	C	B	C	A	B	C	C	I.L.
Formaldehyde	B	B	A	A	A	A	A	A	A	A	A	A	A	A
Formic Acid	I.L.	C	B	B	A	A	A	A	A	C	B	C	C	B
Freon, Wet	B	B	B	A	A	A	A	A	A	A	A	I.L.	I.L.	I.L.
Freon, Dry	B	B	A	A	A	A	A	A	A	A	A	I.L.	I.L.	I.L.
Furfural	A	A	A	A	A	A	A	A	A	A	A	B	B	I.L.
Gasoline, Refined	A	A	A	A	A	A	A	A	A	A	A	A	A	A
Glucose	A	A	A	A	A	A	A	A	A	A	A	A	A	A
Hydrochloric Acid (Aerated)	C	C	C	C	C	C	A	B	C	C	B	C	C	C
Hydrochloric Acid (Air Free)	C	C	C	C	C	C	A	B	C	C	B	C	C	C
Hydrofluoric Acid (Aerated)	B	C	C	B	C	C	A	A	B	C	B	C	C	C
Hydrofluoric Acid (Air Free)	A	C	C	B	C	A	A	A	B	C	I.L.	C	C	I.L.
Hydrogen	A	A	A	A	A	A	A	A	A	A	A	A	A	A
Hydrogen Peroxide	I.L.	A	A	A	C	A	B	B	A	A	I.L.	B	B	I.L.
Hydrogen Sulfide, Liquid	C	C	A	A	C	C	A	A	B	A	A	C	C	I.L.
Magnesium Hydroxide	A	A	A	A	B	A	A	A	A	A	A	A	A	I.L.
Mercury	A	A	A	A	C	B	A	A	A	A	A	A	A	B
Methanol	A	A	A	A	A	A	A	A	A	A	A	A	B	A
Methyl Ethyl Ketone	A	A	A	A	A	A	A	A	A	I.L.	A	A	A	A
Milk	C	C	A	A	A	A	A	A	A	A	A	C	C	C
Natural Gas	A	A	A	A	A	A	A	A	A	A	A	A	A	A
Nitric Acid	C	C	A	B	C	C	C	B	A	A	C	C	C	B

KEY: A - Can be or is being successfully used.
B - Proceed with caution.
C - Should not be used.
I.L.- Information lacking.

Table VIII. cont'd. CORROSION SELECTION

	Carbon Steel	Cast Iron	302 or 304 Stainless Steel	316 Stainless Steel	Bronze	Monel	Hastelloy "B"	Hastelloy "C"	Durimet 20	Titanium	Alloy 6 (co-cr)	416 Stainless Steel	440 C Hard Stainless Steel	17-4 Ph Hard Stainless Steel
Oleic Acid	C	C	A	A	B	A	A	A	A	A	A	A	A	I.L.
Oxalic Acid	C	C	B	B	B	B	A	A	A	B	B	B	B	I.L.
Oxygen	A	A	A	A	A	A	A	A	A	A	A	A	A	A
Petroleum Oils, Refined	A	A	A	A	A	A	A	A	A	A	A	A	A	A
Phosphoric Acid (Aerated)	C	C	A	A	C	C	A	A	A	B	A	C	C	I.L.
Phosphoric Acid (Air Free)	C	C	A	A	C	B	A	A	A	B	A	C	C	I.L.
Phosphoric Acid Vapors	C	C	B	B	C	C	A	I.L.	A	B	C	C	C	I.L.
Picric Acid	C	C	A	A	C	C	A	A	A	I.L.	I.L.	B	B	I.L.
Potassium Chloride	B	B	A	A	B	B	A	A	A	A	I.L.	C	C	I.L.
Potassium Hydroxide	B	B	A	A	B	A	A	A	A	A	I.L.	B	B	I.L.
Propane	A	A	A	A	A	A	A	A	A	A	A	A	A	A
Rosin	B	B	A	A	A	A	A	A	A	I.L.	A	A	A	A
Silver Nitrate	C	C	A	A	C	C	A	A	A	A	B	B	B	I.L.
Sodium Acetate	A	A	B	A	A	A	A	A	A	A	A	A	A	A
Sodium Carbonate	A	A	A	A	A	A	A	A	A	A	A	B	B	A
Sodium Chloride	C	C	B	B	A	A	A	A	A	A	A	B	B	B
Sodium Chromate	A	A	A	A	A	A	A	A	A	A	A	A	A	A
Sodium Hydroxide	A	A	A	A	C	A	A	A	A	A	A	B	B	A
Sodium Hypochloride	C	C	C	C	B-C	B-C	C	A	B	A	I.L.	C	C	I.L.
Sodium Thiosulfate	C	C	A	A	C	C	A	A	A	A	I.L.	B	B	I.L.
Stannous Chloride	B	B	C	A	C	B	A	A	A	A	I.L.	C	C	I.L.
Stearic Acid	A	C	A	A	B	B	A	A	A	A	B	B	B	I.L.

KEY:
A - Can be or is being successfully used.
B - Proceed with caution.
C - Should not be used.
I.L. - Information lacking.

Table VIII cont'd. CORROSION SELECTION

	Carbon Steel	Cast Iron	302 or 304 Stainless Steel	316 Stainless Steel	Bronze	Monel	Hastelloy "B"	Hastelloy "C"	Durimet 20	Titanium	Alloy 6 (co-cr)	416 Stainless Steel	440 C Hard Stainless Steel	17-4 Ph Hard Stainless Steel
Sulfate Liquor (Black)	A	A	A	A	C	A	A	A	A	A	A	I.L.	I.L.	I.L.
Sulfur	A	A	A	A	C	A	A	A	A	A	A	A	A	A
Sulfur Dioxide, Dry	A	A	A	A	A	A	B	A	A	A	A	B	B	I.L.
Sulfur Trioxide, Dry	A	A	A	A	A	A	B	A	A	A	A	B	B	I.L.
Sulfuric Acid (Aerated)	C	C	C	C	C	C	A	A	A	B	B	C	C	C
Sulfuric Acid (Air Free)	C	C	C	C	B	B	A	A	A	B	B	C	C	C
Sulfurous Acid	C	C	B	B	B	C	A	A	A	A	B	C	C	I.L.
Tar	A	A	A	A	A	A	A	A	A	A	A	A	A	A
Trichloroethylene	B	B	B	A	A	A	A	A	A	A	A	B	B	I.L.
Turpentine	B	B	A	A	A	B	A	A	A	A	A	A	A	A
Vinegar	C	C	A	A	B	A	A	A	A	I.L.	A	C	C	A
Water, Boiler Feed	B	C	A	A	C	A	A	A	A	A	A	B	A	A
Water, Distilled	A	A	A	A	A	A	A	A	A	A	A	B	B	I.L.
Water, Sea	B	B	B	B	A	A	A	A	A	A	A	C	C	A
Whiskey and Wines	C	C	A	A	A	B	A	A	A	A	A	C	C	I.L.
Zinc Chloride	C	C	C	C	C	C	A	A	A	A	B	C	C	I.L.
Zinc Sulfate	C	C	A	A	B	A	A	A	A	A	A	B	B	I.L.

KEY:
A - Can be or is being successfully used.
B - Proceed with caution
C - Should not be used.
I.L. Information lacking.

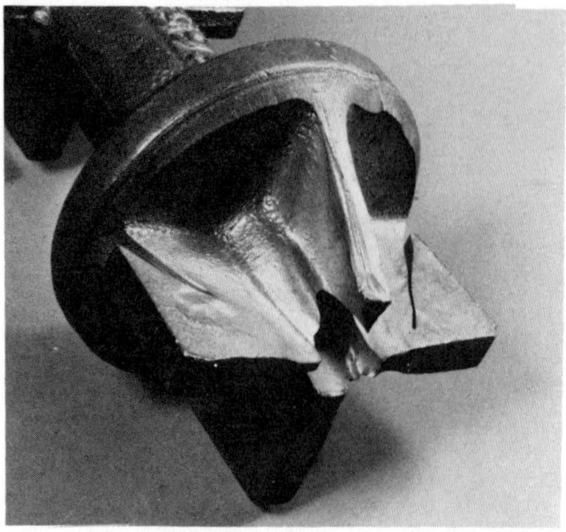

Figure 4. DAMAGE OF VALVE PLUG SEAT BY WIRE-DRAWING.

Figure 5. EROSION DAMAGE OF VALVE PLUG BY CATALYST FINES.

Erosion Resistance

The valve plug and seat ring are probably the control valve components that are most susceptible to damage by erosion caused by impingement of high velocity particles. Figure 4 illustrates valve plug seat damage where wet-steam liquid droplets impinged when the valve plug was nearly closed. This type of damage is commonly referred to as "wire drawing."

Figure 5 illustrates a damaged plug which was handling catalyst fines. The typical appearance of erosion damage is the smooth polished wear areas.

Following is a tabulation of various materials listed in order of increasing relative resistance to erosion (the first material would be rated relatively good and the last relatively poor).

Bronze
Aluminum Bronze
Nickel
Alloy 20
Monel
Hastelloy B & C
Type 316
Type 304
K Monel
17-4PH 40RC
Type 416 40RC
Inconel
Type 304 & Type 316 Alloy 6 faced
Type 440 60RC
Chrome & Tungsten Carbide
Ceramic

Wear and Galling Resistance

Wear and galling of valve plug post and bushing materials are usually encountered in applications where high mass flow rates exist. A factor that may contribute to wear is the instability of the plug caused by buffeting forces. Temperature and lubricating qualities of the flowing media and material selection are contributing factors to galling.

Figure 6 illustrates wear damage and Figure 7 illustrates guide post galling. Material selection for both wear and galling resistance are generally in the same category; and better life can be expected by the use of two different materials between the contact surface areas. If like materials must be used, different hardnesses may be beneficial.

Table IX lists the wear and galling resistance of various combinations of materials. In addition to these materials, plastic lined bushings have been found to be effective when service conditions permit their usage.

The lubricating qualities of the flowing media also affects wear and galling and such media as dry superheated steam and certain heat exchange liquids have a notable tendency toward this type of failure. As would be expected, lubricating oils have the least tendency.

Temperature Requirements

The service temperature is also a factor in selection of trim materials. Table X lists the more common trim materials with recommended upper and lower service temperatures. Note that some platings are listed that are occasionally used on valve guiding surfaces. Elastomers that are used for seat discs and packing are included.

Figure 6. VALVE PLUG WEAR DAMAGE.

Figure 7. GALLING OF GUIDE POST IN CONTROL VALVE.

Elastomers have a variety of usages in a control valve assembly in such applications as actuator diaphragms, piston seals, seat discs, O-ring seals, etc. The choice of materials depends upon service media, pressures and temperatures. Table XI lists the general properties of several materials that are used.

Table XII illustrates maximum and minimum service temperature limitations for several diaphragm materials. Note that the limits are the extreme temperatures. For practical service life, the maximum temperature should be reduced by at least 25° F. and minimum temperatures raised by 20° to 50° F.

Pressures and temperatures exceed the ratings covered by established codes and standards in some control valve applications. In these cases it is recommended that the manufacturer be contacted as these conditions generally warrant control valves of special construction.

REFERENCES

1. Lyman, Taylor, ed., *Metals Handbook,* 8th Edition, Vol. 1, American Society for Metals.
2. Nelson, G.A., ed., 1967, *Corrosion Data Survey,* National Association of Corrosion Engineers, Houston, Texas.
3. Polar, J.P., ed., 1961, *A Guide to Corrosion Resistance,* Climax Molybdenum Company.
4. Uhlig, H.H., ed., 1948, *The Corrosion Handbook,* John Wiley & Sons, Inc., New York.

Table IX. WEAR & GALLING RESISTANCE CHART OF MATERIAL COMBINATIONS

	304 SS	316 SS	BRONZE	INCONEL	MONEL	HASTELLOY B	HASTELLOY C	TITANIUM 75A	NICKEL	ALLOY 20	TYPE 416 HARD	TYPE 440 HARD	17-4 PH	ALLOY 6 (co-cr)	ENC*	Cr PL	AL BRONZE
304 SST	P	P	F	P	P	P	F	P	P	P	F	F	F	F	F	F	F
316 SS	P	P	F	P	P	P	F	P	P	P	F	F	F	F	F	F	F
BRONZE	F	F	S	S	S	S	S	S	S	S	F	F	F	F	F	F	F
INCONEL	P	P	S	P	P	P	F	P	F	F	F	F	F	F	F	F	S
MONEL	P	P	S	P	P	P	F	F	F	F	F	F	F	S	F	F	S
HASTELLOY B	P	P	S	P	P	P	F	F	S	F	F	F	F	S	F	S	S
HASTELLOY C	F	F	S	F	F	F	F	F	F	F	F	F	F	S	F	S	S
TITANIUM 75 A	P	P	S	P	F	F	F	P	F	F	F	F	F	S	F	F	S
NICKEL	P	P	S	F	F	S	F	F	P	P	F	F	F	S	F	F	S
ALLOY 20	P	P	S	F	F	F	F	F	P	P	F	F	F	S	F	F	S
TYPE 416 HARD	F	F	F	F	F	F	F	F	F	F	F	F	F	S	S	S	S
TYPE 440 HARD	F	F	F	F	F	F	F	F	F	F	S	F	S	S	S	S	S
17-4 PH	F	F	F	F	F	F	F	F	F	F	F	S	P	S	S	S	S
ALLOY 6 (co-cr)	F	F	F	F	S	S	S	S	S	S	S	S	S	F	S	S	S
ENC*	F	F	F	F	F	F	F	F	F	F	S	S	S	S	P	S	S
Cr PLATE	F	F	F	F	F	S	S	F	F	F	S	S	S	S	S	P	S
AL BRONZE	F	F	F	S	S	S	S	S	S	S	S	S	S	S	S	S	P

*Electroless Nickel Coating
S - Satisfactory
F - Fair
P - Poor

Table X. VALVE TRIM MATERIAL TEMPERATURE LIMITS DEGREES F.

MATERIAL	LOWER	UPPER
Type 304 Stainless Steel	-450	600
Type 316 Stainless Steel	-450	600
Bronze	-460	450
Inconel	-400	1200
K Monel	-400	900
Monel	-400	900
Hastelloy B		700
Hastelloy C		1000
Titanium		600
Nickel	-325	600
Alloy 20	-50	600
Type 416 Stainless Steel 40RC	-20	800
Type 440 Stainless Steel 60RC	-20	800
17-4 PH	-40	800
Alloy 6 (co-cr)	-460	1500
Electroless Nickel Plating	-450	800
Chrome Plating	-450	1100
Aluminum Bronze	-460	600
Nitrile (Buna N)	-40	200
Fluoroelastomer (Viton[1] and Fluorel[2])	-10	400
TFE	-450	450
Nylon	-100	200
Polyethylene	-100	200
Neoprene	-40	180

1. Registered Trademark - DuPont 2. Registered Trademark - 3M Co.

Table XI. GENERAL PROPERTIES OF ELASTOMERS.

Property	Natural Rubber	Buna-S	Buna-N	Neoprene	Butyl	Thiokol	Silicone	Hypalon	① ② Viton	② Poly-urethane	① Poly-acrylic	③ Ethylene Propylene
Tensile: Pure Gum	3000	400	600	3500	3000	300	200-450	4000	----	----	100	----
Reinforced . . .	4500	3000	4000	3500	3000	1500	1100	4400	2300	6500	1800	2500
Tear Resistance.	Excellent	Poor-Fair	Fair	Good	Good	Fair	Poor-Fair	Excellent	Good	Excellent	Fair	Poor
Abrasion Resistance. . . .	Excellent	Good	Good	Excellent	Fair	Poor	Poor	Excellent	Very Good	Excellent	Good	Good
Aging: Sunlight	Poor	Poor	Poor	Excellent	Excellent	Good	Good	Excellent	Excellent	Excellent	Excellent	Excellent
Oxidation.	Good	Fair	Fair	Good	Good	Good	Very Good	Very Good	Excellent	Excellent	Excellent	Good
Heat (Max. Temp.).	200°F.	200°F.	250°F.	200°F.	200°F.	140°F.	450°F.	300°F.	400°F.	200°F.	350°F.	350°F.
Static (Shelf)	Good	Good	Good	Very Good	Good	Fair	Good	Good	----	----	Good	Good
Flex Cracking Resistance .	Excellent	Good	Good	Excellent	Excellent	Fair	Fair	Excellent	----	Excellent	Good	----
Compression Set Resistance	Good	Good	Very Good	Excellent	Fair	Poor	Good	Poor	Poor	Good	Good	Fair
Solvent Resistance:												
Aliphatic Hydrocarbon. .	Very Poor	Very Poor	Good	Fair	Poor	Excellent	Poor	Fair	Excellent	Very Good	Good	Poor
Aromatic Hydrocarbon . .	Very Poor	Very Poor	Fair	Poor	Very Poor	Good	Very Poor	Poor	Very Good	Fair	Poor	Fair
Oxygenated Solvent . . .	Good	Good	Poor	Fair	Good	Fair	Poor	Poor	Good	Poor	Poor	---
Halogenated Solvent. . .	Very Poor	Very Poor	Very Poor	Very Poor	Poor	Poor	Very Poor	Very Poor	----	----	Poor	Poor
Oil Resistance:												
Low Aniline Mineral Oil.	Very Poor	Very Poor	Excellent	Fair	Very Poor	Excellent	Poor	Fair	Excellent	----	Excellent	Poor
High Aniline Mineral Oil	Very Poor	Very Poor	Excellent	Good	Very Poor	Excellent	Good	Good	Excellent	----	Excellent	Poor
Synthetic Lubricants . .	Very Poor	Very Poor	Fair	Very Poor	Poor	Poor	Fair	Poor	----	----	Fair	Poor
Organic Phosphates . . .	Very Poor	Very Poor	Very Poor	Very Poor	Good	Poor	Poor	Poor	Poor	Poor	Poor	Very Good
Gasoline Resistance:												
Aromatic	Very Poor	Very Poor	Good	Poor	Very Poor	Excellent	Poor	Poor	Good	Fair	Fair	Fair
Non-Aromatic	Very Poor	Very Poor	Excellent	Good	Very Poor	Excellent	Good	Fair	Very Good	Good	Poor	Poor
Acid Resistance:												
Dilute (Under 10%) . . .	Good	Good	Good	Fair	Good	Poor	Fair	Good	Excellent	Fair	Poor	Very Good
▲Concentrated	Fair	Poor	Poor	Fair	Fair	Very Poor	Poor	Good	Very Good	Poor	Poor	Good
Low Temperature Flexibility (Max.) . . .	-65°F.	-50°F.	-40°F.	- 40°F.	-40°F.	-40°F.	-100°F.	-20°F.	-30°F.	-40°F.	-10°F.	-50°F.
Permeability to Gases. . .	Fair	Fair	Fair	Very Good	Very Good	Good	Fair	Very Good	Good	Good	Good	Good
Water Resistance	Good	Very Good	Very Good	Fair	Very Good	Fair	Fair	Fair	Excellent	Fair	Fair	Very Good
Alkali Resistance:												
Dilute (under 10%) . . .	Good	Good	Good	Good	Very Good	Poor	Fair	Good	Excellent	Fair	Poor	Excellent
Concentrated	Fair	Fair	Fair	Good	Very Good	Poor	Poor	Good	Very Good	Poor	Poor	Good
Resilience	Very Good	Fair	Fair	Very Good	Very Good	Poor	Good	Good	Good	Fair	Very Poor	Very Good
Elongation (Max.).	700%	500%	500%	500%	700%	400%	300%	300%	425%	625%	200%	500%

▲ Except Nitric and Sulfuric.

NOTES:

1. Do not use with steam.
2. Do not use with ammonia.
3. Do not use with petroleum base fluids. Use with ester Base (non-flammable) hydraulic oils and low pressure steam applications to 300°F.

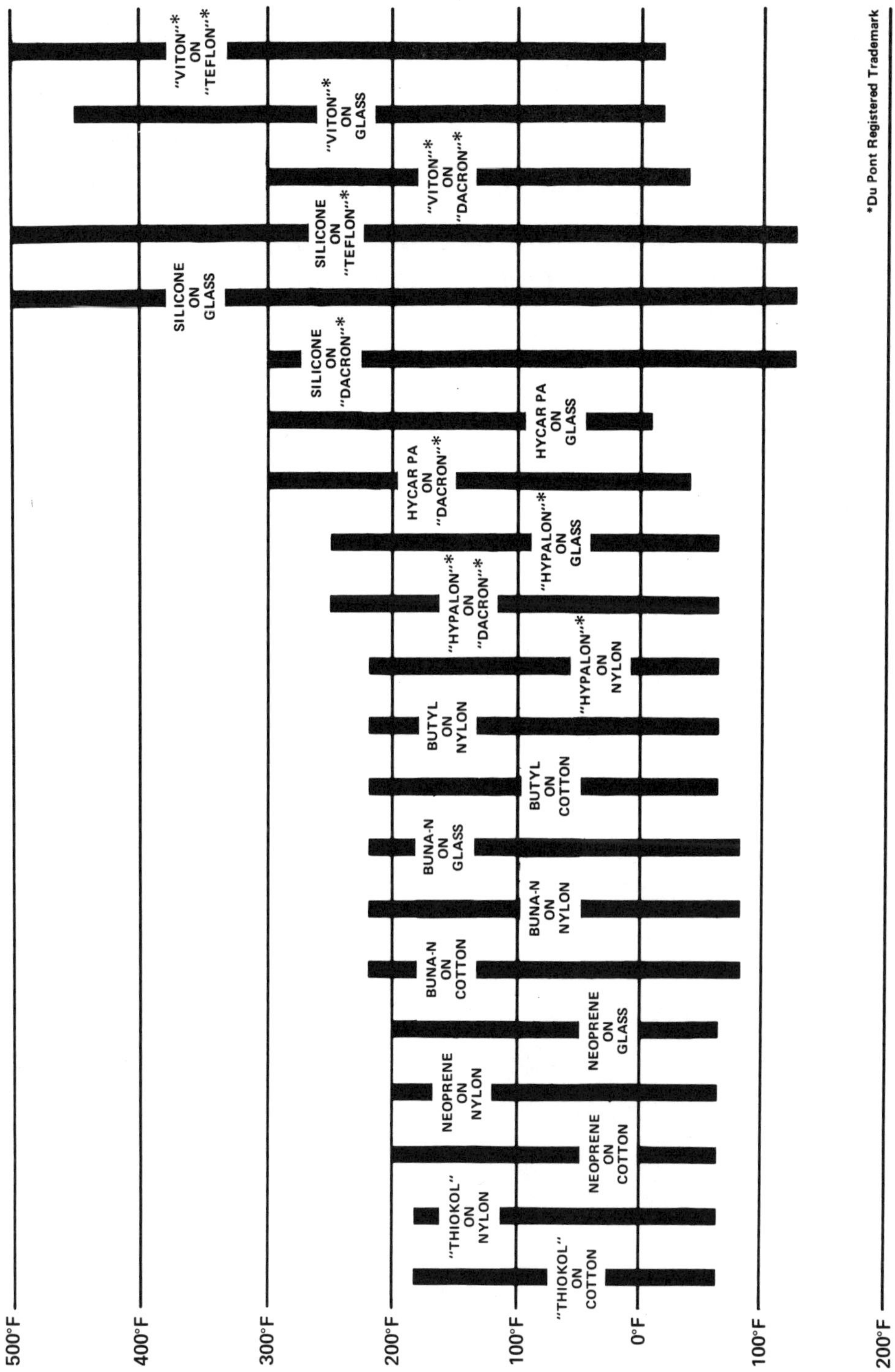

Chart courtesy of E. I. du Pont de Nemours & Co., Fabrics & Finishes Department

Upper serviceable limit is the highest temperature at which the material can be folded 180° back upon itself without cracking after 72 hours' exposure. Low limit is the lowest temperature at which the material can be bent around a 1" mandrel after 1 hour's exposure.

Table XII. SERVICE TEMPERATURE LIMITATIONS FOR GENERAL DIAPHRAGM MATERIALS.

CHAPTER 10

MAINTENANCE OF CONTROL VALVES

J. A. Parker and C. R. Thornton

E. I. Du Pont de Nemours & Co.
Wilmington, Delaware

CONTENTS

INTRODUCTION

The diaphragm control valve is very often the most costly element in a control loop. As a part of the process stream, it is also the element most likely to cause process down-time in the event of malfunction. These reasons alone provide the justification for a conscientious control valve maintenance program.

Control valve maintenance is generally divided into two categories. The first is preventive maintenance, which involves the installation precautions taken and the inline follow-up of the valve's performance and environmental deterioration while in service. The second is the overhaul of a control valve when it is no longer able to perform satisfactorily. The latter phase is generally a shop repair, but in some industries, it is performed in the pipeline, under controlled shut-down conditions.

Control valve manufacturers provide detailed installation and overhaul instructions for every type of control valve they produce. This chapter is not intended to replace these manufacturers' instructions. Instead, it is intended to impart a general knowledge of control valve maintenance requirements.

PREVENTIVE MAINTENANCE

Preventive maintenance is a term used to identify a program of scheduled overhaul, prior to failure. The schedule is usually based on previous knowledge of performance in a given service and is such, that a maximum safe in-time is established to minimize the chance of unexpected breakdown. Repair effected under such a program for control valves is a shop overhaul, and will be described later in this chapter. Because of this, the term *preventive maintenance* is being borrowed here to describe the work, which should be done during installation and operation to prevent the untimely breakdown of a control valve, ahead of schedule.

Prior to Start-Up

The preventive maintenance of a properly specified control valve begins at the time of installation. While the body and connections of a control valve are indeed rugged structures, they are not intended to be the means of aligning improperly fitted pipe. Care must be taken to insure that any stresses caused by improper pipe alignment are relieved elsewhere in the pipe system. Pipeline stresses on a control valve may result in misalignment of the stem, guide, and seat system of the valve and, in the case of a split body valve, may cause separation of body flanges. This may cause hysteresis, leak-through in the seat or flanges and possibly packing leakage.

Every effort should be made to install control valves so that their stems travel in a vertical plane with the diaphragm housing above the body. If a valve must be installed with the stem traveling in a horizontal plane, support braces should be provided to support the diaphragm housing. An unsupported valve may have stem misalignment, resulting in unacceptable hysteresis and packing leakage. Control valves should be located so they are accessible for periodic inspection and adjustment. Access ladders and platforms may be used for this purpose.

Prior to start-up, it should be determined if the control valve is to be the collection point for line trash such as rust, weld slag, and other foreign objects. If so, temporary screens or strainers should be considered for installation immediately upstream. A well-lapped seat may be ruined by one small piece of slag in the valve, when the valve is trying to close. If the process stream may normally contain scale, dirt or other hard foreign matter, the installation of permanent-type strainers or filters should be considered.

Control valve diaphragm housings are not air consumers. They are, however, collectors of liquids and other foreign matter in the air system. A clean, dry, oil-free source of air should be selected to operate the control valve. Valves with positioners or booster relays attached are particularly susceptible to failure when exposed to unclean air systems.

After Start-Up

After start-up, periodic field inspection of the control valve installation should be made. The packing will usually require adjustment after a short period in service and the lubricators will require readjusting. Dusty or dirty conditions may require a plastic or rubber boot to be placed around the stem to protect the packing gland and/or the guides in the bottom of the diaphragm housing. Corrosive fume conditions or process drippings may require additional special protection for the control valve. Plastic bags are often an easy solution to this problem since they may be used to completely enclose a valve without affecting its performance. Installation accessories such as lagging or tracing should be inspected and repaired, because these usually have a major effect on a control valve's performance.

Certainly, it is not possible to describe every installation or in-service potential failure situation in this writing. These more-common situations provide the basis for a good control valve preventive maintenance program and, together with the experience gained in any plant situation, will aid in extending the useful life of a control valve between overhauls.

SHOP OVERHAUL

Decontamination

The control valve that has been removed from the line for overhaul poses a special problem to the maintenance organization. The instrument has been a part of the process piping and may be contaminated with the process fluid. Certainly, a valve in water, steam, air, inert gas, etc., service is no problem and may be overhauled immediately; however, many fluids used in modern processing are corrosive, explosive, radioactive, or, in other ways, hazardous to personnel or equipment. Such contamination must be removed from all process-wetted parts of the valve before it can enter a modern repair shop and before it can be handled by personnel unfamiliar with such process hazards.

The actual decontamination treatment of the wetted parts will vary, depending on the nature of the process fluid. Many times, only a thorough water-wash or steaming of the parts will suffice, while other contaminants may require acid washing or even a special heat treatment of several hundred degree-hours, as is the case with some heavy metal compounds. The required treatment must be determined by someone familiar with the nature of the contaminant. Facilities for control valve decontamination will vary, depending on the type of contaminants involved. Location of the facilities should be chosen so that cross-contamination of other locations or facilities is avoided. This usually means a site remote from the operation and the repair facilities.

Since decontamination applies to the wetted parts of the control valve, disassembly of the body and bonnet components is required. Facilities for this work must be provided. These always include lifting and transporting equipment for handling valves during disassembly. An air supply is essential to stroke the valves to prevent further seat damage during stem and plug removal. Air is also useful for pneumatic tools and, in some cases, is needed for health reasons such as breathing air or chemical suit purging. The need for weather protection will depend upon the need for uninterrupted use of the decontamination facilities. If weather protection is provided, then good lighting and ventilation become essential. The need for tanks or vats, heating facilities, steam hoses and other washing or treating equipment can be determined only after the nature of the contaminant is known. Consideration should be given to maintaining exclusive tool kits and wearing apparel for use at the decontamination site.

Special attention must be given to the selection of the personnel who will do the decontamination service. Because of the generally dirty nature of the work, decontamination is an undesirable job. It is extremely essential that the employee who is disassembling a control valve for decontamination, thoroughly understands the control valve, as well as the nature of the chemical hazards involved, in order to avoid personal injury. It is also important that the details of construction be understood, to avoid further deterioration or damage to the control valve during disassembly. Such things as whether or not bellows seals are involved, whether the plug and stem are removed from the top or bottom of the body, if the seats are threaded into the body or otherwise retained, will affect the actual disassembly procedure. Many expensive bellows seals, for instance, have been destroyed by unsuspecting mechanics trying to improperly remove them. It is further essential that the location of all blind passages in the body or bonnet be known so that they may be thoroughly cleaned. The precision-finished surfaces, such as the stem, plug facing, seat surfaces, and guide surfaces, must be protected from further damage to minimize the overhaul costs. Skill in the use of special tools, such as packing ring pullers, seat ring extractors or flange spreaders, is essential to prevent misuse and resulting damage to the valve.

When all of the process wetted parts of the control valve are free of contamination, the valve should be reassembled, finger-tight, to assure that no components are lost during transit to the repair shop. Prior to shipment to the shop, the valve should be marked in some obvious manner to indicate that it has been decontaminated.

This procedure aids in preventing a contaminated valve from being delivered to the repair shop and worked on by unsuspecting personnel.

Disassembly of the Valve

The decontaminated control valve should arrive at the repair shop accompanied by

No. 17082 **C**

Deliver to Spot #__________

INSTRUMENT SHOP RECORD - CONTROL VALVES

Bldg. Name & No.__________	**Inst. Item No.**__________
Equip No.__________	**Serial No.**__________
Vendor__________	**DV No.**__________
Body Size__________	**Type No.**__________

What is Ailment of Valve__________

Repair Requirements:__________

Special Items Required

Flow Check ( )

Press. Test________**Lbs. Water** **Action Required** **ATO ( )** **ATC ( )** **Change Trim ( )**

Deadend Press. Test at________**Lbs. With**________

Type Packing Required________**Dry Valve Completely**

Stroke Required________**Inches**

Operating Conditions:

Material__________	**Normal Flow**__________
Press.__________	**Sp. Gr.**__________
Press. Drop__________	**Max. Flow**__________
Issued By__________	**Mechanic**__________
Date Issued__________	**Start: Date**_____**Time**_____
Required Date_____**Time**_____	**Finish: Date**_____**Time**_____
Code_____**W. O.**_____	

MU-1932-BR1

Figure 1. A TYPICAL JOB SHOP RECORD CARD FOR CONTROL VALVES.
Courtesy E. I. DuPont DeNemours & Co.

a job card which gives as much information as possible about the valve and its service in the process. Figure 1 is an example of a typical shop record card.

The first step in the overhaul of a control valve is to mark the actuator connection orientation, in relation to the body flanges. A complete disassembly is next. This means, that after separating the actuator from the bonnet, and the bonnet from the body, the actuator is to be disassembled completely. The bonnet and packing gland components are separated; the plug and stem, as well as the blind head flange, are removed from the body. An inspection of all components should be made at this time to determine the extent of repair, reconditioning and replacement that is required.

In many cases, all of the exposed surfaces of the valve components are rusty and unfit for repainting. These parts should be descaled before proceeding. Sand blasting equipment is especially good for this. Care must be taken, before descaling, to protect all of the machined surfaces such as flange faces. This protection can be heavy masking tape for sand-blasting, or it may require a special grease coating if acid washes are to be used.

After the descaling has been completed, the need for any machine work should be determined. The body and trim will most likely require the services of a machinist. All of the flange surfaces should be inspected for damage that would prevent proper gasket seating. Generally, such damage will be in the form of scratches or grooves across the flange face, from the process side to the outside.

Seat rings that are screwed into the body should be given special attention to determine if replacement is necessary. Wire drawing, corrosion and erosion are usually the reason for replacing seat rings, other than to change the capacity of the valve. Small pitting and worn surfaces are usually repaired by machining.

Many manufacturers machine a 60° bevel on the seating edge of the seat ring and a 65° bevel on the seating edge of the plug, as shown in Figure 2. This causes the actual seat contact to be at the top of the bore of the seat ring. This arrangement is satisfactory for a new control valve; however, a control valve that is in need of repair will often have the bore of the seat ring, worn non-uniformly. Restoring concentricity would require enlarging the bore of the seat ring, which will further increase the capacity of the valve, as well as alter its flow characteristic.

One solution to this problem, other than discarding an otherwise reusable seat ring, is to machine a new surface on the face of the seat ring at a 60° bevel, and to change the bevel on the plug from 65° to 59°, as shown in Figure 3. This causes the seating contact to change to the top of the seat ring bevel, which can be re-

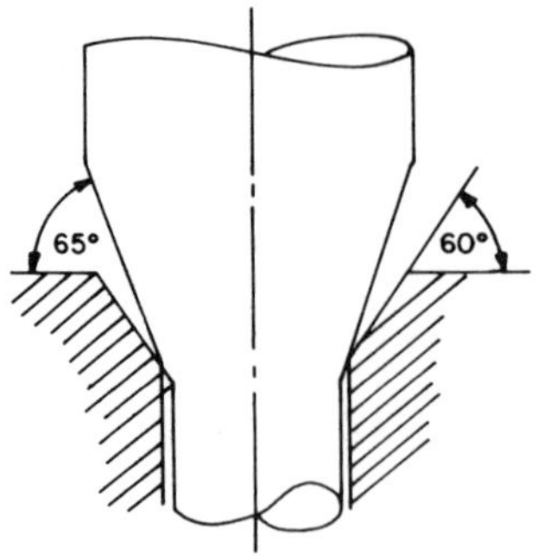

Figure 2. MANUFACTURER'S BEVEL OF 60° ON THE SEAT RING AND 65° ON THE PLUG.

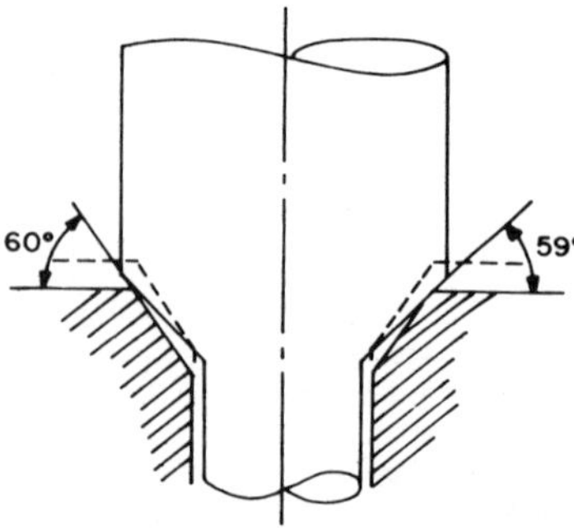

Figure 3. RE-MACHINED SURFACES OF A REPAIRED VALVE, WITH BEVEL OF 60° ON THE SEAT RING AND 59° ON THE PLUG.

machined uniformly without further affecting the valve's capacity or the characteristic of the plug-seat ring contact.

If threaded seat rings are to be replaced, this should be done before any machine work is done on the body. All too often, forces required to tighten a new seat ring will cause distortion of the body. Concentricity must be restored or the valve will never close tightly. Restoration of seat concentricity is best accomplished by using a lathe.

The removal and replacement of threaded seat rings require special tool sets. Most control valve manufacturers produce sets to fit their valves.

Threaded seat rings very often require some type of thread sealing compound to prevent leakage through the threaded area. Not only will such leakage prevent tight shut-off of the flow, but the resulting erosion and wear will also cause irreparable damage to the valve body.

Whenever possible, the valve manufacturer should be consulted for recommendations of sealing compounds for a particular service. If the manufacturer's advice is not available, someone familiar with the process chemistry should be consulted before applying any of the commercially available sealing compounds.

When the necessary machining has been completed, the components are ready for painting. Paint specifications for a control valve should be the same as that for other machinery in the process line. Proper masking will prevent the paint from being applied to gasket surfaces.

Valve plugs, stems, and seat rings represent some of the most precisely machined parts in the field of instrumentation. The finish on control valve stems is so fine that replacement with a factory-supplied part is the only practical solution to a marred stem surface. This high polish is required to give the best possible packing seal with a minimum amount of hysteresis. The plug and seat ring, of course, have to be perfectly matched, first through precise machining and finally through hand grinding.

Grinding or "lapping" the plug and seat of a control valve requires more skill and knowledge of the craftsman than any other phase of the overhaul, except perhaps, any machining that may be required. It must first be determined that the faults in the seating surfaces can be economically corrected by lapping rather than requiring replacement or remachining, if neither has been done. Alignment of the plug to the seat is absolutely essential for uniform seating. This generally requires that all guides be in place before the lapping operation begins.

Most of the commercially available valve grinding compound sets, which contain a coarse and a fine compound, are satisfactory for control valve lapping.

The coarse compound should be used until a fine, continuous line is accomplished around the seating edges of the plug and seat ring. After thoroughly cleaning all of the coarse compound from the valve, the fine compound should be used to polish this seating seal line.

When the lapping operation has been completed, it is absolutely essential that all of the grinding compound be cleaned from the valve.

Satisfactory flow shut-off and smooth operation of the plug and stem are dependent on the alignment of the plug in the seat ring. Guides are provided to assure excellent alignment. Some valves have guides both below and above the seat ring while others are simply top guided. These guides, in order to accomplish their mission, must be machined to a tolerance close to that of the guide shoulder on the plug-stem assembly. Any corrosion or wear that destroys the close fit renders the guide useless.

Some guides are threaded into place while others are pressed into place. Replacement of threaded guides is straightforward, but pressed guides must be forcibly removed. Various methods work well for this operation. Press-type removal equipment is preferable, if available. Where press equipment is not available, differential heat methods, for expansion of the guide holder and contraction of the guide, will work. This latter method must be used with care in order that permanent damage not be done to the guide holder. New guides can usually be pressed in place with a bench vise.

Disassembly of the Actuator

The control valve actuator, after having been completely disassembled, should be carefully inspected, component by component. Cast steel yokes should be examined for any signs of cracking since all of the valve stroking force is transmitted through this component. Corrosion that either weakens or causes misfitting of components is cause for parts replacement. The diaphragm should be examined for any signs of cracking, hardening or abrasion to indicate possible future failure.

The same surface preparation and painting procedure used on the body components should be applied to the actuator components.

All threaded parts should have the threads dressed for smooth assembly and adjustment. It is generally more profitable to discard all diaphragm case bolts and replace them with new ones.

Reverse-acting actuators should have any stem seals replaced with new ones.

Reassembly of the Valve and Actuator

Reassembly of the control valve body is a straightforward operation. The chief points of concern are that gaskets are used of proper size and material and properly centered; and, that bolts or studs are used of proper size, length and material. When tightening the bonnet and the blind-head flanges to the body, it is important to tighten diagonally opposite bolts evenly, to assure an even loading on the gasket, thus preventing flange leakage. The valve manufacturer's recommendations for proper torque loading of the particular bolts or studs should be adhered to when tightening the body bolts or studs. Over-tightening will cause excessive elongation, resulting in a weakened flange joint.

The depth of the packing gland should be measured to determine the proper number of packing rings. Preformed packing rings should be used whenever possible. These are commercially available in most materials and sizes. Their uniformity assures the best possible seal in the gland. If a lantern gland is used, enough rings of packing should be placed beneath it, to assure that it will mate with the lubrication port in the gland housing. Packing rings should be inserted one at a time and each should be lightly tapped into place. *Do not pound packing rings.* Pounding distorts the rings and causes them to lose their resilience. When the gland has been filled with the predetermined number of packing rings, the follower and yoke should be installed, finger tight, for hysteresis testing of the assembled valve. It should be noted that special packings, such as Chevron sets, and special spring-loaded glands would be assembled differently. The manufacturer's instructions for these special packings must be followed if a good gland seal is to be expected.

Reassembly of the actuator should include the application of a suitable lubricant to all friction points. These include the stem where it passes through guides and seals, the spring thrust bearings and retainers, and the spring-adjusting nuts.

Test and Adjustment of Reassembled Valve and Actuator

When the reassembly has been completed, the diaphragm case should be leak-tested. This is most easily accomplished by loading the diaphragm chamber with 45 psig of air *(do not exceed the manufacturer's rating for the case)* and leak-testing around the flange edge with a soap solution. A hysteresis test of the actuator should be made before it is mated with the body assembly, to determine that there is no binding of components within the actuator.

The hysteresis test may be made by attaching a clamp-on arm to the exposed, movable portion of the stem and by attaching a micrometer gauge to the unmovable yoke or spring barrel. The micrometer gauge must, of course, have as much travel capacity as the stem. After zeroing the micrometer gauge to the at-rest position of the stem, record the air-loading pressure required to stroke the actuator at every 10% increment of its travel. Then, while returning the stem to the at-rest position, record the air-loading at every 10% increment of travel. Compare the recorded air-loading pressures in the two directions, for each increment of travel. The differential pressure between the two readings at each increment, should not exceed 0.1 psi. Differential pressures greater than 0.1 psi, indicate an intolerable friction condition, within the actuator assembly, which must be corrected, even if complete disassembly and reassembly of the actuator is required.

The assembled and tested actuator may be mated to the valve body with the hysteresis test equipment attached, because a final-assembly hysteresis test will be made later.

When mating the actuator to the body, it is important that the air connections on the actuator be oriented to the body flanges, exactly as they were before disassembly. This will allow the valve to be reinstalled into the pipeline, without requiring any alterations to the field tubing or piping assembly.

The full-travel loading pressure, required for any control valve, is a function of the diaphragm effective area versus the stem loading (the accumulation of the spring loading and the plug area times the shut-off differential pressure across the plug). Rarely will these factors result in an ideal 15 psi (30 psi for a 6 to 30 psi control system) full-travel, diaphragm pressure.

If the full travel diaphragm pressure requirement is less than 12 psi or more than 18 psi for a 3 to 15 psi control system (24 psi to 36 psi for a 6 to 30 psi control system), and if the valve is not normally equipped with a valve positioner, then consideration should be given to changing either the size of the actuator or the size of the actuator spring.

When the stroke and spring compression have been properly adjusted, a final hysteresis check should be made on the actuator. This is done with the packing loose and with no pressure differential across the valve. This test is to determine that proper alignment of all the components has been obtained.

The test should be run on the complete assembly, just as it was run on the actuator. The hysteresis differential pressure of a properly assembled control valve should not exceed 0.25 psi under these testing conditions. Higher differential pressures indicate that binding is occurring somewhere within the inner valve mechanism. The degree of binding is indicated by the magnitude of the loading-pressure differentials obtained during the test.

Upon completion of a satisfactory hysteresis test, the valve body should be filled with a test fluid with the stem at mid-travel. The test fluid pressure within the body should be raised to the maximum process operating pressure and the packing should be tightened only enough to stop leakage. The stem should be stroked continuously, while the packing is being tightened, to properly seat the packing and to prevent stem binding.

Note: Special packings previously mentioned should be adjusted according to the manufacturer's recommendations. While the packing is being adjusted, the body should be examined for external leaks.

Metal-to-metal seated diaphragm control valves are *not* tight shut-off devices. They will all leak through the seat to some degree. Most manufacturers design and build to a shut-off condition that will allow no more than 0.1% of maximum flow capacity leakage at the shut-off position for a double-seated valve, and will allow no more than 0.01% of maximum flow capacity leakage for a single-seated valve. These leak rates can be duplicated if reasonable care is exercised during the overhaul of a control valve.

Upon the completion of leak testing, any accessories that were a part of the valve assembly should be reconditioned and installed on the valve.

Preparation for Shipment

Preparation of the reconditioned control valve for shipment will vary, depending upon whether the valve is to be reinstalled in the process line from which it was removed or whether it is to be put into storage as a standby spare. To prevent foreign matter from entering, all air connections to the valve should be plugged and the flange openings should be covered. If the valve is to be reinstalled immediately, heavy masking tape will suffice. If the valve is to go to storage, flange guards of wood or heavy cardboard should be fastened to the flange faces. Any protruding accessories such as lubricators or positioner gauges, that can be broken off easily, should be removed from their receptacles and packaged separately to accompany the valve for installation at the field location.

Control valves that are to be stored as standby spares require a clean, dry location for storage. If such facilities are not available, the valve should be completely enveloped in a moisture-proof bag or similar container before storage. The storage area should be such that the valve will not be damaged by other work being performed in the area.

OVERHAUL CHECK-LIST

The step-by-step procedure for control valve overhaul is lengthy. In order to assure that no phase of the overhaul has been overlooked, a careful check-out procedure should be established. The following is a step-by-step check list that one large repair shop foreman has prepared to aid his personnel.

SHOP OVERHAUL DIAPHRAGM VALVES, 1/2 INCH TO 3 INCHES.

Step-by-step procedure and check-list.

1. Obtain job ticket from foreman.
2. Transport valve to work bench or to wash stand if necessary to clean.
3. Install on either control valve vise of repair bench or in standard vise to facilitate disassembly.
4. Mark actuator and body orientation and remove actuator from body. Remove positioner if necessary.
5. Disassemble body by removing bonnet flange, packing, stem and blind flange.
6. Secure actuator, mark for orientation of air connection, back-off the spring adjuster, remove the cap screws from the case assembly with an impact wrench, discard cap screws and nuts.
7. Remove diaphragm, diaphragm plate and actuator stem as a unit.
8. Remove spring adjuster screw nut, spring adaptor screw assembly, spring adjuster, spring seat and spring.
9. Examine diaphragm case assembly and spring barrel case assembly interior; steam-clean if necessary.
10. Remove body from vise; transport body, body bonnets, diaphragm case assembly and spring barrel to sand-blast facility; do general clean-up.
11. Spray-paint body and body components with one coat of silver paint.
12. Spray-paint the diaphragm case assembly and spring barrel exterior with one coat of gray primer. Paint interior of case assembly and spring barrel with one coat of gray primer, if needed. Spray-paint the case and barrel exterior with one coat of orange.
13. Inspect the diaphragm for holes, cracks or excessive wear; replace if needed. Replace all stem seals in reverse actuators; lightly grease the "O" ring before installation.
14. Dress threads on all actuator components; lightly grease.
15. Move parts from the spray booth to the bench. Install the body in a vise.
16. Inspect valve body, flange, ring and seat ring; replace on machine as necessary. Lightly grind with compound to lap-in the seat and ring.
17. Chase the body studs with a die; replace studs as needed; tap-out the stud holes at this time; assemble the flanges to the body; line-up the reference marks for proper orientation.
18. Repack with approved packing; make-up the gland finger tight.
19. Assemble the parts of the actuator barrel; install the diaphragm case, using new cap screws and nuts, and tighten with an impact wrench. Apply air pressure and check for air leaks.
20. Adjust the spring range and stroke of the actuator by applying proper air pressure values.
21. With sufficient air on the actuator to put it at approximately mid-stroke, attach the actuator to the valve body and either bolt or clamp it with a yoke-holding device.
22. Install the actuator and the plug stem connector finger tight; either apply or remove air to adjust the connector for proper seating of the plug. Tighten the jam nuts or the connector. Adjust the travel indicator.
23. Body-test and dead-end test the valve in accordance with information contained on the work card.
24. If the positioner has been removed, reinstall it by first chasing the threads of the positioner bracket or the boss that are on the actuator.
25. Mount the positioner and the positioner arm, and fasten the turnbuckle assembly to the positioner arm.
26. Switch the bypass valve to the instrument and apply sufficient air pressure to the actuator to obtain a mid-stroke position.

27. Adjust the positioner arm turnbuckle to make the front lever parallel to the rear lever of the positioner.
28. Loosen the slider clamp screw and adjust the slider to the approximate actual (in-service) valve travel.
29. Turn the bypass valve back to the positioner and alternately adjust the turnbuckle length and operating spring seat until the valve is at mid-stroke when the applied pressure is at mid-range of the control instrument pressure, in accordance with the information contained on the work card.
30. Check the positioner and the valve for full stroke performance.
31. Complete the necessary paper work in accordance with shop practice; stencil as required.
32. Remove the valve from the vise; mask the valve openings; install proper tag; and prepare it for transport.

Figure 4. A SMALL SAND-BLAST BOOTH.

OVERHAUL FACILITIES

Continuous control valve overhaul work requires a work area that is specifically adapted to the tasks. Real efficiency and cost-savings can be realized from the availability of such facilities as a small sand-blast booth, as shown in Figure 4, for surface preparation work. A steam-solvent cleaning booth, similar to that shown in Figure 5, is particularly useful for degreasing and grit removal after sand-blasting. Suitable spray-paint facilities, as shown in Figure 6, provide for fast repainting, with a minimum of fume and dirt contamination to the adjacent work areas.

Proper lifting equipment is particularly useful in handling larger control valves with a minimum number of personnel, as shown in Figure 7.

A holding device, such as that shown in Figure 8, is particularly useful in reassembly work. The valve body can be mounted on the large universal flange, which in turn, can be rotated 360° by turning the crank handle. The universal flange can be piped to a pneumatically-driven pump, which will pump test fluid to the valve body for seat leakage and hydrostatic testing. The valve need not be dismantled from the holding device until the final assembly and test has been completed. A set of blind and screw flanges is a useful accessory to the holding device. The flanges illustrated in Figure 9 are in sizes of one-half inch through six inches, all of 600-pound rating. The bolt holes have

Figure 5. A STEAM-SOLVENT CLEANING BOOTH.

Figure 6. A SPRAY-PAINTING FACILITY.

been elongated to the respective 150-pound bolt circle. This allows one set of flanges to be used for flange ratings of from 150 pounds through 600 pounds while being nonstandard enough to discourage their removal for other uses.

Proper seat ring removal tools, as shown in Figure 10, are more than justified by the amount of time that can be saved in seat-ring removal. An additional bonus can be realized by the number of seat rings that are removed undamaged and by avoiding injury to personnel, that could result from using improper rigs.

The importance of a machine tool facility, similar to that shown in Figure 11, cannot be overemphasized for repairing otherwise unusable valve trim.

A test panel, adjacent to the valve-holding device, can save a large amount of time during the final checkout of the control valve. The panel shown in Figure 12 provides gages and regulators for the valve positioner air supply and the pneumatic pump-loading air supply. A gage for indicating hydrotest pressures and a manometer and regulator for the instrument signal pressures are also included.

SUMMARY

A well-planned maintenance program, with trained personnel who have proper job aids, will result in overhauled, diaphragm control valves that will have an inline use-period before breakdown that can be equal to that of new valves, but at a fraction of their cost.

Figure 7. LIFTING EQUIPMENT FOR LARGE CONTROL VALVES.

Figure 8. A CONTROL VALVE HOLDING DEVICE.

Figure 9. AN ACCESSORY SET OF UNIVERSAL BLIND AND SCREW FLANGES.

Figure 10. SEAT-RING REMOVAL TOOLS.

Figure 11. MACHINE TOOL FACILITY FOR REPAIRING VALVE TRIM.

Figure 12. AN INSTRUMENT PANEL FOR CHECK-OUT DURING OVERHAUL.

CHAPTER 11

INSTALLATION PRACTICES

S. F. Luna

Westinghouse Nuclear Energy Systems
Pittsburgh, Pennsylvania

CONTENTS

INTRODUCTION

The engineering technique for selecting the best valve for the job has been treated in depth in the preceding chapters. The fruits of this labor may not be realized unless proper installation practices are followed.

There is but one major factor that can motivate the not so glamorous job of engineering the installation and that factor, of course, is *money.*

1. If the valve to be installed should fail, resulting in a system shutdown for say 10 minutes, or one hour, or four hours, or more, how much money in production and maintenance will this cost; $100, $5,000, or $________?
2. If a new plant is being built and the piping installation drawings are incorrect, resulting in an incorrectly piped valve installation, what would it cost to correct the piping? Consider the material cost, cost of labor, drawing changes, overhead, profit, delay charges, etc.
3. Consider the plant start-up situation with the control room full of people; customers, test engineers, standby skilled help, vendor's field personnel, etc., all ready to start-up. If after two to three days of sequentially bringing the plant to a major progress point, the control valve in the next sequence would fail to pass fluid, because it is plugged up with rags, welding rods, left-overs from lunch, etc.; what would this cost? It has been reported that on certain installations the cost during this stage, involving many people, is $2,000 per hour.

The above emphasizes the likelihood of wasting large sums of money, because of the failure to follow good installation practices. There are many other reasons for giving this phase of control valve engineering proper attention—longer valve life, less maintenance, better system performance, etc., but they all boil down to *"money in the hopper, if the installation is proper."*

Good installation practices start with the initial engineering practices, particularly in the preparation of the piping drawings, piping standards and procedures.

Most firms responsible for the installation of control systems have developed control valve installation piping standards and procedures which are either used with specific piping drawings or are used to produce specific piping drawings.

This subject has been covered extensively in the literature. This chapter contains material, suitable for this handbook, which has, for the most part, appeared in the literature.

The considerations for valve installations may be classified into the following six areas, each of which is discussed in the following sections.

1. The safety of people and equipment in all stages of the valve's life; that is, installation, test, operation, and maintenance.
2. The performance of the control valve as a final control element in a control loop.
3. The requirements of the installation to permit intelligent manual operation of the system, if necessary.
4. The accessibility of the control valve for maintenance.
5. The cost of installation, operation and maintenance.
6. The protection of the valve from either mechanical or environmental forces.

SAFETY CONSIDERATIONS

The following problem areas and solutions must be considered in the valve piping design and installation:

Valves may leak, either initially or through use, at the packing glands, flange gaskets, or at pin holes formed at valve faults. It should be assumed that a leak will develop. What damage can conceivably occur, considering the type of fluid passing through the valve, at the most severe operating conditions (temperature and pressure)? Will a corrosive fluid collect in an electrical wire-way? Will a combustible fluid drop onto a hot tank? Will a hot stream of caustic solution injure an operator?

After a shut-down, valves may hold system pressure for an extended period of time. Means must be provided to safely reduce this pressure, if the valve is to be either opened or removed for maintenance. Likewise, means must be provided for draining trapped damaging fluids.

In the design of the valving system, the following should be considered, to reduce control valve pressure:

1. The design tightness of the control valve. Single seated valves can be furnished to provide a leak rate of less than one drop per hour. Double-seated valves rarely close tightly on both seats; therefore, consideration must be given to possible replacement with a single-seated valve.
2. The volume of trapped fluid between the stop valves. Obviously a large volume of trapped fluid at a high pressure can represent a sizable force, capable of seriously injuring a man

who is careless in the manner in which he relieves control valve pressure.

3. The nature of the fluid. Is it a gas, clean fluid, or fluid with suspended solids? What is the pH of the fluid?

For good design one should assume that it won't be possible to avoid shut-down trapping high pressure fluids; therefore, it is recommended that bleed and/or drain valves be installed on either side of the control valve, if the potential energy of the trapped fluid is dangerously high.

For valves with large trapped volumes at elevated pressures, it may be wise to consider double vent and bleed valves. Figure 1 shows such an arrangement. This arrangement traps small volumes between the two valves and enables pressure relief of the system in steps, by repeated pressure build-up and let-down of the volume between the two drain and/or bleed valves. Note that the control valve must be in open position during this procedure.

If the fluid to be relieved is a hazardous gas, the vent lines must be piped to a safe location. If the gas is not combustible, it still must be piped or directed to avoid a blast of gas carrying rust or other projectiles that can injure the person handling the vent valve.

A let-down valve should be selected that will restrict the flow, thus protecting the careless operator who fails to follow the safe practice of slowly opening the let-down valves. Such restricted let-down valves can plug up easily if the fluid contains suspended solids. It may, therefore, be necessary to let-down pressure from valves located on top of the pipe and then complete the draining operation with larger-sized drain valves. In this case it is very important to train personnel on the safe manipulation of the let-down valves. If the fluid to be drained is a dangerous liquid, the piping system designer should consider piping it to a safe location. During plant operation, valves may be very hot; therefore, provisions must be made to prevent heat damage to the personnel.

Valves handling radioactive fluids require extreme methods of care. The rules and regulations covering this area are very extensive and not covered in this text.

PERFORMANCE

The prime function of the control valve must be considered in the design of the valve installation to avoid imposing dynamic conditions that may detract from the factors used in selecting the valve. The process control engineer is generally very careful in selecting a control valve that will have a specified capacity and a specified flow rate vs. valve lift characteristic. He will normally take into consideration the pressure drop in the piping system, when sizing the valve.

The designer must design the piping system to be consistent with the process con-

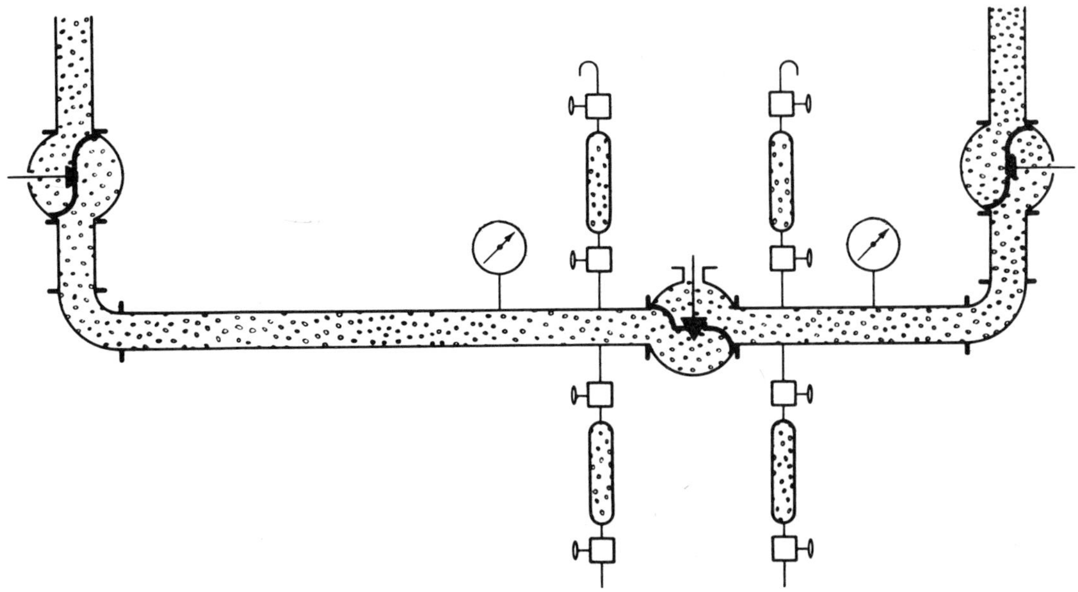

Figure 1. CONTROL VALVE INSTALLATION WITH VENT OR DRAIN CONNECTIONS.

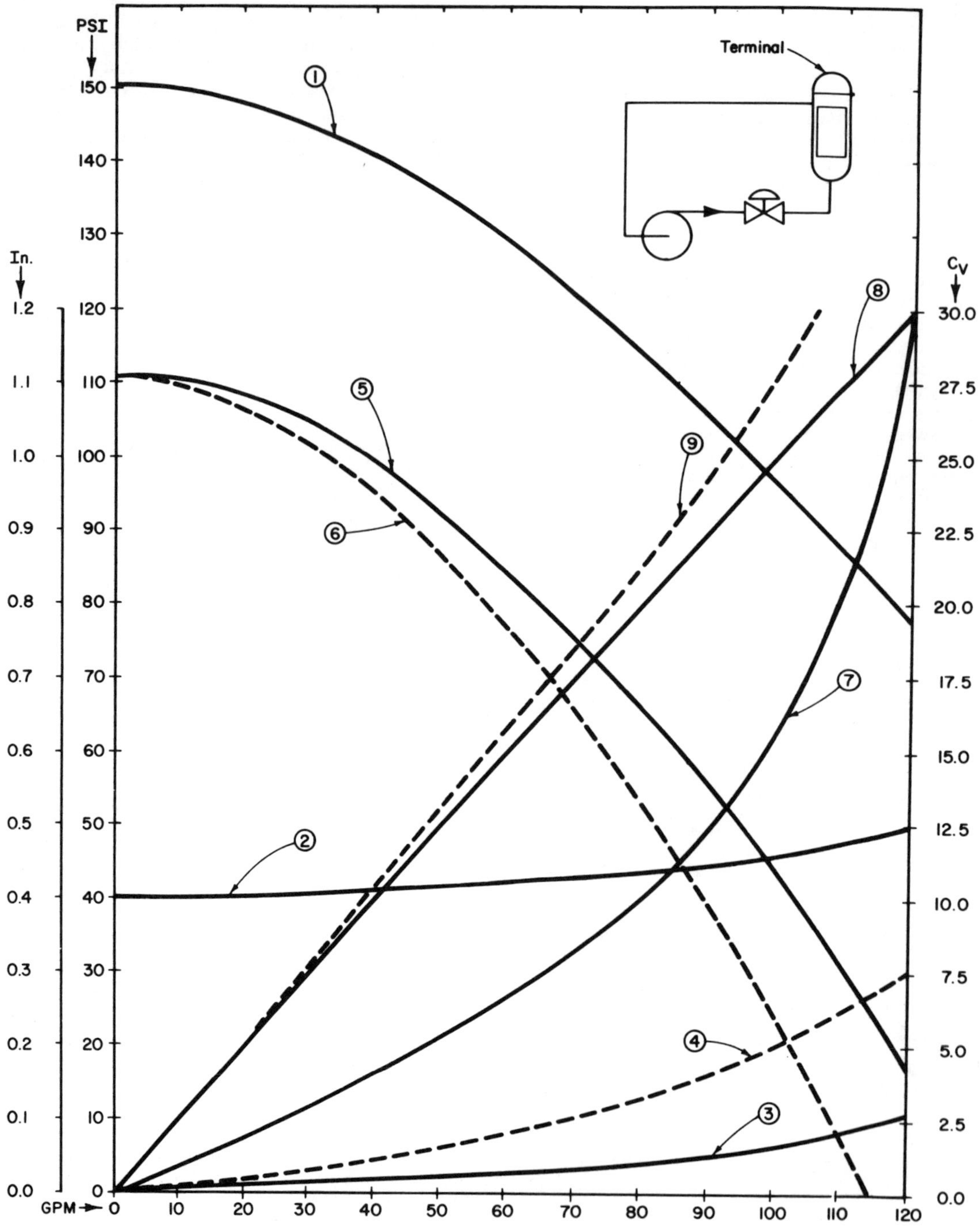

① Pump head vs. flow
② Terminal ΔP vs. flow
③ Design piping ΔP vs. flow
④ Actual piping ΔP vs. flow
⑤ Design available valve ΔP vs. flow
⑥ Actual valve ΔP vs. flow
⑦ Design valve C_V vs. lift
⑧ Design valve lift vs. flow
⑨ Actual valve lift vs. flow

Figure 2. SYSTEM PERFORMANCE CURVES.

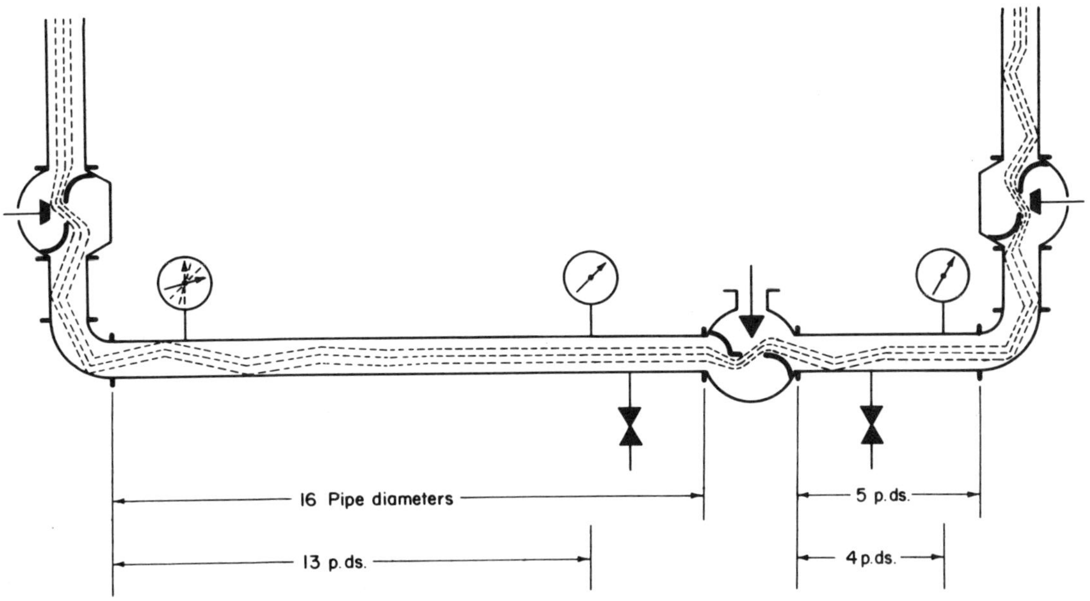

Figure 3. TYPICAL SATISFACTORY ARRANGEMENT OF UPSTREAM AND DOWNSTREAM PIPING.

trol engineer's calculations, or vice versa. Figure 2 shows a method of plotting system pressures to arrive at available pressure drop vs. valve lift. A similar plot will be used by the valve manufacturer to accurately design the valve plug. The piping design engineer should review such plots in designing the piping system. The design friction loss, through the lengths of pipes and fittings selected, can be made consistent with such plots. This calculation should include the friction drop through the fully opened control valve as part of the piping system. The characteristic pump curve is generally available from the pump manufacturer. The terminal pressure drop vs. flow through the process vessel may be more difficult to obtain.

It can be seen from Figure 2 that valve performance—capacity and flow vs. valve lift, will be altered if the actual piping system consumes more pressure drop (curve 4) than that considered in the valve sizing calculation.

In designing the inlet and exit piping at a control valve, a conservative approach is to consider the control valve as a variable orifice; thus, the piping recommendations used to pipe orifice plate assemblies can be applied. The following considerations are recommended:

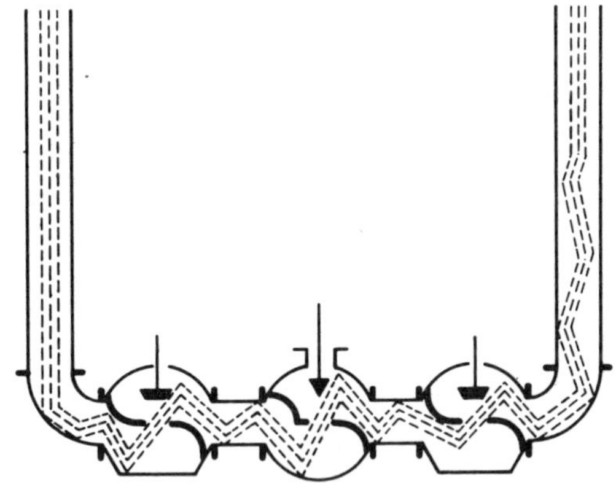

Figure 4. UNSATISFACTORY ARRANGEMENT OF UPSTREAM AND DOWNSTREAM PIPING.

1. Allow the maximum number of feet of straight run inlet piping that is consistent with other piping demands. A good rule of thumb is 10 to 20 pipe diameters. This is easy to obtain in small size lines, but may be impossible to obtain in line sizes larger than nominal 5 inches. Nonetheless the practice of allowing for a maximum straight inlet run will allow better valve performance.
2. Where possible, allow a straight run of 3 to 5 pipe diameters for the outlet piping.
3. The straight run on the inlet to the valve will permit the fluid stream to enter the valve at a steady inlet pressure. Thus, for every position taken by the valve, or in other words, for every new orifice opening, a steady inlet pressure condi-

tion will insure a steady and repeatable flow rate. Figure 3 represents a good flow approach and Figure 4 shows a poor flow approach.

4. It will be noted that the location of the pressure gages in Figure 3 allows a steady, true, pressure head and not pressure head plus and minus the velocity heads due to the non-uniform flow pattern. The accurate measurements of upstream and downstream pressure are particularly important in trouble-shooting fluid system problems. An accurate measurement of control valve pressure drop, plus valve position, can verify if the valve is performing to its design requirements, or if there is internal trouble requiring a valve shut-down and repair.
5. The location of an upstream or downstream tap for a pressure regulating valve should definitely follow the above recommendations to assure sensing accurate and steady pressures undisturbed by perturbations of pipe fittings.

MANUAL OPERATION CONSIDERATIONS

In locating the control valve, consideration should be given to the following factors that will facilitate good manual operation, if required:

1. Is the manually operated valve located where an operator can conveniently manipulate it and at the same time be able to read the indicators displaying the plant parameter he is trying to control?
2. Can the operator see the level gage glass on a tank, if he is manually controlling liquid level?
3. Can the operator determine how much of a change he has made on the valve stem by observing line pressure gages or the valve stem motion indicator?
4. Are indicators of other parameters available so that the operator can anticipate changes? For example, if he is controlling boiler feedwater flow, does he have visual access to a steam pressure gage so he can anticipate the need for more or less feed water if the steam pressure or flow gauge indicates a decrease or increase in steam flow?
5. If the manual operation is performed by the use of a by-pass valve, it is desirable that this valve be selected to have a valve lift vs. flow characteristic that is similar to that of the control valve.

ACCESSIBILITY

There are control valves that have been in operation for 10 to 20 or more years and have never been subject to dissasembly and repair. In fact, some of these valves have been in operation for such a long time that the type of inner valve is unknown. These valves are generally found in service on low pressure systems with non corrosive fluids.

In today's process plants, control valves are required to handle a wide variety of fluids at ever increasing pressures and temperatures; therefore, the piping designer can assume that the control valve will have to be either serviced in-place, or removed, on a routine schedule. The cost for this servicing will depend largely upon the attention given to valve accessibility.

The piping designer should follow a manufacturer's outline drawing that shows the clearances required in the following areas:

1. Above the valve for the removal of the top works with stem and plug as an assembly.
2. Below the valve for removal of the bottom flange and removal of the plug and stem assembly, if required.
3. On the side of the valve for removal of accessories such as manual operators, electric motor drives, solenoid valves, valve positioners, etc.
4. At the body flanges for removal of flange bolts. This dimension is seldom furnished, thus the piping designer must consider this in designing reducers to connect control valves to inlet and outlet piping. (See Figure 18).

Good manufacturer's drawings will permit the preparation of piping drawings that call out clearance requirements. (Figure 20 may be used as a guide in lieu of manufacturer's drawings.) This in itself is not enough since the heating and ventilating designer and the electrical wire routing designer may not avail themselves of the piping drawings. Costly rerouting of heating and ventilating ducts or wire-ways can be avoided if the piping designer advises other group designers of his valve clearance requirements.

CONTROL VALVE MANIFOLDS

The piping arrangement selected for the installation of a control valve involves the consideration of many factors which are tied to the fundamental objective, which is: *Excellent and dependable con-*

trol valve performance at the minimum overall cost over its design life.

The designer must, therefore, address himself to the many considerations given in following check list:

1. How should the inlet and outlet piping to the valve be designed.
2. How much straight run should be used?
3. What line size should be used?
4. Are reducers needed? Does the down stream piping need to be enlarged to allow for expansion of the fluid?
5. Are block valves needed? If so, what size and type?
6. Is a by-pass valve required? If so, what size and type?
7. Where should the valve be located in the line relative to sensing elements?
8. At what elevation shall the control valve be located to avoid obstruction and to provide access.
9. Will the valve be accessible for maintenance and for manual operation? Will the valve stem motion indicator be visible during normal operation?
10. Will the valve be near equipment or indicators that are required for manual operation?
11. Will the valve present a hazard to people or equipment in any phase of its operation?
12. Are vent or drain valves required? If so, how many, what size and type, and where shall they be located? Should these be piped to a safe location?
13. Are pressure gauges required? If so, how many, where should they be located and how should they be piped?
14. Are pressure relieving devices required? If so, what size and where should they be located?
15. Are line strainers needed? If so, for initial plant construction or for reinsertion after a plant shutdown and wet layup?
16. What type of fittings will be needed?
17. What materials of construction should be used?
18. What codes must the installation meet?
19. Will it be necessary to insulate the installation?
20. Will it be necessary to heat trace the installation?
21. If the valve cannot be installed in a vertical position, what orientation should be selected?
22. How should the control valve be supported?
23. What are the requirements for the pneumatic or hydraulic tubing and auxiliaries?
24. If the control valve were removed, are means available to isolate the pneumatic or hydraulic tubing and auxiliaries and are they independently supported? Electrical hardware associated with electrical driven control valves must be given the same consideration.
25. Are steam traps required? If so, where should they be located?
26. Will pressure taps be required— If so, where shall they locate them, and what size and type of tap valve should be used?
27. Will the valve be subject to damage by moving vehicles? If so, how should it be protected?

The design of a control valve manifold starts with an examination of the overall piping design. Piping is generally selected to meet pressure codes, corrosion and wear requirements, noise level requirements and plant expansion requirements. The following fluid velocities are typical: liquids - 5 fps; steam - 100-125 fps; and gases - 250-400 fps.

The size of the pipe selected is greatly influenced by the cost of material and by the cost of fabrication. A high pressure stainless steel piping system requiring welding, stress relieving, and radiography, may be selected to tolerate higher line velocities. Steam flows greater than 400 fps and feed water flows of 40 fps are reported in the literature for high pressure, high temperature power plants.

In many cases the pipe selected is oversized to allow for future plant expansion; thus, the control valve required for initial plant operation may be one or two sizes smaller than the pipe size into which it must be fitted.

There are two alternatives in selecting the control valve, either:

1. a line-sized control valve with reduced trim (for example, a 4 in. valve body with 2 in. trim); or
2. a minimum-sized control valve (for example, a 2 in. valve body with full sized trim).

The selection is generally determined by a cost analysis considering the difference in cost of the valves, piping and pressure drop requirements related to the pumping cost.

The selection of a line size valve with reduced trim is recommended unless the cost analysis shows a significant advantage for the minimum size valve. It is common practice in many process industries to automatically select line size control valves for line sizes of 2 in. and smaller.

With the process piping fixed and the control valve sized, manifold design can proceed. Fortunately, control valve body dimensions have been standardized. (ISA RP 4.1, "Uniform Face to Face Dimensions for Flanged Control Valve Bodies.")

ISA RP 4.2 "Standard Control Valve Manifold Designs" and API RP 550 "Manual on Installation of Refinery Instruments and Control Systems" can be useful in designing a particular manifold, but will probably require some modification to meet specific requirements. Other manifold arrangements and recommendations are given in a series of diagrams appearing later in this chapter. The specific requirements are first determined by a review of a check list as noted above. Modification of a suitable reference design can then be made accordingly.

It is common practice in many processes to install block valves and by-pass valves. These are installed to facilitate continuous operation. The block valves are used to isolate the control valve and the by-pass valve is used while the control valve is being repaired.

There are exceptions where the added expense of one or more block valves precludes their use. Batch processes may be piped so that one block valve is all that is required to isolate a control valve for repair. Certain critical processes cannot tolerate the inadvertent opening of a by-pass valve or the closing of a block valve.

Most companies have adopted recommended minimum sizes similar to Table 1, because of the cost factor. Typical examples of arrangements of the control valves, block valves and by-pass valves are illustrated throughout the Figures in Appendix I of this chapter.

PLANT INSTALLATION PRACTICES

Good plant installation practices start with the proper handling of the control valve from the time it is received at the site.

It is most important that the personnel who will be handling the control valve at all stages of its installation realize that control valves are precision devices with finished surfaces that range down to 1 - 2 RMS (roughness in micro-inches). Furthermore, all personnel handling the control valves should be made aware of the cost and lead-time required to secure control valves. Specially designed control valves can take from 6-14 months to deliver after order placement. Careless handling of valves may cause considerable delay in plant startup.

The degree of inspection of a control valve upon receipt will vary with the facility and expertise of personnel handling it. The range of care and degree of inspection varies widely from one purchaser to another. It is not uncommon on a new plant construction site to find control valves unboxed and exposed to the elements. At the other extreme the so called "expert", armed with an assembly drawing and a garden variety of tools, will attempt a complete disassembly of the control valve in order to check the dimensions of all the parts. Somewhere between these extremes is the proper degree of receipt inspection.

Many firms provide for thorough inspection, by the purchaser, at the place of manufacture. In this case, the receipt inspection consists of the minimum effort required to determine that the valve was not damaged in shipment.

The following is considered the minimum inspection criteria for factory assembled

Table I. RECOMMENDED MINIMUM BLOCK AND BYPASS VALVE SIZING

(All Sizes in Inches)

Line Size:	½		¾		1		1½		2		3		4		6		8		10		12	
	Valve		Valve		Valve		Valve		Valve		Valve		Valve		Valve		Valve		Valve		Valve	
Control Valve Size	Block	Bypass	Block	Bypass	Block	Bypass	Block	Bypass	Block	Bypass	Block	Bypass	Block	Bypass	Block	Bypass	Block	Bypass	Block	Bypass	Block	Bypass
½	½	½	¾	¾	1	1	1½	1½														
¾			¾	¾	1	1	1½	1½	2	2												
1					1	1	1½	1½	2	2	2	2										
1½							1½	1½	2	2	2	2	3	3								
2									2	2	3	2	3	3	4	4						
3											3	3	4	3	4	4	6	6				
4													4	4	6	4	6	6	8	8		
6															6	6	8	6	8	8	10	10
8																	8	8	10	8	10	10
10																			10	10	12	10
12																					12	12

control valves:

1. visual examination, using the assembly drawing;
2. hydrostatic test;
3. leakage test;
4. hysteresis check;
5. valve-travel check by the operator;
6. operational check of all accessories - limit switches, valve positioners, etc.
7. electrical tests (Megger and Hipot) for electrical devices; and
8. packaging and shipping checks to specifications and procedures.

If these tests are properly performed at the factory the receipt inspection will only require visual examination. If these these tests were not performed at the factory, or if there is reason to suspect problems, these tests should be performed on site, as required.

After the receipt inspection, the control valve should be prepared for shipment either to the plant area for installation, or to storage. The valve body's air or electrical connections should be plugged to keep-out dirt. If the control valve is to be stored for any length of time it should be packed for protection against the elements. Refer to Chapter 10 for a more detailed discussion on preparing a valve for storage.

Summary of Installation Practices

The following detailed items are recommended for consideration in every installation:

1. The air supply used to operate pneumatic control valves should be free from oil and moisture.
2. Follow the manufacturer's recommendations for the hydraulic system to power hydraulically driven control valves.
3. Follow the manufacturer's recommendations for the electronic system for electrically or electro-hydraulically driven control valves.
4. Inlet piping to control valves with small passageways, should be fitted with appropriate filters to eliminate internal valve damage from foreign matter in the piping system.
5. Do not put excessive stress on valve bodies when installing in the system. This is particularly important for split body valves.
6. Before initial start-up and after a maintenance shut-down, install screens ahead of the control valve to collect pipe scale, rust, and other debris. Whenever possible, the piping system should be fitted with a spool piece and flushed out prior to control valve installation.
7. If the valve is to operate in a dusty atmosphere, install a rubber or plastic boot around the stem to protect its polished surface from damage.
8. Be sure to follow all of the manufacturer's instructions for adjustments and switch positions for the accessories. For example, do not leave the valve positioner by-pass switch in the by-pass position.
9. If the control valve is to be removed from the system after installation be sure that all block valves are closed and tagged. If the control valve contains damaging fluids or contaminants, it should be tagged accordingly, for proper cleaning, prior to disassembly.
10. Be sure the valve is installed with the flow direction arrow in the proper direction. Cases have been reported where the manufacturer furnished a valve with the arrow pointing in the wrong direction.
11. Review all of the control valve manufacturer's specific instructions prior to installation. The first page of an instruction sheet provided by the Jordan Valve Division of Richards Industries, Inc. (I & M Procedure #60) is in Figure 5, as an example of the kind of material available from manufacturers.

REFERENCES

1. API RP550, Second Edition, 1965, *Manual on Installation of Refinery Instruments and Control Systems,* Part I, Process Instrumentation and Control, 126 pp. American Petroleum Institute, New York.
2. ISA RP4.1, 1951, *Uniform Face to Face Dimensions for Flanged Control Valve Bodies,* 2 pp, Instrument Society of America, Pittsburgh.
3. ISA RP4.2, 1956, *Standard Control Valve Manifold Designs* (Carbon Steel Valves), 20 pp, Instrument Society of America, Pittsburgh.
4. Masek, John A., January, 1967-September, 1969, "Flow Meter and Control Valve Manifold Piping," a collection of reprints from *Heating, Piping and Air Conditioning,* Reinhold Publishing Company, New York.
5. VC-1006-A, *Piping Pointers,* Crane Company, Chicago.

JORDAN VALVE™

DIVISION OF RICHARDS INDUSTRIES, INC.
HELEN & BLADE STS. CINCINNATI, OHIO 45216
Telephone (513) 242-3100

60 I & M

INSTRUCTIONS FOR INSTALLATION AND MAINTENANCE

SERIES 60 PRESSURE REGULATORS 1/4" to 2" (PRV)

INSTALLATION

YOUR SERIES 60 PRESSURE REGULATOR WILL PROVIDE YOU WITH LONG, TROUBLE-FREE SERVICE IF IT IS CORRECTLY INSTALLED. A FEW MINUTES OF YOUR TIME SPENT READING THESE INSTRUCTIONS MAY SAVE HOURS OF TROUBLE AND DOWNTIME LATER.

VALVE INSTALLATION (READ CAREFULLY)

1. To protect the regulator from grit, scale, thread chips, and other foreign matter, all pipe lines and piping components should be blown out and thoroughly cleaned before the regulator is installed.

2. Shut-off valves, pressure gages, and by-pass piping should be installed as indicated in the diagram to provide easier adjustment, operation, and testing.

3. In preparing threaded pipe connections, care should be exercised to prevent pipe sealing compound from getting into the pipe lines. Pipe sealing compound should be used sparingly, leaving the two lead threads clean. Jordan uses, and recommends, Seyco #2415 thread sealer Teflon ribbon.

4. **A LINE STRAINER SHOULD BE INSTALLED ON THE INLET SIDE OF THE REGULATOR TO PROTECT IT FROM GRIT, SCALE AND OTHER FOREIGN MATTER. A .033 PERFORATED SCREEN IS USUALLY SUITABLE. LINE STRAINERS ARE AVAILABLE FROM JORDAN VALVE.**

5. Install the regulator in the highest horizontal line of piping to provide drainage for inlet and outlet piping, to prevent water hammer, and to obtain faster regulation.

6. The flow arrow on the regulator body must be pointed in the direction of flow. The regulator may be installed vertically or horizontally without affecting its operation.

7. For best control, 3' - 0" straight sections of pipe should be installed on either side of the regulator.

8. In hot vapor lines, upstream and downstream piping near the regulator should be insulated to minimize condensation.

9. If possible, install a relief valve downstream from the regulator. Set at 15 psi above the control point of the regulator.

10. Expand the outlet piping at least one pipe size if the controlled pressure (downstream) is 25% of the inlet pressure or less. A standard tapered expander connected to the outlet of the regulator is recommended.

11. Where surges are severe, a piping accumulator is recommended.

IDEAL INSTALLATION

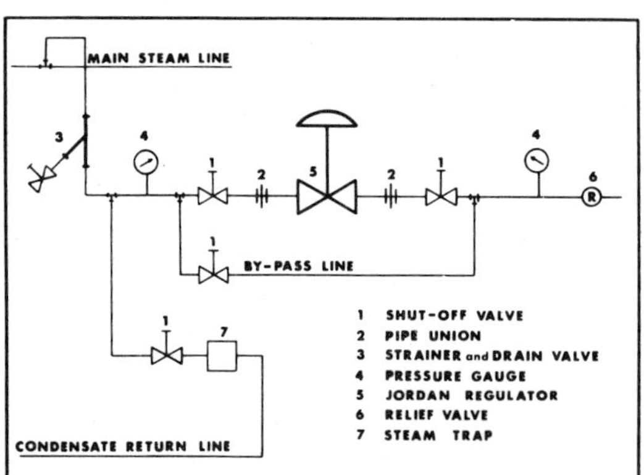

TROUBLE SHOOTING

TROUBLE	POSSIBLE CAUSE & CURE
Erratic Control	Oversizing causes cycling and hunting and reduces the rangeability of the valve. Make certain that your sizing is correct.
	Steam traps downstream may need attention.
	Safety valve may be jammed open. Repair as necessary.
	Excessive foreign matter on seats. Clean them.
	Valve stroke out of adjustment. Check and readjust if necessary.
	Valve disc may not be moving freely. Check disc guide clearance.
Will not Operate	Diaphragm ruptured. Replace.
	Adjusting spring broken. Replace.
	Improper spring setting. Reset.

PROTECT VALVES WITH LINE STRAINERS

Figure 5. SAMPLE INSTALLATION INSTRUCTION SHEET AVAILABLE FROM MANUFACTURERS.
Courtesy Jordan Valve Division

APPENDIX I

CONTROL VALVE MANIFOLD AND PIPING ARRANGEMENTS

Manifold arrangements illustrative of various typical situations are presented in a series of diaphrams on the following pages. These may be quite suitable for use as shown, or may be made suitable for specific requirements by minor modifications.

Figures 6 through 17 are reproduced by permission of American Petroleum Institute, from RP-550, *Manual on Installation of Refinery Instruments and Control Systems;* Part I, Process Instrumentation and and Control, 1965.

Figures 18 through 35 are reproduced by permission of *Heating, Piping and Air Conditioning Journal* from the reprint Collection by John A. Masek, "Flow Meter and Control Valve Piping," 1967, 96 pp.

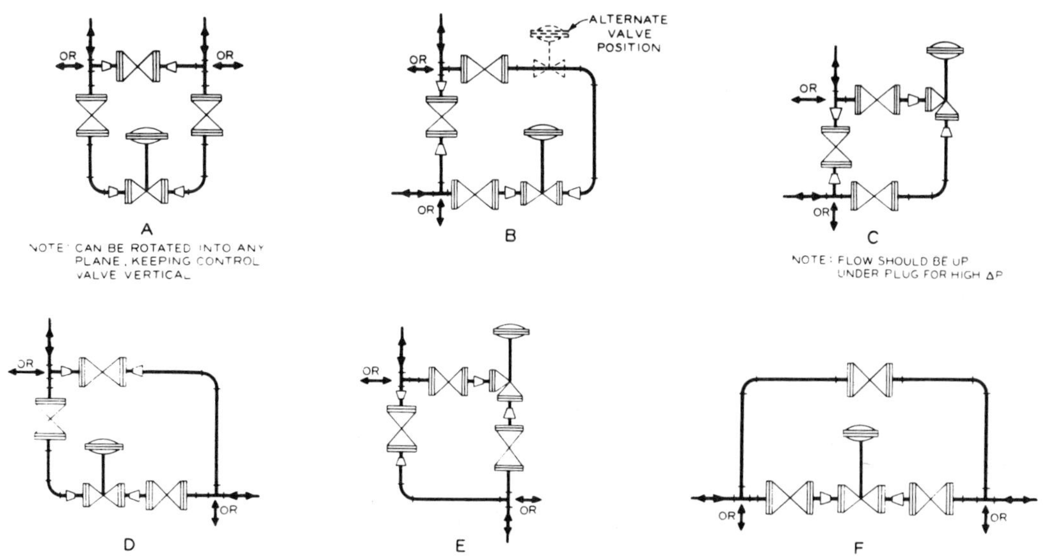

Arrangement A (ISA RP 4.2, Type 1) is preferred because manifold is compact, control valve is readily accessible for maintenance, and the assembly is easily drained.

Arrangement B is preferred because control valve is more readily accessible.

Arrangement C is often used with angle valves. Control valve is self-draining.

Arrangement D (ISA RP 4.2, Type 3) is preferred because the control valve is readily accessible. Bypass is self-draining.

Arrangement E results in a compact manifold, but control valve may not be too accessible.

Arrangement F is preferred because bypass is self-draining; however, requires greater space.

Note: Block and bypass valves should be installed close to tees, as shown, to minimize pockets. Drain and vents on either side of the control valve are not shown, nor are the supports.

Figure 6. CONTROL VALVE MANIFOLD ARRANGEMENTS.

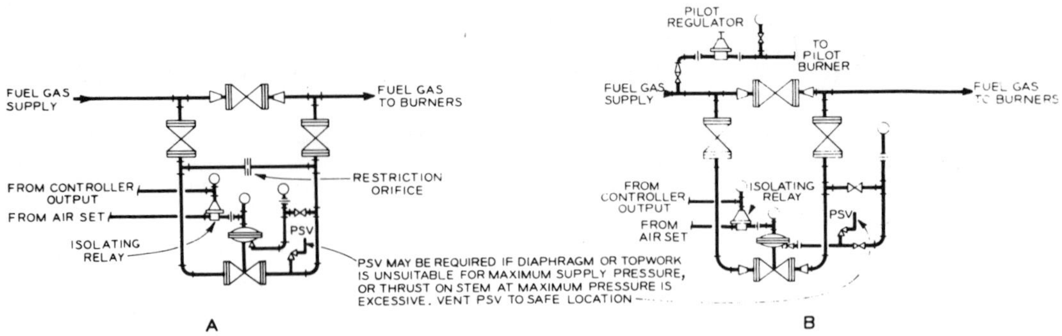

Arrangement A shows pressure-balanced valve with isolating relay and restriction orifice for maintaining minimum fires. Restriction orifice is sized for the manufacturer's recommended minimum turndown capacity for the burners. To prevent the control valve from closing completely, travel stops may be used in lieu of the restriction orifice.

Arrangement B shows a pilot regulator which is suggested where minimum fire maintained by restriction would cause furnace tube failure on process flow stoppage. Isolating relay is used to prevent gas flow into controller on diaphragm leakage or breakage. Ratio relay may be used to permit air loading to match fuel gas burner pressure.

Figure 7. PIPING ARRANGEMENTS FOR PRESSURE-BALANCED VALVE AT PROCESS HEATERS.

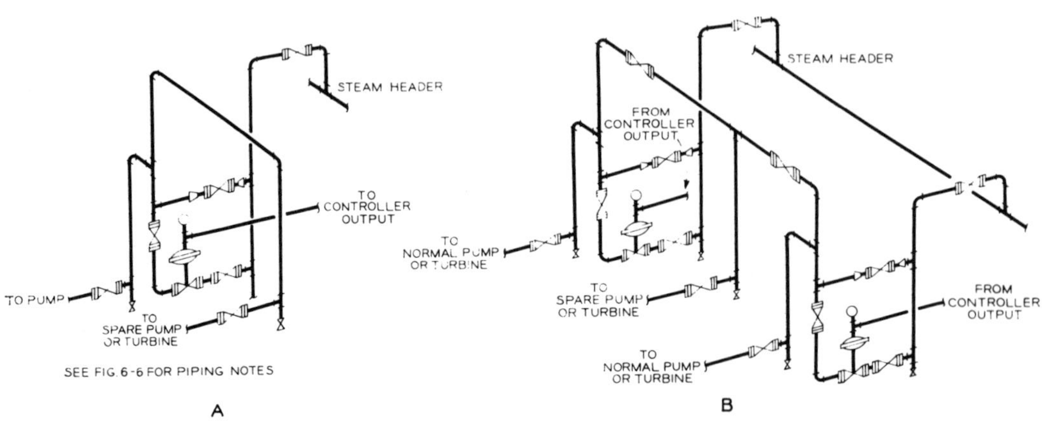

Arrangement A is used where one control valve is employed for normal pump, or turbine, and spare.

Arrangement B is used where two normal pumps have a common spare.

Figure 8. STEAM PIPING TO SPARE PUMP OR TURBINE.

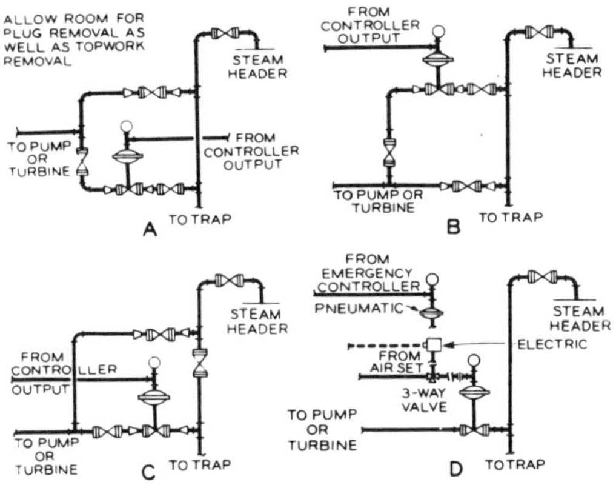

Arrangement A makes the control valve readily accessible; the piping is self-draining. Steam inlet may be above minimum distance required for plug removal.

Arrangement B may be used where the bypass can be installed at turbine or pump inlet. Control valve may be too high for easy access; piping is self-draining.

Arrangement C is often used to make control valve more accessible because the control valve can be located at the same elevation as turbine or pump steam inlet.

Arrangement D is often used for emergency start, or stop, of pump or turbine.

Notes:

1. Bypass and bypass valve should not be so located as to form a pocket; they should be self-draining.
2. Location of separators, strainers, and traps not shown because company standards differ on use and placement. The control valve should be located as low as possible for easy access.

Figure 9. CONTROL VALVE ARRANGEMENTS FOR STEAM TO PUMPS OR TURBINES.

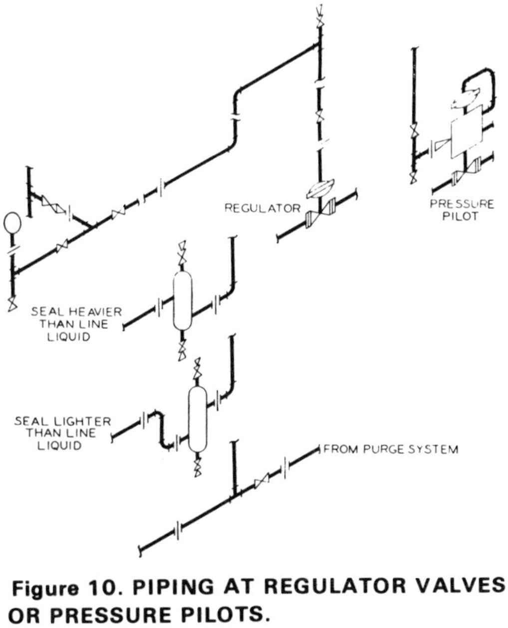

Figure 10. PIPING AT REGULATOR VALVES OR PRESSURE PILOTS.

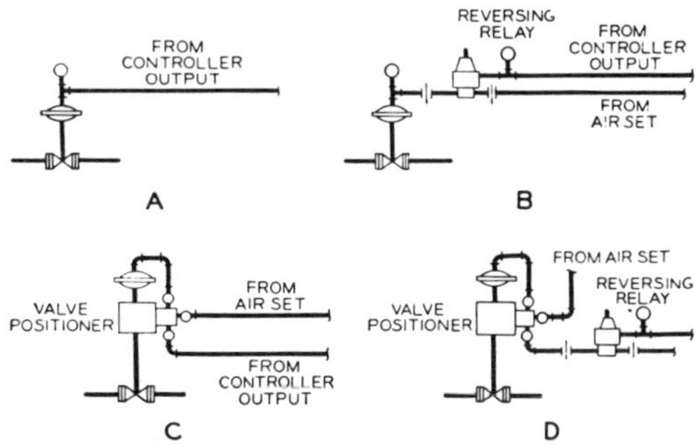

Arrangement A shows nominal control valve piping with gage for reading diaphragm pressure.

Arrangement B shows control valve with reversing relay. The reversing relay is used to reverse controller output to valve, where two valves with different action are connected to one controller, or is used to provide the same action when on automatic or manual control.

Arrangement C shows control valve with positioner.

Arrangement D shows control valve with positioner and reversing relay. The reversing relay is the same as used in B with standard control valve; in addition, it allows the positioner to be bypassed. A reverse-acting positioner may be used but it may not be bypassed.

Figure 11. ARRANGEMENTS OF PNEUMATIC PIPING OR TUBING, TO CONTROL VALVES.

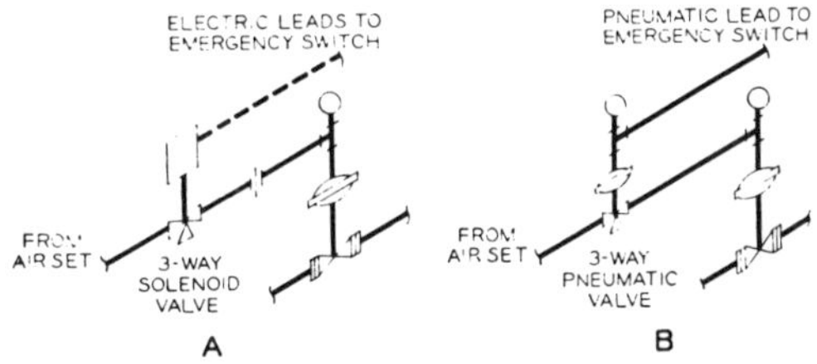

Arrangement A shows control valve with emergency solenoid actuator.

Arrangement B shows control valve with emergency pneumatic actuator.

Figure 12. CONTROL VALVE PIPING WITH EMERGENCY CONTROL ACTION.

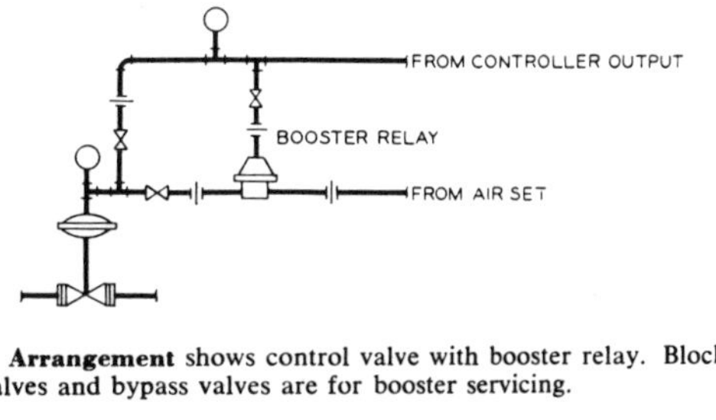

Arrangement shows control valve with booster relay. Block valves and bypass valves are for booster servicing.

Figure 13. BOOSTER RELAY PIPING.

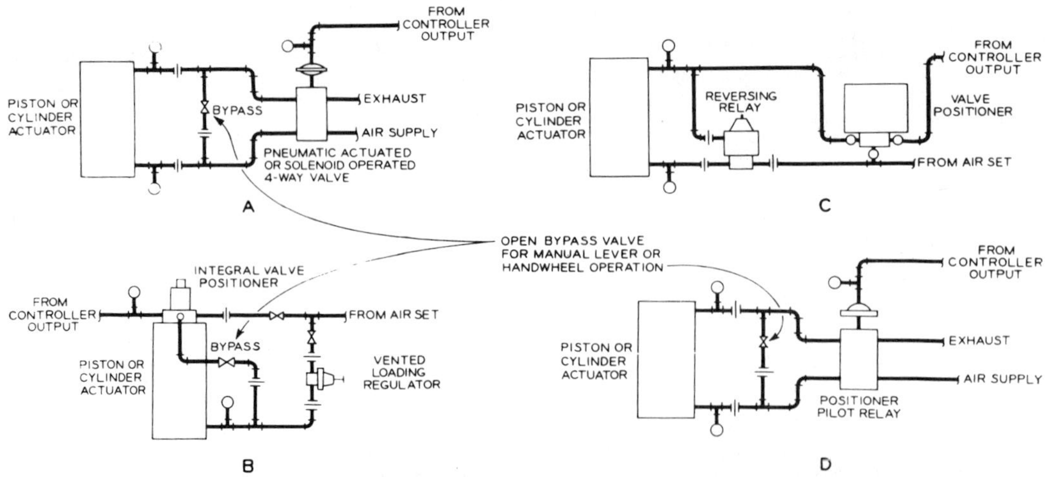

Arrangement A shows the actuator used with four-way valve in two-position service.

Arrangement B shows the actuator with constant pressure loading on one side.

Arrangement C shows the actuator with reversing relay.

Arrangement D shows the actuator with pilot relay for throttling service.

Figure 14. AIR CYLINDER OR PISTON ACTUATORS.

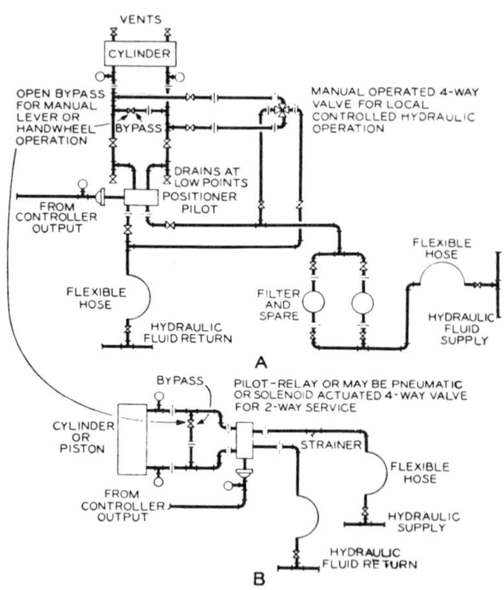

Arrangement A shows the cylinder actuator with provision for local manual hydraulic operation.

Arrangement B shows the cylinder or piston actuator without local hydraulic operation.

Figure 15. HYDRAULIC CYLINDER OR PISTON ACTUATORS.

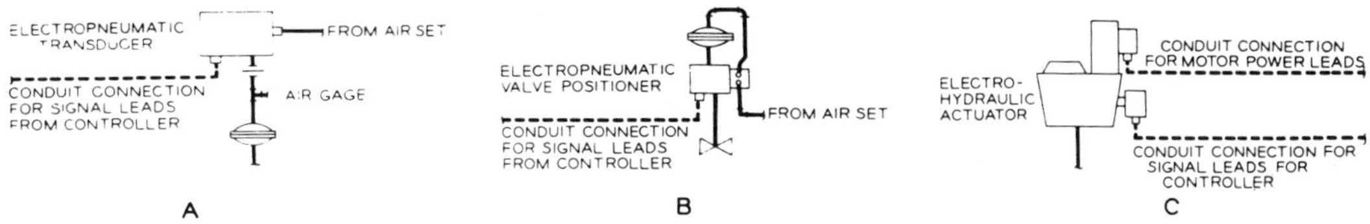

Arrangement A shows the electropneumatic transducer mounted at pneumatic control valve.

Arrangement B shows the electropneumatic valve positioner on valve.

Arrangement C shows the electrohydraulic valve actuator.

Note: Provide seals in electrical leads as required by electrical specifications for area in which actuator is installed.

Figure 16. ELECTROPNEUMATIC AND ELECTROHYDRAULIC ACTUATORS.

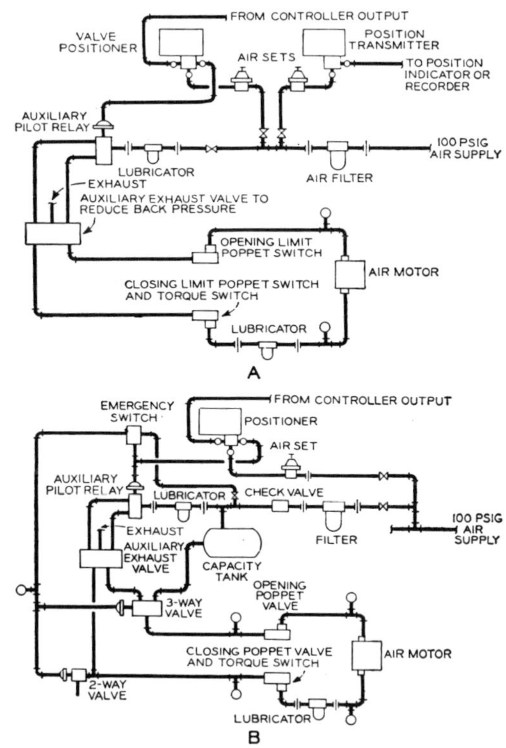

Arrangement A shows air motor actuator with position transmitter.

Arrangement B shows air motor actuator with auxiliary equipment to close valve on emergency.

Figure 17. AIR MOTOR ACTUATORS.

a) Weld neck flange and eccentric reducer; configuraation can prevent stud removal.
b) 4 in. long pipe spool piece welded between flange and reducer, allowing bolt removal space and drain connections.
c) Butt weld reducing tee, providing bolt removal space and drain connections.
d) 6 in. control valve flange with 8 in. eccentric reducer; stud bolts can be removed.
e) same as c) except slip-on flange is used with appropriate length pipe.

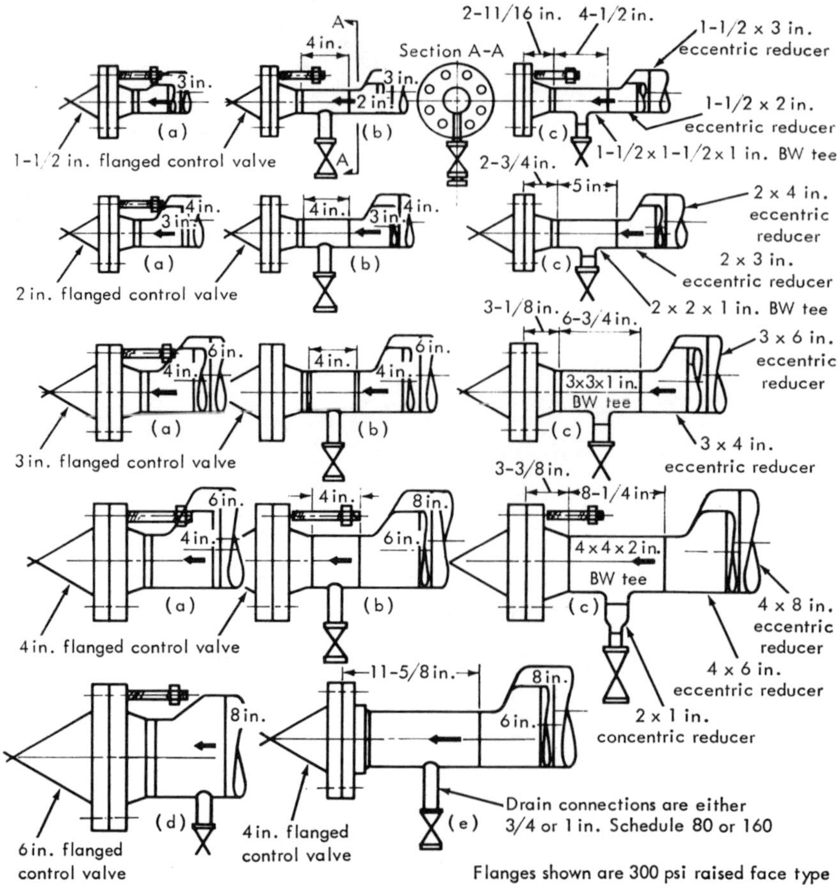

FIGURE 18. TYPICAL CONNECTIONS AT CONTROL VALVE IN MANIFOLD ASSEMBLY.

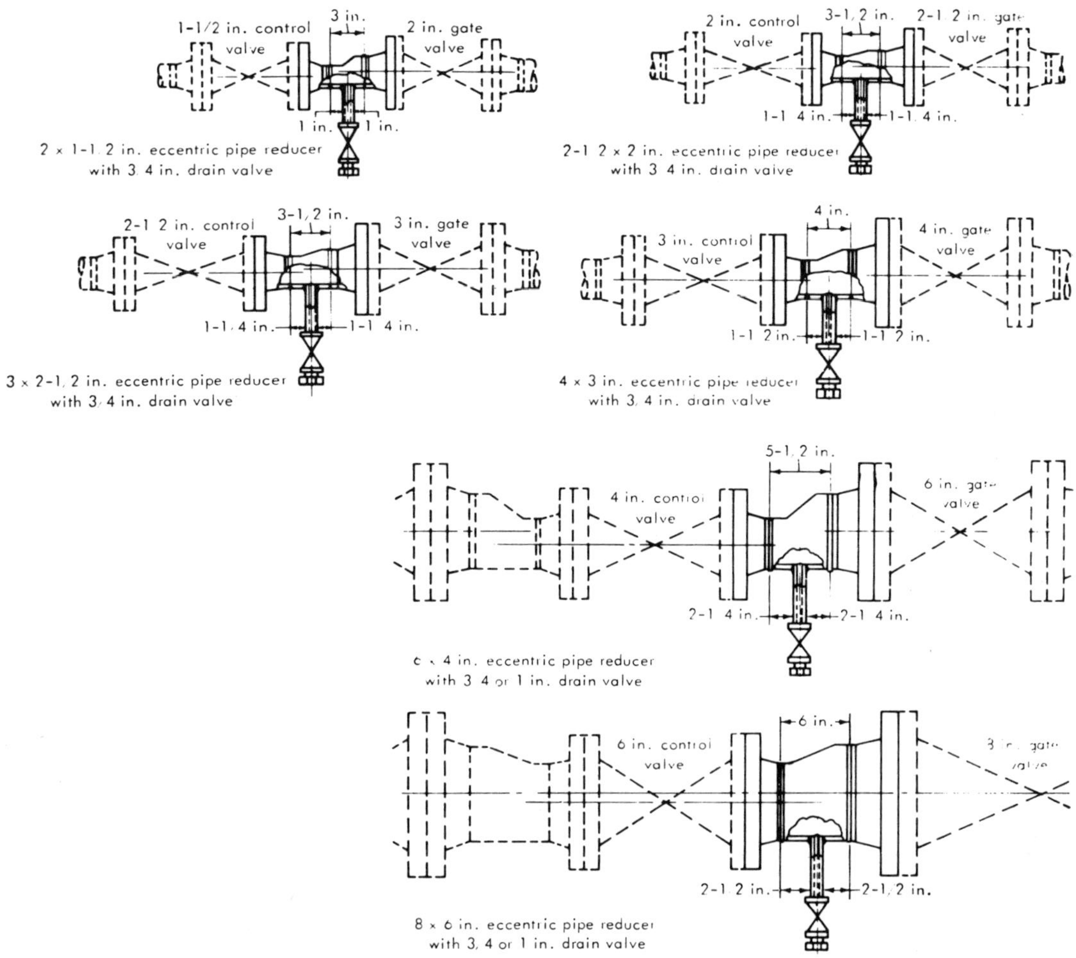

Figure 19. DRAIN VALVES INSTALLED IN BOTTOM OF PIPE REDUCERS ADJACENT TO CONTROL VALVES.

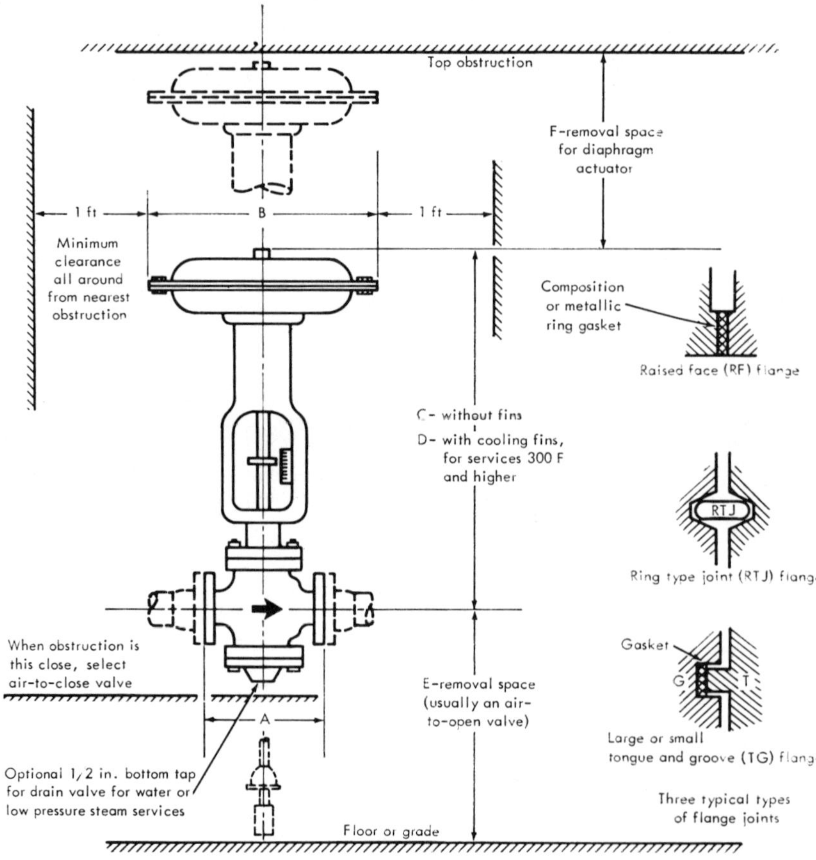

Figure 20. CLEARANCE REQUIREMENTS FOR CONTROL VALVE MANIFOLDS.

Control valve body size, in.	A (face-to-face dimensions) 150 psi raised face flanges	300 psi raised face flanges	400 psi raised face flanges	B	C	D	E*	F*
2**	10	10½	11¼	15½	33	39	21	18
2½	10⅞	11½	12¼	15½	34	40	21	20
3	11¾	12½	13¼	15½	35	40	21	23
4	13⅞	14½	15½	15½	36	42	21	24
6	17¾	18⅝	19½	19⅛	40	43	30	54
8	21⅜	22⅜	23⅜	20¾	46	50	30	54

*Dimensions for stem and trim removal and body removal are minimum; increase if possible.

**Standard screwed control valves should have same clearance dimensions as shown for 2 in. plugged control valves.

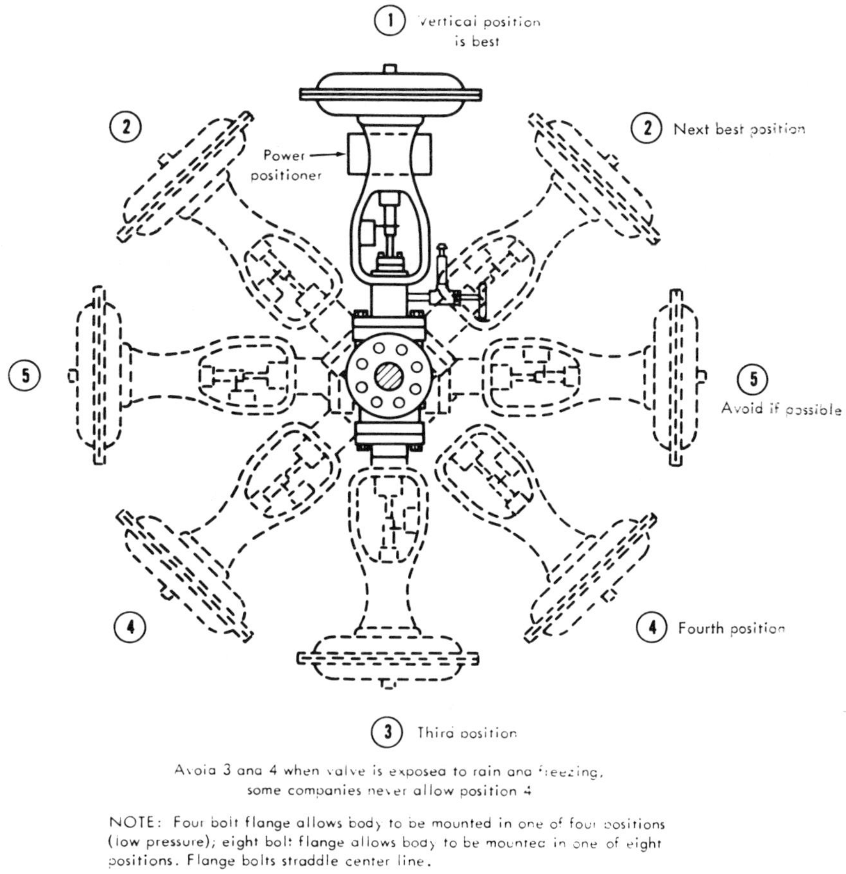

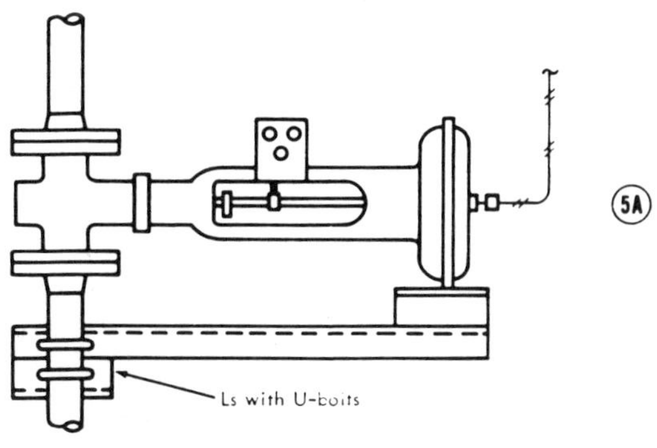

Figure 21. ALTERNATE ORIENTATIONS FOR DIAPHRAGM CONTROL VALVES, POSITION 1 IS PREFERRED.

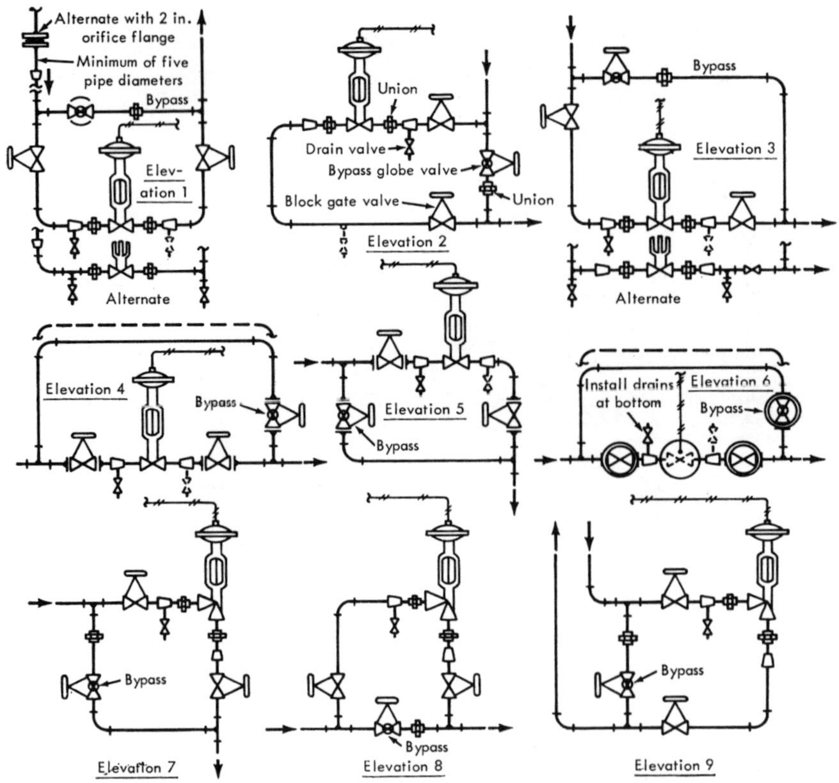

Figure 22. NINE TYPICAL CONTROL VALVE MANIFOLD ARRANGEMENTS FOR THREADED END GLOBE AND ANGLE BODY CONTROL VALVES.

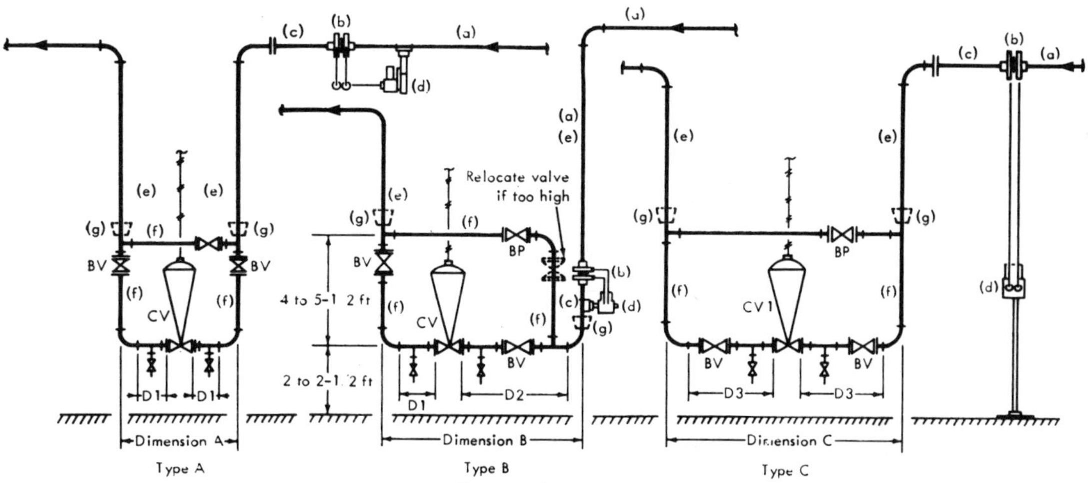

Three typical elevations

Reducer size (g), in.		Valve sizes, in.*			Type A manifold				Type B manifold		Type C manifold		
						Straight pipe			Straight pipe			Straight pipe	
Main size (e), 2 in.	Manifold size (f), in.	CV	BV	BP	Dimension A, in.	D1, in.	D1, pipe diameters	Dimension B, in.	D2, in.	D2, pipe diameters	Dimension C, in.	D3, in.	D3, pipe diameters
2	2	1½	1½	1½	28¼	19	6	45¾	24	16	54¼	20	13
3	2	1½	1½	1½	28¼	19	6	45¾	24	16	54¼	20	13
3	2	2	2	2	32⅛	22	4	51¼	26	12	60⅜	25	12
4	2	2	2	2	32⅛	22	4	51¼	26	12	60⅜	25	12
4	3	3	3	3	41⅜	28	3	65½	37	11	76⅜	32	10
6	3	3	3	3	41⅜	28	3	65½	37	11	76⅜	32	10
6	4	4	4	4	49⅞	33	3	77	39	9	87⅝	37	9
8	4	4	4	4	49⅞	33	3	77	39	9	87⅝	37	9
8	6	6	6	6	73	50	2	101¼	50	6	114½	48	8

**All valves are flanged, raised face, 300 psi USASI rating; CV = control valve, BV = block valves (gate), and BP = bypass valve (globe).*

(1) Control valve body (with reduced trim) and manifold all in process line size
(2) Manifold piping all in control valve body size but smaller than process line size

(a) Process line upstream from orifice plate, minimum of 15 nominal pipe diameters of straight pipe
(b) Orifice plate in orifice flange assembly
(c) Process line downstream from orifice plate, minimum of 5 to 10 diameters of straight pipe
(d) Differential pressure flow transmitter
(e) Process line size
(f) Control valve manifold, control valve body pipe size, process line size, or smaller
(g) Pipe reducer (optional when manifold is smaller than process line size)
CV-Diaphragm control valve, CV-1 — Control valve with split body
BV-Block gate valve, control valve size
BP-Bypass globe valve, control valve size
D -Straight pipe section upstream and downstream of control valve

Figure 23. THREE TYPICAL CONTROL VALVE MANIFOLDS.

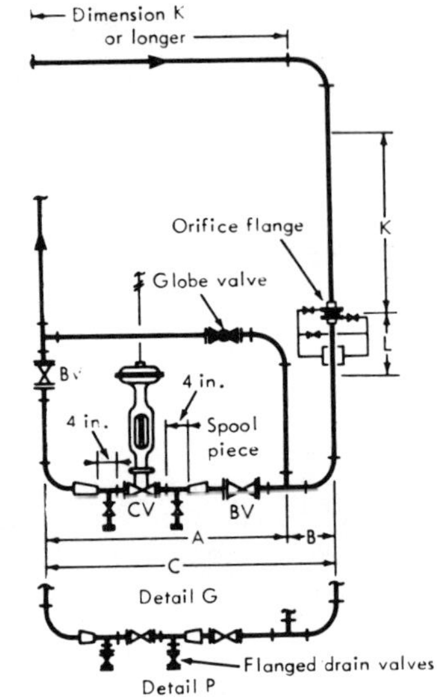

Main size, in.	Dimension K, in.[1]	Dimension L, in.[2]	Valve sizes, in.[3] BV	CV	Detail G dimensions, in.[4] A	B	C	Detail H dimensions, in.[5] E	B	F
2	28 to 40	10	2	1½	48⅛	5½	53⅝	49⅛	5½	54⅝
3	42 to 60	15	3	1½	54⅞	7⅞	62¾	55⅞	7⅞	63¾
3	42 to 72	15	3	2	56¼	7⅞	64⅛	61¾	7⅞	69⅝
4	56 to 80	20	4	2	60⅞	10⅛	71	66⅜	10⅛	76½
4	56 to 80	20	4	3	63⅝	10⅛	73¾	72⅛	10⅛	82¼
6	84 to 128	30	6	3	76	14⅝	90⅝	84½	14⅝	99⅛
6	84 to 128	30	6	4	76½	14⅝	93⅛	93	14⅝	107⅝
8	112 to 160	40	8	4	83½	19	104½	100	19	119
8	112 to 160	40	8	6	90⅝	19	109⅝	110⅝	19	129⅝

[1]*Upstream straight pipe length based on 0.70 or 0.80 beta ratio.*
[2]*Downstream straight pipe length based on 5 nominal pipe diameters.*
[3]*Valves are flanged, raised face. 300 psi USASI rating; BV = block valves (gate), CV = control valve.*
[4]*Pipe spool piece 4 in. long on each side of control valve.*
[5]*Butt weld tee with dimension M (see table below) on each side of control valve.*

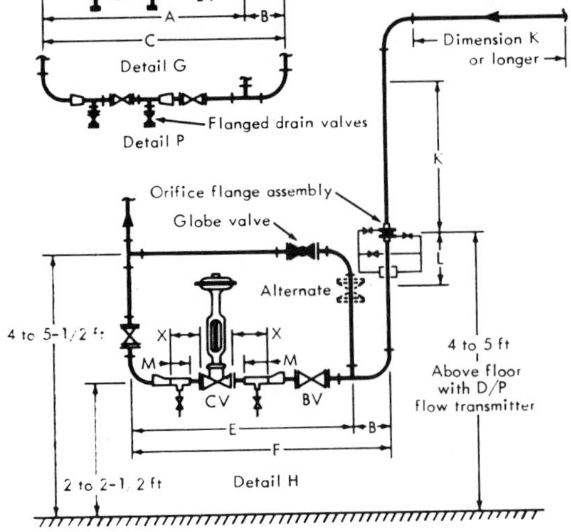

Control valve flange size, in.	Butt weld tee size, in.*	BW tee dimension M, in.	Approximate straight pipe length, X, nominal pipe diameters
1½	1½×1½×1	4½	4+
2	2 ×2 ×1	5	3+
3	3 ×3 ×1	6¾	3+
4	4 ×4 ×1½	8¼	2+
6	6 ×6 ×2	11¼	2+
8	8 ×8 ×3	14	2+

* *With 1 in. drain valve at bottom.*

Figure 24. FLANGED CONTROL VALVE MANIFOLDS WITH ORIFICE FLANGE ASSEMBLIES.

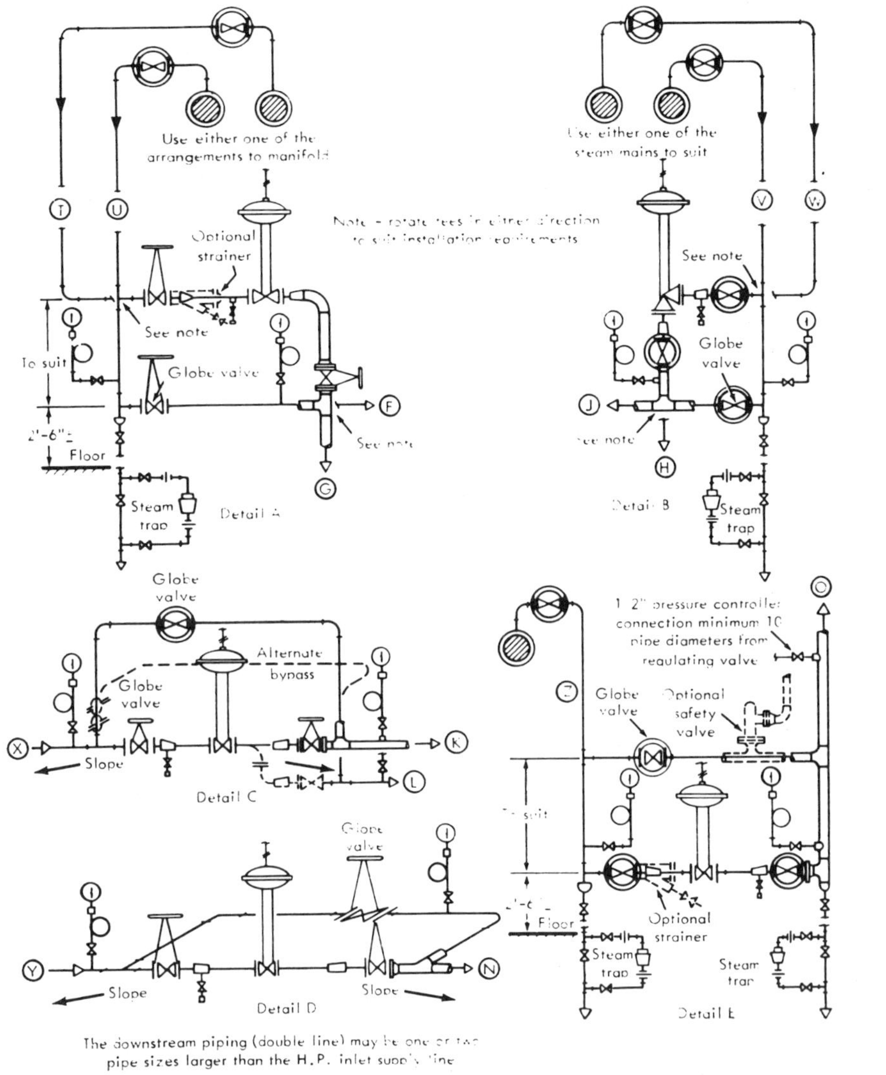

Figure 25. MANIFOLD DESIGN ARRANGEMENTS FOR FLOW AND PRESSURE REDUCING CONTROL VALVES IN LOW PRESSURE STEAM SERVICE USING FLANGED FITTINGS.

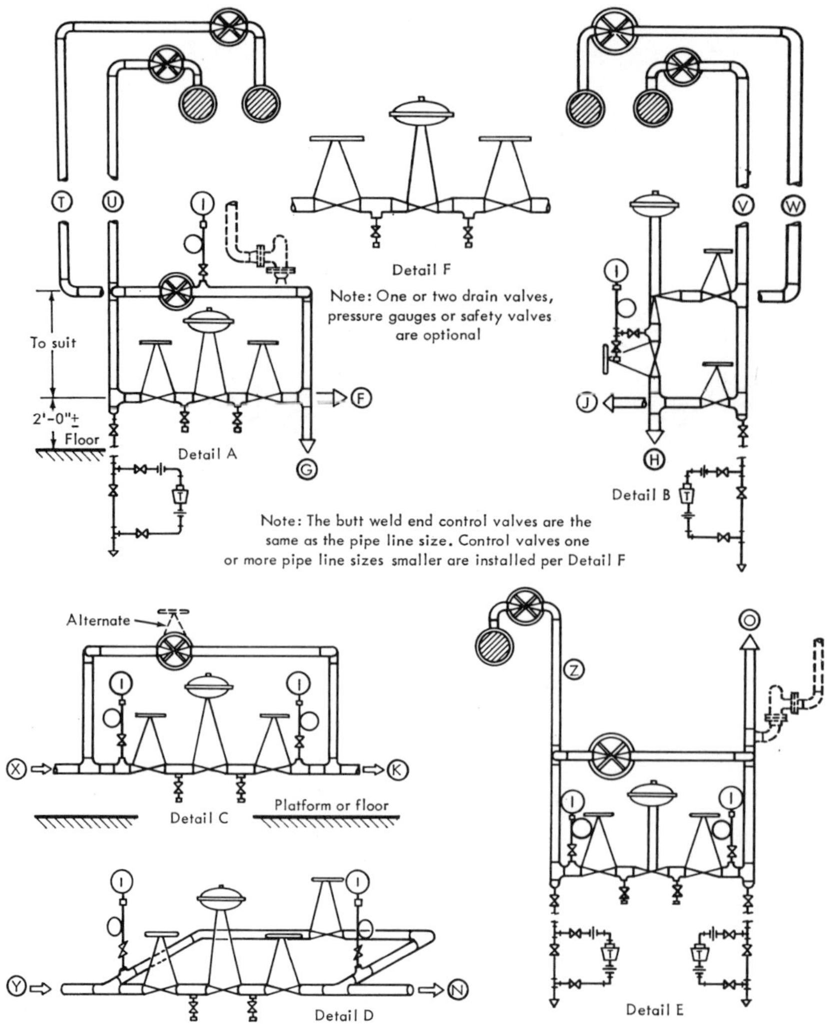

Figure 26. CONTROL VALVE MANIFOLD USING WELDED CONNECTIONS FOR UPPER PRESSURE RANGE SERVICES ON STEAM AND PROCESS STREAMS.

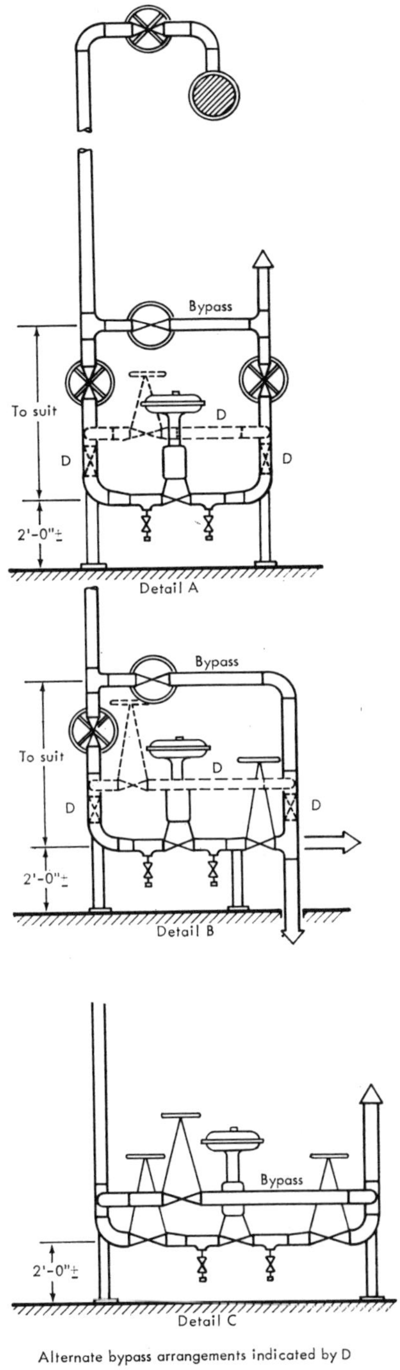

Figure 27. BUTT WELD END TYPE CONTROL, BLOCK, AND BYPASS VALVES IN PIPING MANIFOLDS PROVIDING STEAM OR OTHER FLUID FLOW REGULATION.

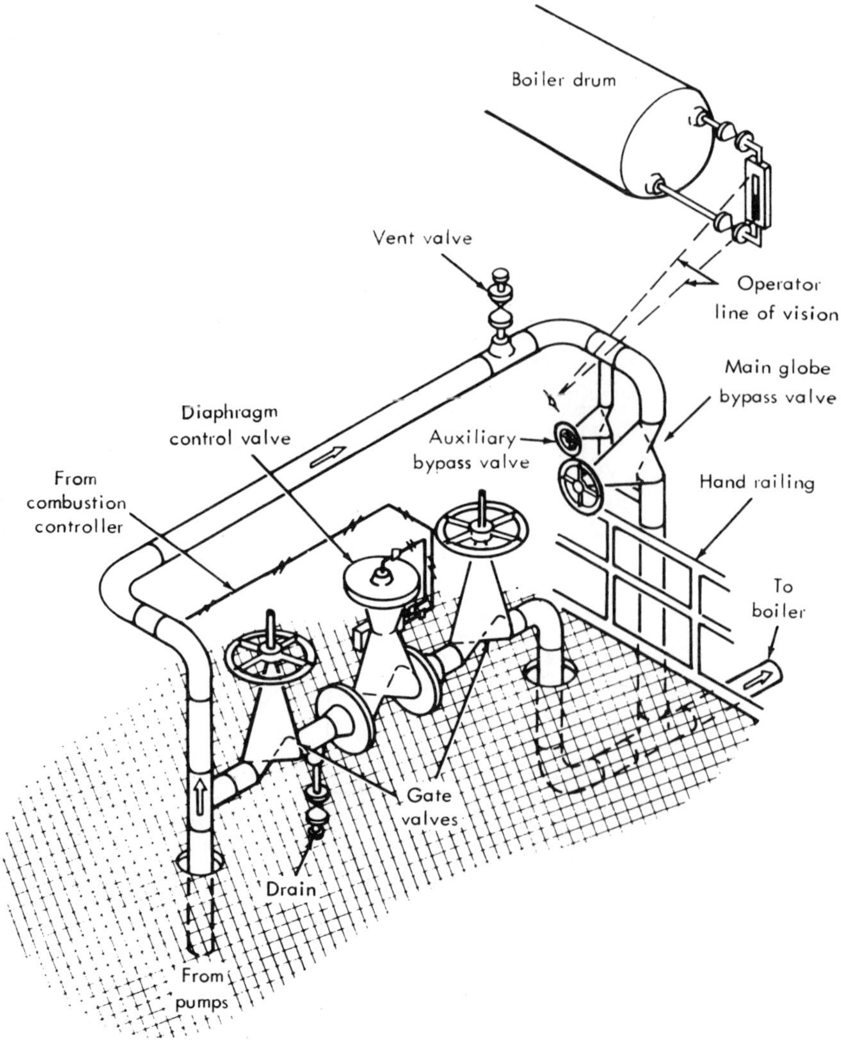

Figure 28. CONTROL VALVE MANIFOLD WITH BYPASS PIPING EXTENDING BEYOND ACCESS PLATFORM.

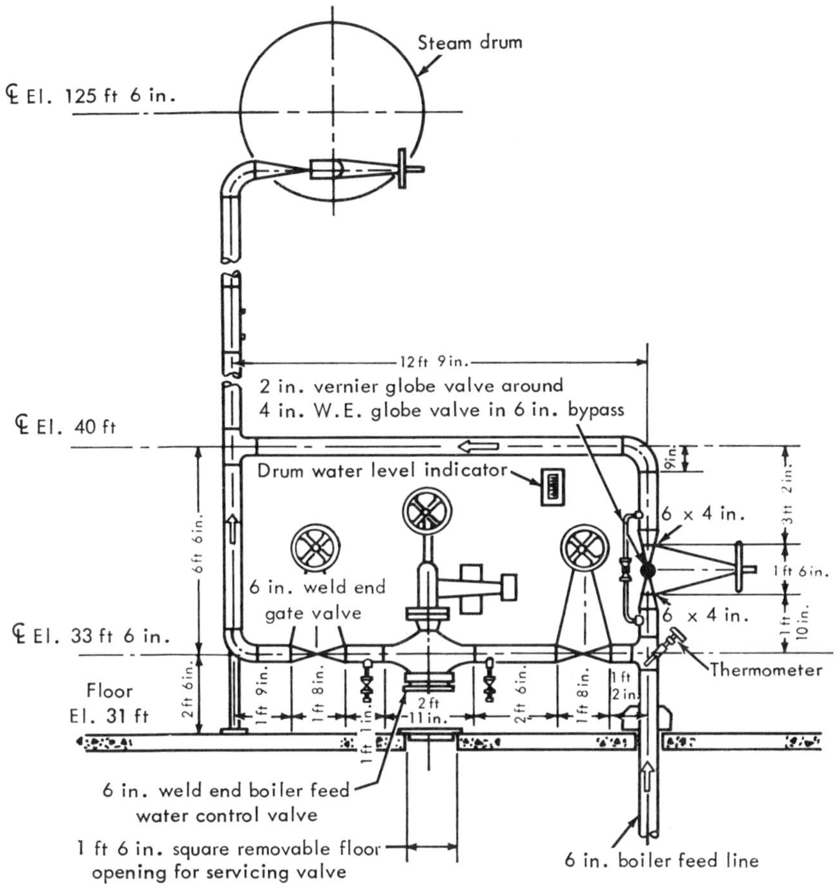

Figure 29. FEED WATER CONTROL VALVE MANIFOLD BELOW INLET TO BOILER STEAM DRUM.

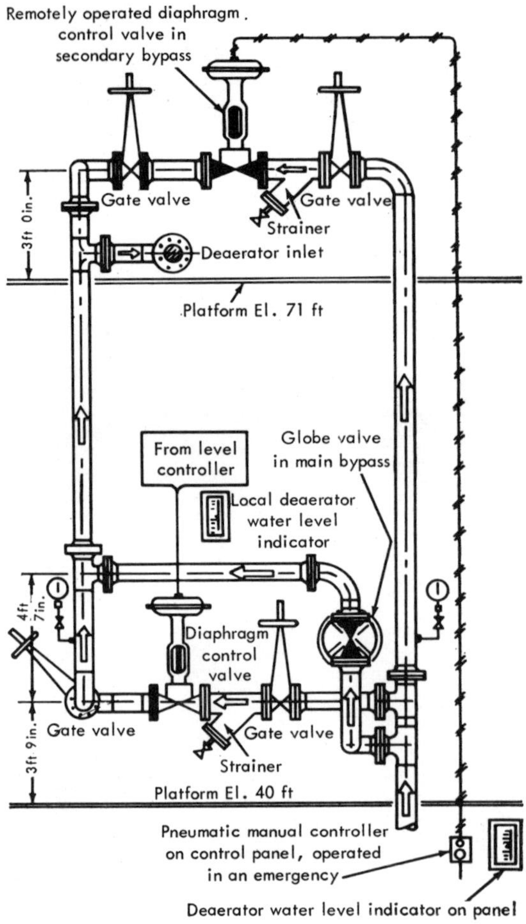

Figure 30. MAKEUP WATER CONTROL VALVE MANIFOLD COMBINED WITH AN AUXILIARY VALVE MANIFOLD, REMOTELY CONTROLLED.

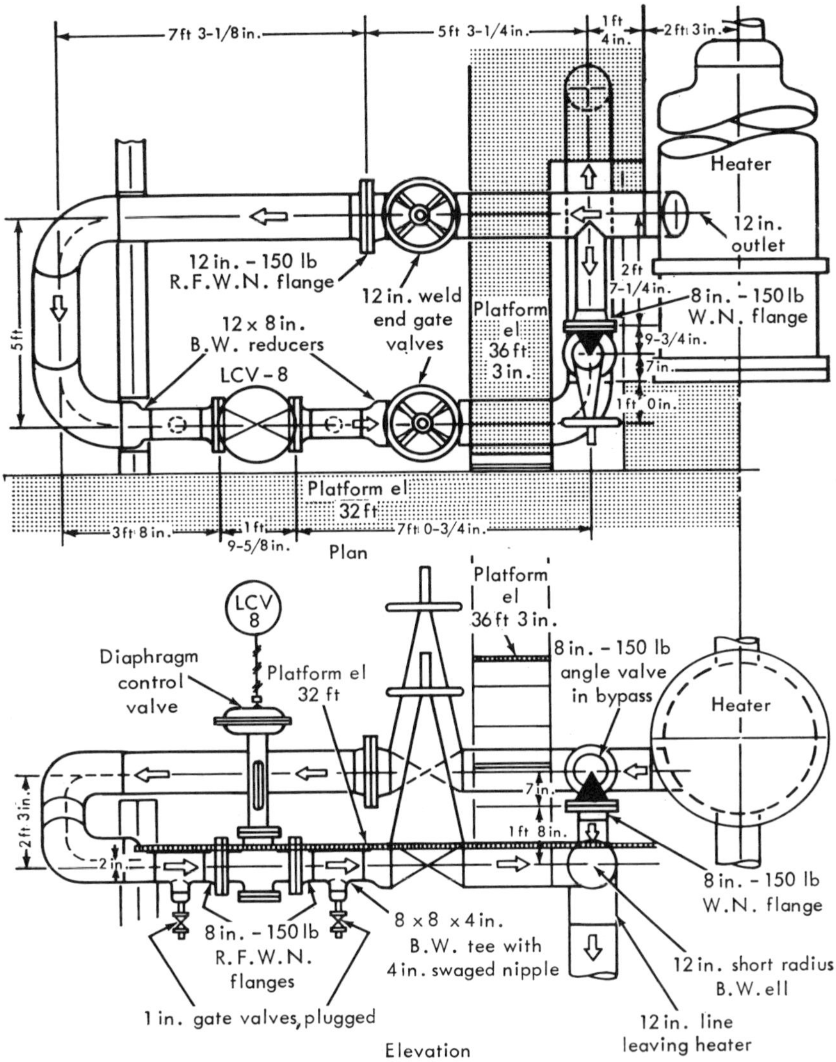

Figure 31. CONTROL VALVE MANIFOLD LOOP AT HEATER OUTLET.

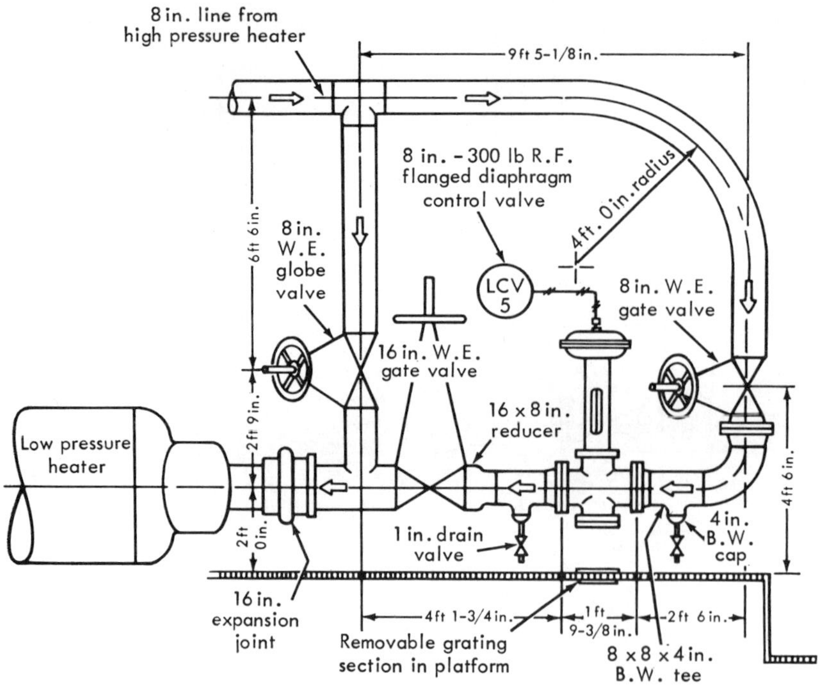

Figure 32. MANIFOLDING OF CONTROL VALVE FOR LOOPED ENTRY TO HEATER FROM OVERHEAD SUPPLY LINE.

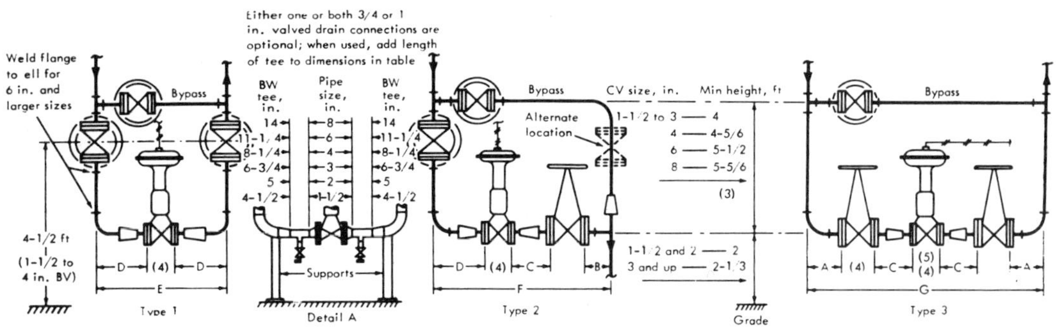

Figure 33. DIMENSIONS FOR FLANGED CONTROL VALVE MANIFOLDS.

Valve sizes, in.		Dimensions, in., with swaged nipples							Dimensions, in., with weld reducers				
BV	CV	A	B	C	D	E	F	G	C	D	E	F	G
150 psi block and control valves—													
2	1 1/2	5 1/2	5	11 7/16	11 15/16	32 3/4	44 3/8	57	7 15/16	8 1/16	—[2]	—[2]	50 3/8
3	1 1/2	7 1/4	6 1/8	13 3/16	14 15/16	38 3/4	51 1/4	66	8 1/2	10 7/16	—[2]	42 1/4	57
3	2	7 1/4	6 1/8	13 1/4	15	40 1/8	52 5/8	67 3/8	8 3/4	10 1/2	—[2]	43 5/8	58 3/8
4	2	9	7 1/8	14 1/2	17 1/2	45 1/8	58 3/8	75 3/8	9 1/2	12 1/2	35 1/8	48 3/8	65 3/8
4	3	9	7 1/8	14 3/4	17 3/4	47 3/8	60 5/8	77 5/8	9 3/4	12 3/4	37 3/8	50 5/8	67 5/8
6	3	12 1/2	9 1/8	18 1/4	23 3/4	59 3/8	—[1]	—[1]	11 3/4	17 1/4	46 3/8	60 5/8	81 5/8
6	4	12 1/2	9 1/8	18 1/2	24	62	—[1]	—[1]	12	17 1/2	49	63 1/4	84 1/4
8	4	16	11	20	28	70	—[1]	—[1]	13	21	56	70 5/8	95 1/4
8	6	16	11	20 1/2	28 1/2	74 7/8	—[1]	—[1]	13 1/2	21 1/2	60 7/8	75 1/2	100 1/8
10	6	19	12 1/2	22 1/2	33 1/2	84 7/8	—[1]	—[1]	14 1/2	25 1/2	68 7/8	83 1/2	111 1/8
10	8	19	12 1/2	23	34	89 1/2	—[1]	—[1]	15	26	73 1/2	88 1/8	115 3/4
150 psi block valve, 300 psi control valve—													
2	1 1/2			11 11/16			45 3/8	58	8 3/16			—[2]	51
3	1 1/2			13 7/16			52 1/4	67	8 15/16			43 1/4	58
3	2			13 1/2			53 5/8	68 3/8	9			44 1/8	59 3/8
4	2			14 3/4			59 3/8	76 3/8	9 3/4			48 7/8	66 3/8
4	3			15 1/8			62 1/8	79 1/8	10 1/8			52 1/8	69 1/8
6	3			18 3/8			—[1]	—[1]	12 1/8			62 1/8	83 1/8
6	4			18 7/8			—[1]	—[1]	12 3/8			64 5/8	85 5/8
8	4			20 3/8			—[1]	—[1]	13 3/8			72	96 5/8
8	6			20 7/8			—[1]	—[1]	13 7/8			77 1/8	101 3/4
10	6			22 7/8			—[1]	—[1]	14 7/8			85 1/8	112 3/4
10	8			23 3/8			—	—	15 3/8			89 7/8	117 1/2
300 psi block and control valves—													
2	1 1/2	5 3/4	5 1/4	11 13/16	12 3/16	33 3/4	47 3/8	62	8 7/16	8 11/16	—[2]	—[2]	55
3	1 1/2	7 5/8	6 1/2	13 13/16	15 3/16	39 3/4	56 1/8	74 3/4	9 5/16	10 11/16	—[2]	47 1/8	65 3/4
3	2	7 5/8	6 1/2	13 7/8	15 1/4	41 1/8	57 1/2	76 1/8	9 3/8	10 3/4	—[2]	48 1/2	67 1/8
4	2	9 3/8	7 1/2	15 1/8	17 3/4	46 1/8	65 1/8	83 7/8	10 1/8	12 3/4	36 1/8	53 1/8	73 7/8
4	3	9 3/8	7 1/2	15 1/2	18 1/8	48 7/8	65 7/8	86 5/8	10 1/2	13 1/8	38 7/8	53 7/8	76 5/8
6	3	12 7/8	9 1/2	19	24 1/8	60 7/8	—[1]	—[1]	12 1/2	17 5/8	47 7/8	68 1/4	95 3/8
6	4	12 7/8	9 1/2	19 1/4	24 3/8	63 5/8	—[1]	—[1]	12 3/4	17 7/8	50 3/8	70 3/4	97 7/8
8	4	16 3/8	11 3/8	20 3/4	28 3/8	71 3/8	—[1]	—[1]	13 3/4	21 3/8	57 3/8	77 3/4	108 1/8
8	6	16 3/8	11 3/8	21 1/4	28 7/8	76 1/2	—[1]	—[1]	14 1/4	21 7/8	62 1/2	82 7/8	113 1/4
10	6	19 5/8	13 1/8	23 1/2	33 7/8	86 1/2	—[1]	—[1]	15 1/2	25 7/8	70 1/2	91 3/8	125 1/4
10	8	19 5/8	13 1/8	24	34 3/8	91 1/4	—[1]	—[1]	16	26 3/8	75 1/4	96 1/8	130
600 psi block and control valves—													
2	1 1/2	6 1/8	5 5/8	12 5/8	12 1/2	35 1/8	52 5/8	71 1/8	9 1/8	9	—[2]	—[2]	64 1/8
3	1 1/2	8	6 7/8	14 1/2	15 1/2	41 1/8	61 1/4	83 5/8	10	11	—[2]	52 1/4	74 5/8
3	2	8	6 7/8	14 5/8	15 5/8	42 3/4	62 3/4	85 1/4	10 1/8	11 1/8	—[2]	53 7/8	76 1/4
4	2	10 1/4	8 3/8	16 3/8	18 1/8	47 3/4	71 5/8	99 1/4	11 3/8	13 1/8	37 3/4	61 5/8	88 1/4
4	3	10 1/4	8 3/8	16 3/4	18 1/2	50 1/2	74 3/8	102	11 3/4	13 1/2	40 1/2	64 3/8	92
6	3	13 7/8	10 1/2	20 3/8	24 1/2	62 1/2	—[1]	—[1]	13 7/8	18	49 1/2	78 1/8	113 1/2
6	4	13 7/8	10 1/2	21 1/8	25 1/4	66 1/4	—[1]	—[1]	14 5/8	18 3/4	53 1/4	81 7/8	117 1/4
8	4	17 1/2	12 1/2	22 3/4	29 1/4	74 1/4	—[1]	—[1]	15 3/4	22 1/4	60 1/4	92 1/2	134 3/4
8	6	17 1/2	12 1/2	23 3/8	29 7/8	80	—[1]	—[1]	16 3/8	22 7/8	66	98 1/4	140 1/2
10	6	21 1/4	14 3/4	26 1/8	34 7/8	90	—[1]	—[1]	18 1/8	26 7/8	74	111 1/4	161 1/2
10	8	21 1/4	14 3/4	26 3/4	35 1/2	95 1/4	—[1]	—[1]	18 3/4	27 1/2	79 1/4	116 1/2	166 3/4

[1]Use butt weld reducers.
[2]Use butt weld swaged pipe nipples.
[3]Add 6 in. for air fins.
[4]Gasket thickness is 1/16 in. for 150 and 300 psi valves; 1/8 in. for 600 psi valves.
[5]Split body valve optional.

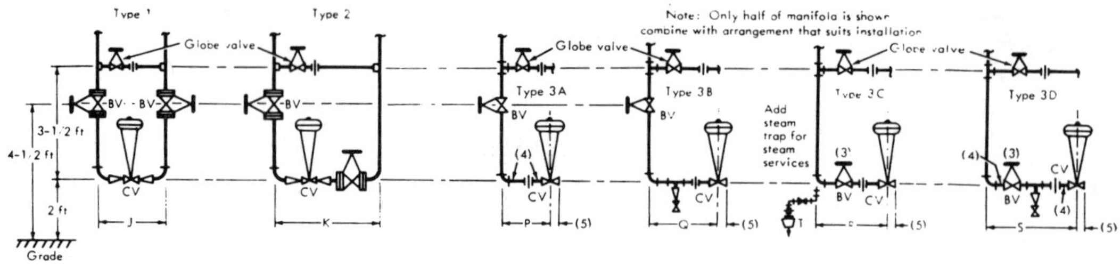

Valve sizes, in.		Dimensions, in., for flanged block valve, screwed control valve				Dimensions, in., for screwed block and control valves (and manifold type)														
			K, for ratings, psi, of:																	
BV	CV	J	150	300	600	P	Q	R	S	2P (1)	2Q (1)	2R (3)	2S (3)	P+Q (1)	P+R (2)	P+S (2)	Q+R (2)	Q+S (2)	R+S (3)
¾	½	—	—	—	—	11 1/16	16 15/16	17 7/16	23 5/16	22⅛	33⅞	34⅞	46⅝	28	28½	34⅜	34⅜	40¼	40¾
1	½	—	—	—	—	11⅝	17¾	18⅜	24½	23¼	35½	36¾	49	29⅜	30	36⅛	36⅛	42¼	42⅞
1	¾	—	—	—	—	11 9/16	17 11/16	18 5/16	24 7/16	23⅛	35⅜	36⅝	48⅞	29¼	29⅞	36	36	42⅛	42¾
1½	¾	24	36	37	40	13¾	21⅛	20⅞	28¼	27½	42¼	41¾	56½	34⅞	34⅝	42	42	49⅜	49⅛
1½	1	24	36	37	40	14 7/16	21 13/16	21 9/16	28 13/16	28⅞	43⅝	43⅛	57⅞	36¼	36	43⅜	43⅜	50¾	50½
2	1	24¼	36⅜	38⅜	42¼	16¾	24¼	24½	32	33½	48½	49	64	41	41¼	48¾	48¾	56¼	56½
2	1½	25⅛	37¼	38¼	43⅛	17 3/16	24 11/16	24 15/16	32⅞	34⅜	49⅞	49⅞	64⅞	41⅞	42⅛	49⅝	49⅝	57⅛	57⅞

Basic dimensions, in., used to establish dimensions P through S																			
Pipe size of item	3000 psi ell and tee, c/l to end	3000 psi union, end to end	600 psi gate valve, end to end[3]	4 in. nipple less threads[4]	Control valve, c/l to end[5]	Swaged nipple less engagement for small end size of:				Pipe nipples less thread engagement, in., for nipple length, in., of:									
						½	¾	1	1½	1½	2	2½	3	3½	4	6	8	10
½	—	—	—	—	2½	—	—	—	—	½	1	1½	2	2½	3	5	7	9
¾	1½	2¼	3½	2⅞	2½	1 5/16	—	—	—	⅜	⅞	1⅜	1⅞	2¾	2⅞	4⅞	6⅞	8⅞
1	1¾	2 7/16	4⅛	2⅝	3 5/16	2 5/16	2¼	—	—	—	⅝	1⅛	1⅝	2⅛	2⅝	4⅜	6⅝	8⅝
1½	2⅜	3	4½	2⅝	3¾	—	3¼	3⅛	—	—	⅝	1⅛	1⅝	2⅛	2⅝	4⅜	6⅝	8⅝
2	2½	3⅜	5¼	2½	4½	—	—	5 1/16	5 1/16	—	—	1	1½	2	2½	4½	6½	8½

[3,4,5] Refer to drawings above; also, for Note 5, see control valve manufacturer's catalog for exact end-to-end dimension.

Figure 34. DIMENSIONS FOR SCREWED CONTROL VALVE MANIFOLDS.

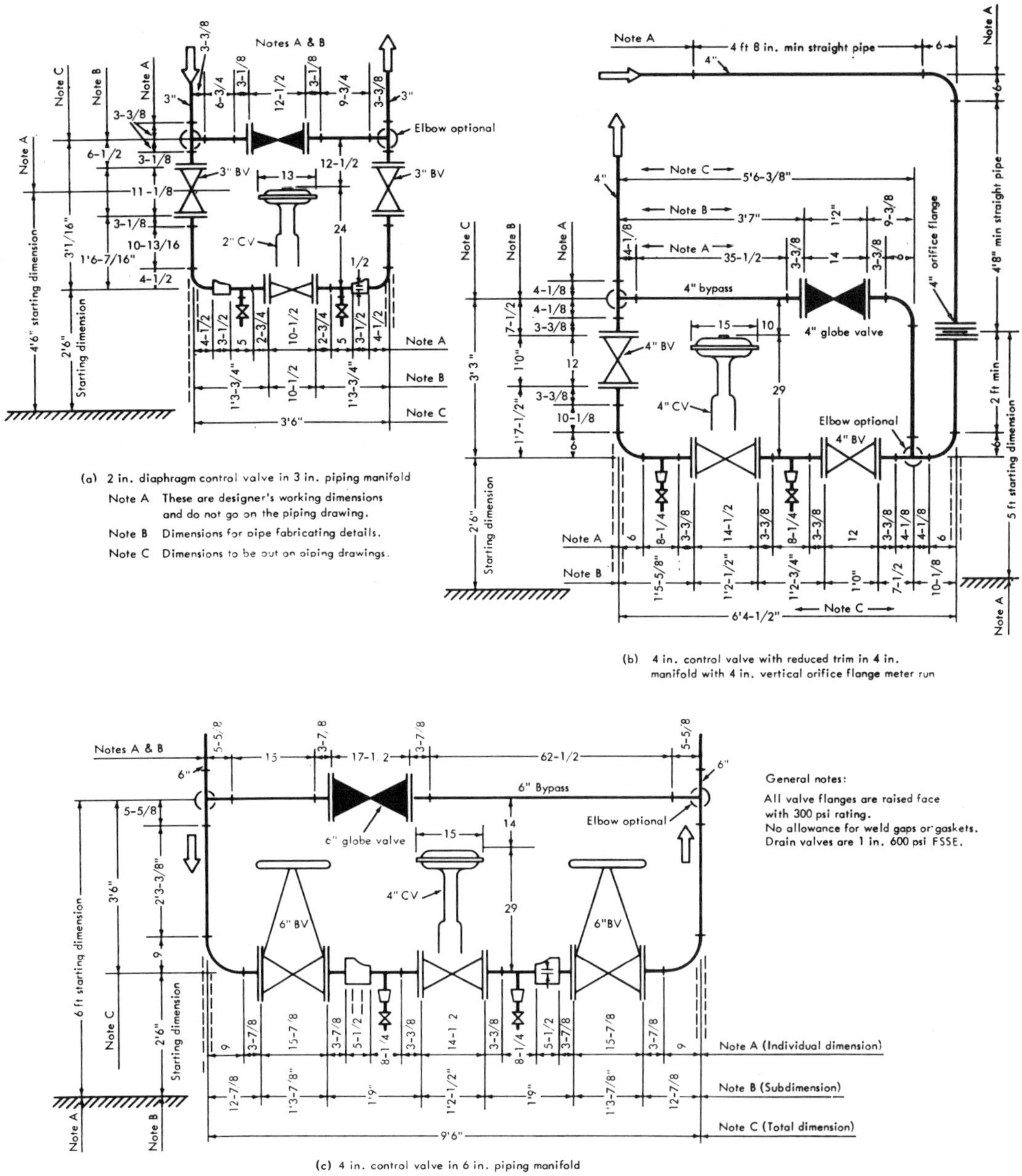

(a) 2 in. diaphragm control valve in 3 in. piping manifold

Note A These are designer's working dimensions and do not go on the piping drawing.

Note B Dimensions for pipe fabricating details.

Note C Dimensions to be put on piping drawings.

(b) 4 in. control valve with reduced trim in 4 in. manifold with 4 in. vertical orifice flange meter run

General notes:

All valve flanges are raised face with 300 psi rating.

No allowance for weld gaps or gaskets.

Drain valves are 1 in. 600 psi FSSE.

(c) 4 in. control valve in 6 in. piping manifold

Figure 35. METHODS FOR DIMENSIONING CONTROL VALVE MANIFOLD PIPING. ALL UNDESIGNATED DIMENSIONS ARE IN INCHES.

APPENDIX II
HELPFUL HINTS ON HANDLING TOOLS AND EQUIPMENT

The following piping hints and diagrams were reproduced by permission from Crane Company's Bulletin VC-1006-A, *Piping Pointers.*

They are included as a helpful supplement since it is felt that in many cases, field problems arise because of failure to observe such simple and obvious practices and procedures.

know your wrenches

easy does it with the proper tools

Good handling of piping materials begins with knowing the proper tools to use and how to use them. Wrenches are important tools on every piping job. When used improperly, however, they can cause considerable damage. Two simple rules regarding wrenches can keep you out of trouble:

Rule 1: choose the right type.
Rule 2: choose the right size.

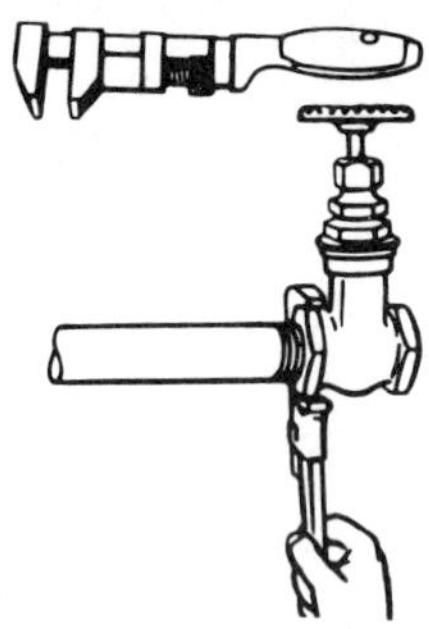

hex and monkey wrench

Your hexagonal or monkey wrench with its smooth, square jaws—is always your best bet for hex end valves and fittings. It gives a better fit on the part to be turned, and doesn't have the crushing effect of a pipe wrench. Because of the shape of their jaws, hex or monkey wrenches obviously cannot be used on pipe or other round objects.

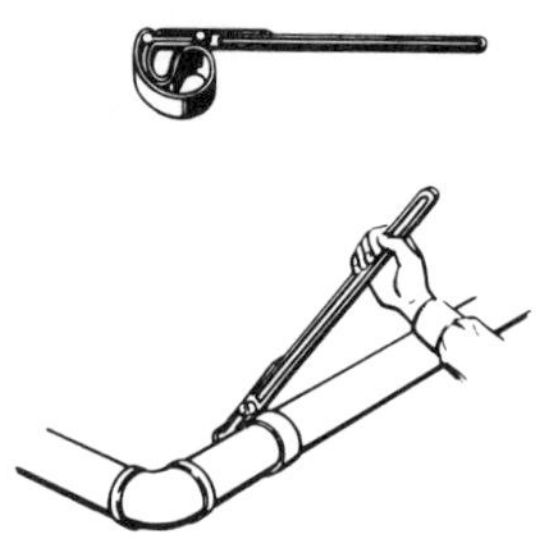

strap wrench

Strap wrenches are used mainly when working with plated or polished materials to avoid marring the surface. They also come in handy in tight places where you can't get in with a pipe wrench.

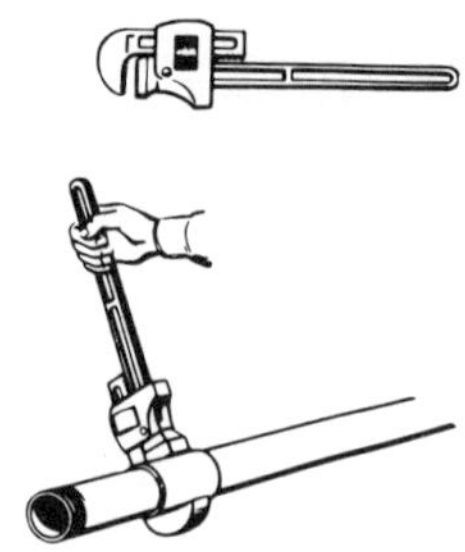

pipe wrench

Designed for use on pipe and screwed end fittings only. Not as efficient as a monkey wrench on hexagonal ends. The harder you pull a pipe wrench, the tighter it squeezes. This squeezing action can do great damage to a valve. It may cripple or even crush its body.

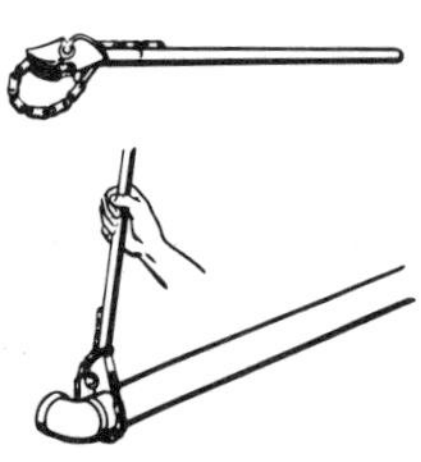

pipe tongs or chain pipe wrench

Pipe tongs are wrenches of the chain strap and lever type. They are generally used for larger sizes of pipe—about 3-inches and above—although they are available for sizes down to 1/8-inch.

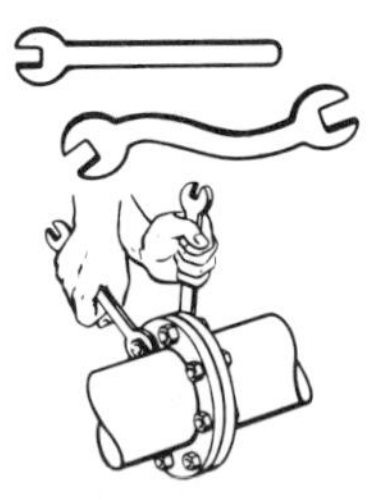

open end wrenches

Open end or socket wrenches are most suitable and fastest to work with when pulling up flange bolts and nuts. Getting the right size is the most important consideration here. Slippage of the wrench can cause bruised knuckles and can wear nut and bolt heads round.

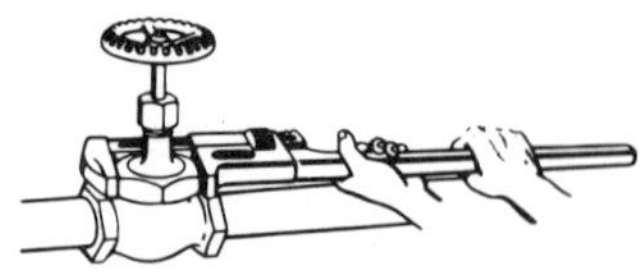

too much leverage is dangerous

Avoid bearing down on the joint with an oversize wrench or a wrench with an extension handle. It often gives you more leverage than you need, and may result in pulling up a joint so tight that the fitting will be cracked, the valve twisted out of shape, or the pipe run clear into the seats.

how to make up a screwed joint

The screwed joint (threaded male and female ends screwed together) is the most common method of joining pipe. A high degree of standardization is maintained in the threading of piping materials. American Pipe Thread Standards guide the piping industry.

The chart below shows the normal engagement male and female threads for making up screwed joints.

use a wire brush

Since dirt can make a good fit impossible, the first thing to do is to wipe both male and female threads clean. Wire brushing is recommended, especially if threaded pipe has been exposed to weather. Running a tap or die over the threads will usually straighten any that may be damaged and will help to assure a good metal-to-metal joint.

lubricant is good - but not inside a line

Next, use a bit of good thread lubricant or "pipe dope," but put it on the *male threads only.* This prevents any excess from squeezing into the pipe and causing harm to valve seats or other mechanism. Lubricant reduces friction when pulling up a joint. It is not expected to seal the joint, or to compensate for poor workmanship.

With lubricant applied, start the joint by hand. If the thread engagement feels "right," turn it up as far as it will go. If the threading job is up to standard . . . and if the threads are free of dirt . . . the joint is now started right. Further turns with a wrench will complete the job.

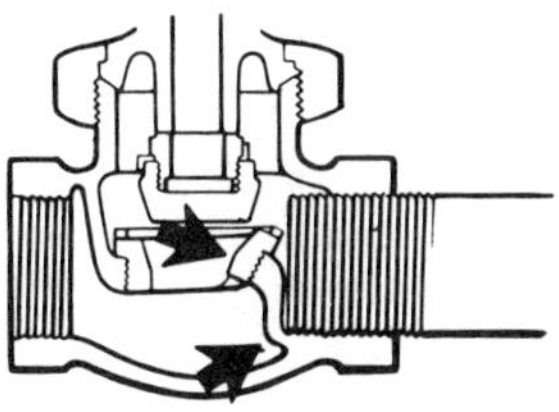

pipe run in too far

But be careful with the wrench. Never use an oversize one that will tempt you to "lean on the joint." Too much pull up can cause damage, especially in the case of valves. Don't try to run all the male threads into the joint. The lead of the die always leaves a few imperfect and unusable male threads which should be left exposed.

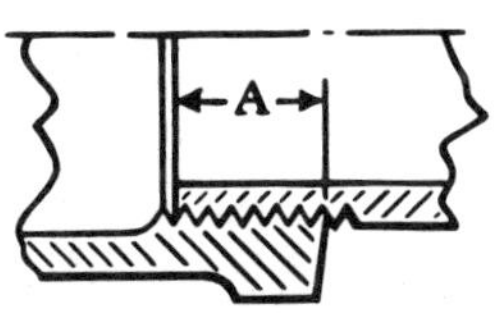

American Standard and A.P.I. line pipe threads

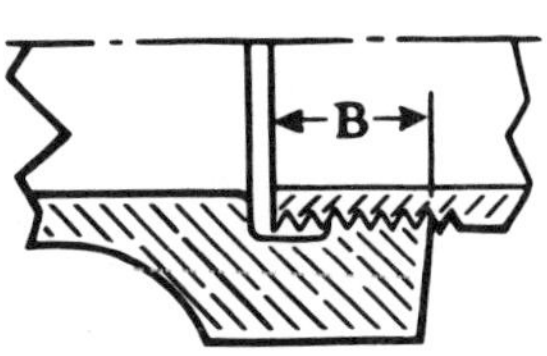

shoulder type drainage fitting treads

normal engagement between male and female threads to make tight joints

DIMENSIONS, IN INCHES

Dimensions given do not allow for variations in tapping or threading.

Size	1/8	1/4	3/8	1/2	3/4	1	1-1/4	1-1/2	2	2-1/2
A	1/4	3/8	3/8	1/2	9/16	11/16	11/16	11/16	3/4	15/16
B*						9/16	5/8	5/8	5/8	7/8

Size	3	3-1/2	4	5	6	8	10	12	14
A	1	1-1/16	1-1/8	1-1/4	1-5/16	1-7/16	1-5/8	1-3/4	
B*	15/16	1	1-1/16	1-3/16	1-1/4	1-3/8	1-9/16	1-11/16	1-7/8

*Using American Standard Taper Male Thread with Crane shoulder type Drainage Fittings. The external thread, however, should not be threaded small to gauge and not more than one turn large.

how to assemble a flanged joint

Flanged joints are made up by bolting two flanges together with a gasket between their machined faces to assure a tight seal. The proper steps in assembling a flanged joint are shown below.

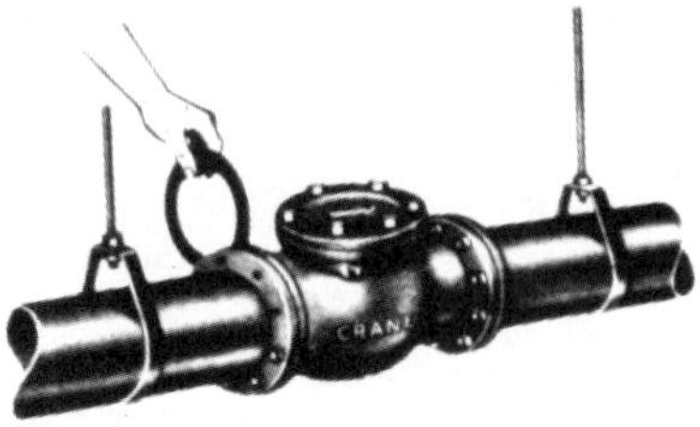

good!

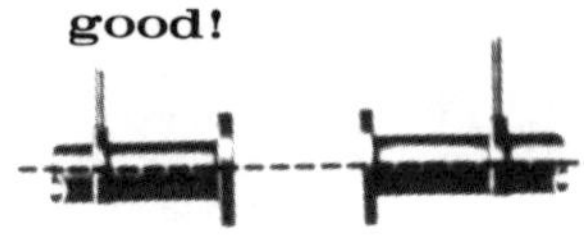

bad!

clean all parts

As with screwed joints, all parts should be clean to assure best results. Use a solvent-soaked rag to remove the rust-preventing grease put on flanges at the factory. Next, clean off all dirt and grit particles. Then wipe off the gasket just to be sure it is clean.

align and support pipe

When the pipe is in place, be sure it is properly supported. For example, a valve cannot hold up an unsupported length of pipe without undergoing great strain. The flanges should be accurately aligned—by checking with a spirit level, both horizontally along the pipe and vertically across the flange faces. It is then ready to bė bolted tight.

insert gasket

With the flanges secured in position, slip in half the bolts at the bottom. They will hold the gasket in place. Gaskets should be coated with a little graphite and oil or other recommended lubricant before they are inserted. They are then easier to remove if the joint is opened later. Finally, slip in the rest of the bolts, apply a little thread lubricant to each, and then turn up all nuts by hand, far as they'll go.

it's best to tighten bolts in this order

Now start pulling up with a wrench—not in rotation, but by the cross-over method to load the bolts evenly and eliminate any concentrated stresses. Keep repeating by going over and across until the joint is uniformly tight.

flange facings should be alike

Do not attempt to get a tight flanged joint with flange faces which are not alike. Steel valves and fittings normally are manufactured with raised flange faces. Piping materials in iron and bronze usually have plain or flat flange faces.

As shown, the proper procedure is to join a plain face to a plain face—or a raised face to a raised face. The proper gaskets will assure a tight joint. But never should a raised face be joined to a flat or plain face.

When joining 125-pound iron or any bronze flanges to steel flanges, be sure to remove the raised face on the steel flange. A full faced gasket is preferred for such a joint.

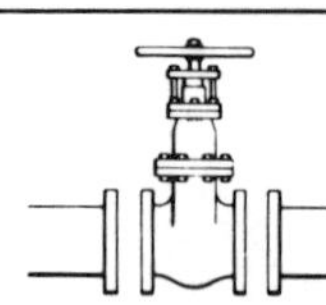

plain faces

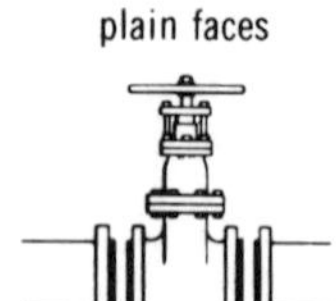

raised faces

APPENDIX

Numerous tables, charts, graphs, and standards are presented in the Appendix. It was impractical to include every item of information that an engineer may need; however, the attempt was made to select the most commonly referred-to data that would be required in applying the contents of the foregoing Chapters.

CONTENTS

ISA – RP4.1

1950

RECOMMENDED PRACTICE

UNIFORM FACE TO FACE DIMENSIONS FOR FLANGED CONTROL VALVE BODIES

Sponsor

INSTRUMENT SOCIETY of AMERICA
400 Stanwix Street
Pittsburgh, Pennsylvania 15222

Published September 1950
Reaffirmed September 1959
(With Tentative Status Removed)

Printed in the U.S.A.

FOREWORD

This Recommended Practice has been prepared as a part of the service of the Instrument Society of America toward a goal of uniformity in the field of instrumentation. To be of real value this report should not be static, but should be subjected to periodic review. Toward this end the Society welcomes all comments and criticisms, and asks that they be addressed to Secretary, Standards & Practice Board, Instrument Society of America, Penn Sheraton Hotel, 530 William Penn Place, Pittsburgh 19, Pennsylvania.

This report was prepared by the Sub-Committee on Control Valve and By-pass Manifolds (RP-4).

C. W. Bates, Chairman 1949-50–Humble Oil & Refining Co.
W. H. Fortney, Chairman 1947-49–Humble Oil & Refining Co.
M. K. Anderson–Carbide & Carbon Chemicals Corp.
H. C. Givens–LaGloria Corporation
J. L. Henderson–Champion Paper & Fibre Company
J. M. Jones, Jr.–The Texas Company
R. H. Koehler–Pan American Refining Company
E. D. Mattix–Cities Service Refining Corporation
J. A. Parker–Shell Chemical Corporation
R. R. Proctor–The Pure Oil Company
H. T. Wicker–Standard Oil Company

The assistance of the Fluid Controls Institute in developing this Recommended Practice is gratefully acknowledged.

PURPOSE

This report is presented as a means of giving official endorsement to the control valve face-to-face dimensions shown herein. These dimensions were developed by the majority of control valve manufacturers with the assistance of the National Steam Specialty Club, when it was called to their attention as a result of the committee's initial questionnaire, that the user industry was 100% in agreement that uniform face to face dimensions for control valves were desirable. The adoption of these dimensions by all control valve manufacturers will allow piping design, where control valves are involved, without giving special consideration to the make of equipment to be used.

SCOPE

The sizes shown are those most widely used throughout industry. As indicated in the title only flanged valves are covered. The flange ratings are those most frequently used in industry. Only valves which are in the category of diaphragm actuated control valves of the plug type are included.

UNIFORM FACE TO FACE DIMENSIONS FOR FLANGED CONTROL VALVE BODIES

PIPE SIZE INCHES	125 PSI IRON 150 PSI STEEL SEE NOTE 1	250 PSI IRON 300 PSI STEEL SEE NOTE 2	600 PSI STEEL SEE NOTE 3
½	....	7½	8
¾	....	7⅝	8⅛
1	7¼	7¾	8¼
1½	8¾	9¼	9⅞
2	10	10½	11¼
2½	10⅞	11½	12¼
3	11¾	12½	13¼
4	13⅞	14½	15½
6	17¾	18⅝	20
8	21⅜	22⅜	24

NOTE 1—Cast iron bodies with 125 PSI ASA B16A-1939 flanges, and steel bodies with 150 PSI raised face ASA-B16E-1939 flanges.
NOTE 2—Cast iron bodies with 250 PSI ASA B16B-1944 flanges, and steel bodies with 300 PSI raised face B16E-1939 flanges.
NOTE 3—Steel bodies with 600 PSI raised face ASA B16E-1939 flanges.

AMERICAN PIPE FLANGE DIMENSIONS
Diameter of Bolt Circle - Inches

Nominal Pipe Size	125 lb. Cast Iron	ANSI 150 lb.	250 lb. Cast Iron	ANSI 300 lb.	ANSI 400 lb.	ANSI 600 lb.	ANSI 900 lb.	ANSI 1500 lb.	ANSI 2500 lb.
1	3-1/8	3-1/8	3-1/2	3-1/2	3-1/2	3-1/2	4	4	4-1/4
1-1/4	3-1/2	3-1/2	3-7/8	3-7/8	3-7/8	3-7/8	4-3/8	4-3/8	5-1/8
1-1/2	3-7/8	3-7/8	4-1/2	4-1/2	4-1/2	4-1/2	4-7/8	4-7/8	5-3/4
2	4-3/4	4-3/4	5	5	5	5	6-1/2	6-1/2	6-3/4
2-1/2	5-1/2	5-1/2	5-7/8	5-7/8	5-7/8	5-7/8	7-1/2	7-1/2	7-3/4
3	6	6	6-5/8	6-5/8	6-5/8	6-5/8	7-1/2	8	9
4	7-1/2	7-1/2	7-7/8	7-7/8	7-7/8	8-1/2	9-1/4	9-1/2	10-3/4
5	8-1/2	8-1/2	9-1/4	9-1/4	9-1/4	10-1/2	11	11-1/2	12-3/4
6	9-1/2	9-1/2	10-5/8	10-5/8	10-5/8	11-1/2	12-1/2	12-1/2	14-1/2
8	11-3/4	11-3/4	13	13	13	13-3/4	15-1/2	15-1/2	17-1/4
10	14-1/4	14-1/4	15-1/4	15-1/4	15-1/4	17	18-1/2	19	21-1/4
12	17	17	17-3/4	17-3/4	17-3/4	19-1/4	21	22-1/2	24-3/8
16	21-1/4	21-1/4	22-1/2	22-1/2	22-1/2	23-3/4	24-1/4	27-3/4	-
18	22-3/4	22-3/4	24-3/4	24-3/4	24-3/4	25-3/4	27	30-1/2	-
20	25	25	27	27	27	28-1/2	29-1/2	32-3/4	-
24	29-1/2	29-1/2	32	32	32	33	35-1/2	39	-

AMERICAN PIPE FLANGE DIMENSIONS
Number and Diameter of Stud Bolts - Inches

Nominal Pipe Size	125 lb. Cast Iron		ANSI 150 lb.		250 lb. Cast Iron		ANSI 300 lb.		ANSI 400 lb.		ANSI 600 lb.		ANSI 900 lb.		ANSI 1500 lb.		ANSI 2500 lb.	
	No.	Dia.	No.	Dia.	No.	Dia.	No.	Dia.	No.	Dia.	No.	Dia.	No.	Dia.	No.	Dia.	No.	Dia.
1	4	1/2	4	1/2	4	5/8	4	5/8	4	5/8	4	5/8	4	7/8	4	7/8	4	7/8
1-1/4	4	1/2	4	1/2	4	5/8	4	5/8	4	5/8	4	5/8	4	7/8	4	7/8	4	1
1-1/2	4	1/2	4	1/2	4	3/4	4	3/4	4	3/4	4	3/4	4	1	4	1	4	1-1/8
2	4	5/8	4	5/8	8	5/8	8	5/8	8	5/8	8	5/8	8	7/8	8	7/8	8	1
2-1/2	4	5/8	4	5/8	8	3/4	8	3/4	8	3/4	8	3/4	8	1	8	1	8	1-1/8
3	4	5/8	4	5/8	8	3/4	8	3/4	8	3/4	8	3/4	8	7/8	8	1-1/8	8	1-1/4
4	8	5/8	8	5/8	8	3/4	8	3/4	8	7/8	8	7/8	8	1-1/8	8	1-1/4	8	1-1/2
5	8	3/4	8	3/4	8	3/4	8	3/4	8	7/8	8	1	8	1-1/4	8	1-1/2	8	1-3/4
6	8	3/4	8	3/4	12	3/4	12	3/4	12	7/8	12	1	12	1-1/8	12	1-3/8	8	2
8	8	3/4	8	3/4	12	7/8	12	7/8	12	1	12	1-1/8	12	1-3/8	12	1-5/8	12	2
10	12	7/8	12	7/8	16	1	16	1	16	1-1/8	16	1-1/4	16	1-3/8	12	1-7/8	12	2-1/2
12	12	7/8	12	7/8	16	1-1/8	16	1-1/8	16	1-1/4	20	1-1/4	20	1-3/8	16	2	12	2-3/4
16	16	1	16	1	20	1-1/4	20	1-1/4	20	1-3/8	20	1-1/2	20	1-5/8	16	2-1/2	-	-
18	16	1-1/8	16	1-1/8	24	1-1/4	24	1-1/4	24	1-3/8	20	1-5/8	20	1-7/8	16	2-3/4	-	-
20	20	1-1/8	20	1-1/8	24	1-1/4	24	1-1/4	24	1-1/2	24	1-5/8	20	2	16	3	-	-
24	20	1-1/4	20	1-1/4	24	1-1/2	24	1-1/2	24	1-3/4	24	1-7/8	20	2-1/2	16	3-1/2	-	-

Reprinted from "Control Valve Handbook," 1965, Fisher Controls Company, Marshalltown, Iowa.

AMERICAN PIPE FLANGE DIMENSIONS
Flange Thickness - Inches

Nominal Pipe Size	125 lb. Cast Iron	ANSI 150 lb.		250 lb. Cast Iron	ANSI 300 lb.		ANSI 400 lb.		ANSI 600 lb.		ANSI 900 lb.		ANSI 1500 lb.		ANSI 2500 lb.	
	FF	RF	RTJ	RF	RF	RTJ	RF	RTJ	RF	RTJ	RF	RTJ	RF	RTJ	RF	RTJ
1	7/16	7/16	11/16	11/16	11/16	15/16	15/16	15/16	15/16	15/16	1-3/8	1-3/8	1-3/8	1-3/8	1-5/8	1-5/8
1-1/4	1/2	1/2	3/4	3/4	3/4	1	1-1/16	1-1/16	1-1/16	1-1/16	1-3/8	1-3/8	1-3/8	1-3/8	1-3/4	1-13/16
1-1/2	9/16	9/16	13/16	13/16	13/16	1-1/16	1-1/8	1-1/8	1-1/8	1-1/8	1-1/2	1-1/2	1-1/2	1-1/2	2	2-1/16
2	5/8	5/8	7/8	7/8	7/8	1-3/16	1-1/4	1-5/16	1-1/4	1-5/16	1-3/4	1-13/16	1-3/4	1-13/16	2-1/4	2-5/16
2-1/2	11/16	11/16	15/16	1	1	1-5/16	1-3/8	1-7/16	1-3/8	1-7/16	1-7/8	1-15/16	1-7/8	1-15/16	2-1/2	2-5/8
3	3/4	3/4	1	1-1/8	1-1/8	1-7/16	1-1/2	1-9/16	1-1/2	1-9/16	1-3/4	1-13/16	2-1/8	2-3/16	2-7/8	3
4	15/16	15/16	1-3/16	1-1/4	1-1/4	1-9/16	1-5/8	1-11/16	1-3/4	1-13/16	2	2-1/16	2-3/8	2-7/16	3-1/4	3-7/16
5	15/16	15/16	1-3/16	1-3/8	1-3/8	1-11/16	1-3/4	1-13/16	2	2-1/16	2-1/4	2-5/16	3-1/8	3-3/16	3-7/8	4-1/8
6	1	1	1-1/4	1-7/16	1-7/16	1-3/4	1-7/8	1-15/16	2-1/8	2-3/16	2-7/16	2-1/2	3-1/2	3-5/8	4-1/2	4-3/4
8	1-1/8	1-1/8	1-3/8	1-5/8	1-5/8	1-15/16	2-1/8	2-3/16	2-7/16	2-1/2	2-3/4	2-13/16	3-7/8	4-1/16	5-1/4	5-9/16
10	1-3/16	1-3/16	1-7/16	1-7/8	1-7/8	2-3/16	2-3/8	2-7/16	2-3/4	2-13/16	3	3-1/16	4-1/2	4-11/16	6-3/4	7-3/16
12	1-1/4	1-1/4	1-1/2	2	2	2-5/16	2-1/2	2-9/16	2-7/8	2-15/16	3-3/8	3-7/16	5-1/8	5-7/16	7-1/2	7-15/16
16	1-7/16	1-7/16	1-11/16	2-1/4	2-1/4	2-9/16	2-3/4	2-13/16	3-1/4	3-5/16	3-3/4	3-15/16	6	6-7/16	-	-
18	1-9/16	1-9/16	1-13/16	2-3/8	2-3/8	2-11/16	2-7/8	2-15/16	3-1/2	3-9/16	4-1/4	4-1/2	6-5/8	7-1/16	-	-
20	1-11/16	1-11/16	1-15/16	2-1/2	2-1/2	2-7/8	3	3-1/8	3-3/4	3-7/8	4-1/2	4-3/4	7-1/4	7-11/16	-	-
24	1-7/8	1-7/8	2-1/8	2-3/4	2-3/4	3-3/16	3-1/4	3-7/16	4-1/4	4-7/16	5-3/4	6-1/8	8-1/4	8-13/16	-	-

Abbreviations Used: FF = Flat Face; RF = Raised Face; RTJ = Ring Type Joint.

AMERICAN PIPE FLANGE DIMENSIONS
Flange Diameter - Inches

Nominal Pipe Size	125 lb. Cast Iron	ANSI 150 lb.	250 lb. Cast Iron	ANSI 300 lb.	ANSI 400 lb.	ANSI 600 lb.	ANSI 900 lb.	ANSI 1500 lb.	ANSI 2500 lb.
1	4-1/4	4-1/4	4-7/8	4-7/8	4-7/8	4-7/8	5-7/8	5-7/8	6-1/4
1-1/4	4-5/8	4-5/8	5-1/4	5-1/4	5-1/4	5-1/4	6-1/4	6-1/4	7-1/4
1-1/2	5	5	6-1/8	6-1/8	6-1/8	6-1/8	7	7	8
2	6	6	6-1/2	6-1/2	6-1/2	6-1/2	8-1/2	8-1/2	9-1/4
2-1/2	7	7	7-1/2	7-1/2	7-1/2	7-1/2	9-5/8	9-5/8	10-1/2
3	7-1/2	7-1/2	8-1/4	8-1/4	8-1/4	8-1/4	9-1/2	10-1/2	12
4	9	9	10	10	10	10-3/4	11-1/2	12-1/4	14
5	10	10	11	11	11	13	13-3/4	14-3/4	16-1/2
6	11	11	12-1/2	12-1/2	12-1/2	14	15	15-1/2	19
8	13-1/2	13-1/2	15	15	15	16-1/2	18-1/2	19	21-3/4
10	16	16	17-1/2	17-1/2	17-1/2	20	21-1/2	23	26-1/2
12	19	19	20-1/2	20-1/2	20-1/2	22	24	26-1/2	30
16	23-1/2	23-1/2	25-1/2	25-1/2	25-1/2	27	27-3/4	32-1/2	-
18	25	25	28	28	28	29-1/4	31	36	-
20	27-1/2	27-1/2	30-1/2	30-1/2	30-1/2	32	33-3/4	38-3/4	-
24	32	32	36	36	36	37	41	46	-

Reprinted from "Control Valve Handbook," 1965, Fisher Controls Company, Marshalltown, Iowa.

Commercial Wrought Steel Pipe Data

Schedule Wall Thickness—Per ASA B36.10-1950*

Schedule	Nominal Pipe Size	Outside Diameter	Thickness	Inside Diameter		Inside Diameter Functions (In Inches)				Transverse Internal Area	
				d	D	d^2	d^3	d^4	d^5	a	A
	Inches	Inches	Inches	Inches	Feet					Sq. In.	Sq. Ft.
Schedule 10	14	14	0.210	13.58	1.132	184.42	2504.4	34010.	461856.	144.75	1.005
	16	16	0.250	15.5	1.291	240.25	3723.9	57720.	894660.	188.69	1.310
	18	18	0.250	17.5	1.4583	306.25	5359.4	93789.	1641309.	240.53	1.670
	20	20	0.250	19.5	1.625	380.25	7414.9	144590.	2819500.	298.65	2.074
	24	24	0.250	23.5	1.958	552.25	12977.	304980.	7167030.	433.74	3.012
	30	30	0.312	29.376	2.448	862.95	25350.	744288.	21864218.	677.76	4.707
Schedule 20	8	8.625	0.250	8.125	0.6771	66.02	536.38	4359.3	35409.	51.85	0.3601
	10	10.75	0.250	10.25	0.8542	105.06	1076.9	11038.	113141.	82.52	0.5731
	12	12.75	0.250	12.25	1.021	150.06	1838.3	22518.	275855.	117.86	0.8185
	14	14	0.312	13.376	1.111	178.92	2393.2	32012.	428185.	140.52	0.9758
	16	16	0.312	15.376	1.281	236.42	3635.2	55894.	859442.	185.69	1.290
	18	18	0.312	17.376	1.448	301.92	5246.3	91156.	1583978.	237.13	1.647
	20	20	0.375	19.250	1.604	370.56	7133.3	137317.	2643352.	291.04	2.021
	24	24	0.375	23.25	1.937	540.56	12568.	292205.	6793832.	424.56	2.948
	30	30	0.500	29.00	2.417	841.0	24389.	707281.	20511149.	660.52	4.587
Schedule 30	8	8.625	0.277	8.071	0.6726	65.14	525.75	4243.2	34248.	51.16	0.3553
	10	10.75	0.307	10.136	0.8447	102.74	1041.4	10555.	106987.	80.69	0.5603
	12	12.75	0.330	12.09	1.0075	146.17	1767.2	21366.	258304.	114.80	0.7972
	14	14	0.375	13.25	1.1042	175.56	2326.2	30821.	408394.	137.88	0.9575
	16	16	0.375	15.25	1.2708	232.56	3546.6	54084.	824801.	182.65	1.268
	18	18	0.438	17.124	1.4270	293.23	5021.3	85984.	1472397.	230.30	1.599
	20	20	0.500	19.00	1.5833	361.00	6859.0	130321.	2476099.	283.53	1.969
	24	24	0.562	22.876	1.9063	523.31	11971.	273853.	6264703.	411.00	2.854
	30	30	0.625	28.75	2.3958	826.56	23764.	683201.	19642160.	649.18	4.508
Schedule 40	1/8	0.405	0.068	0.269	0.0224	0.0724	0.0195	0.005242	0.00141	0.057	0.00040
	1/4	0.540	0.088	0.364	0.0303	0.1325	0.0482	0.01756	0.00639	0.104	0.00072
	3/8	0.675	0.091	0.493	0.0411	0.2430	0.1198	0.05905	0.02912	0.191	0.00133
	1/2	0.840	0.109	0.622	0.0518	0.3869	0.2406	0.1497	0.09310	0.304	0.00211
	3/4	1.050	0.113	0.824	0.0687	0.679	0.5595	0.4610	0.3799	0.533	0.00371
	1	1.315	0.133	1.049	0.0874	1.100	1.154	1.210	1.270	0.864	0.00600
	1 1/4	1.660	0.140	1.380	0.1150	1.904	2.628	3.625	5.005	1.495	0.01040
	1 1/2	1.900	0.145	1.610	0.1342	2.592	4.173	6.718	10.82	2.036	0.01414
	2	2.375	0.154	2.067	0.1722	4.272	8.831	18.250	37.72	3.355	0.02330
	2 1/2	2.875	0.203	2.469	0.2057	6.096	15.051	37.161	91.75	4.788	0.03322
	3	3.500	0.216	3.068	0.2557	9.413	28.878	88.605	271.8	7.393	0.05130
	3 1/2	4.000	0.226	3.548	0.2957	12.59	44.663	158.51	562.2	9.886	0.06870
	4	4.500	0.237	4.026	0.3355	16.21	65.256	262.76	1058.	12.730	0.08840
	5	5.563	0.258	5.047	0.4206	25.47	128.56	648.72	3275.	20.006	0.1390
	6	6.625	0.280	6.065	0.5054	36.78	223.10	1352.8	8206.	28.891	0.2006
	8	8.625	0.322	7.981	0.6651	63.70	508.36	4057.7	32380.	50.027	0.3474
	10	10.75	0.365	10.02	0.8350	100.4	1006.0	10080.	101000.	78.855	0.5475
	12	12.75	0.406	11.938	0.9965	142.5	1701.3	20306.	242470.	111.93	0.7773
	14	14.0	0.438	13.124	1.0937	172.24	2260.5	29666.	389340.	135.28	0.9394
	16	16.0	0.500	15.000	1.250	225.0	3375.0	50625.	759375.	176.72	1.2272
	18	18.0	0.562	16.876	1.4063	284.8	4806.3	81111.	1368820.	223.68	1.5533
	20	20.0	0.593	18.814	1.5678	354.0	6659.5	125320.	2357244.	278.00	1.9305
	24	24.0	0.687	22.626	1.8855	511.9	11583.	262040.	5929784.	402.07	2.7921
Schedule 60	8	8.625	0.406	7.813	0.6511	61.04	476.93	3725.9	29113.	47.94	0.3329
	10	10.75	0.500	9.750	0.8125	95.06	926.86	9036.4	88110.	74.66	0.5185
	12	12.75	0.562	11.626	0.9688	135.16	1571.4	18268.	212399.	106.16	0.7372
	14	14.0	0.593	12.814	1.0678	164.20	2104.0	26962.	345480.	128.96	0.8956
	16	16.0	0.656	14.688	1.2240	215.74	3168.8	46544.	683618.	169.44	1.1766
	18	18.0	0.750	16.500	1.3750	272.25	4492.1	74120.	1222982.	213.83	1.4849
	20	20.0	0.812	18.376	1.5313	337.68	6205.2	114028.	2095342.	265.21	1.8417
	24	24.0	0.968	22.064	1.8387	486.82	10741.	236994.	5229036.	382.35	2.6552
Schedule 80	1/8	0.405	0.095	0.215	0.0179	0.0462	0.00994	0.002134	0.000459	0.036	0.00025
	1/4	0.540	0.119	0.302	0.0252	0.0912	0.0275	0.008317	0.002513	0.072	0.00050
	3/8	0.675	0.126	0.423	0.0353	0.1789	0.0757	0.03200	0.01354	0.141	0.00098
	1/2	0.840	0.147	0.546	0.0455	0.2981	0.1628	0.08886	0.04852	0.234	0.00163
	3/4	1.050	0.154	0.742	0.0618	0.5506	0.4085	0.3032	0.2249	0.433	0.00300
	1	1.315	0.179	0.957	0.0797	0.9158	0.8765	0.8387	0.8027	0.719	0.00499
	1 1/4	1.660	0.191	1.278	0.1065	1.633	2.087	2.6667	3.409	1.283	0.00891

(continued on the next page)

Courtesy Crane Co., Reproduced from Technical Paper No. 410, "Flow of Fluids."

*The latest revision of this standard is ANSI B36.10-1970; therefore, the schedule wall thickness used to derive the given data may vary slightly from the current standard in some cases.

Commercial Wrought Steel Pipe Data

Schedule Wall Thickness—Per ASA B36.10-1950 *

Schedule	Nominal Pipe Size (Inches)	Outside Diameter (Inches)	Thickness (Inches)	Inside Diameter d (Inches)	Inside Diameter D (Feet)	d^2	d^3	d^4	d^5	Transverse Internal Area a (Sq. In.)	Transverse Internal Area A (Sq. Ft.)
Schedule 80—cont.	1½	1.900	0.200	1.500	0.1250	2.250	3.375	5.062	7.594	1.767	0.01225
	2	2.375	0.218	1.939	0.1616	3.760	7.290	14.136	27.41	2.953	0.02050
	2½	2.875	0.276	2.323	0.1936	5.396	12.536	29.117	67.64	4.238	0.02942
	3	3.5	0.300	2.900	0.2417	8.410	24.389	70.728	205.1	6.605	0.04587
	3½	4.0	0.318	3.364	0.2803	11.32	38.069	128.14	430.8	8.888	0.06170
	4	4.5	0.337	3.826	0.3188	14.64	56.006	214.33	819.8	11.497	0.07986
	5	5.563	0.375	4.813	0.4011	23.16	111.49	536.38	2583.	18.194	0.1263
	6	6.625	0.432	5.761	0.4801	33.19	191.20	1101.6	6346.	26.067	0.1810
	8	8.625	0.500	7.625	0.6354	58.14	443.32	3380.3	25775.	45.663	0.3171
	10	10.75	0.593	9.564	0.7970	91.47	874.82	8366.8	80020.	71.84	0.4989
	12	12.75	0.687	11.376	0.9480	129.41	1472.2	16747.	190523.	101.64	0.7058
	14	14.0	0.750	12.500	1.0417	156.25	1953.1	24414.	305176.	122.72	0.8522
	16	16.0	0.843	14.314	1.1928	204.89	2932.8	41980.	600904.	160.92	1.1175
	18	18.0	0.937	16.126	1.3438	260.05	4193.5	67626.	1090518.	204.24	1.4183
	20	20.0	1.031	17.938	1.4948	321.77	5771.9	103536.	1857248.	252.72	1.7550
	24	24.0	1.218	21.564	1.7970	465.01	10027.	216234.	4662798.	365.22	2.5362
Schedule 100	8	8.625	0.593	7.439	0.6199	55.34	411.66	3062.	22781.	43.46	0.3018
	10	10.75	0.718	9.314	0.7762	86.75	807.99	7526.	69357.	68.13	0.4732
	12	12.75	0.843	11.064	0.9220	122.41	1354.4	14985.	165791.	96.14	0.6677
	14	14.0	0.937	12.126	1.0105	147.04	1783.0	21621.	262173.	115.49	0.8020
	16	16.0	1.031	13.938	1.1615	194.27	2707.7	37740.	526020.	152.58	1.0596
	18	18.0	1.156	15.688	1.3057	246.11	3861.0	60572.	950250.	193.30	1.3423
	20	20.0	1.281	17.438	1.4532	304.08	5302.6	92467.	1612438.	238.83	1.6585
	24	24.0	1.531	20.938	1.7448	438.40	9179.2	192195.	4024179.	344.32	2.3911
Schedule 120	4	4.50	0.438	3.624	0.302	13.133	47.595	172.49	625.1	10.315	0.07163
	5	5.563	0.500	4.563	0.3802	20.82	95.006	433.5	1978.	16.35	0.1136
	6	6.625	0.562	5.501	0.4584	30.26	166.47	915.7	5037.	23.77	0.1650
	8	8.625	0.718	7.189	0.5991	51.68	371.54	2671.	19202.	40.59	0.2819
	10	10.75	0.843	9.064	0.7553	82.16	744.66	6750.	61179.	64.53	0.4481
	12	12.75	1.000	10.750	0.8959	115.56	1242.3	13355.	143563.	90.76	0.6303
	14	14.0	1.093	11.814	0.9845	139.57	1648.9	19480.	230137.	109.62	0.7612
	16	16.0	1.218	13.564	1.1303	183.98	2495.5	33849.	459133.	144.50	1.0035
	18	18.0	1.375	15.250	1.2708	232.56	3546.6	54086.	824804.	182.66	1.2684
	20	20.0	1.500	17.000	1.4166	289.00	4913.0	83521.	1419857.	226.98	1.5762
	24	24.0	1.812	20.376	1.6980	415.18	8459.7	172375.	3512313.	326.08	2.2645
Schedule 140	8	8.625	0.812	7.001	0.5834	49.01	343.15	2402.	16819.	38.50	0.2673
	10	10.75	1.000	8.750	0.7292	76.56	669.92	5862.	51291.	60.13	0.4176
	12	12.75	1.125	10.500	0.8750	110.25	1157.6	12155.	127628.	86.59	0.6013
	14	14.0	1.250	11.500	0.9583	132.25	1520.9	17490.	201136.	103.87	0.7213
	16	16.0	1.438	13.124	1.0937	172.24	2260.5	29666.	389340.	135.28	0.9394
	18	18.0	1.562	14.876	1.2396	221.30	3292.0	48972.	728502.	173.80	1.2070
	20	20.0	1.750	16.5	1.3750	272.25	4492.1	74120.	1222981.	213.82	1.4849
	24	24.0	2.062	19.876	1.6563	395.06	7852.1	156069.	3102022.	310.28	2.1547
Schedule 160	½	0.840	0.187	0.466	0.0388	0.2172	0.1012	0.04716	0.02197	0.1706	0.00118
	¾	1.050	0.218	0.614	0.0512	0.3770	0.2315	0.1421	0.08726	0.2961	0.00206
	1	1.315	0.250	0.815	0.0679	0.6642	0.5413	0.4412	0.3596	0.5217	0.00362
	1¼	1.660	0.250	1.160	0.0966	1.346	1.561	1.811	2.100	1.057	0.00734
	1½	1.900	0.281	1.338	0.1115	1.790	2.395	3.205	4.288	1.406	0.00976
	2	2.375	0.343	1.689	0.1407	2.853	4.818	8.138	13.74	2.241	0.01556
	2½	2.875	0.375	2.125	0.1771	4.516	9.596	20.39	43.33	3.546	0.02463
	3	3.50	0.438	2.624	0.2187	6.885	18.067	47.41	124.4	5.408	0.03755
	4	4.50	0.531	3.438	0.2865	11.82	40.637	139.7	480.3	9.283	0.06447
	5	5.563	0.625	4.313	0.3594	18.60	80.230	346.0	1492.	14.61	0.1015
	6	6.625	0.718	5.189	0.4324	26.93	139.72	725.0	3762.	21.15	0.1469
	8	8.625	0.906	6.813	0.5677	46.42	316.24	2155.	14679.	36.46	0.2532
	10	10.75	1.125	8.500	0.7083	72.25	614.12	5220.	44371.	56.75	0.3941
	12	12.75	1.312	10.126	0.8438	102.54	1038.3	10514.	106461.	80.53	0.5592
	14	14.0	1.406	11.188	0.9323	125.17	1400.4	15668.	175292.	98.31	0.6827
	16	16.0	1.593	12.814	1.0678	164.20	2104.0	26961.	345482.	128.96	0.8956
	18	18.0	1.781	14.438	1.2032	208.45	3009.7	43454.	627387.	163.72	1.1369
	20	20.0	1.968	16.064	1.3387	258.05	4145.3	66590.	1069715.	202.67	1.4074
	24	24.0	2.343	19.314	1.6095	373.03	7204.7	139152.	2687582.	292.98	2.0346

Courtesy Crane Co., Reproduced from Technical Paper No. 410, "Flow of Fluids."

*The latest revision of this standard is ANSI B36.10-1970; therefore, the schedule wall thickness used to derive the given data may vary slightly from the current standard in some cases.

Commercial Wrought Steel Pipe Data

(Per ASA B36.10-1950) *

Nominal Pipe Size	Outside Diameter	Thickness	Inside Diameter		Inside Diameter Functions (In Inches)				Transverse Internal Area	
			d	D	d^2	d^3	d^4	d^5	a	A
Inches	Inches	Inches	Inches	Feet					Sq. In.	Sq. Ft.
Standard Wall Pipe										
1/8	0.405	0.068	0.269	0.0224	0.0724	0.0195	0.00524	0.00141	0.057	0.00040
1/4	0.540	0.088	0.364	0.0303	0.1325	0.0482	0.01756	0.00639	0.104	0.00072
3/8	0.675	0.091	0.493	0.0411	0.2430	0.1198	0.05905	0.02912	0.191	0.00133
1/2	0.840	0.109	0.622	0.0518	0.3869	0.2406	0.1497	0.0931	0.304	0.00211
3/4	1.050	0.113	0.824	0.0687	0.679	0.5595	0.4610	0.3799	0.533	0.00371
1	1.315	0.133	1.049	0.0874	1.100	1.154	1.210	1.270	0.864	0.00600
1 1/4	1.660	0.140	1.380	0.1150	1.904	2.628	3.625	5.005	1.495	0.01040
1 1/2	1.900	0.145	1.610	0.1342	2.592	4.173	6.718	10.82	2.036	0.01414
2	2.375	0.154	2.067	0.1722	4.272	8.831	18.250	37.72	3.355	0.02330
2 1/2	2.875	0.203	2.469	0.2057	6.096	15.051	37.161	91.75	4.788	0.03322
3	3.500	0.216	3.068	0.2557	9.413	28.878	88.605	271.8	7.393	0.05130
3 1/2	4.000	0.226	3.548	0.2957	12.59	44.663	158.51	562.2	9.886	0.06870
4	4.500	0.237	4.026	0.3355	16.21	65.256	262.76	1058.	12.730	0.08840
5	5.563	0.258	5.047	0.4206	25.47	128.56	648.72	3275.	20.006	0.1390
6	6.625	0.280	6.065	0.5054	36.78	223.10	1352.8	8206.	28.891	0.2006
8	8.625	0.277	8.071	0.6725	65.14	525.75	4243.0	34248.	51.161	0.3553
	8.625S	0.322	7.981	0.6651	63.70	508.36	4057.7	32380.	50.027	0.3474
10	10.75	0.279	10.192	0.8493	103.88	1058.7	10789.	109876.	81.585	0.5666
	10.75	0.307	10.136	0.8446	102.74	1041.4	10555.	106987.	80.691	0.5604
	10.75S	0.365	10.020	0.8350	100.4	1006.0	10080.	101000.	78.855	0.5475
12	12.75	0.330	12.090	1.0075	146.17	1767.2	21366.	258300.	114.80	0.7972
	12.75S	0.375	12.000	1.000	144.0	1728.0	20736.	248800.	113.10	0.7854
Extra Strong Pipe										
1/8	0.405	0.095	0.215	0.0179	0.0462	0.00994	0.002134	0.000459	0.036	0.00025
1/4	0.540	0.119	0.302	0.0252	0.0912	0.0275	0.008317	0.002513	0.072	0.00050
3/8	0.675	0.126	0.423	0.0353	0.1789	0.0757	0.03201	0.01354	0.141	0.00098
1/2	0.840	0.147	0.546	0.0455	0.2981	0.1628	0.08886	0.04852	0.234	0.00163
3/4	1.050	0.154	0.742	0.0618	0.5506	0.4085	0.3032	0.2249	0.433	0.00300
1	1.315	0.179	0.957	0.0797	0.9158	0.8765	0.8387	0.8027	0.719	0.00499
1 1/4	1.660	0.191	1.278	0.1065	1.633	2.087	2.6667	3.409	1.283	0.00891
1 1/2	1.900	0.200	1.500	0.1250	2.250	3.375	5.062	7.594	1.767	0.01225
2	2.375	0.218	1.939	0.1616	3.760	7.290	14.136	27.41	2.953	0.02050
2 1/2	2.875	0.276	2.323	0.1936	5.396	12.536	29.117	67.64	4.238	0.02942
3	3.500	0.300	2.900	0.2417	8.410	24.389	70.728	205.1	6.605	0.04587
3 1/2	4.000	0.318	3.364	0.2803	11.32	38.069	128.14	430.8	8.888	0.06170
4	4.500	0.337	3.826	0.3188	14.64	56.006	214.33	819.8	11.497	0.07986
5	5.563	0.375	4.813	0.4011	23.16	111.49	536.6	2583.	18.194	0.1263
6	6.625	0.432	5.761	0.4801	33.19	191.20	1101.6	6346.	26.067	0.1810
8	8.625	0.500	7.625	0.6354	58.14	443.32	3380.3	25775.	45.663	0.3171
10	10.75	0.500	9.750	0.8125	95.06	926.86	9036.4	88110.	74.662	0.5185
12	12.75	0.500	11.750	0.9792	138.1	1622.2	19072.	223970.	108.434	0.7528
Double Extra Strong Pipe										
1/2	0.840	0.294	0.252	0.0210	0.0635	0.0160	0.004032	0.00102	0.050	0.00035
3/4	1.050	0.308	0.434	0.0362	0.1884	0.0817	0.03549	0.01540	0.148	0.00103
1	1.315	0.358	0.599	0.0499	0.3588	0.2149	0.1287	0.07711	0.282	0.00196
1 1/4	1.660	0.382	0.896	0.0747	0.8028	0.7193	0.6445	0.5775	0.630	0.00438
1 1/2	1.900	0.400	1.100	0.0917	1.210	1.331	1.4641	1.611	0.950	0.00660
2	2.375	0.436	1.503	0.1252	2.259	3.395	5.1031	7.670	1.774	0.01232
2 1/2	2.875	0.552	1.771	0.1476	3.136	5.554	9.8345	17.42	2.464	0.01710
3	3.500	0.600	2.300	0.1917	5.290	12.167	27.984	64.36	4.155	0.02885
3 1/2	4.000	0.636	2.728	0.2273	7.442	20.302	55.383	151.1	5.845	0.04059
4	4.500	0.674	3.152	0.2627	9.935	31.315	98.704	311.1	7.803	0.05419
5	5.563	0.750	4.063	0.3386	16.51	67.072	272.58	1107.	12.966	0.09006
6	6.625	0.864	4.897	0.4081	23.98	117.43	575.04	2816.	18.835	0.1308
8	8.625	0.875	6.875	0.5729	47.27	324.95	2234.4	15360.	37.122	0.2578

Courtesy Crane Co., Reproduced from Technical Paper No. 410, "Flow of Fluids."

***The latest revision of this standard is ANSI B36.10-1970; therefore, the schedule wall thickness used to derive the given data may vary slightly from the current standard in some cases.**

Stainless Steel Pipe Data

Schedule Wall Thickness — Per ASA B36.19-1957 *

Nominal Pipe Size	Outside Diameter	Thickness	Inside Diameter		Inside Diameter Functions (In Inches)				Transverse Internal Area	
			d	D	d^2	d^3	d^4	d^5	a	A
Inches	Inches	Inches	Inches	Feet					Sq. In.	Sq. Ft.
Schedule 5 S										
1/2	0.840	0.065	0.710	0.0592	0.504	0.358	0.254	0.1804	0.396	0.00275
3/4	1.050	0.065	0.920	0.0767	0.846	0.779	0.716	0.659	0.664	0.00461
1	1.315	0.065	1.185	0.0988	1.404	1.664	1.972	2.337	1.103	0.00766
1 1/4	1.660	0.065	1.530	0.1275	2.341	3.582	5.480	8.384	1.839	0.01277
1 1/2	1.900	0.065	1.770	0.1475	3.133	5.545	9.815	17.37	2.461	0.01709
2	2.375	0.065	2.245	0.1871	5.040	11.31	25.40	57.03	3.958	0.02749
2 1/2	2.875	0.083	2.709	0.2258	7.339	19.88	53.86	145.9	5.764	0.04003
3	3.500	0.083	3.334	0.2778	11.12	37.06	123.6	411.9	8.733	0.06065
3 1/2	4.000	0.083	3.834	0.3195	14.70	56.36	216.1	828.4	11.545	0.08017
4	4.500	0.083	4.334	0.3612	18.78	81.41	352.8	1529.	14.750	0.1024
5	5.563	0.109	5.345	0.4454	28.57	152.7	816.2	4363.	22.439	0.1558
6	6.625	0.109	6.407	0.5339	41.05	263.0	1685.	10796.	32.241	0.2239
8	8.625	0.109	8.407	0.7006	70.68	594.2	4995.	41996.	55.512	0.3855
10	10.750	0.134	10.482	0.8375	109.9	1152.	12072.	126538.	86.315	0.5994
12	12.750	0.156	12.438	1.0365	154.7	1924.	23933.	297682.	121.50	0.8438
Schedule 10 S										
1/8	0.405	0.049	0.307	0.0256	0.0942	0.0289	0.00888	0.00273	0.074	0.00051
1/4	0.540	0.065	0.410	0.0342	0.1681	0.0689	0.02826	0.01159	0.132	0.00092
3/8	0.675	0.065	0.545	0.0454	0.2970	0.1619	0.08822	0.04808	0.233	0.00162
1/2	0.840	0.083	0.674	0.0562	0.4543	0.3062	0.2064	0.1391	0.357	0.00248
3/4	1.050	0.083	0.884	0.0737	0.7815	0.6908	0.6107	0.5398	0.614	0.00426
1	1.315	0.109	1.097	0.0914	1.203	1.320	1.448	1.589	0.945	0.00656
1 1/4	1.660	0.109	1.442	0.1202	2.079	2.998	4.324	6.235	1.633	0.01134
1 1/2	1.900	0.109	1.682	0.1402	2.829	4.759	8.004	13.46	2.222	0.01543
2	2.375	0.109	2.157	0.1798	4.653	10.04	21.65	46.69	3.654	0.02538
2 1/2	2.875	0.120	2.635	0.2196	6.943	18.30	48.21	127.0	5.453	0.03787
3	3.500	0.120	3.260	0.2717	10.63	34.65	112.9	368.2	8.347	0.05796
3 1/2	4.000	0.120	3.760	0.3133	14.14	53.16	199.9	751.5	11.11	0.07712
4	4.500	0.120	4.260	0.3550	18.15	77.31	329.3	1403.	14.26	0.09899
5	5.563	0.134	5.295	0.4413	28.04	148.5	786.1	4162.	22.02	0.1529
6	6.625	0.134	6.357	0.5298	40.41	256.9	1633.	10382.	31.74	0.2204
8	8.625	0.148	8.329	0.6941	69.37	577.8	4813.	40083.	54.48	0.3784
10	10.750	0.165	10.420	0.8683	108.6	1131.	11789.	122840.	85.29	0.5923
12	12.750	0.180	12.390	1.0325	153.5	1902.	23566.	291982.	120.6	0.8372
Schedule 40 S										
1/8 to 12	Values are the same, size for size, as those shown on the facing page for Standard Wall Pipe (heaviest weight on 8, 10, and 12-inch sizes).									
Schedule 80 S										
1/8 to 12	Values are the same, size for size, as those shown on the facing page for Extra Strong Pipe.									

Courtesy Crane Co., Reproduced from Technical Paper No. 410, "Flow of Fluids."

*** The latest revision of this standard is ANSI B36.19-1965; therefore, the schedule wall-thickness used to derive the given data may vary slightly from the current standard in some cases.**

DIN STANDARDS*

DIN Pressure-Temperature Ratings Carbon Steel Flanges and Pipes

Nominal Pressure	Maximum Permissible Working Pressure: I (W) For Fluids, Gases and Vapours up to 120° C.	II (G) For Fluids, Gases and Vapours up to 300° C.	III (H) For Fluids, Gases and Vapours up to 400° C.		Hydraulic Test Pressure
	Flanges & Pipes	Flanges & Pipes	Flanges	Pipes	
16	16	13	13	10	25
25	25	20	20	16	40
40	40	32	32	25	60
64	64	50	40	40	96
100	100	80	64	64	150
160	160	125	100	100	240
250	250	200	160	160	375
320	320	250	200	200	480
400	400	320	250	250	600

DIN Cast Steel Flange Standard Nenndruck 16

(Nominal Pressure 16 Kgf/cm²)

Nominal Bore	Pipe: Outside Dia.	Pipe: Thickness	Flange: Outside Dia.	Flange: Thickness	Flange: Bolt Circle Dia.	Bolting: No. of Bolts	Bolting: Thread	Bolting: Bolt Hole Dia.
10	22	6	90	16	60	4	M12	14
15	25	6	93	16	65	4	M12	14
20	33	6.5	105	18	75	4	M12	14
25	39	7	115	18	85	4	M12	14
32	46	7	140	16	100	4	M16	18
40	55	7.5	150	16	110	4	M16	18
50	65	7.5	165	18	125	4	M16	18
65	86	8	185	18	145	4	M16	18
80	97	8.5	200	20	160	8	M16	18
100	119	9.5	220	20	180	8	M16	18
125	145	10	250	22	210	8	M16	18
150	172	11	285	22	240	8	M20	23
175	199	12	315	24	270	8	M20	23
200	224	12	340	24	295	12	M20	23
250	278	14	405	26	355	12	M24	27
300	330	15	460	28	410	12	M24	27

DIN Cast Steel Flange Standard Nenndruck 25

(Nominal Pressure 25 Kgf/cm²)

Nominal Bore	Pipe: Outside Dia.	Pipe: Thickness	Flange: Outside Dia.	Flange: Thickness	Flange: Bolt Circle Dia.	Bolting: No. of Bolts	Bolting: Thread	Bolting: Bolt Hole Dia.
10	22	6	90	16	60	4	M12	14
15	25	6	95	16	65	4	M12	14
20	33	6.5	105	18	75	4	M12	14
25	39	7	115	18	85	4	M12	14
32	46	7	140	18	100	4	M16	18
40	55	7.5	150	18	110	4	M16	18
50	65	7.5	165	20	125	4	M16	18
65	86	8	185	22	145	8	M16	18
80	97	8.5	200	24	160	8	M16	18
100	119	9.5	235	24	190	8	M20	23
125	145	10	270	26	220	8	M24	27
150	172	11	300	28	250	8	M24	27
175	199	12	330	28	280	12	M24	27
200	224	12	360	30	310	12	M24	27
250	278	14	425	32	370	12	M27	30
300	330	15	485	34	430	16	M27	30

DIN Cast Steel Flange Standard Nenndruck 40

(Nominal Pressure 40 Kgf/cm²)

Nominal Bore	Pipe: Outside Dia.	Pipe: Thickness	Flange: Outside Dia.	Flange: Thickness	Flange: Bolt Circle Dia.	Bolting: No. of Bolts	Bolting: Thread	Bolting: Bolt Hole Dia.
10	22	6	90	16	60	4	M12	14
15	25	6	95	16	65	4	M12	14
20	33	6.5	105	18	75	4	M12	14
25	39	7	115	18	85	4	M12	14
32	46	7	140	18	100	4	M16	18
40	55	7.5	150	18	110	4	M16	18
50	66	8	165	20	125	4	M16	18
65	87	8.5	185	22	145	8	M16	18
80	98	9	200	24	160	8	M16	18
100	120	10	235	24	190	8	M20	23
125	147	11	270	26	220	8	M24	27
150	174	12	300	28	250	8	M24	27
175	201	13	350	32	295	12	M27	30
200	228	14	375	34	320	12	M27	30
250	282	16	450	38	385	12	M30	33
300	334	17	515	42	450	16	M30	33

All dimensions in mm.

All pressures in Kgf/cm^2.

*German flange standards that are widely used in Europe.

Reprinted from "Control Valve Handbook," 1965, Fisher Controls Company, Marshalltown, Iowa.

DIN STANDARDS

DIN Cast Steel Flange Standard Nenndruck 64

(Nominal Pressure 64 Kgf/cm²)

Nominal Bore	Pipe Thickness	Flange: Outside Dia.	Flange: Thickness	Flange: Bolt Circle Dia.	Bolting: No. of Bolts	Bolting: Thread	Bolting: Bolt Hole Dia.
10	8	100	20	70	4	M12	14
15	8	105	20	75	4	M12	14
25	9	140	24	100	4	M16	18
40	9	170	26	125	4	M20	23
50	10	180	26	135	4	M20	23
65	10	205	26	160	8	M20	23
80	11	215	28	170	8	M20	23
100	12	250	30	200	8	M24	27
125	13	295	34	240	8	M27	30
150	14	345	36	280	8	M30	33
175	15	375	40	310	12	M30	33
200	16	415	42	345	12	M33	36
250	19	470	46	400	12	M33	36
300	21	530	52	460	16	M33	36

DIN Cast Steel Flange Standard Nenndruck 100

(Nominal Pressure 100 Kgf/cm²)

Nominal Bore	Pipe Thickness	Flange: Outside Dia.	Flange: Thickness	Flange: Bolt Circle Dia.	Bolting: No. of Bolts	Bolting: Thread	Bolting: Bolt Hole Dia.
10	10	100	20	70	4	M12	14
15	10	105	20	75	4	M12	14
25	11	140	24	100	4	M16	18
40	11	170	26	125	4	M20	23
50	12	195	28	145	4	M24	27
65	13	220	30	170	8	M24	27
80	14	230	32	180	8	M24	27
100	15	265	36	210	8	M27	30
125	16	315	40	250	8	M30	33
150	18	355	44	290	12	M30	33
175	20	385	48	320	12	M30	33
200	21	430	52	360	12	M33	36
250	25	505	60	430	12	M36	39
300	29	585	68	500	16	M39	42

DIN Cast Steel Flange Standard Nenndruck 160

(Nominal Pressure 160 Kgf/cm²)

Nominal Bore	Pipe Thickness	Flange: Outside Dia.	Flange: Thickness	Flange: Bolt Circle Dia.	Bolting: No. of Bolts	Bolting: Thread	Bolting: Bolt Hole Dia.
10	10	100	20	70	4	M12	14
15	10	105	20	75	4	M12	14
25	10	140	24	100	4	M16	18
40	10	170	28	125	4	M20	23
50	10	195	30	145	4	M24	27
65	11	220	34	170	8	M24	27
80	12	230	36	180	8	M24	27
100	14	265	40	210	8	M27	30
125	16	315	44	250	8	M30	33
150	18	355	50	290	12	M30	33
175	19	390	54	320	12	M33	36
200	21	430	60	360	12	M33	36
250	24	515	68	430	12	M39	42
300	28	585	78	500	16	M39	42

DIN Cast Steel Flange Standard Nenndruck 250

(Nominal Pressure 250 Kgf/cm²)

Nominal Bore	Pipe Thickness	Flange: Outside Dia.	Flange: Thickness	Flange: Bolt Circle Dia.	Bolting: No. of Bolts	Bolting: Thread	Bolting: Bolt Hole Dia.
10	10	125	24	85	4	M16	18
15	10	130	26	90	4	M16	18
25	11	150	28	105	4	M20	23
40	13	185	34	135	4	M24	27
50	13	200	38	150	8	M24	27
65	14	230	42	180	8	M24	27
80	16	255	46	200	8	M27	30
100	19	300	54	235	8	M30	33
125	22	340	60	275	12	M30	33
150	25	390	68	320	12	M33	36
175	29	430	74	355	12	M36	39
200	32	485	82	400	12	M39	42
250	38	585	100	490	16	M45	48
300	47	690	120	590	16	M48	52

All dimensions in mm.

Reprinted from "Control Valve Handbook," 1965, Fisher Controls Company, Marshalltown, Iowa.

DIN STANDARDS

DIN Cast Steel Flange Standard Nenndruck 320

(Nominal Pressure 320 Kgf/cm^2)

Nominal Bore	Pipe Thickness	Flange			Bolting		
		Outside Dia.	Thickness	Bolt Circle Dia.	No. of Bolts	Thread	Bolt Hole Dia.
10	11	125	24	85	4	M16	18
15	11	130	26	90	4	M16	18
25	11	160	34	115	4	M20	23
40	14	195	38	145	4	M24	27
50	15	210	42	160	8	M24	27
65	18	255	51	200	8	M27	30
80	19	275	55	220	8	M27	30
100	24	335	65	265	8	M33	36
125	27	380	75	310	12	M33	36
150	32	425	84	350	12	M36	39
175	35	485	95	400	12	M39	42
200	38	525	103	440	16	M39	42
250	49	640	125	540	16	M48	52

All dimensions in mm.

DIN Cast Steel Flange Standard Nenndruck 400

(Nominal Pressure 400 Kgf/cm^2)

Nominal Bore	Pipe Thickness	Flange			Bolting		
		Outside Dia.	Thickness	Bolt Circle Dia.	No. of Bolts	Thread	Bolt Hole Dia.
10	11	125	28	85	4	M16	18
15	11	145	30	100	4	M20	23
20	11	160	34	115	4	M20	23
25	12	180	38	130	4	M24	27
32	14	205	44	150	4	M27	30
40	15	220	48	165	4	M27	30
50	18	235	52	180	8	M27	30
65	22	290	64	225	8	M30	33
80	25	305	68	240	8	M30	33
100	30	370	80	295	8	M36	39
125	36	415	92	340	12	M36	39
150	41	475	105	390	12	M39	42
175	47	545	120	450	12	M45	48
200	53	585	130	490	16	M45	48

Reprinted from "Control Valve Handbook," 1965, Fisher Controls Company, Marshalltown, Iowa.

VALVE SIZING DATA
TYPICAL VALVE FACTORS

	Size Range (inches)	F_L	x_T	C_d
Single Port Globe				
wing guided plug	0.5 - 4	0.90	0.75	11
V-port skirt	0.5 - 6	0.90	0.75	9
turned plug	1 - 4	0.90	0.75	11
V-plug	1 - 6	0.90	0.75	9.5
Double Port Globe				
wing guided plug	1 - 8	0.90	0.75	14
V-port skirt	1 - 4	0.90	0.75	13
turned plug	1 - 8	0.85	0.72	13
V-plug	1 - 8	0.88	0.70 - 0.85	12.5
Angle Body - Turned Plug				
flow-to-close				
straight orifice	1 - 4	0.8	0.65	20
straight reduced port	1 - 4	0.7	0.55	--
tapered orifice 2:1	1 - 8	0.45	0.15	--
flow-to-open				
full port	1 - 4	0.9	0.7	17
reduced port	1 - 10	0.95	0.8	--
Ball Valve				
standard	4 - 16	0.55	0.15	30 - 40
Butterfly				
60-degree Open	2 - 72	0.75	0.38	16 - 18
90-degree Open	2 - 72	0.55	0.20	20 - 45

The factors given above are the average values applicable over the size range shown. Values for specific valves may vary somewhat from the average particularly in sizes outside the given size range.

F_L - Pressure loss factor.
X_T - Terminal or ultimate valve of X, used to establish expansion factor, Y.
C_d - Relative Capacity (at rated Cv).

Courtesy of L. R. Driskell, author of Part I, Chapter 6. Refer to pp. 124-140 for additional explanation and application of this data.

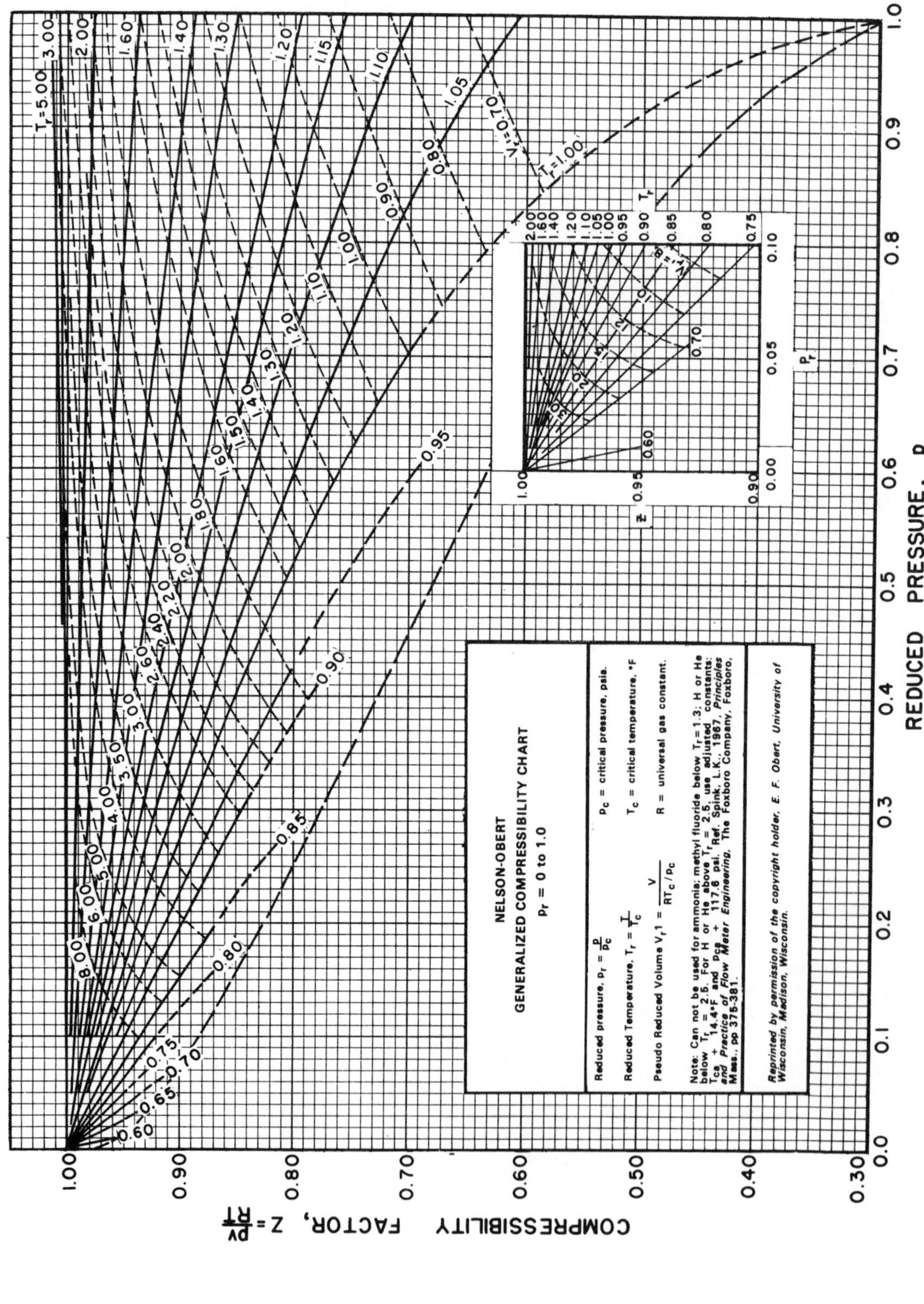
NELSON-OBERT
GENERALIZED COMPRESSIBILITY CHART
p_r = 0 to 1.0
Reduced pressure, $p_r = \frac{p}{p_c}$
p_c = critical pressure, psia.
Reduced Temperature, $T_r = \frac{T}{T_c}$
T_c = critical temperature, °F
Pseudo Reduced Volume $V_r 1 = \frac{V}{RT_c / p_c}$
R = universal gas constant.
Note: Can not be used for ammonia; methyl fluoride below T_r = 1.3; H or He below T_r = 2.5. For H or He above T_r = 2.5, use adjusted constants: T_{ca} + 14.4°F and p_{ca} + 117.6 psi. Ref. Spink, L.K., 1967, *Principles and Practice of Flow Meter Engineering*, The Foxboro Company, Foxboro, Mass., pp 375-381.
Reprinted by permission of the copyright holder, E. F. Obert, University of Wisconsin, Madison, Wisconsin.
COMPRESSIBILITY FACTOR, $Z = \frac{pv}{RT}$
REDUCED PRESSURE, p_r

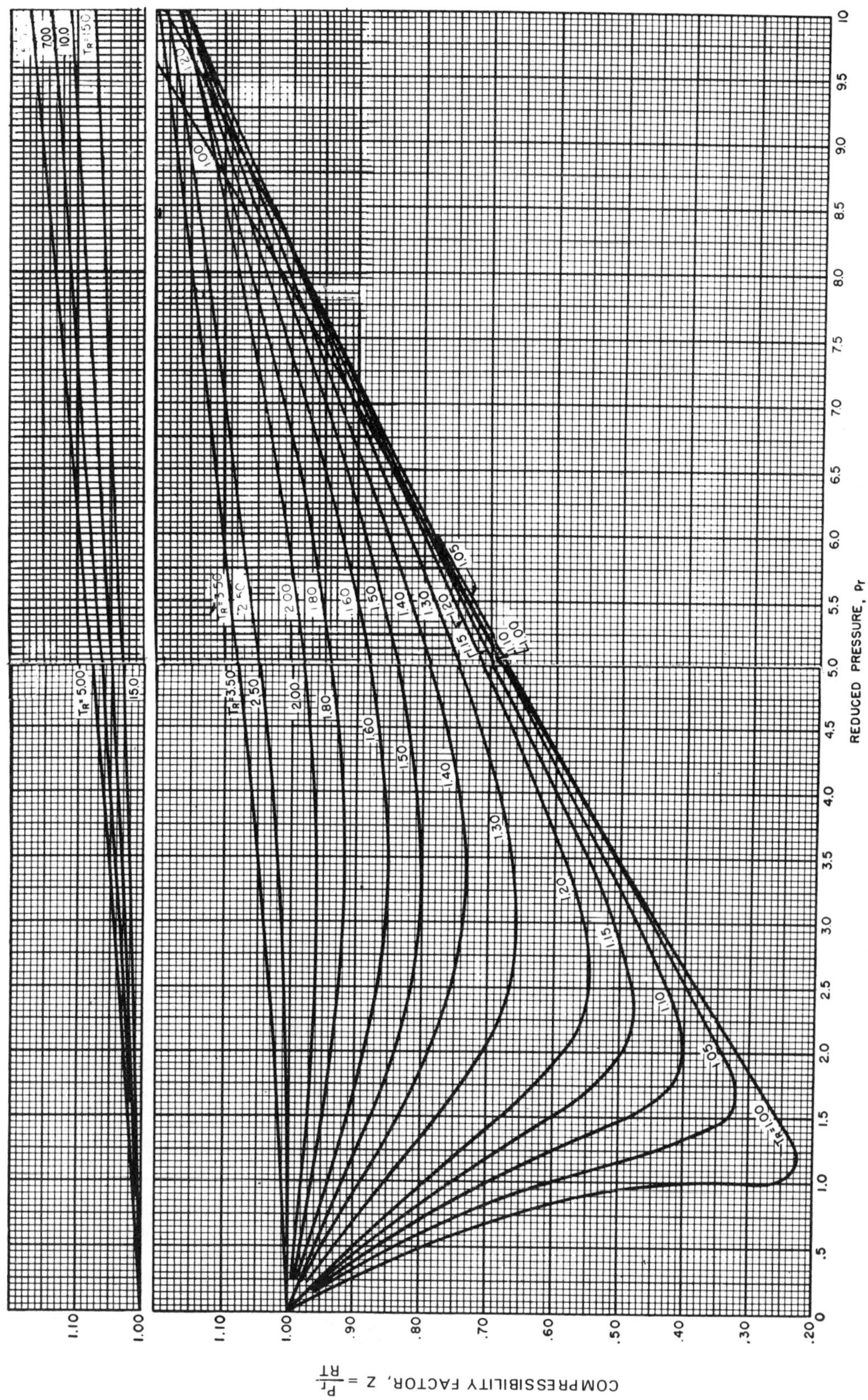

NELSON-OBERT GENERALIZED COMPRESSIBILITY CHART, p_r = 0 to 10.

Note: Refer to page 272 for explanation of p_r, T_r and V_r1 and for limitations with certain gases.

Reprinted by permission of the copyright holder, E. F. Obert, University of Wisconsin, Madison, Wisconsin.

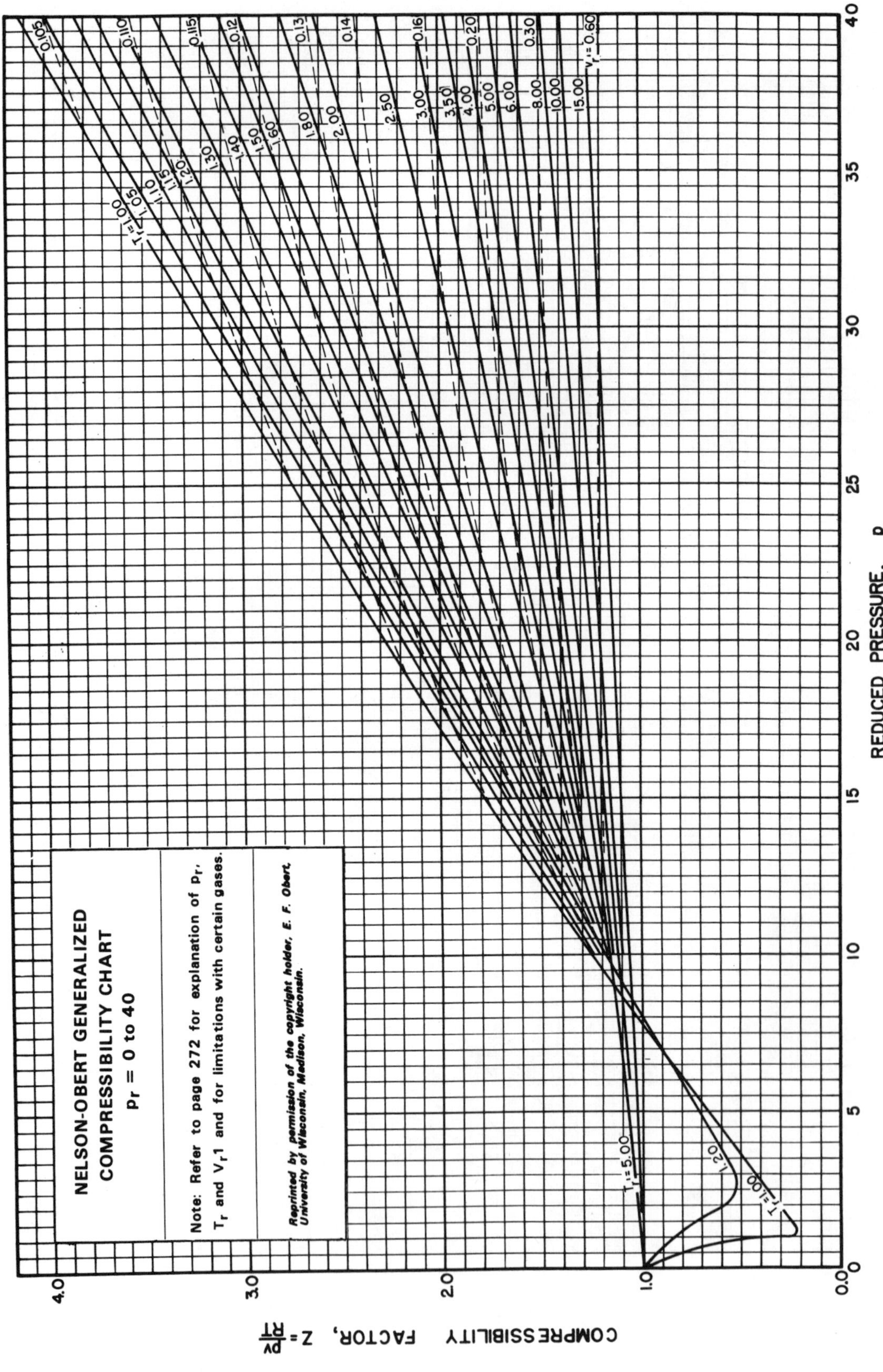
NELSON-OBERT GENERALIZED
COMPRESSIBILITY CHART
p_r = 0 to 40
Note: Refer to page 272 for explanation of p_r, T_r and V_r1 and for limitations with certain gases.
Reprinted by permission of the copyright holder, E. F. Obert, University of Wisconsin, Madison, Wisconsin.
REDUCED PRESSURE, p_r
COMPRESSIBILITY FACTOR, $Z = \frac{pv}{RT}$

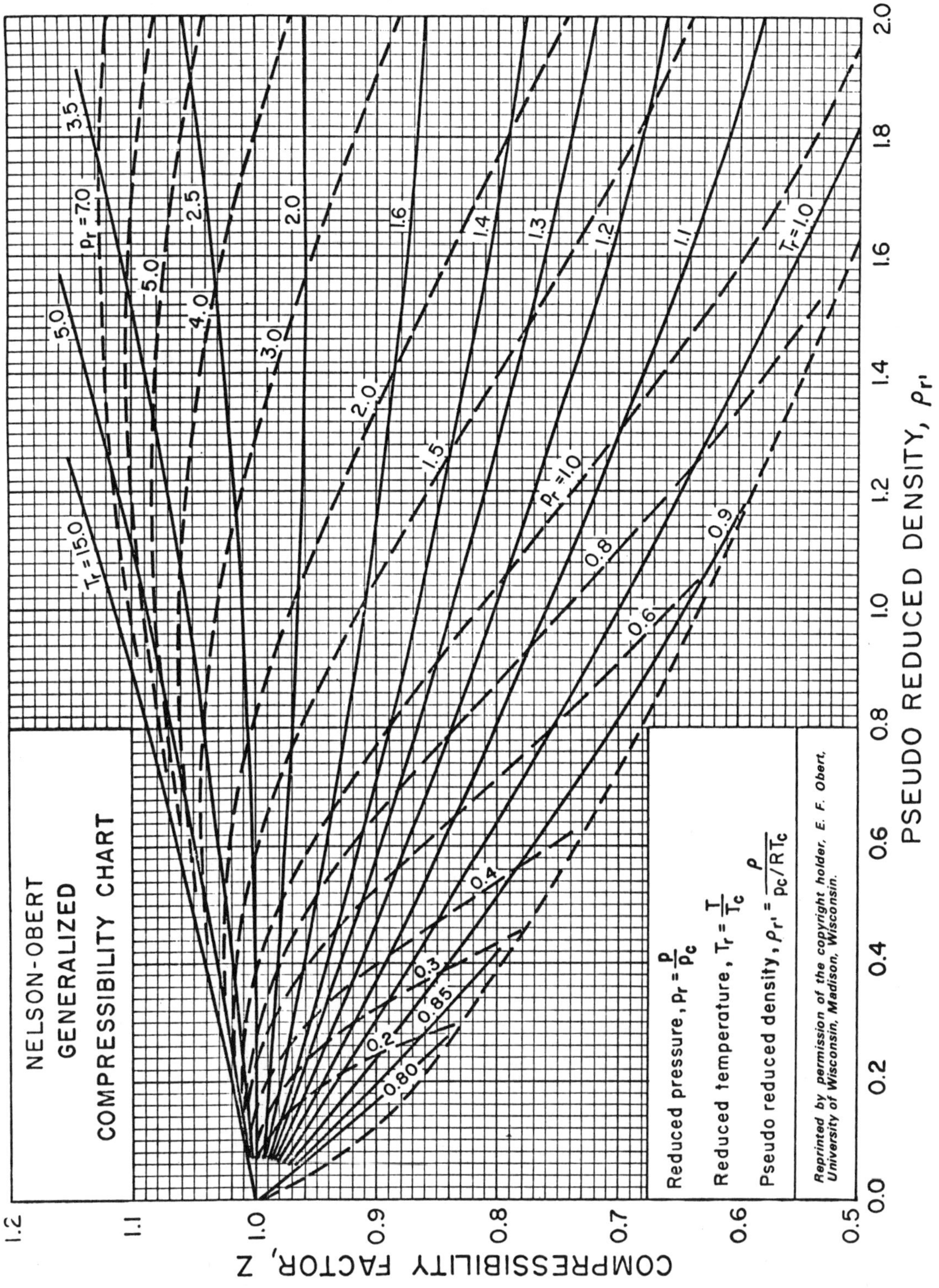
NELSON-OBERT
GENERALIZED
COMPRESSIBILITY CHART
Reduced pressure, $p_r = \frac{p}{p_c}$
Reduced temperature, $T_r = \frac{T}{T_c}$
Pseudo reduced density, $\rho_{r'} = \frac{\rho}{p_c/RT_c}$
Reprinted by permission of the copyright holder, E. F. Obert, University of Wisconsin, Madison, Wisconsin.
PSEUDO REDUCED DENSITY, $\rho_{r'}$
COMPRESSIBILITY FACTOR, Z
$T_r = 15.0$
$T_r = 1.0$
$p_r = 7.0$
$p_r = 1.0$

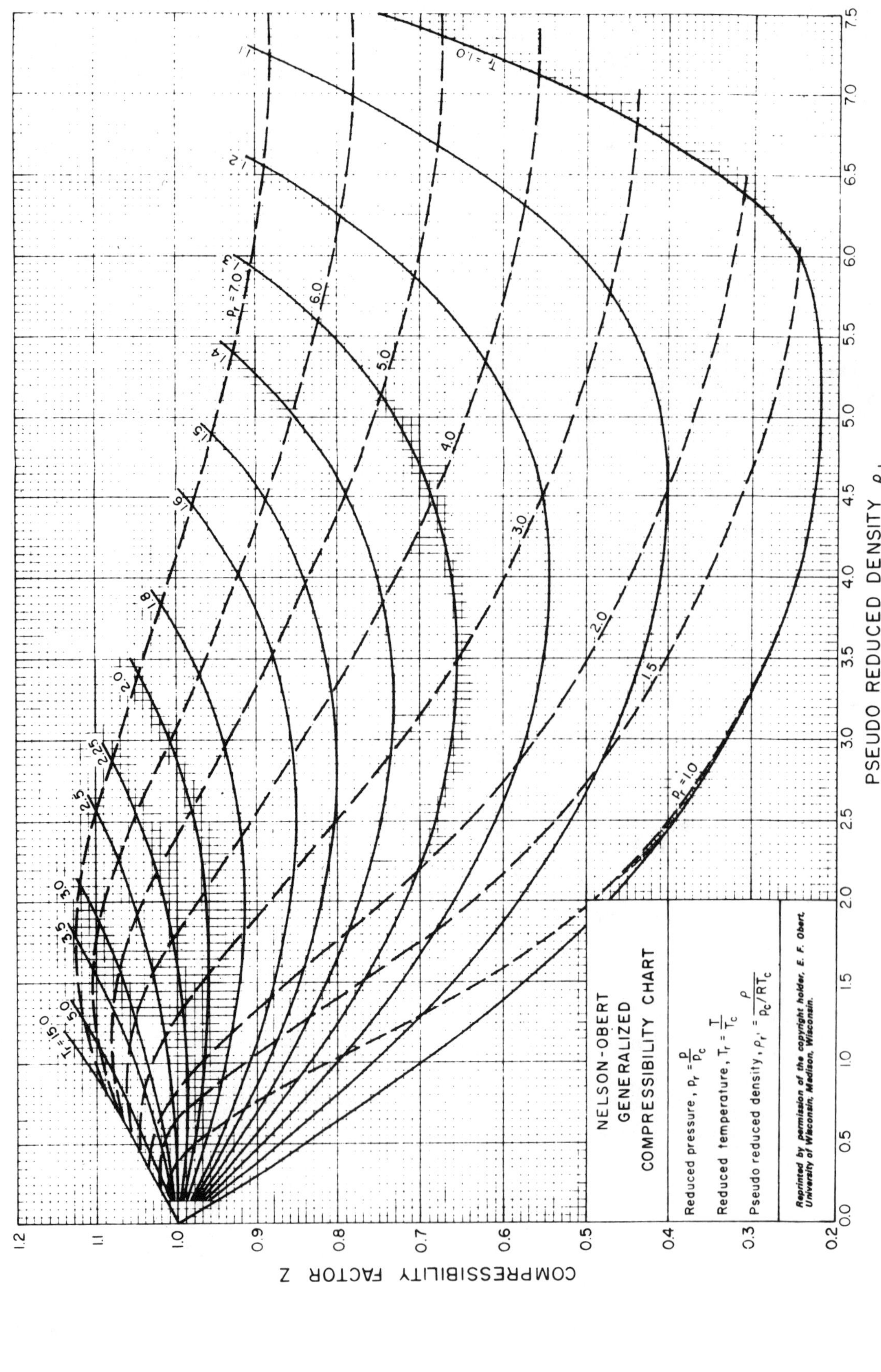
NELSON-OBERT
GENERALIZED
COMPRESSIBILITY CHART
Reduced pressure, $p_r = \frac{p}{p_c}$
Reduced temperature, $T_r = \frac{T}{T_c}$
Pseudo reduced density, $\rho_r' = \frac{\rho}{p_c/RT_c}$
Reprinted by permission of the copyright holder, E. F. Obert, University of Wisconsin, Madison, Wisconsin.
COMPRESSIBILITY FACTOR Z
PSEUDO REDUCED DENSITY ρ_r'

COEFFICIENTS FOR CONTRACTION LOSSES IN FLUID FLOW

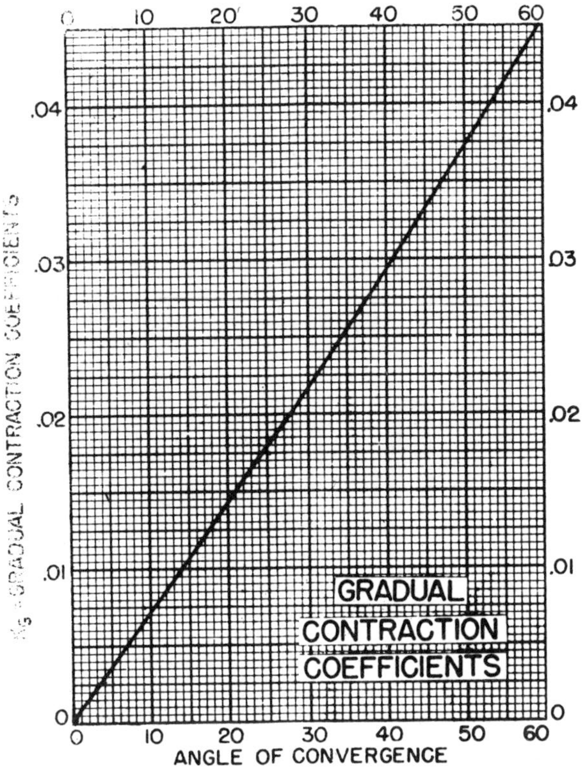

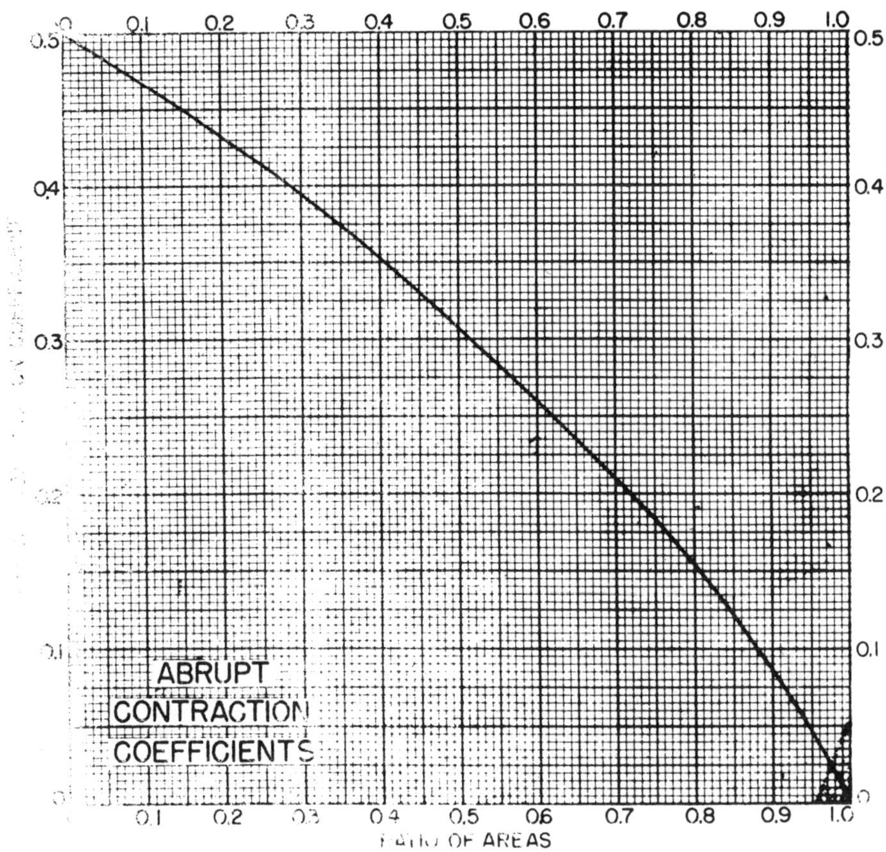

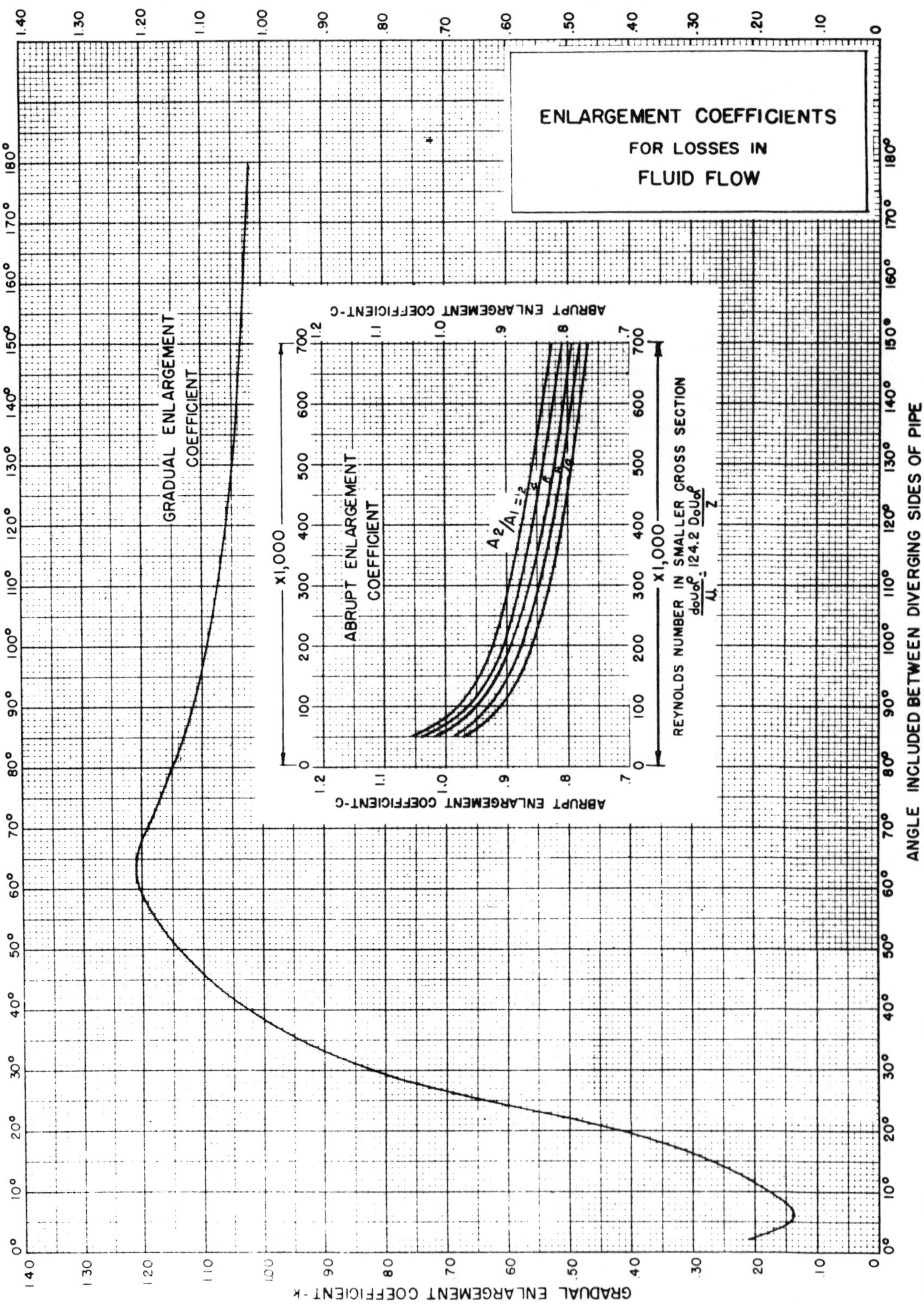
ENLARGEMENT COEFFICIENTS
FOR LOSSES IN
FLUID FLOW
GRADUAL ENLARGEMENT COEFFICIENT
GRADUAL ENLARGEMENT COEFFICIENT-K
ANGLE INCLUDED BETWEEN DIVERGING SIDES OF PIPE
ABRUPT ENLARGEMENT COEFFICIENT
ABRUPT ENLARGEMENT COEFFICIENT-C
A_2/A_1 =
X1,000
REYNOLDS NUMBER IN SMALLER CROSS SECTION
$\frac{d_0 U_0 \rho}{\mu} = 124.2 \frac{D_0 U_0 \rho}{Z}$

Specific Gravity and Viscosity of Liquids

Liquid	Boiling point at atm press °F	Specific gravity		Kinematic viscosity			Coordinates chart, pg 81	
		Temp °F	Water at 4°C=1	Temp °F	Centistokes	SSU	a	b
Acetaldehyde	69	68	.782	68	.295	—	3.0	0
Acetic acid	245	68	1.049	68	1.17	—	3.3	5.1
Acetone	133	68	.79	68	.41	—	2.8	1.5
Aero motor oil (typical)	—	60	.895	100	—	1100		
	—	—	——	210	—	95	2.7	12.2
Alcohol, allyl-n	207	68	.855	68	1.60	—	3.9	5.6
Alcohol, butyl-n	243	68	.81	68	3.64	38	3.3	7.3†
	243	158	.78	158	1.17	—	4.65	7.05††
Alcohol, ethyl-n (grain)	172	68	.789	68	1.52	—	3.35	5.67
Alcohol, methyl-n (wood)	151	68	.79	59	.74	—	3.8	3.8
Alcohol, propyl-n	207	68	.804	68	2.8	35	3.2	6.95
Ammonia (liquid)	-28	0	.662	0	.30	—	—	—
Aniline	363	68	1.022	68	4.37	40	4.2	7.15
Automobile crankcase oils,								
SAE 10	—	60	.88-94	130	—	90-120	3.0	10.7
SAE 20	—	60	.88-94	130	—	120-185	3.0	11.3
SAE 3	—	60	.88-94	130	—	185-225	3.0	11.65
SAE 40	—	60	.88-94	130	—	225-400	2.9	12.0
SAE 50	—	60	.88-94	210	—	80-105	2.8	12.2
SAE 60	—	60	.88-94	210	—	105-125	2.65	12.65
SAE 70	—	60	.88-94	210	—	125-150	2.6	12.7
Automobile transmission lub,								
SAE 80	—	60	.88-94	0	—	100,000 max	—	—
SEA 90	—	60	.88-94	100	—	800-1500	—	—
SAE 140	—	60	.88-94	210	—	120-200	—	—
SAE 250	—	60	.88-94	210	—	200 min	—	—
Beer	—	60	1.01	68	—	32 (est)	—	—
Benzol (Benzene)	176	68	.879	68	.744	—	3.6	3.85
Brine, calcium chloride, 25%	—	60	1.23	60	—	39	—	—
Brine, sodium chloride, 25%	—	60	1.19	60	—	34	—	—
Bromine	142	68	2.9	68	.34	—	—	—
Butyric acid-n	316	68	.959	68	1.61	—	3.0	5.95
Carbolic acid (phenol)	360	65	1.08	65	11.83	65	4.5	7.9
Carbon disulphide	115	68	1.263	68	.298	—	-.9	8.9*
Carbon tetrachloride	170	68	1.594	68	.612	—	4.0	3.25
Castor Oil	—	68	.96	68	—	4600	2.75	12.2
Chloroform	142	68	1.489	68	.38	—	3.05	1.2
Compounded steam cyl oil (5% tallow)	—	60	.90	110	—	1800	2.4	12.8
				210	—	145		
Decane-n	343	68	.73	68	1.24	—	2.95	5.35
Diethyl ether	94.4	68	.714	68	.32	—	3.0	.6
Ethyl acetate	171	68	.90	68	.49	—	3.7	2.5
Ethyl bromide	101	59	1.45	68	.27	—	-.2	8.3*
Ethylene bromide	269	68	2.18	68	.787	—	3.6	3.95
Ethylene chloride	183	68	1.246	68	.668	—	3.5	3.5
Formic acid	213	68	1.221	68	1.48	—	3.9	5.5

†Accurate for the range between 32 and 158°F.
††Accurate for the range between 139 and 158°F.
*Divide figure obtained with these coordinates by 10 to obtain viscosity in centistokes.

Reprinted from "Cameron Hydraulics Data," 1958, Cameron Pump Division, Ingersoll Rand Company, Phillipsburg, New Jersey.

Specific Gravity and Viscosity of Liquids

Liquid	Boiling point at atm press °F	Specific gravity		Kinematic viscosity			Coordinates chart, pg 81	
		Temp °F	Water at 4°C—1	Temp °F	Centi stokes	SSU	a	b
Freon 11	—	70	1.49	70	.32	—	—	—
Freon 12	—	70	1.33	70	.21	—	—	—
Freon 21	—	70	1.37	70	.27	—	—	—
Fuel Oil, No. 1	—	60	.82–95	100	—	30–40	—	—
No. 2	—	60	.82–95	100	—	35-50	—	—
No. 3	—	60	.82–95	100	—	55 max	—	—
No. 5	—	60	.82–95	100	—	60–450	—	—
No. 6	—	60	.82–95	122	—	430–2900	—	—
Gasoline, typical (a)	—	6	.74	60	.88	—	3.15	4.15
(b)	—	6	.72	60	.64	—	3.7	3.2
(c)	—	6	.68	60	.46	—	2.6	1.7
Glycerine, 100%	554	68	1.26	68.6	—	2950	3.25	11.5
Glycerine and water, 50%	—	68	1.13	68	—	43	—	—
Glycol, Ethylene	—	68	1.125	70	17.8	88.4		
Heptane-n	209	68	.684	68	.60	—	2.9	3.1
Hexane-n	156	68	.66	68	.49	—	3.0	2.3
Hydrochloric acid, 31.5%	—	68	1.05	68	1.9	—	—	—
Kerosene	—	60	.78–.82	68	—	35	—	—
Lard oil	—	60	.91–.92	100	—	180–200	—	—
Linseed oil (raw)	538	60	.92–.94	100	—	143	2.0	11.1
Marine engine oil (20% blown rape)	—	60	.94	210	—	70	2.5	11.9
Methy acetate	135	68	.93	68	.44	—	3.1	1.6
Methy iodide	108	68	2.28	68	.213	—	.35	7.5*
Milk	—	68	1.02-1.04	68	—	32 est	—	—
Naphthelene	424	68	1.145	176	.9	—	—	—
Neatsfoot oil	—	60	.91–.92	60	107-123	—	—	—
Nitric acid, 60%	—	68	1.37	68	1.87	—	—	—
Nitrobenzene	412	68	1.203	68	1.67	—	3.1	6.0
Nonane-n	302	68	.718	68	.97	—	1.6	4.7
Octane-n	258	68	.70	68	.77	—	3.0	3.95
Olive oil	(570)	68	.91	68	93.0	—	2.4	11.4
Pentane-n	97	68	.63	68	.37	—	2.8	1.1
Petroleum either (benzine)	—	60	.64	60	—	31 est	—	—
Propionic acid	286	68	.99	68	1.13	—	3.2	5.05
Quenching oil (typical)	—	60	.86–89	100	—	100–120	—	—
Rapeseed oil	—	68	.91	68	180	—	2.6	11.5
Soya bean oil	—	60	.924	60	86	—	2.1	11.3
Sperm oil	(209)	77	.88	100	21	—	2.3	10.7
Sugar, 20%	—	68	1.08	68		33.5	—	—
40%	—	68	1.18	68	5.3	43	—	—
60%	—	68	1.29	68	44	202	—	—
Sulfuric acid, 100%	640	68	1.83	68	14.6	76	—	
95%	—	68	1.83	68	14.5	75		
60%	—	68	1.50	68	4.4	41	—	
Turbine oil (typical medium)	—	60	.91	100	—	270	—	
Turpentine	320	60	.86–87	60	1.83	—	2.9	6.1
Water (fresh)	212	60	1.0	60	1.1	31.5	5.0	4.5
Water (sea)	—	60	1.03	60	—	31.5	—	—
Xyolene-o	287	68	87	68	93	—	2.8	4.5

*Divide figure obtained with these coordinates by 10 to obtain viscosity in centistokes.

Reprinted from "Cameron Hydraulics Data," 1958, Cameron Pump Division, Ingersoll Rand Company, Phillipsburg, New Jersey.

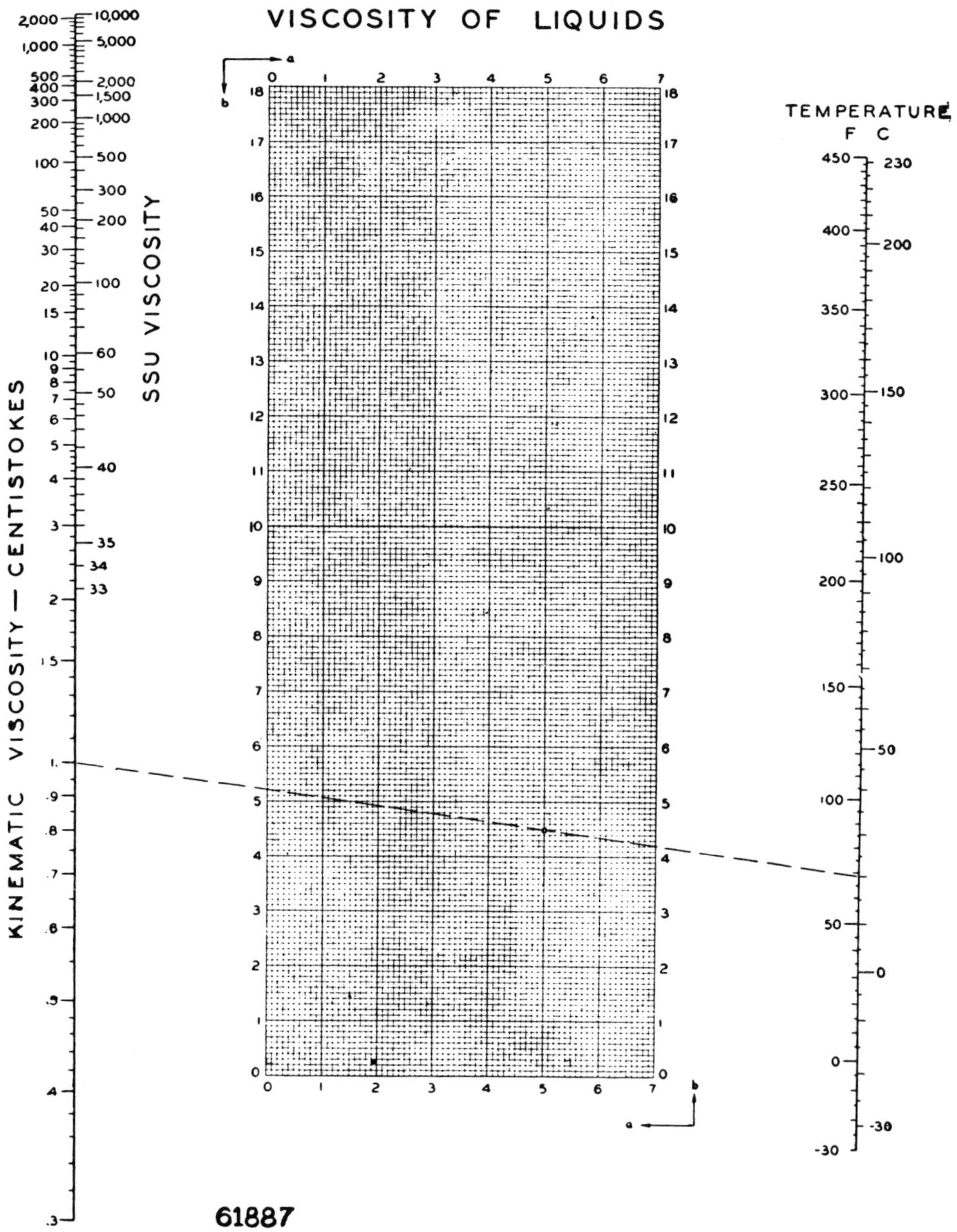

If viscosity is known at two temperatures, plot these points on the graph. The intersection of the lines from temperature to viscosity will give a point. Lines through this point will give viscosity at any other temperature. The point shown is for water. Other points are given in the tables pages 78 and 79. These data are approximate. However, coordinates have been given only for those liquids upon which consistent data over the normal temperature range was available.

Reprinted from "Cameron Hydraulics Data," 1958, Cameron Pump Division, Ingersoll Rand Company, Phillipsburg, New Jersey.

Physical Properties of Water

Temperature of Water t Degrees Fahrenheit	Saturation Pressure P' Pounds per Square Inch Absolute	Specific Volume $\overline{V}$ Cubic Feet Per Pound	Weight Density ρ Pounds per Cubic Foot	Weight Pounds Per Gallon
32	0.08859	0.016022	62.414	8.3436
40	0.12163	0.016019	62.426	8.3451
50	0.17796	0.016023	62.410	8.3430
60	0.25611	0.016033	62.371	8.3378
70	0.36292	0.016050	62.305	8.3290
80	0.50683	0.016072	62.220	8.3176
90	0.69813	0.016099	62.116	8.3037
100	0.94924	0.016130	61.996	8.2877
110	1.2750	0.016165	61.862	8.2698
120	1.6927	0.016204	61.7132	8.2498
130	2.2230	0.016247	61.550	8.2280
140	2.8892	0.016293	61.376	8.2048
150	3.7184	0.016343	61.188	8.1797
160	4.7414	0.016395	60.994	8.1537
170	5.9926	0.016451	60.787	8.1260
180	7.5110	0.016510	60.569	8.0969
190	9.340	0.016572	60.343	8.0667
200	11.526	0.016637	60.107	8.0351
210	14.123	0.016705	59.862	8.0024
212	14.696	0.016719	59.812	7.9957
220	17.186	0.016775	59.613	7.9690
240	24.968	0.016926	59.081	7.8979
260	35.427	0.017089	58.517	7.8226
280	49.200	0.017264	57.924	7.7433
300	67.005	0.01745	57.307	7.6608
350	134.604	0.01799	55.586	7.4308
400	247.259	0.01864	53.648	7.1717
450	422.55	0.01943	51.467	6.8801
500	680.86	0.02043	48.948	6.5433
550	1045.43	0.02176	45.956	6.1434
600	1543.2	0.02364	42.301	5.6548
650	2208.4	0.02674	37.397	4.9993
700	3094.3	0.03662	27.307	3.6505

Specific gravity of water at 60 F = 1.00

Weight per gallon is based on 7.48052 gallons per cubic foot.

All data on volume and pressure are abstracted from ASME Steam Tables (1967), with permission of publisher, The American Society of Mechanical Engineers, 345 East 47th Street, New York, N.Y. 10017.

Courtesy Crane Co., Reproduced from Technical Paper No. 410, "Flow of Fluids."

PROPERTIES OF SATURATED STEAM

Absolute Pressure		Vacuum Inches of Hg	Temperature	Heat of the Liquid	Latent Heat of Evaporation	Total Heat of Steam	Specific Volume
Lbs. Per Sq. In. P'	Inches of Hg		t Degrees F.	Btu/lb.	Btu/lb.	Hg Btu/lb.	$\overline{V}$ Cu. Ft. per Lb.
0.20	0.41	29.51	53.14	21.21	1063.8	1085.0	1526.0
0.25	0.51	29.41	59.30	27.36	1060.3	1087.7	1235.3
0.30	0.61	29.31	64.47	32.52	1057.4	1090.0	1039.5
0.35	0.71	29.21	68.93	36.97	1054.9	1091.9	898.5
0.40	0.81	29.11	72.86	40.89	1052.7	1093.6	791.9
0.45	0.92	29.00	76.38	44.41	1050.7	1095.1	708.5
0.50	1.02	28.90	79.58	47.60	1048.8	1096.4	641.4
0.60	1.22	28.70	85.21	53.21	1045.7	1098.9	540.0
0.70	1.43	28.49	90.08	58.07	1042.9	1101.0	466.9
0.80	1.63	28.29	94.38	62.36	1040.4	1102.8	411.7
0.90	1.83	28.09	98.24	66.21	1038.3	1104.5	368.4
1.0	2.04	27.88	101.74	69.70	1036.3	1106.0	333.6
1.2	2.44	27.48	107.92	75.87	1032.7	1108.6	280.9
1.4	2.85	27.07	113.26	81.20	1029.6	1110.8	243.0
1.6	3.26	26.66	117.99	85.91	1026.9	1112.8	214.3
1.8	3.66	26.26	122.23	90.14	1024.5	1114.6	191.8
2.0	4.07	25.85	126.08	93.99	1022.2	1116.2	173.73
2.2	4.48	25.44	129.62	97.52	1020.2	1117.7	158.85
2.4	4.89	25.03	132.89	100.79	1018.3	1119.1	146.38
2.6	5.29	24.63	135.94	103.83	1016.5	1120.3	135.78
2.8	5.70	24.22	138.79	106.68	1014.8	1121.5	126.65
3.0	6.11	23.81	141.48	109.37	1013.2	1122.6	118.71
3.5	7.13	22.79	147.57	115.46	1009.6	1125.1	102.72
4.0	8.14	21.78	152.97	120.86	1006.4	1127.3	90.63
4.5	9.16	20.76	157.83	125.71	1003.6	1129.3	81.16
5.0	10.18	19.74	162.24	130.13	1001.0	1131.1	73.52
5.5	11.20	18.72	166.30	134.19	998.5	1132.7	67.24
6.0	12.22	17.70	170.06	137.96	996.2	1134.2	61.98
6.5	13.23	16.69	173.56	141.47	994.1	1135.6	57.50
7.0	14.25	15.67	176.85	144.76	992.1	1136.9	53.64
7.5	15.27	14.65	179.94	147.86	990.2	1138.1	50.29
8.0	16.29	13.63	182.86	150.79	988.5	1139.3	47.34
8.5	17.31	12.61	185.64	153.57	986.8	1140.4	44.73
9.0	18.32	11.60	188.28	156.22	985.2	1141.4	42.40
9.5	19.34	10.58	190.80	158.75	983.6	1142.3	40.31
10.0	20.36	9.56	193.21	161.17	982.1	1143.3	38.42
11.0	22.40	7.52	197.75	165.73	979.3	1145.0	35.14
12.0	24.43	5.49	201.96	169.96	976.6	1146.6	32.40
13.0	26.47	3.45	205.88	173.91	974.2	1148.1	30.06
14.0	28.50	1.42	209.56	177.61	971.9	1149.5	28.04

Reprinted from "Control Valve Handbook," 1965, Fisher Controls Company, Marshalltown, Iowa.

PROPERTIES OF SATURATED STEAM

Pressure Lbs. per Sq. In. Absolute P'	Gauge P	Temperature t Degrees F.	Heat of the Liquid Btu/lb.	Latent Heat of Evaporation Btu/lb.	Total Heat of Steam Hg Btu/lb.	Specific Volume $\overline{V}$ Cu. Ft. per Lb.
14.696	0.0	212.00	180.07	970.3	1150.4	26.80
15.0	0.3	213.03	181.11	969.7	1150.8	26.29
16.0	1.3	216.32	184.42	967.6	1152.0	24.75
17.0	2.3	219.44	187.56	965.5	1153.1	23.39
18.0	3.3	222.41	190.56	963.6	1154.2	22.17
19.0	4.3	225.24	193.42	961.9	1155.3	21.08
20.0	5.3	227.96	196.16	960.1	1156.3	20.089
21.0	6.3	230.57	198.79	958.4	1157.2	19.192
22.0	7.3	233.07	201.33	956.8	1158.1	18.375
23.0	8.3	235.49	203.78	955.2	1159.0	17.627
24.0	9.3	237.82	206.14	953.7	1159.8	16.938
25.0	10.3	240.07	208.42	952.1	1160.6	16.303
26.0	11.3	242.25	210.62	950.7	1161.3	15.715
27.0	12.3	244.36	212.75	949.3	1162.0	15.170
28.0	13.3	246.41	214.83	947.9	1162.7	14.663
29.0	14.3	248.40	216.86	946.5	1163.4	14.189
30.0	15.3	250.33	218.82	945.3	1164.1	13.746
31.0	16.3	252.22	220.73	944.0	1164.7	13.330
32.0	17.3	254.05	222.59	942.8	1165.4	12.940
33.0	18.3	255.84	224.41	941.6	1166.0	12.572
34.0	19.3	257.58	226.18	940.3	1166.5	12.226
35.0	20.3	259.28	227.91	939.2	1167.1	11.898
36.0	21.3	260.95	229.60	938.0	1167.6	11.588
37.0	22.3	262.57	231.26	936.9	1168.2	11.294
38.0	23.3	264.16	232.89	935.8	1168.7	11.015
39.0	24.3	265.72	234.48	934.7	1169.2	10.750
40.0	25.3	267.25	236.03	933.7	1169.7	10.498
41.0	26.3	268.74	237.55	932.6	1170.2	10.258
42.0	27.3	270.21	239.04	931.6	1170.7	10.029
43.0	28.3	271.64	240.51	930.6	1171.1	9.810
44.0	29.3	273.05	241.95	929.6	1171.6	9.601
45.0	30.3	274.44	243.36	928.6	1172.0	9.401
46.0	31.3	275.80	244.75	927.7	1172.4	9.209
47.0	32.3	277.13	246.12	926.7	1172.9	9.025
48.0	33.3	278.45	247.47	925.8	1173.3	8.848
49.0	34.3	279.74	248.79	924.9	1173.7	8.678
50.0	35.3	281.01	250.09	924.0	1174.1	8.515
51.0	36.3	282.26	251.37	923.0	1174.4	8.359
52.0	37.3	283.49	252.63	922.2	1174.8	8.208
53.0	38.3	284.70	253.87	921.3	1175.2	8.062
54.0	39.3	285.90	255.09	920.5	1175.6	7.922
55.0	40.3	287.07	256.30	919.6	1175.9	7.787
56.0	41.3	288.23	257.50	918.8	1176.3	7.656
57.0	42.3	289.37	258.67	917.9	1176.6	7.529
58.0	43.3	290.50	259.82	917.1	1176.9	7.407
59.0	44.3	291.61	260.96	916.3	1177.3	7.289
60.0	45.3	292.71	262.09	915.5	1177.6	7.175
61.0	46.3	293.79	263.20	914.7	1177.9	7.064
62.0	47.3	294.85	264.30	913.9	1178.2	6.957
63.0	48.3	295.90	265.38	913.1	1178.5	6.853
64.0	49.3	296.94	266.45	912.3	1178.8	6.752

Reprinted from "Control Valve Handbook," 1965, Fisher Controls Company, Marshalltown, Iowa.

PROPERTIES OF SATURATED STEAM

Pressure Lbs. per Sq. In.		Temperature	Heat of the Liquid	Latent Heat of Evaporation	Total Heat of Steam	Specific Volume
Absolute P'	Gauge P	t Degrees F.	Btu/lb.	Btu/lb.	H_g Btu/lb.	$\overline{V}$ Cu. Ft. per lb.
65.0	50.3	297.97	267.50	911.6	1179.1	6.655
66.0	51.3	298.99	268.55	910.8	1179.4	6.560
67.0	52.3	299.99	269.58	910.1	1179.7	6.468
68.0	53.3	300.98	270.60	909.4	1180.0	6.378
69.0	54.3	301.96	291.61	908.7	1180.3	6.291
70.0	55.3	302.92	272.61	907.9	1180.6	6.206
71.0	56.3	303.88	273.60	907.2	1180.8	6.124
72.0	57.3	304.83	274.57	906.5	1181.1	6.044
73.0	58.3	305.76	275.54	905.8	1181.3	5.966
74.0	59.3	306.68	276.49	905.1	1181.6	5.890
75.0	60.3	307.60	277.43	904.5	1181.9	5.816
76.0	61.3	308.50	278.37	903.7	1182.1	5.743
77.0	62.3	309.40	279.30	903.1	1182.4	5.673
78.0	63.3	310.29	280.21	902.4	1182.6	5.604
79.0	64.3	311.16	281.12	901.7	1182.8	5.537
80.0	65.3	312.03	282.02	901.1	1183.1	5.472
81.0	66.3	312.89	282.91	900.4	1183.3	5.408
82.0	67.3	313.74	283.79	899.7	1183.5	5.346
83.0	68.3	314.59	284.66	899.1	1183.8	5.285
84.0	69.3	315.42	285.53	898.5	1184.0	5.226
85.0	70.3	316.25	286.39	897.8	1184.2	5.168
86.0	71.3	317.07	287.24	897.2	1184.4	5.111
87.0	72.3	317.88	288.08	896.5	1184.6	5.055
88.0	73.3	318.68	288.91	895.9	1184.8	5.001
89.0	74.3	319.48	289.74	895.3	1185.1	4.948
90.0	75.3	320.27	290.56	894.7	1185.3	4.896
91.0	76.3	321.06	291.38	894.1	1185.5	4.845
92.0	77.3	321.83	292.18	893.5	1185.7	4.796
93.0	78.3	322.60	292.98	892.9	1185.9	4.747
94.0	79.3	323.36	293.78	892.3	1186.1	4.699
95.0	80.3	324.12	294.56	891.7	1186.2	4.652
96.0	81.3	324.87	295.34	891.1	1186.4	4.606
97.0	82.3	325.61	296.12	890.5	1186.6	4.561
98.0	83.3	326.35	296.89	889.9	1186.8	4.517
99.0	84.3	327.08	297.65	889.4	1187.0	4.474
100.0	85.3	327.81	298.40	888.8	1187.2	4.432
101.0	86.3	328.53	299.15	888.2	1187.4	4.391
102.0	87.3	329.25	299.90	887.6	1187.5	4.350
103.0	88.3	329.96	300.64	887.1	1187.7	4.310
104.0	89.3	330.66	301.37	886.5	1187.9	4.271
105.0	90.3	331.36	302.10	886.0	1188.1	4.232
106.0	91.3	332.05	302.82	885.4	1188.2	4.194
107.0	92.3	332.74	303.54	884.9	1188.4	4.157
108.0	93.3	333.42	304.26	884.3	1188.6	4.120
109.0	94.3	334.10	304.97	883.7	1188.7	4.084
110.0	95.3	334.77	305.66	883.2	1188.9	4.049
111.0	96.3	335.44	306.37	882.6	1189.0	4.015
112.0	97.3	336.11	307.06	882.1	1189.2	3.981
113.0	98.3	336.77	307.75	881.6	1189.4	3.947
114.0	99.3	337.42	308.43	881.1	1189.5	3.914

Reprinted from "Control Valve Handbook," 1965, Fisher Controls Company, Marshalltown, Iowa.

PROPERTIES OF SATURATED STEAM

Pressure Lbs. per Sq. In.		Temperature	Heat of the Liquid	Latent Heat of Evaporation	Total Heat of Steam	Specific Volume
Absolute P'	Gauge P	t Degrees F.	Btu/lb.	Btu/lb.	H_g Btu/lb.	V Cu. Ft. per lb.
115.0	100.3	338.07	309.11	880.6	1189.7	3.882
116.0	101.3	338.72	309.79	880.0	1189.8	3.850
117.0	102.3	339.36	310.46	879.5	1190.0	3.819
118.0	103.3	339.99	311.12	879.0	1190.1	3.788
119.0	104.3	340.62	311.78	878.4	1190.2	3.758
120.0	105.3	341.25	312.44	877.9	1190.4	3.728
121.0	106.3	341.88	313.10	877.4	1190.5	3.699
122.0	107.3	342.50	313.75	876.9	1190.7	3.670
123.0	108.3	343.11	314.40	876.4	1190.8	3.642
124.0	109.3	343.72	315.04	875.9	1190.9	3.614
125.0	110.3	344.33	315.68	875.4	1191.1	3.587
126.0	111.3	344.94	316.31	874.9	1191.2	3.560
127.0	112.3	345.54	316.94	874.4	1191.3	3.533
128.0	113.3	346.13	317.57	873.9	1191.5	3.507
129.0	114.3	346.73	318.19	873.4	1191.6	3.481
130.0	115.3	347.32	318.81	872.9	1191.7	3.455
131.0	116.3	347.90	319.43	872.5	1191.9	3.430
132.0	117.3	348.48	320.04	872.0	1192.0	3.405
133.0	118.3	349.06	320.65	871.5	1192.1	3.381
134.0	119.3	349.64	321.25	871.0	1192.2	3.357
135.0	120.3	350.21	321.85	870.6	1192.4	3.333
136.0	121.3	350.78	322.45	870.1	1192.5	3.310
137.0	122.3	351.35	323.05	869.6	1192.6	3.287
138.0	123.3	351.91	323.64	869.1	1192.7	3.264
139.0	124.3	352.47	324.23	868.7	1192.9	3.242
140.0	125.3	353.02	324.82	868.2	1193.0	3.220
141.0	126.3	353.57	325.40	867.7	1193.1	3.198
142.0	127.3	354.12	325.98	867.2	1193.2	3.177
143.0	128.3	354.67	326.56	866.7	1193.3	3.155
144.0	129.3	355.21	327.13	866.3	1193.4	3.134
145.0	130.3	355.76	327.70	865.8	1193.5	3.114
146.0	131.3	356.29	328.27	865.3	1193.6	3.094
147.0	132.3	356.83	328.83	864.9	1193.8	3.074
148.0	133.3	357.36	329.39	864.5	1193.9	3.054
149.0	134.3	357.89	329.95	864.0	1194.0	3.034
150.0	135.3	358.42	330.51	863.6	1194.1	3.015
152.0	137.3	359.46	331.61	862.7	1194.3	2.977
154.0	139.3	360.49	332.70	861.8	1194.5	2.940
156.0	141.3	361.52	333.79	860.9	1194.7	2.904
158.0	143.3	362.53	334.86	860.0	1194.9	2.869
160.0	145.3	363.53	335.93	859.2	1195.1	2.834
162.0	147.3	364.53	336.98	858.3	1195.3	2.801
164.0	149.3	365.51	338.02	857.5	1195.5	2.768
166.0	151.3	366.48	339.05	856.6	1195.7	2.736
168.0	153.3	367.45	340.07	855.7	1195.8	2.705
170.0	155.3	368.41	341.09	854.9	1196.0	2.675
172.0	157.3	369.35	342.10	854.1	1196.2	2.645
174.0	159.3	370.29	343.10	853.3	1196.4	2.616
176.0	161.3	371.22	344.09	852.4	1196.5	2.587
178.0	163.3	372.14	345.06	851.6	1196.7	2.559

Reprinted from "Control Valve Handbook," 1965, Fisher Controls Company, Marshalltown, Iowa.

PROPERTIES OF SATURATED STEAM

Pressure Lbs. per Sq. In.		Temperature	Heat of the Liquid	Latent Heat of Evaporation	Total Heat of Steam	Specific Volume
Absolute P'	Gauge P	t Degrees F.	Btu/lb.	Btu/lb.	H_g Btu/lb.	V Cu. Ft. per lb.
180.0	165.3	373.06	346.03	850.8	1196.9	2.532
182.0	167.3	373.96	347.00	850.0	1197.0	2.505
184.0	169.3	374.86	347.96	849.2	1197.2	2.479
186.0	171.3	375.75	348.92	848.4	1197.3	2.454
188.0	173.3	376.64	349.86	847.6	1197.5	2.429
190.0	175.3	377.51	350.79	846.8	1197.6	2.404
192.0	177.3	378.38	351.72	846.1	1197.8	2.380
194.0	179.3	379.24	352.64	845.3	1197.9	2.356
196.0	181.3	380.10	353.55	844.5	1198.1	2.333
198.0	183.3	380.95	354.46	843.7	1198.2	2.310
200.0	185.3	381.79	355.36	843.0	1198.4	2.288
205.0	190.3	383.86	357.58	841.1	1198.7	2.234
210.0	195.3	385.90	359.77	839.2	1199.0	2.183
215.0	200.3	387.89	361.91	837.4	1199.3	2.134
220.0	205.3	389.86	364.02	835.6	1199.6	2.087
225.0	210.3	391.79	366.09	833.8	1199.9	2.0422
230.0	215.3	393.68	368.13	832.0	1200.1	1.9992
235.0	220.3	395.54	370.14	830.3	1200.4	1.9579
240.0	225.3	397.37	372.12	828.5	1200.6	1.9183
245.0	230.3	399.18	374.08	826.8	1200.9	1.8803
250.0	235.3	400.95	376.00	825.1	1201.1	1.8438
255.0	240.3	402.70	377.89	823.4	1201.3	1.8086
260.0	245.3	404.42	379.76	821.8	1201.5	1.7748
265.0	250.3	406.11	381.60	820.1	1201.7	1.7422
270.0	255.3	407.78	383.42	818.5	1201.9	1.7107
275.0	260.3	409.43	385.21	816.9	1202.1	1.6804
280.0	265.3	411.05	386.98	815.3	1202.3	1.6511
285.0	270.3	412.65	388.73	813.7	1202.4	1.6228
290.0	275.3	414.23	390.46	812.1	1202.6	1.5954
295.0	280.3	415.79	392.16	810.5	1202.7	1.5689
300.0	285.3	417.33	393.84	809.0	1202.8	1.5433
320.0	305.3	423.29	400.39	803.0	1203.4	1.4485
340.0	325.3	428.97	406.66	797.1	1203.7	1.3645
360.0	345.3	434.40	412.67	791.4	1204.1	1.2895
380.0	365.3	439.60	418.45	785.8	1204.3	1.2222
400.0	385.3	444.59	424.0	780.5	1204.5	1.1613
420.0	405.3	449.39	429.4	775.2	1204.6	1.1061
440.0	425.3	454.02	434.6	770.0	1204.6	1.0556
460.0	445.3	458.50	439.7	764.9	1204.6	1.0094
480.0	465.3	462.82	444.6	759.9	1204.5	0.9670
500.0	485.3	467.01	449.4	755.0	1204.4	0.9278
520.0	505.3	471.07	454.1	750.1	1204.2	0.8915
540.0	525.3	475.01	458.6	745.4	1204.0	0.8578
560.0	545.3	478.85	463.0	740.8	1203.8	0.8265
580.0	565.3	482.58	467.4	736.1	1203.5	0.7973
600.0	585.3	486.21	471.6	731.6	1203.2	0.7698
620.0	605.3	489.75	475.7	727.2	1202.9	0.7440
640.0	625.3	493.21	479.8	722.7	1202.5	0.7198
660.0	645.3	496.58	483.8	718.3	1202.1	0.6971
680.0	665.3	499.88	487.7	714.0	1201.7	0.6757

Reprinted from "Control Valve Handbook," 1965, Fisher Controls Company, Marshalltown, Iowa.

PROPERTIES OF SATURATED STEAM

Pressure Lbs. per Sq. In.		Temperature	Heat of the Liquid	Latent Heat of Evaporation	Total Heat of Steam	Specific Volume
Absolute P'	Gauge P	t Degrees F.	Btu/lb.	Btu/lb.	H_g Btu/lb.	$\overline{V}$ Cu.Ft.per lb.
700.0	685.3	503.10	491.5	709.7	1201.2	0.6554
720.0	705.3	506.25	495.3	705.4	1200.7	0.6362
740.0	725.3	509.34	499.0	701.2	1200.2	0.6180
760.0	745.3	512.36	502.6	697.1	1199.7	0.6007
780.0	765.3	515.33	506.2	692.9	1199.1	0.5843
800.0	785.3	518.23	509.7	688.9	1198.6	0.5687
820.0	805.3	521.08	513.2	684.8	1198.0	0.5538
840.0	825.3	523.88	516.6	680.8	1197.4	0.5396
960.0	845.3	526.63	520.0	676.8	1196.8	0.5260
880.0	865.3	529.33	523.3	672.8	1196.1	0.5130
900.0	885.3	531.98	526.6	668.8	1195.4	0.5006
920.0	905.3	534.59	529.8	664.9	1194.7	0.4886
940.0	925.3	537.16	533.0	661.0	1194.0	0.4772
960.0	945.3	539.68	536.2	657.1	1193.3	0.4663
980.0	965.3	542.17	539.3	653.3	1192.6	0.4557
1000.0	985.3	544.61	542.4	649.4	1191.8	0.4456
1050.0	1035.3	550.57	550.0	639.9	1189.9	0.4218
1100.0	1085.3	556.31	557.4	630.4	1187.8	0.4001
1150.0	1135.3	561.86	564.6	621.0	1185.6	0.3802
1200.0	1185.3	567.22	571.7	611.7	1183.4	0.3619
1250.0	1235.3	572.42	578.6	602.4	1181.0	0.3450
1300.0	1285.3	577.46	585.4	593.2	1178.6	0.3293
1350.0	1335.3	582.35	592.1	584.0	1176.1	0.3148
1400.0	1385.3	587.10	598.7	574.7	1173.4	0.3012
1450.0	1435.3	591.73	605.2	565.5	1170.7	0.2884
1500.0	1485.3	596.23	611.6	556.3	1167.9	0.2765
1600.0	1585.3	604.90	624.1	538.0	1162.1	0.2548
1700.0	1685.3	613.15	636.3	519.6	1155.9	0.2354
1800.0	1785.3	621.03	648.3	501.1	1149.4	0.2179
1900.0	1885.3	628.58	660.1	482.4	1142.4	0.2021
2000.0	1985.3	635.82	671.7	463.4	1135.1	0.1878
2100.0	2085.3	642.77	683.3	444.1	1127.4	0.1746
2200.0	2185.3	649.46	694.8	424.4	1119.2	0.1625
2300.0	2285.3	655.91	706.5	403.9	1110.4	0.1513
2400.0	2385.3	662.12	718.4	382.7	1101.1	0.1407
2500.0	2485.3	668.13	730.6	360.5	1091.1	0.1307
2600.0	2585.3	673.94	743.0	337.2	1080.2	0.1213
2700.0	2685.3	679.55	756.2	312.1	1068.3	0.1123
2800.0	2785.3	684.99	770.1	284.7	1054.8	0.1035
2900.0	2885.3	690.26	785.4	253.6	1039.0	0.0947
3000.0	2985.3	695.36	802.5	217.8	1020.3	0.0858
3100.0	3085.3	700.31	825.0	168.1	993.1	0.0753
3200.0	3185.3	705.11	872.4	62.0	934.4	0.0580
3206.2	3191.5	705.40	902.7	0.0	902.7	0.0503

Reprinted from "Control Valve Handbook," 1965, Fisher Controls Company, Marshalltown, Iowa.

PROPERTIES OF SUPERHEATED STEAM

$\bar{V}$ = specific volume, cubic feet per pound
h_g = total heat of steam, Btu per pound

Pressure Lbs. per Sq. In.		Sat. Temp.		Total Temperature - Degrees Fahrenheit (t)										
Absolute P'	Gauge P	t		360°	400°	440°	480°	500°	600°	700°	800°	900°	1000°	1200°
14.696	0.0	212.00	$\bar{V}$	33.03	34.68	36.32	37.96	38.78	42.86	46.94	51.00	55.07	59.13	67.25
			h_g	1221.1	1239.9	1258.8	1277.6	1287.1	1334.8	1383.2	1432.3	1482.3	1533.1	1637.5
20.0	5.3	227.96	$\bar{V}$	24.21	25.43	26.65	27.86	28.46	31.47	34.47	37.46	40.45	43.44	49.41
			h_g	1220.3	1239.2	1258.2	1277.1	1286.6	1334.4	1382.9	1432.1	1482.1	1533.0	1637.4
30.0	15.3	250.33	$\bar{V}$	16.072	16.897	17.714	18.528	18.933	20.95	22.96	24.96	26.95	28.95	32.93
			h_g	1218.6	1237.9	1257.0	1276.2	1285.7	1333.8	1382.4	1431.7	1481.8	1532.7	1637.2
40.0	25.3	267.25	$\bar{V}$	12.001	12.628	13.247	13.862	14.168	15.688	17.198	18.702	20.20	21.70	24.69
			h_g	1216.9	1236.5	1255.9	1275.2	1284.8	1333.1	1381.9	1431.3	1481.4	1532.4	1637.0
50.0	35.3	281.01	$\bar{V}$	9.557	10.065	10.567	11.062	11.309	12.532	13.744	14.950	16.152	17.352	19.747
			h_g	1215.2	1235.1	1254.7	1274.2	1283.9	1332.5	1381.4	1430.9	1481.1	1532.1	1636.8
60.0	45.3	292.71	$\bar{V}$	7.927	8.357	8.779	9.196	9.403	10.427	11.441	12.449	13.452	14.454	16.451
			h_g	1213.4	1233.6	1253.5	1273.2	1283.0	1331.8	1380.9	1430.5	1480.8	1531.9	1636.6
70.0	55.3	302.92	$\bar{V}$	6.762	7.136	7.502	7.863	8.041	8.924	9.796	10.662	11.524	12.383	14.097
			h_g	1211.5	1232.1	1252.3	1272.2	1282.0	1331.1	1380.4	1430.1	1480.5	1531.6	1636.3
80.0	65.3	312.03	$\bar{V}$	5.888	6.220	6.544	6.862	7.020	7.797	8.562	9.322	10.077	10.830	12.332
			h_g	1209.7	1230.7	1251.1	1271.1	1281.1	1330.5	1379.9	1429.7	1480.1	1531.3	1636.2
90.0	75.3	320.27	$\bar{V}$	5.208	5.508	5.799	6.084	6.225	6.920	7.603	8.279	8.952	9.623	10.959
			h_g	1207.7	1229.1	1249.8	1270.1	1280.1	1329.8	1379.4	1429.3	1479.8	1531.0	1635.9
100.0	85.3	327.81	$\bar{V}$	4.663	4.937	5.202	5.462	5.589	6.218	6.835	7.446	8.052	8.656	9.860
			h_g	1205.7	1227.6	1248.6	1269.0	1279.1	1329.1	1378.9	1428.9	1479.5	1530.8	1635.7
120.0	105.3	341.25	$\bar{V}$	3.844	4.081	4.307	4.527	4.636	5.165	5.683	6.195	6.702	7.207	8.212
			h_g	1201.6	1224.4	1246.0	1266.9	1277.2	1327.7	1377.8	1428.1	1478.8	1530.2	1635.3
140.0	125.3	353.02	$\bar{V}$	3.258	3.468	3.667	3.860	3.954	4.413	4.861	5.301	5.738	6.172	7.035
			h_g	1197.3	1221.1	1243.3	1264.7	1275.2	1326.4	1376.8	1427.3	1478.2	1529.7	1634.9
160.0	145.3	363.53	$\bar{V}$	...	3.008	3.187	3.359	3.443	3.849	4.244	4.631	5.015	5.396	6.152
			h_g	...	1217.6	1240.6	1262.4	1273.1	1325.0	1375.7	1426.4	1477.5	1529.1	1634.5
180.0	165.3	373.06	$\bar{V}$	...	2.649	2.813	2.969	3.044	3.411	3.764	4.110	4.452	4.792	5.466
			h_g	...	1214.0	1237.8	1260.2	1271.0	1323.5	1374.7	1425.6	1476.8	1528.6	1634.1
200.0	185.3	381.79	$\bar{V}$	...	2.361	2.513	2.656	2.726	3.060	3.380	3.693	4.002	4.309	4.917
			h_g	...	1210.3	1234.9	1257.8	1268.9	1322.1	1373.6	1424.8	1476.2	1528.0	1633.7
220.0	205.3	389.86	$\bar{V}$	...	2.125	2.267	2.400	2.465	2.772	3.066	3.352	3.634	3.913	4.467
			h_g	...	1206.5	1231.9	1255.4	1266.7	1320.7	1372.6	1424.0	1475.5	1527.5	1633.3
240.0	225.3	397.37	$\bar{V}$	...	1.9276	2.062	2.187	2.247	2.533	2.804	3.068	3.327	3.584	4.093
			h_g	...	1202.5	1228.8	1253.0	1264.5	1319.2	1371.5	1423.2	1474.8	1526.9	1632.9
260.0	245.3	404.42	$\bar{V}$	...	...	1.8882	2.006	2.063	2.330	2.582	2.827	3.067	3.305	3.776
			h_g	...	...	1225.7	1250.5	1262.3	1317.7	1370.4	1422.3	1474.2	1526.3	1632.5
280.0	265.3	411.05	$\bar{V}$	...	...	1.7388	1.8512	1.9047	2.156	2.392	2.621	2.845	3.066	3.504
			h_g	...	...	1222.4	1247.9	1260.0	1316.2	1369.4	1421.5	1473.5	1525.8	1632.1
300.0	285.3	417.33	$\bar{V}$	...	...	1.6090	1.7165	1.7675	2.005	2.227	2.442	2.652	2.859	3.269
			h_g	...	...	1219.1	1245.3	1257.6	1314.7	1368.3	1420.6	1472.8	1525.2	1631.7
320.0	305.3	423.29	$\bar{V}$	...	...	1.4950	1.5985	1.6472	1.8734	2.083	2.285	2.483	2.678	3.063
			h_g	...	...	1215.6	1242.6	1255.2	1313.2	1367.2	1419.8	1472.1	1524.7	1631.3
340.0	325.3	428.97	$\bar{V}$	...	...	1.3941	1.4941	1.5410	1.7569	1.9562	2.147	2.334	2.518	2.881
			h_g	...	...	1212.1	1239.9	1252.8	1311.6	1366.1	1419.0	1471.5	1524.1	1630.9
360.0	345.3	434.40	$\bar{V}$	...	...	1.3041	1.4012	1.4464	1.6533	1.8431	2.025	2.202	2.376	2.719
			h_g	...	...	1208.4	1237.1	1250.3	1310.1	1365.0	1418.1	1470.8	1523.5	1630.5

Reprinted from "Control Valve Handbook," 1965, Fisher Controls Company, Marshalltown, Iowa.

PROPERTIES OF SUPERHEATED STEAM

$\bar{V}$ = specific volume, cubic feet per pound
h_g = total heat of steam, Btu per pound

Pressure. Lbs. per Sq. In.		Sat. Temp.		Total Temperature - Degrees Fahrenheit (t)										
Absolute P'	Gauge P	t		500°	540°	600°	640°	660°	700°	740°	800°	900°	1000°	1200°
380.0	365.3	439.60	$\bar{V}$	1.3616	1.4444	1.5605	1.6345	1.6707	1.7419	1.8118	1.9149	2.083	2.249	2.575
			h_g	1247.7	1273.1	1308.5	1331.0	1342.0	1363.8	1385.3	1417.3	1470.1	1523.0	1630.0
400.0	385.3	444.59	$\bar{V}$	1.2851	1.3652	1.4770	1.5480	1.5827	1.6508	1.7177	1.8161	1.9767	2.134	2.445
			h_g	1245.1	1271.0	1306.9	1329.6	1340.8	1362.7	1384.3	1416.4	1469.4	1522.4	1629.6
420.0	405.3	449.39	$\bar{V}$	1.2158	1.2935	1.4014	1.4697	1.5030	1.5684	1.6324	1.7267	1.8802	2.031	2.327
			h_g	1242.5	1268.9	1305.3	1328.3	1339.5	1361.6	1383.3	1415.5	1468.7	1521.9	1629.2
440.0	425.3	454.02	$\bar{V}$	1.1526	1.2282	1.3327	1.3984	1.4306	1.4934	1.5549	1.6454	1.7925	1.9368	2.220
			h_g	1239.8	1266.7	1303.6	1326.9	1338.2	1360.4	1382.3	1414.7	1468.1	1521.3	1628.8
460.0	445.3	458.50	$\bar{V}$	1.0948	1.1685	1.2698	1.3334	1.3644	1.4250	1.4842	1.5711	1.7124	1.8508	2.122
			h_g	1237.0	1264.5	1302.0	1325.4	1336.9	1359.3	1381.3	1413.8	1467.4	1520.7	1628.4
480.0	465.3	462.82	$\bar{V}$	1.0417	1.1138	1.2122	1.2737	1.3038	1.3622	1.4193	1.5031	1.6390	1.7720	2.033
			h_g	1234.2	1262.3	1300.3	1324.0	1335.6	1358.2	1380.3	1412.9	1466.7	1520.2	1628.0
500.0	485.3	467.01	$\bar{V}$	0.9927	1.0633	1.1591	1.2188	1.2478	1.3044	1.3596	1.4405	1.5715	1.6996	1.9504
			h_g	1231.3	1260.0	1298.6	1322.6	1334.2	1357.0	1379.3	1412.1	1466.0	1519.6	1627.6
520.0	505.3	471.07	$\bar{V}$	0.9473	1.0166	1.1101	1.1681	1.1962	1.2511	1.3045	1.3826	1.5091	1.6326	1.8743
			h_g	1228.3	1257.7	1296.9	1321.1	1332.9	1355.8	1378.2	1411.2	1465.3	1519.0	1627.2
540.0	525.3	475.01	$\bar{V}$	0.9052	0.9733	1.0646	1.1211	1.1485	1.2017	1.2535	1.3291	1.4514	1.5707	1.8039
			h_g	1225.3	1255.4	1295.2	1319.7	1331.5	1354.6	1377.2	1410.3	1464.6	1518.5	1626.8
560.0	545.3	478.85	$\bar{V}$	0.8659	0.9330	1.0224	1.0775	1.1041	1.1558	1.2060	1.2794	1.3978	1.5132	1.7385
			h_g	1222.2	1253.0	1293.4	1318.2	1330.2	1353.5	1376.1	1409.4	1463.9	1517.9	1626.4
580.0	565.3	482.58	$\bar{V}$	0.8291	0.8954	0.9830	1.0368	1.0627	1.1331	1.1619	1.2331	1.3479	1.4596	1.6776
			h_g	1219.0	1250.5	1291.7	1316.7	1328.8	1352.3	1375.1	1408.6	1463.2	1517.3	1626.0
600.0	585.3	486.21	$\bar{V}$	0.7947	0.8602	0.9463	0.9988	1.0241	1.0732	1.1207	1.1899	1.3013	1.4096	1.6208
			h_g	1215.7	1248.1	1289.9	1315.2	1327.4	1351.1	1374.0	1407.7	1462.5	1516.7	1625.5
620.0	605.3	489.75	$\bar{V}$	0.7624	0.8272	0.9118	0.9633	0.9880	1.0358	1.0821	1.1494	1.2577	1.3628	1.5676
			h_g	1212.4	1245.5	1288.1	1313.7	1326.0	1349.9	1373.0	1406.8	1461.8	1516.2	1625.1
640.0	625.3	493.21	$\bar{V}$	0.7319	0.7963	0.8795	0.9299	0.9541	1.0008	1.0459	1.1115	1.2168	1.3190	1.5178
			h_g	1209.0	1243.0	1286.2	1312.2	1324.6	1348.6	1371.9	1405.9	1461.1	1515.6	1624.7
660.0	645.3	496.58	$\bar{V}$	0.7032	0.7670	0.8491	0.8985	0.9222	0.9679	1.0119	1.0759	1.1784	1.2778	1.4709
			h_g	1205.4	1240.4	1284.4	1310.6	1323.2	1347.4	1370.8	1405.0	1460.4	1515.0	1624.3
680.0	665.3	499.88	$\bar{V}$	0.6759	0.7395	0.8205	0.8690	0.8922	0.9369	0.9800	1.0424	1.1423	1.2390	1.4269
			h_g	1201.8	1237.7	1282.5	1309.1	1321.7	1346.2	1369.8	1404.1	1459.7	1514.5	1623.9
700.0	685.3	503.10	$\bar{V}$	...	0.7134	0.7934	0.8411	0.8639	0.9077	0.9498	1.0108	1.1082	1.2024	1.3853
			h_g	...	1235.0	1280.6	1307.5	1320.3	1345.0	1368.7	1403.2	1459.0	1513.9	1623.5
750.0	735.3	510.86	$\bar{V}$	...	0.6540	0.7319	0.7778	0.7996	0.8414	0.8813	0.9391	1.0310	1.1196	1.2912
			h_g	...	1227.9	1275.7	1303.5	1316.6	1341.8	1366.0	1400.9	1457.2	1512.4	1622.4
800.0	785.3	518.23	$\bar{V}$	...	0.6015	0.6779	0.7223	0.7433	0.7833	0.8215	0.8763	0.9633	1.0470	1.2088
			h_g	...	1220.5	1270.7	1299.4	1312.9	1338.6	1363.2	1398.6	1455.4	1511.0	1621.4
850.0	835.3	525.26	$\bar{V}$	...	0.5546	0.6301	0.6732	0.6934	0.7320	0.7685	0.8209	0.9037	0.9830	1.1360
			h_g	...	1212.7	1265.5	1295.2	1309.0	1335.4	1360.4	1396.3	1453.6	1509.5	1620.4
900.0	885.3	531.98	$\bar{V}$	...	0.5124	0.5873	0.6294	0.6491	0.6863	0.7215	0.7716	0.8506	0.9262	1.0714
			h_g	...	1204.4	1260.1	1290.9	1305.1	1332.1	1357.5	1393.9	1451.8	1508.1	1619.3
950.0	935.3	538.42	$\bar{V}$	...	0.4740	0.5489	0.5901	0.6092	0.6453	0.6793	0.7275	0.8031	0.8753	1.0136
			h_g	...	1195.5	1254.6	1286.4	1301.1	1328.7	1354.7	1391.6	1450.0	1506.6	1618.3
1000.0	985.3	544.61	$\bar{V}$	...	...	0.5140	0.5546	0.5733	0.6084	0.6413	0.6878	0.7604	0.8294	0.9615
			h_g	...	...	1248.8	1281.9	1297.0	1325.3	1351.7	1389.2	1448.2	1505.1	1617.3

Reprinted from "Control Valve Handbook," 1965, Fisher Controls Company, Marshalltown, Iowa.

PROPERTIES OF SUPERHEATED STEAM

$\bar{V}$ = specific volume, cubic feet per pound
h_g = total heat of steam, Btu per pound

Pressure Lbs. per Sq. In. Absolute P'	Gauge P	Sat. Temp. t		Total Temperature - Degrees Fahrenheit (t) 660°	700°	740°	760°	780°	800°	860°	900°	1000°	1100°	1200°
1100.0	1085.3	556.31	$\bar{V}$	0.5110	0.5445	0.5755	0.5904	0.6049	0.6191	0.6601	0.6866	0.7503	0.8117	0.8716
			h_g	1288.5	1318.3	1345.8	1358.9	1371.7	1384.3	1420.8	1444.5	1502.2	1558.8	1615.2
1200.0	1185.3	567.22	$\bar{V}$	0.4586	0.4909	0.5206	0.5347	0.5484	0.5617	0.6003	0.6250	0.6843	0.7412	0.7967
			h_g	1279.6	1311.0	1339.6	1353.2	1366.4	1379.3	1416.7	1440.7	1499.2	1556.4	1613.1
1300.0	1285.3	577.46	$\bar{V}$	0.4139	0.4454	0.4739	0.4874	0.5004	0.5131	0.5496	0.5728	0.6284	0.6816	0.7333
			h_g	1270.2	1303.4	1333.3	1347.3	1361.0	1374.3	1412.5	1437.0	1496.2	1553.9	1611.0
1400.0	1385.3	587.10	$\bar{V}$	0.3753	0.4062	0.4338	0.4468	0.4593	0.4714	0.5061	0.5281	0.5805	0.6305	0.6789
			h_g	1260.3	1295.5	1326.7	1341.3	1355.4	1369.1	1408.2	1433.1	1493.2	1551.4	1608.9
1500.0	1485.3	596.23	$\bar{V}$	0.3413	0.3719	0.3989	0.4114	0.4235	0.4352	0.4684	0.4893	0.5390	0.5862	0.6318
			h_g	1249.8	1287.2	1320.0	1335.2	1349.7	1363.8	1403.9	1429.3	1490.1	1548.9	1606.8
1600.0	1585.3	604.90	$\bar{V}$	0.3112	0.3417	0.3682	0.3804	0.3921	0.4034	0.4353	0.4553	0.5027	0.5474	0.5906
			h_g	1238.7	1278.7	1313.0	1328.8	1343.9	1358.4	1399.5	1425.3	1487.0	1546.4	1604.6
1700.0	1685.3	613.15	$\bar{V}$	0.2842	0.3148	0.3410	0.3529	0.3643	0.3753	0.4061	0.4253	0.4706	0.5132	0.5542
			h_g	1226.8	1269.7	1305.8	1322.3	1337.9	1352.9	1395.0	1421.4	1484.0	1543.8	1602.5
1800.0	1785.3	621.03	$\bar{V}$	0.2597	0.2907	0.3166	0.3284	0.3395	0.3502	0.3801	0.3986	0.4421	0.4828	0.5218
			h_g	1214.0	1260.3	1298.4	1315.5	1331.8	1347.2	1390.4	1417.4	1480.8	1541.3	1600.4
1900.0	1885.3	628.58	$\bar{V}$	0.2371	0.2688	0.2947	0.3063	0.3173	0.3277	0.3568	0.3747	0.4165	0.4556	0.4929
			h_g	1200.2	1250.4	1290.6	1308.6	1325.4	1341.5	1385.8	1413.3	1477.7	1538.8	1598.2
2000.0	1985.3	635.82	$\bar{V}$	0.2161	0.2489	0.2748	0.2863	0.2972	0.3074	0.3358	0.3532	0.3935	0.4311	0.4668
			h_g	1184.9	1240.0	1282.6	1301.4	1319.0	1335.5	1381.2	1409.2	1474.5	1536.2	1596.1
2100.0	2085.3	642.77	$\bar{V}$	0.1962	0.2306	0.2567	0.2682	0.2789	0.2890	0.3167	0.3337	0.3727	0.4089	0.4433
			h_g	1167.7	1229.0	1274.3	1294.0	1312.3	1329.5	1376.4	1405.0	1471.4	1533.6	1593.9
2200.0	2185.3	649.46	$\bar{V}$	0.1768	0.2135	0.2400	0.2514	0.2621	0.2721	0.2994	0.3159	0.3538	0.3887	0.4218
			h_g	1147.8	1217.4	1265.7	1286.3	1305.4	1323.3	1371.5	1400.8	1468.2	1531.1	1591.8
2300.0	2285.3	655.91	$\bar{V}$	0.1575	0.1978	0.2247	0.2362	0.2468	0.2567	0.2835	0.2997	0.3365	0.3703	0.4023
			h_g	1123.8	1204.9	1256.7	1278.4	1298.4	1316.9	1366.6	1396.5	1464.9	1528.5	1589.6
2400.0	2385.3	662.12	$\bar{V}$	...	0.1828	0.2105	0.2221	0.2327	0.2425	0.2689	0.2848	0.3207	0.3534	0.3843
			h_g	...	1191.5	1247.3	1270.2	1291.1	1310.3	1361.6	1392.2	1461.7	1525.9	1587.4
2500.0	2485.3	668.13	$\bar{V}$	...	0.1686	0.1973	0.2090	0.2196	0.2294	0.2555	0.2710	0.3061	0.3379	0.3678
			h_g	...	1176.8	1237.6	1261.8	1283.6	1303.6	1356.5	1387.8	1458.4	1523.2	1585.3
2600.0	2585.3	673.94	$\bar{V}$	...	0.1549	0.1849	0.1967	0.2074	0.2172	0.2431	0.2584	0.2926	0.3236	0.3526
			h_g	...	1160.6	1227.3	1252.9	1275.8	1296.8	1351.4	1383.4	1455.1	1520.6	1583.1
2700.0	2685.3	679.55	$\bar{V}$	...	0.1415	0.1732	0.1853	0.1960	0.2059	0.2315	0.2466	0.2801	0.3103	0.3385
			h_g	...	1142.5	1216.5	1243.8	1267.9	1289.7	1346.1	1378.9	1451.8	1518.0	1580.9
2800.0	2785.3	684.99	$\bar{V}$	...	0.1281	0.1622	0.1745	0.1854	0.1953	0.2208	0.2356	0.2685	0.2979	0.3254
			h_g	...	1121.4	1205.1	1234.2	1259.6	1282.4	1340.8	1374.3	1448.5	1515.4	1578.7
2900.0	2885.3	690.26	$\bar{V}$	...	0.1143	0.1517	0.1644	0.1754	0.1853	0.2108	0.2254	0.2577	0.2864	0.3132
			h_g	...	1095.9	1193.0	1224.3	1251.1	1274.9	1335.3	1369.7	1445.1	1512.7	1576.5
3000.0	2985.3	695.36	$\bar{V}$	...	0.0984	0.1416	0.1548	0.1660	0.1760	0.2014	0.2159	0.2476	0.2757	0.3018
			h_g	...	1060.7	1180.1	1213.8	1242.2	1267.2	1329.7	1365.0	1441.8	1510.0	1574.3
3100.0	3085.3	700.31	$\bar{V}$	...	...	0.1320	0.1456	0.1571	0.1672	0.1926	0.2070	0.2382	0.2657	0.2911
			h_g	...	...	1166.2	1202.9	1233.0	1259.3	1324.1	1360.3	1438.4	1507.4	1572.1
3200.0	3185.3	705.11	$\bar{V}$	...	...	0.1226	0.1369	0.1486	0.1589	0.1843	0.1986	0.2293	0.2563	0.2811
			h_g	...	...	1151.1	1191.4	1223.5	1251.1	1318.3	1355.5	1434.9	1504.7	1569.9
3206.2	3191.5	705.40	$\bar{V}$	...	...	0.1220	0.1363	0.1480	0.1583	0.1838	0.1981	0.2288	0.2557	0.2806
			h_g	...	...	1150.2	1190.6	1222.9	1250.5	1317.9	1355.2	1434.7	1504.5	1569.8

Reprinted from "Control Valve Handbook," 1965, Fisher Controls Company, Marshalltown, Iowa.

PHYSICAL CONSTANTS OF VARIOUS FLUIDS

Fluid	Formula	Molecular Weight	Boiling Point °F. at 14.696 psia	Specific Gravity	
				Liquid 60/60 °F.	Gas
Acetic Acid	$HC_2H_3O_2$	60.05	245	1.05	
Acetone	C_3H_6O	58.08	133	.79	2.01
Air	N_2O_2	28.97	-317	.86	1.0
Alcohol, Ethyl	C_2H_6O	46.07	173	.79	1.59
Alcohol, Methyl	CH_4O	32.04	148	.79	1.11
Ammonia	NH_3	17.03	-28	.62	.59
Ammonium Chloride*	NH_4Cl			1.07	
Ammonium Hydroxide*	NH_4OH			.91	
Ammonium Sulfate*	$(NH_4)_2SO_4$			1.15	
Aniline	C_6H_7N	93.12	365	1.02	
Argon	A	39.94	-302	1.65	1.38
Beer				1.01	
Bromine	Br_2	159.84	138	2.93	5.52
Calcium Chloride*	$CaCl_2$			1.23	
Carbon Dioxide	CO_2	44.01	-109		1.52
Carbon Disulfide	CS_2	76.1	115	1.29	2.63
Carbon Monoxide	CO	28.01	-314	.80	.97
Carbon Tetrachloride	CCl_4	153.84	170	1.59	5.31
Chlorine	Cl_2	70.91	-30	1.42	2.45
Chromic Acid	H_2CrO_4	118.03		1.21	
Citric Acid	$C_6H_8O_7$	192.12		1.54	
Copper Sulfate*	$CuSO_4$			1.17	
Ether	$(C_2H_5)_2O$	74.12	34	.74	2.55
Ferric Chloride*	$FeCl_3$			1.23	
Fluorine	F_2	38.00	-305	1.11	1.31
Formaldehyde	H_2CO	30.03	-6	.82	1.08
Formic Acid	HCO_2H	46.03	214	1.23	
Furfural	$C_5H_4O_2$	96.08	324	1.16	
Glycerine	$C_3H_8O_3$	92.09	554	1.26	
Glycol	$C_2H_6O_2$	62.07	387	1.11	
Helium	He	4.003	-454	.18	.14
Hydrochloric Acid	HCl	36.47	-115	1.64	

* Aqueous Solution - 25% by weight of compound.

Reprinted from "Control Valve Handbook," 1965, Fisher Controls Company, Marshalltown, Iowa.

PHYSICAL CONSTANTS OF VARIOUS FLUIDS

Fluid	Formula	Molecular Weight	Boiling Point °F. at 14.696 psia	Specific Gravity	
				Liquid 60/60 °F.	Gas
Hydrofluoric Acid	HF	20.01	66	.92	
Hydrogen	H_2	2.016	-422	.07	.07
Hydrogen Chloride	HCl	36.47	-115		1.26
Hydrogen Sulfide	H_2S	34.07	-76	.79	1.17
Isopropyl Alcohol	C_3H_8O	60.09	180	.78	2.08
Linseed Oil			538	.93	
Magnesium Chloride*	$MgCl_2$			1.22	
Mercury	Hg	200.61	670	13.6	6.93
Methyl Bromide	CH_3Br	94.95	38	1.73	3.27
Methyl Chloride	CH_3Cl	50.49	-11	.99	1.74
Naphthalene	$C_{10}H_8$	128.16	424	1.14	4.43
Nitric Acid	HNO_3	63.02	187	1.5	
Nitrogen	N_2	28.02	-320	.81	.97
Oil, Vegetable				.91 - .94	
Oxygen	O_2	32	-297	1.14	1.105
Phosgene	$COCl_2$	98.92	47	1.39	3.42
Phosphoric Acid	H_3PO_4	98.00	415	1.83	
Potassium Carbonate*	K_2CO_3			1.24	
Potassium Chloride*	KCl			1.16	
Potassium Hydroxide*	KOH			1.24	
Sodium Chloride*	NaCl			1.19	
Sodium Hydroxide*	NaOH			1.27	
Sodium Sulfate*	Na_2SO_4			1.24	
Sodium Thiosulfate*	$Na_2S_2O_3$			1.23	
Starch	$(C_6H_{10}O_5)_x$			1.50	
Sugar Solutions*	$C_{12}H_{22}O_{11}$			1.10	
Sulfuric Acid	H_2SO_4	98.08	626	1.83	
Sulfur Dioxide	SO_2	64.6	14	1.39	2.21
Turpentine			320	.87	
Water	H_2O	18.016	212	1.00	.62
Zinc Chloride*	$ZnCl_2$			1.24	
Zinc Sulfate*	$ZnSO_4$			1.31	

* Aqueous Solution - 25% by weight of compound.

Reprinted from "Control Valve Handbook," 1965, Fisher Controls Company, Marshalltown, Iowa.

PHYSICAL CONSTANTS OF COMMON HYDROCARBONS

No.	Compound	Formula	Molecular Weight	Boiling Point °F., 14.696 psi. abs.	Specific Gravity Liquid 60/60 °F.	Specific Gravity Gas
1	Methane	CH_4	16.042	-258.69	...	0.555
2	Ethane	C_2H_6	30.068	-127.53	...	1.046
3	Propane	C_3H_8	44.094	-43.73	0.5077	1.547
4	n-Butane	C_4H_{10}	58.120	31.10	0.5844	2.071
5	2-Methylpropane (isobutane)	C_4H_{10}	58.120	10.89	0.5631	2.067
6	n-Pentane	C_5H_{12}	72.146	96.933	0.63116	2.4906
7	2-Methylbutane (isopentane)	C_5H_{12}	72.146	82.134	0.62476	2.4906
8	2,2-Dimethylpropane (neopentane)	C_5H_{12}	72.146	49.105	0.601	2.4906
9	n-Hexane	C_6H_{14}	86.172	155.736	0.66405	2.9749
10	2-Methylpentane	C_6H_{14}	86.172	140.488	0.65790	2.9749
11	3-Methylpentane	C_6H_{14}	86.172	145.908	0.66902	2.9749
12	2,2-Dimethylbutane (neohexane)	C_6H_{14}	86.172	121.534	0.65399	2.9749
13	2,3-Dimethylbutane	C_6H_{14}	86.172	136.378	0.66639	2.9749
14	n-Heptane	C_7H_{16}	100.198	209.169	0.68819	3.4591
15	2-Methylhexane	C_7H_{16}	100.198	194.094	0.68299	3.4591
16	3-Methylhexane	C_7H_{16}	100.198	197.330	0.69151	3.4591
17	3-Ethylpentane	C_7H_{16}	100.198	200.251	0.70257	3.4591
18	2,2-Dimethylpentane	C_7H_{16}	100.198	174.560	0.67833	3.4591
19	2,4-Dimethylpentane	C_7H_{16}	100.198	176.900	0.67723	3.4591
20	3,3-Dimethylpentane	C_7H_{16}	100.198	186.917	0.69767	3.4591
21	2,2,3-Trimethylbutane (triptane)	C_7H_{16}	100.198	177.584	0.69454	3.4591
22	n-Octane	C_8H_{18}	114.224	258.197	0.70677	3.9432
23	2,5-Dimethylhexane (diisobutyl)	C_8H_{18}	114.224	228.385	0.69795	3.9432
24	2,2,4-Trimethylpentane ("isooctane")	C_8H_{18}	114.224	210.628	0.69625	3.9432
25	n-Nonane	C_9H_{20}	128.250	303.436	0.72171	4.4275
26	n-Decane	$C_{10}H_{22}$	142.276	345.2	0.73413	4.9118
27	Cyclopentane	C_5H_{10}	70.130	120.672	0.75048	2.4211
28	Methylcyclopentane	C_6H_{12}	84.156	161.262	0.75354	2.9053
29	Cyclohexane	C_6H_{12}	84.156	177.328	0.78344	2.9053
30	Methylcyclohexane	C_7H_{14}	98.182	213.681	0.77398	3.3896
31	Ethene (ethylene)	C_2H_4	28.052	-154.68	...	0.9684
32	Propene	C_3H_6	42.078	-53.86	0.5218	1.4526
33	1-Butene	C_4H_8	56.104	20.73	0.6011	1.9368
34	cis-2-Butene	C_4H_8	56.104	38.70	0.6272	1.9368
35	trans-2-Butene	C_4H_8	56.104	33.58	0.6100	1.9368
36	2-Methylpropene (isobutene)	C_4H_8	56.104	19.68	0.6002	1.9368
37	1-Pentene	C_5H_{10}	70.130	85.95	0.6461	2.4210
38	1,2-Butadiene	C_4H_6	54.088	50.5	0.658	1.8673
39	1,3-Butadiene	C_4H_6	54.088	24.06	0.6272	1.8673
40	2-Methyl-1,3-butadiene (isoprene)	C_5H_8	68.114	93.34	0.6861	2.3515
41	Ethyne (acetylene)	C_2H_2	26.036	-119	0.615	0.8988
42	Benzene	C_6H_6	78.108	176.185	0.88458	2.6965
43	Toluene	C_7H_8	92.134	231.121	0.87190	3.1808
44	Ethylbenzene	C_8H_{10}	106.160	277.137	0.87175	3.6650
45	1,2-Dimethylbenzene (o-xylene)	C_8H_{10}	106.160	291.95	0.88482	3.6650
46	1,3-Dimethylbenzene (m-xylene)	C_8H_{10}	106.160	282.38	0.86880	3.6650
47	1,4-Dimethylbenzene (p-xylene)	C_8H_{10}	106.160	281.03	0.86576	3.6650
48	Styrene (Phenyl Ethylene)	C_8H_8	104.144	293.4	0.91112	3.5954
49	Isopropylbenzene (cumene)	C_9H_{12}	120.186	306.31	0.86642	4.1492

Reprinted from "Control Valve Handbook," 1965, Fisher Controls Company, Marshalltown, Iowa.

Properties of Cryogenic Fluids

Gas	He	H_2	CH_4	N_2	O_2	F_2
Density, NTP* lb./cu. ft.	.0103	.0052	.0415	.0724	.0827	.0982
Boiling Point 1 atm. °F.	-452.1	-423.2	-258.7	-320.4	-297.4	-306.6
Vapor Density at B.P. lb./cu. ft.	1.06	.084	.111	.288	.296	---
Liquid Density at B.P. lb./cu.ft.	7.62	4.37	26.46	50.41	71.27	94.2
Heat of Vaporization at B.P. Btu/lb.	8.8	193	219.2	85.2	91.7	74.1
Critical Temperature °F.	-450.3	-400.3	-116.5	-232.8	-181.1	-200.2
Critical Pressure, psia	33.2	187.7	673.1	492.3	736.9	808.3

NTP = Normal Temperature and Pressure, 70°F. and 14.696 psia.

Area Equivalents

Convert From \ Convert To	Square Meters	Square Inches	Square Feet	Square Yards	Square Miles	Square Kilometers
Square Meters	1	1549.99	10.7639	1.19599	3.861×10^{-7}	1×10^{-6}
Square Inches	0.0006452	1	6.944×10^{-7}	7.716×10^{-4}	2.491×10^{-10}	6.452×10^{-10}
Square Feet	0.0929	144	1	0.11111	3.587×10^{-8}	9.29×10^{-8}
Square Yards	0.83613	1296	9	1	3.228×10^{-7}	8.361×10^{-7}
Square Miles	2,589,999	-	27,878,400	3,097,600	1	2.59
Square Kilometers	1,000,000	-	10,763,867	1,195,985	0.3861	1

1 square meter = 100 square decimeters = 10,000 square centimeters

1 square millimeter = 0.01 square centimeter = 0.00155 square inches

Reprinted from "Control Valve Handbook," 1965, Fisher Controls Company, Marshalltown, Iowa.

Volume Equivalents

Convert From \ Convert To	Cubic Decimeters	Cubic Inches	Cubic Feet	U. S. Quart	U. S. Gallon	Imperial Gallon	U. S.* Barrel
Cubic Decimeters	1	61.0234	0.03531	1.05668	0.264178	0.220083	0.00629
Cubic Inches	0.01639	1	5.787×10^{-4}	0.01732	0.004329	0.003606	0.000103
Cubic Feet	28.317	1728	1	29.9221	7.48055	6.22888	0.1781
U.S. Quart	0.94636	57.75	0.03342	1	0.25	0.2082	0.00595
U.S. Gallon	3.78543	231	0.13368	4	1	0.833	0.02381
Imperial Gallon	4.54374	277.274	0.16054	4.80128	1.20032	1	0.02877
U.S.* Barrel	158.98	9,702	5.6146	168	42	34.973	1

1 cubic meter = 1000 cubic decimeters = 1,000,000 cubic centimeters
1 liter = 10 deciliters = 100 centiliters = 1000 milliliters = 1000 cubic centimeters
1 liter = 1 cubic decimeter.
* Petroleum

Volume Rate Equivalents

Convert From \ Convert To	Liters Per Minute	Cubic Meters Per Hour	Cubic Feet Per Hour	Imperial Gallon Per Minute	U.S. Gallon Per Minute	U.S. Barrel Per Day
Liters Per Minute	1	0.06	2.1189	0.21997	0.264178	9.057
Cubic Meters Per Hour	16.667	1	35.314	3.667	4.403	151
Cubic Feet Per Hour	0.4719	0.028317	1	0.1038	0.1247	4.2746
Imperial Gallon Per Minute	4.546	0.2727	9.6325	1	1.20032	41.1
U.S. Gallon Per Minute	3.785	0.2273	8.0208	0.8326	1	34.28
U.S. Barrel Per Day	0.1104	0.006624	0.23394	0.02428	0.02917	1

Reprinted from "Control Valve Handbook," 1965, Fisher Controls Company, Marshalltown, Iowa.

Pressure Equivalents

Convert From \ Convert To	Kg. Per Sq. Cm.	Lb. Per Sq. In.	Atm.	Meters of Hg.	In. of Hg.	Meters of Water	In. of Water	Ft. of Water
Kg. Per Sq. Cm.	1	14.22	0.9678	0.7355	28.96	10.01	394.05	32.84
Lb. Per Sq. In.	0.07031	1	0.06804	0.05171	2.036	0.7037	27.7	2.309
Atm.	1.0332	14.696	1	0.76	29.92	10.34	407.14	33.93
Meters of Hg.	1.3596	19.34	1.316	1	39.37	13.61	535.7	44.64
In. of Hg.	0.03453	0.4912	0.03342	0.0254	1	0.3456	13.61	1.134
Meters of Water	0.09991	1.421	0.0967	0.07349	2.893	1	39.37	3.281
In. of Water	0.002538	0.0361	0.002456	0.001867	0.07349	0.0254	1	0.0833
Ft. of Water	0.03045	0.4332	0.02947	0.0224	0.8819	0.3048	12	1

1 ounce/sq.inch = 0.0625 lbs./sq.inch

Reprinted from "Control Valve Handbook," 1965, Fisher Controls Company, Marshalltown, Iowa.

TEMPERATURE CONVERSION FORMULAS

To Convert From	To	Substitute in Formula
Degrees Centigrade	Degrees Fahrenheit	(°C x 9/5) + 32
Degrees Centigrade	Degrees Kelvin	(°C - 273.16)
Degrees Fahrenheit	Degrees Centigrade	(°F - 32) x 5/9
Degrees Fahrenheit	Degrees Rankine	(°F - 459.69)

Temperature Interpolation Factors

°C.	Degree	°F.
0.56	1	1.8
1.11	2	3.6
1.67	3	5.4
2.22	4	7.2
2.78	5	9.0
3.33	6	10.8
3.89	7	12.6
4.44	8	14.4
5.00	9	16.2

Equivalents of Degrees API, Degrees Baumé, Specific Gravity, Weight Density, and Pounds Per Gallon at 60 F/60 F

Degrees on API or Baumé Scale	Values for API Scale			Values for Baumé Scale					
	Oil			Liquids Lighter Than Water			Liquids Heavier Than Water		
	Specific Gravity S	Weight Density, Lb/Ft^3 ρ	Pounds per Gallon	Specific Gravity S	Weight Density, Lb/Ft^3 ρ	Pounds per Gallon	Specific Gravity S	Weight Density, Lb/Ft^3 ρ	Pounds per Gallon
0	...	...	...	...	...	...	1.0000	62.36	8.337
2	...	...	...	...	...	...	1.0140	63.24	8.454
4	...	...	...	...	...	...	1.0284	64.14	8.574
6	...	...	...	...	...	...	1.0432	65.06	8.697
8	...	...	...	...	...	...	1.0584	66.01	8.824
10	1.0000	62.36	8.337	1.0000	62.36	8.337	1.0741	66.99	8.955
12	0.9861	61.50	8.221	0.9859	61.49	8.219	1.0902	67.99	9.089
14	0.9725	60.65	8.108	0.9722	60.63	8.105	1.1069	69.03	9.228
16	0.9593	59.83	7.998	0.9589	59.80	7.994	1.1240	70.10	9.371
18	0.9465	59.03	7.891	0.9459	58.99	7.886	1.1417	71.20	9.518
20	0.9340	58.25	7.787	0.9333	58.20	7.781	1.1600	72.34	9.671
22	0.9218	57.87	7.736	0.9211	57.44	7.679	1.1789	73.52	9.828
24	0.9100	56.75	7.587	0.9091	56.70	7.579	1.1983	74.73	9.990
26	0.8984	56.03	7.490	0.8974	55.97	7.482	1.2185	75.99	10.159
28	0.8871	55.32	7.396	0.8861	55.26	7.387	1.2393	77.29	10.332
30	0.8762	54.64	7.305	0.8750	54.57	7.295	1.2609	78.64	10.512
32	0.8654	53.97	7.215	0.8642	53.90	7.205	1.2832	80.03	10.698
34	0.8550	53.32	7.128	0.8537	53.24	7.117	1.3063	81.47	10.891
36	0.8448	52.69	7.043	0.8434	52.60	7.031	1.3303	82.96	11.091
38	0.8348	52.06	6.960	0.8333	51.97	6.947	1.3551	84.51	11.297
40	0.8251	51.46	6.879	0.8235	51.36	6.865	1.3810	86.13	11.513
42	0.8155	50.86	6.799	0.8140	50.76	6.786	1.4078	87.80	11.737
44	0.8063	50.28	6.722	0.8046	50.18	6.708	1.4356	89.53	11.969
46	0.7972	49.72	6.646	0.7955	49.61	6.632	1.4646	91.34	12.210
48	0.7883	49.16	6.572	0.7865	49.05	6.557	1.4948	93.22	12.462
50	0.7796	48.62	6.499	0.7778	48.51	6.484	1.5263	95.19	12.725
52	0.7711	48.09	6.429	0.7692	47.97	6.413	1.5591	97.23	12.998
54	0.7628	47.57	6.359	0.7609	47.45	6.344	1.5934	99.37	13.284
56	0.7547	47.07	6.292	0.7527	46.94	6.275	1.6292	101.60	13.583
58	0.7467	46.57	6.225	0.7447	46.44	6.209	1.6667	103.94	13.895
60	0.7389	46.08	6.160	0.7368	45.95	6.143	1.7059	106.39	14.222
62	0.7313	45.61	6.097	0.7292	45.48	6.079	1.7470	108.95	14.565
64	0.7238	45.14	6.034	0.7216	45.00	6.016	1.7901	111.64	14.924
66	0.7165	44.68	5.973	0.7143	44.55	5.955	1.8354	114.46	15.302
68	0.7093	44.23	5.913	0.7071	44.10	5.895	1.8831	117.44	15.699
70	0.7022	43.79	5.854	0.7000	43.66	5.836	1.9333	120.57	16.118
72	0.6953	43.36	5.797	0.6931	43.22	5.778	...	...	...
74	0.6886	42.94	5.741	0.6863	42.80	5.722	...	...	...
76	0.6819	42.53	5.685	0.6796	42.38	5.666	...	...	...
78	0.6754	42.12	5.631	0.6731	41.98	5.612	...	...	...
80	0.6690	41.72	5.577	0.6667	41.58	5.558	...	...	...
82	0.6628	41.33	5.526	0.6604	41.19	5.506	...	...	...
84	0.6566	40.95	5.474	0.6542	40.80	5.454	...	...	...
86	0.6506	40.57	5.424	0.6482	40.42	5.404	...	...	...
88	0.6446	40.20	5.374	0.6422	40.05	5.354	...	...	...
90	0.6388	39.84	5.326	0.6364	39.69	5.306	...	...	...
92	0.6331	39.48	5.278	0.6306	39.33	5.257	...	...	...
94	0.6275	39.13	5.231	0.6250	38.98	5.211	...	...	...
96	0.6220	38.79	5.186	0.6195	38.63	5.165	...	...	...
98	0.6166	38.45	5.141	0.6140	38.29	5.119	...	...	...
100	0.6112	38.12	5.096	0.6087	37.96	5.075	...	...	...

Courtesy Crane Co., Reproduced from Technical Paper No. 410, "Flow of Fluids."

OTHER USEFUL CONVERSIONS

To Convert From	To	Multiply By
Cu. Ft. (Methane)	B.T.U.	1000 (approx.)
Cu. Ft. of Water	Lbs. of Water	62.4
Degrees	Radians	0.01745
Gals.	Lbs. of Water	8.336
Grams	Ounces	0.0352
Horsepower (mech.)	Ft. Lbs. per Min.	33,000
Horsepower (elec.)	Watts	746
Kg.	Lbs.	2.205
Kg. per Cu. Meter	Lbs. per Cu. Ft.	0.06243
Kilowatts	Horsepower	1.34
Lbs.	Kg.	0.4536
Lbs. of Air (14.7 psia and 60° F.)	Cu. Ft. of Air	13.1
Lbs. per Cu. Ft.	Kg. per Cu. Meter	16.0184
Lbs. per Hr. (Gas)	Std. Cu. Ft. per Hr.	$\frac{13.1}{\text{Specific Gravity}}$
Lbs. per Hr. (Water)	Gals. per Min.	0.002
Lbs. per Sec. (Gas)	Std. Cu. Ft. per Hr.	$\frac{46{,}160}{\text{Specific Gravity}}$
Radians	Degrees	57.3
Scfh Air	Scfh Propane	0.81
Scfh Air	Scfh Butane	0.71
Scfh Air	Scfh 0.6 Natural Gas	1.29

PREFIXES

Prefix	Meaning	Numerical Value	
Micromicro-	One-trillionth	0.000,000,000,001 or	10^{-12}
Millimicro-	One-billionth	0.000,000,001	10^{-9}
Micro-	One-millionth	0.000,001	10^{-6}
Milli-	One-thousandth	0.001	10^{-3}
Centi-	One-hundredth	0.01	10^{-2}
Deci-	One-tenth	0.1	10^{-1}
Deka-	Ten	10	10
Hecto-	One hundred	100	10^{2}
Kilo-	One thousand	1,000	10^{3}
Mega-	One million	1,000,000	10^{6}

Reprinted from "Control Valve Handbook," 1965, Fisher Controls Company, Marshalltown, Iowa.

ABOUT THE AUTHORS

Ernest E. Allen is presently a research engineer with Fisher Controls Company, Marshalltown, Iowa. At Fisher, he has directed an extensive research study on valve noise. A member of the ISA Standards and Practices Valve Sizing Committee SP-39, Mr. Allen also belongs to the Acoustical Society of America. A graduate of Iowa State University, Mr. Allen has presented technical papers at meetings of the American Industrial Hygiene Association, the American Gas Association and ISA.

Earl A. Bake is currently Supervisory Engineer for new product development with Rockwell Manufacturing Company, Pittsburgh, Pennsylvania. His previous experience included four years as research product engineer at Rockwell Manufacturing, developing experimental and analytical information on valve flow and cavitation and evolving methods for design to optimize flow performance. Prior to that he worked at Westinghouse Electric Corporation for twelve years as technical administrator of research and development on valves for nuclear power plants. Mr. Bake, who received his BSME from Case Western Reserve, has published an ASME article on high velocity gas flow in reduced port valves.

Hans D. Baumann is currently Vice President of Engineering with Masoneilan International, Inc. Here, he is responsible for all of the company's U.S. and worldwide engineering activities. Prior to coming to Masoneilan, Mr. Baumann served as Chief Engineer with Welland and Tuxhorn of Germany. A Senior member of ISA, Mr. Baumann also belongs to ASME and holds more than 30 U.S. and foreign patents. He has authored or coauthored more than a dozen technical articles in leading trade journals and is a contributing author to both the *Instrument Engineers Handbook* and the *ISA Handbook of Control Valves*. Educated in Germany, Mr. Baumann has taken graduate courses at Case Western Reserve and Northeastern University.

Everett L. Dixson is employed by the Fisher Controls Company as a research engineer. At Fisher, he is responsible for the investigation and testing of such nonmetallic materials as elastomers, plastics, adhesives, etc., with special emphasis on the practical application of these materials to service conditions in assemblies. Previously, Mr. Dixson served as a design and development engineer with Lennox Industries, Marshalltown, Iowa. A member of the American Society of Mechanical Engineers, Mr. Dixson holds a BS degree in Mechanical Engineering from Iowa State University.

Leslie R. Driskell is currently with Blaw-Knox Chemical Plants, Inc., Pittsburgh, Pa. Here, he directs the activities of the Instrument Engineering Department. His previous professional associations include E. I. DuPont de Nemours Co., Riggs Distler and Company, Purdue University and Seagram-Calvert, Inc. An ISA Fellow Member, Mr. Driskell is the author of numerous technical papers; in 1968, he won the Instrumentation Technology Award for an article published in that journal. Mr. Driskell holds a BS degree in Mechanical Engineering from the University of Louisville.

Matthew L. Freeman is Manager of Engineering with Masoneilan International, Inc., Norwood, Massachusetts. Here, he is responsible for transferring product development design into production practice for both United States and foreign operations. His previous professional associations include Rockwell Manufacturing Company, Honeywell Valve Division, McEvoy Company and the Atlantic-Richfield Continental Oil Company. A Senior Member of ISA, Mr. Freeman has held several ISA offices and is also a member of the American Society of Mechanical Engineers. A graduate of Mississippi State University, Mr. Freeman holds ten valve patents and is the co-author of one technical publication.

D. J. Gaffney is currently Manager of Valve Engineering with the Fisher Controls Co., Marshalltown, Iowa. Here, he is responsible for product design and improvement, materials, codes and standards. Mr. Gaffney spent thirty-two years previous to his association with Fisher in design engineering. Chairman of the PMC Section of SAMA, Mr. Gaffney is the holder of a BS Degree in Mechanical Engineering from the University of Nebraska. He also holds five U.S. patents.

Melville W. Hott, Jr., is Executive Vice-President of the Jordan Valve Division of Richards Industries, Inc., Cincinnati, Ohio. He is responsible for the management of the marketing, engineering and manufacturing departments. An ISA Member, Mr. Hott also belongs to the American Society of Civil Engineers and is Secretary of the Fluid Controls Institute. Mr. Hott holds several patents in valve design.

James W. Hutchison is currently a member of the staff of the Honeywell Education Center. Here, he conducts sales training and is responsible for customer training in instrumentation and control equipment. Mr. Hutchison's previous professional associations include the Allied Chemical Corporation, E. I. duPont deNemours & Company and Powers Regulator Company. As an ISA Senior Member, Mr. Hutchison has served as Editor-in-Chief of the *ISA Handbook of Control Valves,* and as both Secretary and Education Chairman of the Philadelphia Section. He is a registered Pennsylvania professional engineer.

Charles F. King is currently a technical sales consultant with the Fisher Controls Company, Marshalltown, Iowa. Here, he acts in an advisory capacity on control valve applications and new products. Mr. King has served the Fisher Controls Company in various capacities—as project engineer, research engineer and director of research. An ISA Senior Member, he is Chairman of Standards and Practices for District VI. The author of several publications, Mr. King holds a BSME from Iowa State University.

Orval P. Lovett is Section Supervisor of Instrumentation for E. I. DuPont de Nemours & Company, Wilmington, Delaware. He is in charge of instrumentation design for DuPont's expanded and new plant facilities. Prior to his association with DuPont, Mr. Lovett was with the Stone & Webster Engineering Corporation. An ISA Fellow Member, Mr. Lovett received a Special Services Award from ISA's Wilmington Section. The author or co-author of four technical papers on control valves published in *Instrumentation Technology,* Mr. Lovett is also a contributor to the *Control Engineers Handbook.* He holds a BS degree in Chemical Engineering from Drexel University.

Sal F. Luna has been associated with Atomic Power Divisions of Westinghouse since 1955. He is primarily concerned with the application of advanced technology to a wide variety of reactor plant control systems. He is presently responsible for the Westinghouse Nuclear Energy Systems. Previously, he was a principal instrument engineer with the Catalytic Construction Company of Philadelphia. Mr. Luna has authored a chapter in the *ISA Transducer Compendium* and has co-authored several technical papers. He has received numerous patent disclosure awards. An ISA Senior Member, he is presently serving on the ISA President's Advisory Committee. Mr. Luna holds a BS degree in Chemistry from Niagara University.

Ralph L. Moore is presently an Instrument Engineer with E. I. DuPont de Nemours & Company, Inc. His responsibilities include the simulation of physical and chemical processes by mathematical modeling and the verification of operation of the proposed instrumentation. Prior to his association with DuPont, Mr. Moore was an instructor in instrumentation at Case Western Reserve. An ISA Senior Member, he is Director of the Education Committee. He is also a member of the National Society of Professional Engineers and the Delaware Society of Professional Engineers. The author or co-author of several articles, Mr. Moore received the ISA Instrumentation Technology Award for the article, "The Computer / Team Approach to Control-System Design". Mr. Moore holds Masters Degrees from both Drexel University and Case Western Reserve.

Joseph O'Connor, Jr. is President of Kieley & Mueller, Inc. of Middletown, New York. His previous experience includes two years with Fluid Controls Company, Philadelphia and since 1953 with Kieley & Mueller in several capacities from sales applications, through Vice President, to his present position as President. Mr. O'Connor, who received his BSME from Princeton University, is an ISA Senior Member.

John A. Parker is presently an Instrument Specialist Engineer with E. I. DuPont de Nemours & Company, Inc., Wilmington, Delaware. Here, he is responsible for the design, procurement, and start-up of instrument systems for chemical facilities. An ISA Member, Mr. Parker has authored papers for the First ISA Final Control Elements Symposium and the Fisher Controls Company's 1970 Process Seminar.

Gordon F. Stiles is currently the Manager of Applied Research at the Fisher Controls Co., Marshalltown, Iowa. Mr. Stiles is the author of several articles on flashing, cavitation and control valve sizing. He holds a BS degree in Mechanical Engineering from Iowa State University.

Edward L. Vollmer, III, presently handles technical writing and editing for Honeywell, Inc., Fort Washington, Pa. He is responsible for initiating, writing and editing technical literature. Previously, Mr. Vollmer spent five years as a valve sales engineer and seven years as an instructor in control theory and instrumentation. He is the co-author of *Introduction to Process Control Dynamics,* a text used at the Honeywell Education Center. Mr. Vollmer holds a BS degree in Science, Mathematics and Education from West Chester State College.

Charles R. Thornton is currently with E. I. DuPont deNemours & Company, Inc., as instrument engineer. Here, he is responsible for the design and maintenance of instrument systems for a continuous stream, integrated multi-product, chemical complex. Prior to his association with DuPont, Mr. Thornton spent fourteen years in instrument maintenance in the chemical and petroleum industries and ten years in the design of instrument systems including hardware evaluation and selection. Mr. Thornton is an ISA Senior Member.

Rolf B. Weil served as Product Manager of Control Valves for the Jamesbury Corporation from 1966-1971. Prior to coming to Jamesbury, Mr. Weil was a sales engineer for the Milton Roy Company. An ISA Member and a member of ISA's Valve Sizing Committee (SP39). Mr. Weil holds a BS degree from the University of the Witwatersrand, Johannesburg, South Africa.

Paul Wing is currently Manager of International Engineering for Masoneilan International, Inc., Norwood, Massachusetts. Here, he is in charge of both USA and overseas operations. Mr. Wing has served 30 years with Masoneilan in various key engineering positions. An ISA Senior Member, Mr. Wing is the author of numerous articles on control valve design and application. He holds a BS degree in Chemical Engineering from Massachusetts Institute of Technology.

ALPHABETICAL INDEX TO MAJOR CHARTS, GRAPHS AND TABLES

The following list of the most significant charts, graphs, and tables is given to provide a quick reference to frequently used items. The numerous illustrations and diagrams used by authors to augment the chapters are not included in this list, because they are meaningful primarily within the context of the subject matter.

SUBJECT INDEX